Probability and Mathematical Statistics (Continued)

SEBER • Linear Regression Analysis
SEBER • Multivariate Observations
SEN • Sequential Nonparametrics: Invariance Principles and Statistical Inference
SERFLING • Approximation Theorems of Mathematical Statistics
TJUR • Probability Based on Radon Measures
WILLIAMS • Diffusions, Markov Processes, and Martingales, Volume I: Foundations
ZACKS • Theory of Statistical Inference

Applied Probability and Statistics

ABRAHAM and LEDOLTER • Statistical Methods for Forecasting
AGRESTI • Analysis of Ordinal Categorical Data
AICKIN • Linear Statistical Analysis of Discrete Data
ANDERSON, AUQUIER, HAUCK, OAKES, VANDAELE, and WEISBERG • Statistical Methods for Comparative Studies
ARTHANARI and DODGE • Mathematical Programming in Statistics
BAILEY • The Elements of Stochastic Processes with Applications to the Natural Sciences
BAILEY • Mathematics, Statistics and Systems for Health
BARNETT • Interpreting Multivariate Data
BARNETT and LEWIS • Outliers in Statistical Data
BARTHOLOMEW • Stochastic Models for Social Processes, *Third Edition*
BARTHOLOMEW and FORBES • Statistical Techniques for Manpower Planning
BECK and ARNOLD • Parameter Estimation in Engineering and Science
BELSLEY, KUH, and WELSCH • Regression Diagnostics: Identifying Influential Data and Sources of Collinearity
BHAT • Elements of Applied Stochastic Processes
BLOOMFIELD • Fourier Analysis of Time Series: An Introduction
BOX • R. A. Fisher, The Life of a Scientist
BOX and DRAPER • Evolutionary Operation: A Statistical Method for Process Improvement
BOX, HUNTER, and HUNTER • Statistics for Experimenters: An Introduction to Design, Data Analysis, and Model Building
BROWN and HOLLANDER • Statistics: A Biomedical Introduction
BROWNLEE • Statistical Theory and Methodology in Science and Engineering, *Second Edition*
CHAMBERS • Computational Methods for Data Analysis
CHATTERJEE and PRICE • Regression Analysis by Example
CHOW • Analysis and Control of Dynamic Economic Systems
CHOW • Econometric Analysis by Control Methods
COCHRAN • Sampling Techniques, *Third Edition*
COCHRAN and COX • Experimental Designs, *Second Edition*
CONOVER • Practical Nonparametric Statistics, *Second Edition*
CONOVER and IMAN • Introduction to Modern Business Statistics
CORNELL • Experiments with Mixtures: Designs, Models and The Analysis of Mixture Data
COX • Planning of Experiments
DANIEL • Biostatistics: A Foundation for Analysis in the Health Sciences, *Third Edition*
DANIEL • Applications of Statistics to Industrial Experimentation
DANIEL and WOOD • Fitting Equations to Data: Computer Analysis of Multifactor Data, *Second Edition*
DAVID • Order Statistics, *Second Edition*
DAVISON • Multidimensional Scaling
DEMING • Sample Design in Business Research
DILLON and GOLDSTEIN • Multivariate Analysis: Methods and Applications
DODGE and ROMIG • Sampling Inspection Tables, *Second Edition*
DOWDY and WEARDEN • Statistics for Research

continued on back

LINEAR STATISTICAL MODELS AND RELATED METHODS

LINEAR STATISTICAL MODELS AND RELATED METHODS

With Applications to Social Research

JOHN FOX

Associate Professor
Department of Sociology
York University, Toronto

JOHN WILEY & SONS

New York • Chichester • Brisbane • Toronto • Singapore

Library of Congress Cataloging in Publication Data:

Fox, John, 1947–
Linear statistical models and related methods.

Includes bibliographical references and indexes.
1. Linear models (Statistics) 2. Regression analysis. 3. Analysis of variance. 4. Social sciences--Statistical methods. I. Title.
QA276.F6435 1984 519.5 83-23278
ISBN 0-471-09913-9

Printed in the United States of America

10 9 8 7 6 5 4 3 2 1

TO BONNIE
AND MY PARENTS

"He who loves practice without theory is like the sailor who boards ship without a rudder and compass and never knows where he may be cast."

Leonardo da Vinci, 1452–1519

Preface

Linear models, their variants, and extensions are among the most useful and widely used statistical tools for social research. This book aims to provide an in-depth, modern treatment of linear models and related methods. The book should be of interest to students and researchers in the social sciences; and while the specific choice of methods and examples reflects this audience, I expect that the book will also prove useful in other disciplines that employ linear models for data analysis, and in courses on applied linear models where the subject-matter of applications is not of special concern.

My major premise in writing this text is that the teaching of social statistics should combine statistical theory, critical application, and methodology. Too often, statistical methods are presented to social scientists either as abstract formalisms or as recipes for research. I feel that there also is an unhealthy tendency to oppose statistical theory and data analysis. The quotation from Leonardo da Vinci on the relationship of theory to practice, reproduced at the front of the book, is therefore doubly apt: First, statistical methods in application must be properly related to substantive theory and research concerns. It is difficult in a statistics text to demonstrate concretely the connection between substance and statistical method, but I have attempted to select illustrations that make this point in an elementary fashion, and the connection is raised explicitly at several critical junctures. Second, I think that the sensitive use of statistical methods for data analysis requires some grounding in statistical theory. Thus, although the style of presentation in this book is nonrigorous, the statistical theory of linear models is developed alongside (or, better, underneath) its applications—few results are presented without informal derivation or intuitive justification.

Throughout the book, general approaches and principles are employed to emphasize the conceptual unity of the techniques covered. For example, the maximum-likelihood method is frequently used to derive estimators and tests, and the vector geometry of linear models is employed to clarify a variety of topics. In addition, a critical approach is adopted to the use of statistical models through explanation of (1) the assumptions underlying the models, (2) diagnostic procedures designed to assess model adequacy, and (3) means of respecifying models so that they more adequately represent the data at hand.

The book assumes that readers have been exposed to the elements of statistical inference and probability. It also assumes some knowledge of elementary calculus and matrix algebra. With the exception of elementary calculus, these areas are reviewed and extended in the appendices. Special attention is paid to topics that are frequently slighted in introductory social-statistics courses (such as probability distributions and properties of estimators), and to extensions of material presumed to be familiar (e.g., vector geometry, matrix calculus).

Chapter 1 is devoted to regression analysis, which examines the relationship of a quantitative dependent variable to one or more quantitative independent variables. Much of the statistical theory of linear models is developed in this chapter.

Chapter 2 extends linear models to include qualitative independent variables. The treatment of analysis of variance in this chapter emphasizes unbalanced (i.e., unequal-cell-frequencies) data.

Chapter 3 presents a variety of material on diagnosing and correcting linear-model problems. The problems examined include collinearity, outliers and influential data, nonlinearity, heteroscedasticity, and nonnormality. The chapter contains a discussion of data transformations and an introduction to nonlinear models.

Chapter 4 takes up structural-equation models, which are systems of linear equations representing the causal relations among sets of variables, some of which may exert mutual influence on each other. Attention is paid not only to the direct application of structural-equation models, but also to the general data-analytic principles that these models embody. The chapter ends with an introductory treatment of models that contain specific measurement-error components and that include multiple indicators of latent variables.

Chapter 5 describes logit models for qualitative dependent variables and log-linear models for contingency tables, stressing the similarity of these models to the linear models of earlier chapters. The relationship between logit and log-linear models is also developed. The chapter includes a discussion of diagnostic methods for logit models.

Most of the material in this book can be covered in a two-semester course, the first several weeks of which teach basic linear algebra and elementary differential calculus (using, e.g., Kleppner and Ramsey's *Quick Calculus*). Chapters 1 through 3 can be covered in a one-semester course that omits some specialized topics.

In learning statistics, it is important for the reader of a text to participate actively, both by working through the arguments presented in the book, and by applying methods to data. Reworking of examples is a good place to start, and I have presented illustrations in such a manner as to make re-analysis and further analysis possible.

Nearly all of the examples employ real data from the social sciences, many of them previously analyzed and published. The exercises that involve data analysis also almost all use real data drawn from a variety of areas of application. A word of warning to readers (and instructors): Many of these exercises are time consuming.[1]

Computational matters are occasionally commented on in passing, but the book generally ignores the finer points of statistical computing in favor of methods that are computationally simple. I feel that this approach facilitates learning; once basic techniques are learned, an experienced data analyst has recourse to carefully designed programs for statistical computations. Similarly, I think that it is a mistake to tie a general discussion of linear models too closely to particular programs or packages. In fact, although the marvelous proliferation of statistical software has routinized the computations for most of the methods described in this book, the workings of computer programs are not sufficiently accessible to promote learning. Consequently, I find it useful to teach APL as part of a course on linear models.

APL is an interactive programming language with powerful operators and functions, including those for common matrix operations. Similar computational facilities are available elsewhere, as in the MATRIX procedure of the SAS statistical program package and in the MINITAB package. Using APL, students are able to write their own programs for the computationally simpler methods. They may be provided with subprograms (e.g., for latent roots and vectors) with which they can construct more complex applications, and for functions such as plotting and data management.

Many of the data-analytic examples in this book are accompanied by computer-generated plots and graphs prepared on a typewriter terminal. Although these figures are less elegant than hand-drawn or computer-drawn pen-and-ink graphs, facilities for conveniently obtaining line-printer and printing-terminal plots are more widely available: I feel that it is most useful to display data as they are likely to appear to readers in their own work. The introduction of sophisticated graphics capabilities into general statistical packages (such as SAS and SPSS) and the increasing availability of flexible graphics hardware suggest that this situation will change in the future.

Because the selection of material for a text of this sort must be more than a matter of personal taste, I would like to comment briefly on some topics that are omitted from the book. First, although some attention is paid to experimentation, experimental design is not developed systematically and comprehensively: There is no discussion of nested designs, random- and mixed-effects models, variance components, and so on. These omissions reflect my judgment

[1]Starred exercises (*) are more difficult and are generally theoretical; exercises marked with a dagger (†) substantially extend the treatment in the text—readers may wish to examine these problems, even if they are not worked out in detail; and exercises marked with a pound sign (#) are intended for hand solution (i.e., with a pocket calculator rather than on a computer).

of which methods are most useful to a general social-scientific readership as well as the availability of good books on linear models for designed experiments.

Second, I have intentionally avoided a discussion of correlated errors in times-series regression. I feel that an adequate treatment of time-series data requires methods beyond the scope of this book. Time-series data are employed, however, to illustrate the problem of collinearity and in an example of a nonlinear model.

Finally, there is no systematic treatment in this book of multivariate statistical methods, although principal-components analysis is introduced in the context of collinearity, and the structural-equation models of Chapter 4 include more than one dependent variable. A serious development of multivariate statistics requires book-length treatment at a higher level of mathematical sophistication. Also, an understanding of multivariate methods is predicated on some background in univariate linear models.

JOHN FOX

Toronto, Canada
March 1984

Acknowledgments

During the 1980–1981 academic year, when the first draft of this book was written, I was partially supported by a leave fellowship from the Social Sciences and Humanities Research Council of Canada. I am grateful to the Inter-University Consortium for Political and Social Research, where I have taught for the past several summers, for providing a stimulating atmosphere in which to work.

A number of individuals have read parts of this book in draft form, and I am indebted to them for criticism and suggestions. I particularly wish to thank Ken Bollen of Dartmouth College, Gene Denzel of York University, Shirley Dowdy of the University of West Virginia, and Douglas Rivers of Harvard University for their valuable comments. Draft chapters of the book were used in courses at York University and at the ICPSR Summer Program. Students at both institutions gracefully accepted their role as guinea pigs, and their critical responses to the text—both positive and negative—helped me in the process of revision.

I wish to thank the researchers, many of them colleagues and friends, who generously made their data available to me for the examples and exercises of this book. I am also grateful to the publishers and authors who have granted me permission to reprint or adapt copyrighted material.

Finally, I wish to thank Beatrice Shube at John Wiley for her encouragement and confidence in this project.

J.F.

Contents

CHAPTER 2. DUMMY-VARIABLE REGRESSION AND ANALYSIS OF VARIANCE 73

CHAPTER 3. DIAGNOSING AND TREATING LINEAR-MODEL PROBLEMS 137

LINEAR STATISTICAL MODELS AND RELATED METHODS

1

Linear Regression Analysis

This chapter develops the theory of regression analysis, a method of wide applicability and a natural point of departure for a treatment of linear models. The first section deals with simple regression, which examines the linear relationship between two numerical variables, one of which is thought to affect, influence, or predict the other. The second section takes up multiple regression, which expresses one numerical variable as a linear function of several others. The remaining sections of the chapter develop topics in linear regression analysis: random regressors (Section 1.3), model specification error (Section 1.4), and standardized regression coefficients (Section 1.5). Most of the theory described in the chapter applies generally to linear models.

When we refer to numerical variables, we mean quantitative variables measured on an interval or ratio scale. The distinction between numerical and qualitative (i.e., nominal) data is often confused with the distinction between continuous and discrete data. A numerical variable may be either continuous (such as temperature) or discrete (such as family size); a qualitative variable (sex, for example) is necessarily discrete. In data analysis, the discrete–continuous distinction becomes, for practical purposes, a distinction between variables that take on relatively few values (e.g., an individual's number of surviving grandparents) and those that take on relatively many (e.g., dollar income). Ordinal quantitative data are something of an embarrassment in linear models, and are frequently dealt with by assigning arbitrary numerical scores to their values. (This practice is discussed further in Problem 3.12, Chapter 3.)

1.1. SIMPLE REGRESSION

Simple regression analysis is a method of limited usefulness. It is, nevertheless, of critical importance to us because (1) the limitations of the method suggest directions in which it may usefully be extended, and (2) many general properties of regression analysis and linear models may be introduced in the context of simple regression.

In this section, least-squares simple regression is treated first as a method for fitting a straight line to a scatter of points that represent a set of observations for two variables. After this descriptive treatment, we introduce a statistical model for which the least-squares regression coefficients may serve as estimators. Upon developing some of the properties of the least-squares estimators, we show how these estimators may be employed for testing hypotheses about population regression coefficients. We then introduce the idea of correlation as a means of assessing the fit of the simple-regression model to data. Finally, we develop the vector geometry of simple regression analysis to provide us with a powerful conceptual tool for exploring the properties of linear models.

1.1.1. Fitting a Least-Squares Line to a Scatter of Points

Suppose that we have a set of n scores for two numerical variables X and Y. Further suppose that we regard X as a cause of Y, or at least that we wish to employ X to predict Y. Using practical terms, suppose that Y is dollar income and that X is years of education, each measured for individuals in a sample of wage earners. If X and Y were perfectly related, we could then write Y as a function of X: $Y = f(X)$. Borrowing from mathematical usage, we term X the *independent variable* and Y the *dependent variable*. Although some statisticians object to this terminology, it is quite firmly entrenched in the literature. We must be careful, however, to avoid attributing broad meaning to the terms "independent" and "dependent". For example, when we introduce several independent variables in Section 1.2, we do so without implying that these variables are *statistically* independent of one another. On occasion, we shall also refer to X as a *regressor*.

In virtually all interesting cases in the social sciences, it is unrealistic to expect that X is the sole cause of Y, or alternatively, that Y is perfectly predictable from X alone. Nevertheless, X may be one of a number of causes of Y, or Y may be partially predictable from X. There are, for instance, many causes of income beyond education: factors such as sex, race, occupation, and experience in the labor force come easily to mind. If this is the case, then we cannot expect to represent Y as an exact function of X: Two observations that are identical in their independent-variable values will in general differ in their dependent-variable values. We hope, however, to specify a function that captures the systematic relationship between the variables: $\hat{Y} = f(X)$, where $\hat{Y}$ (read "Y-hat") is called a *fitted value*. E, the difference between Y and $\hat{Y}$

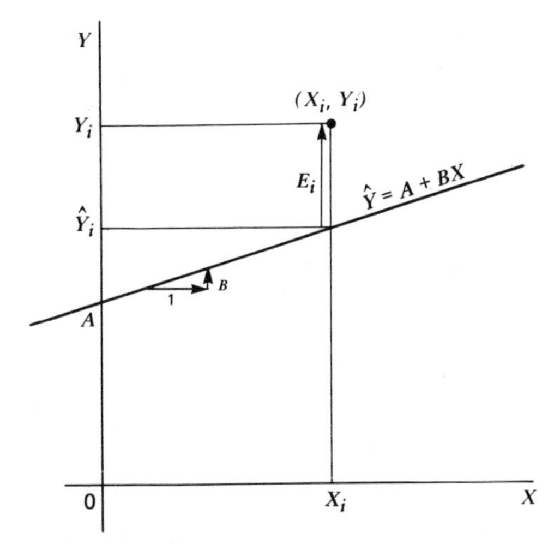

FIGURE 1.1. Linear regression of Y on X.

$(E \equiv Y - \hat{Y})$, represents that portion of the dependent variable not predictable from the independent variable; E is called a *residual*. Thus, $Y = f(X) + E$. A common characteristic of the linear models developed in this text is that they decompose dependent-variable values into fitted and residual components.

Thus far, we have left the functional relation $f(X)$ between Y and X unspecified. We now specify that this relation is linear: $\hat{Y} = A + BX$; or equivalently, $Y = A + BX + E$. For the ith of our n observations, we have

$$Y_i = A + BX_i + E_i \tag{1.1}$$

Here, A and B are respectively the *Y-intercept* (or *constant*) and *slope* of the *regression line* relating Y to X, as illustrated in Figure 1.1.

Although it would be naive to suppose that the relationship between two variables is necessarily linear, straight-line regression is, for several reasons, a natural point of departure. First, it is useful to start an examination of linear models with what is reasonably regarded as the simplest interesting case.[1] Second, many relationships are in fact linear or approximately linear, or may be rendered linear by transforming the data. Finally, in the social sciences, we frequently expect one variable to increase or decrease with another without being able to specify more precisely the functional relationship between the two. It seems sensible, in these instances, to entertain linear relationships because of their simplicity. We should stress at the outset, however, that it is foolish to blindly fit relations—linear or otherwise—to data. We shall take up these issues in detail in Chapter 3.

[1] Even simpler linear models are (1) $Y = BX + E$ (regression through the origin), and (2) $Y = A + E$ (no relationship between Y and X).

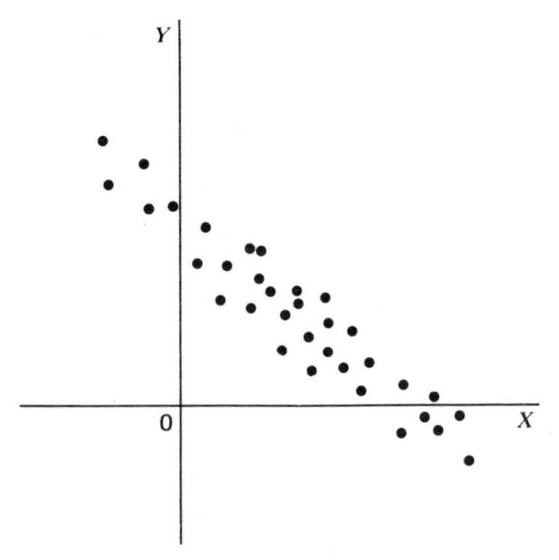

FIGURE 1.2. Scatterplot showing a strong linear relationship between X and Y.

Having decided to fit a straight line to the relationship between X and Y, we need to develop a method for determining the line. It is natural to think about this problem graphically, as in Figure 1.2, representing each of our n observations as a point in X, Y space. A graph of this sort, called a *scatterplot*, provides us with an invaluable tool for examining the relationship between two variables. A scatterplot, for example, can help us decide whether it is reasonable to use a straight line to summarize the relationship between X and Y. This and other data-analytic uses of plots in regression analysis will be discussed at length in Chapter 3. Our present purpose of fitting a linear relation to X, Y data will be well served by finding a line that comes as close as possible to the points in the scatterplot.

An illustrative scatterplot is shown in Figure 1.3. The data plotted in this figure are recorded in Table 1.1. These data, describing agricultural production in a primitive community, were compiled by Sahlins (1972) from information presented in Scudder's (1962) report on the Gwemba valley of Central Africa. The independent variable (X) is the ratio of consumers to productive individuals ("consumers/gardener") in each household, making suitable adjustments for the consumption requirements of different household members. The dependent variable (Y) is a measure of domestic-labor intensity in terms of acres cultivated by the household ("acres/gardener").[2]

In a community in which the social product is not redistributed (except, perhaps, by market exchange), labor intensity would be roughly proportional to the consumer/gardener ratio; that is, the linear regression of Y on X should have a positive slope and should pass through the origin (as shown in Figure 1.4(a)). Sahlins, however, argues that in primitive communities there is some

[2]The use of "ratio" variables such as these in regression analysis is the subject of controversy, especially when different variables share a denominator. For contrasting views of the ratio-variable debate, see Bollen and Ward (1979) and Long (1979).

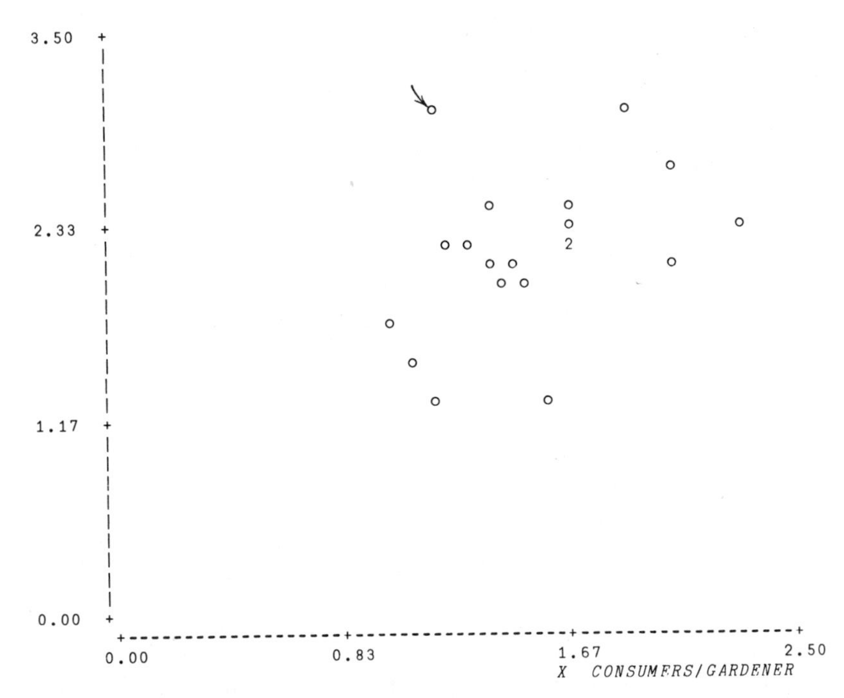

FIGURE 1.3. Scatterplot for Sahlins's data.

TABLE 1.1. Sahlins's Data on Agricultural Production in Mazulu Village

Household	Consumers/ Gardener X_i	Acres/ Gardener Y_i	$\hat{Y}_i$	E_i
1	1.00	1.71	1.892	-0.182
2	1.08	1.52	1.933	-0.413
3	1.15	1.29	1.969	-0.679
4	1.15	3.09	1.969	1.121
5	1.20	2.21	1.995	0.215
6	1.30	2.26	2.047	0.213
7	1.37	2.40	2.083	0.317
8	1.37	2.10	2.083	0.017
9	1.43	1.96	2.114	-0.154
10	1.46	2.09	2.129	-0.039
11	1.52	2.02	2.160	-0.140
12	1.57	1.31	2.186	-0.876
13	1.65	2.17	2.228	-0.058
14	1.65	2.28	2.228	0.052
15	1.65	2.41	2.228	0.182
16	1.66	2.23	2.233	-0.003
17	1.87	3.04	2.341	0.699
18	2.03	2.06	2.424	-0.364
19	2.05	2.73	2.434	0.296
20	2.30	2.36	2.563	-0.203

Source: Reprinted with permission from Sahlins (1972: Table 3.1), copyright Marshall Sahlins, 1972; after Scudder (1962: 258–261), first published in 1962 by Manchester University Press on behalf of the Rhodes-Livingstone Institute; reprinted on behalf of the Institute for African Studies in 1975.

redistribution in favor of households with relatively weak productive capacity. If this is the case, then we should expect a regression line with positive intercept (as in Figure 1.4(b)). In the extreme case, where production is proportional to productive capacity and where the social product is redistributed according to need, the regression of Y on X would be horizontal.

Though the pattern of points is not tight (as in Figure 1.2, for example), the scatterplot of Figure 1.3 suggests that domestic-labor intensity increases with the consumer/producer ratio. The data seem consistent with a linear relationship in which the intercept and slope are both positive. One data point (that for the fourth household, marked with an arrow in the figure) seems unusual, however, in that it combines a small X-value with a large Y-value.

Before we can analytically determine a well-fitting line for data such as those in Figure 1.3, there are two questions that must be answered: (1) How are we to measure the distance between a line and a data point? (2) How are we to cumulate these distances over all n observations? A straightforward answer to the first question follows from the functional relation between Y and X specified in equation (1.1). The residuals, $E_i = Y_i - \hat{Y}_i$, correspond to the signed vertical distances between the line and the data points, as shown in Figure 1.1. We therefore wish to make these vertical distances, which represent errors in predicting Y from X, as small as possible. Note that the asymmetry in roles of the independent and dependent variables implies an asymmetry in determining a well-fitting line; we might choose a different line to predict X from Y.

While minimizing vertical distances is implied by the nature of the putative relation of X and Y, how these distances should be combined to yield an overall index of fit is not so simply determined. It is clear at the outset that we need a criterion that cumulates residuals in some manner, for a line coming close to certain points may be far away from others. We might attempt to determine a regression line visually, but subjectively fit lines have certain deficiencies: (1) unless the linear relationship between X and Y is very strong, it may be difficult to fit a line with any subjective certainty; and (2) we require a method that is generalizable to more complex situations, and that may be used conveniently for statistical inference.

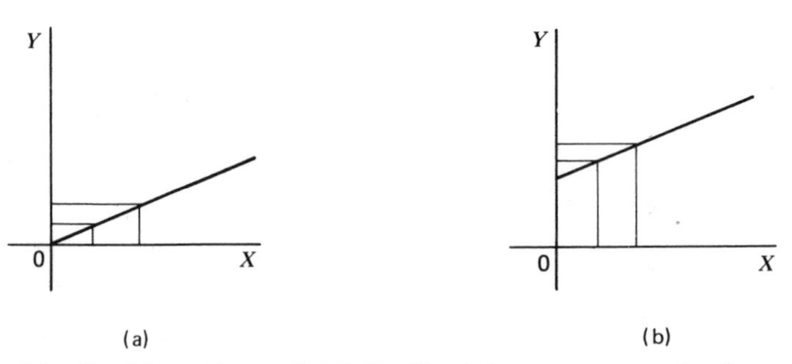

FIGURE 1.4. Possible patterns of relationship between consumers/gardener (X) and acres/gardener (Y).

Although we need to combine residuals for all observations, a simple sum $\Sigma_{i=1}^{n} E_i$ will not do. Since positive residuals can offset negative ones, a line that fits quite poorly can have a sum of residuals close to zero. Indeed, it is easily shown that, regardless of its orientation, any line passing through the means of the variables (the point $(\overline{X}, \overline{Y})$) has $\Sigma E_i = 0$. Such a line satisfies the equation $\overline{Y} = A + B\overline{X}$. Subtracting this equation from equation (1.1), we obtain

$$Y_i - \overline{Y} = B(X_i - \overline{X}) + E_i$$

Summing over all observations produces

$$\sum_{i=1}^{n} E_i = \sum (Y_i - \overline{Y}) - B\sum (X_i - \overline{X}) = 0 - B(0) = 0 \qquad (1.2)$$

A straightforward solution to our problem is to transform the residuals so that they all are positive, which may be accomplished by taking absolute values or by squaring. That is, to find a well-fitting line, we might choose values of A and B that minimize one or the other of the sums

$$\sum |E_i| = 0 \qquad (1.3)$$

$$\sum E_i^2 = 0 \qquad (1.4)$$

The minimum-absolute-deviation criterion (1.3), while intuitively attractive, is relatively unwieldy algebraically. The *least-squares criterion* (1.4), in contrast, is algebraically tractable and leads to a profound and widely applicable method of fitting statistical models to data. As we shall see, coefficients calculated according to the least-squares criterion have certain desirable statistical properties.

Nevertheless, our preference at this point for the least-squares criterion is primarily for expedience; we should be aware that, in certain circumstances, least-squares regression produces potentially misleading results. Because residuals are squared, atypical observations that deviate from a straight-line pattern implied by the body of the data can have disproportionate influence on the fitted line. Rather than accentuating the effects of discrepant data values, we may very well wish to discount them. We shall return to this issue in Section 3.2.3.

Finding the line that satisfies the least-squares criterion is a problem in minimization. We may regard the sum of squared residuals as a function of the coefficients A and B, since each choice of values for these coefficients determines a value for the sum of squares:

$$S(A, B) \equiv \sum_{i=1}^{n} E_i^2 = \sum (Y_i - A - BX_i)^2$$

We want to find the values of A and B that minimize the sum-of-squares function, treating the data X_i, Y_i as given. Differentiating $S(A, B)$ with respect to the regression coefficients, we get

$$\frac{\partial S(A, B)}{\partial A} = \sum(-1)(2)(Y_i - A - BX_i)$$

$$\frac{\partial S(A, B)}{\partial B} = \sum(-X_i)(2)(Y_i - A - BX_i)$$

Setting these partial derivatives to zero yields simultaneous linear equations for A and B, the so-called *normal equations* for simple regression:

$$An + B\sum X_i = \sum Y_i$$

$$A\sum X_i + B\sum X_i^2 = \sum X_i Y_i \qquad (1.5)$$

It is clear without formal proof that the values of A and B satisfying equations (1.5) minimize $S(A, B)$: (1) the sum-of-squares function is quadratic and thus has either a minimum or a maximum; and (2) a stationary point of this function cannot be a maximum, for A and B can be chosen arbitrarily poorly to make $S(A, B)$ as large as we wish.

Solving the normal equations produces the least-squares coefficients

$$A = \overline{Y} - B\overline{X}$$

$$B = \frac{n\Sigma X_i Y_i - \Sigma X_i \Sigma Y_i}{n\Sigma X_i^2 - (\Sigma X_i)^2} = \frac{\Sigma(X_i - \overline{X})(Y_i - \overline{Y})}{\Sigma(X_i - \overline{X})^2} \qquad (1.6)$$

The formula for A implies that the least-squares line passes through the means of the two variables. By equation (1.2), therefore, the least-squares residuals sum to zero. We may use the second normal equation to show that $\Sigma X_i E_i = 0$:

$$\sum X_i E_i = \sum X_i(Y_i - A - BX_i) = \sum X_i Y_i - A\sum X_i - B\sum X_i^2 = 0$$

Similarly, $\Sigma \hat{Y}_i E_i = 0$:

$$\sum \hat{Y}_i E_i = \sum(A + BX_i)(Y_i - A - BX_i)$$

$$= A\sum Y_i - nA^2 - AB\sum X_i + B\sum X_i Y_i - AB\sum X_i - B^2\sum X_i^2$$

$$= A\left(\sum Y_i - B\sum X_i\right) - nA^2 + B\left(\sum X_i Y_i - A\sum X_i - B\sum X_i^2\right)$$

$$= nA^2 - nA^2 + B(0) = 0$$

We shall refer to these properties of the least-squares residuals later in the present section.

It is clear from equations (1.6) that the least-squares coefficients are uniquely defined so long as the independent-variable values are not all identical, for when there is no variation in X, the denominator of B vanishes. This result is intuitively plausible: Only if the independent-variable scores are spread out can we hope to fit a line to the X, Y scatter.

We illustrate the least-squares calculations using Sahlins's data.

$$\overline{Y} = \frac{43.24}{20} = 2.162$$

$$\overline{X} = \frac{30.46}{20} = 1.523$$

$$\sum(X_i - \overline{X})(Y_i - \overline{Y}) = 1.183$$

$$\sum(X_i - \overline{X})^2 = 2.290$$

$$B = \frac{1.183}{2.290} = 0.5166$$

$$A = 2.162 - 0.5166(1.523) = 1.375$$

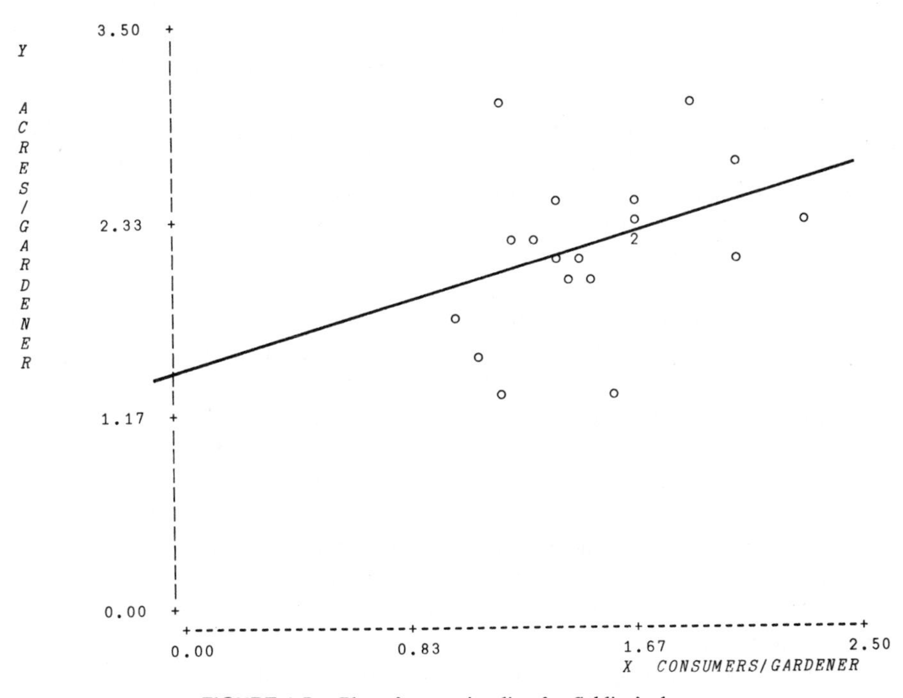

FIGURE 1.5. Plot of regression line for Sahlins's data.

Thus, the least-squares regression equation is $\hat{Y} = 1.38 + 0.517X$. Fitted values and residuals are given in Table 1.1 along with the data, and the regression line is graphed on the scatterplot in Figure 1.5. It is apparent from the graph that the fourth household, which we pointed to earlier as apparently deviant from the remainder of the data, has had an influence on the least-squares fit, increasing the intercept and decreasing the slope of the regression line.

Interpretation of the least-squares regression coefficients is straightforward. The value of $B = 0.517$ indicates that a unit increase in the number of consumers per gardener is associated on average with an increase of 0.517 in the number of acres tended by each gardener. Since the data are not longitudinal, the phrase "a unit increase" here implies not a literal change over time, but rather a static comparison between two households that differ by one in their X-values. Ordinarily, we may interpret the intercept A as the fitted value associated with $X = 0$. Here, however, the ratio of consumers to gardeners cannot fall below one (for each gardener is simultaneously a consumer). Still, the fact that A is positive implies that increases in X yield less-than-proportionate increases in Y, a finding consistent with Sahlins's substantive argument.

1.1.2. The Simple-Regression Model

We have thus far developed least-squares regression as a descriptive method for summarizing the relationship between two numerical variables. In this section, we shall specify a statistical model for simple regression analysis. The assumptions underlying the model provide a basis for model fitting and statistical inference.

Suppose that the procedure generating our X, Y data is repeatable, at least hypothetically. More specifically, imagine that independent-variable values remain fixed across replications of the study, but that the dependent-variable value associated with each observation varies from replication to replication. The stipulation that X is fixed corresponds realistically to experimental research, where the independent variable is manipulated by the experimenter and need not be changed when the experiment is repeated. In observational research, it may be possible to sample within independent-variable values, but to do so would be unusual. Section 1.3 introduces random independent variables; quite surprisingly, as we shall see, the theory developed for fixed X applies virtually without modification when X is a random variable. It is therefore advantageous to treat regression analysis in the simpler fixed-independent-variable context.

To say that X is fixed means that the independent-variable value for each of our n observations is specified before data are collected: $x_1, x_2, \ldots, x_n$. These values need not be distinct, and indeed there is a certain advantage to observing independent-variable values more than once. In the sample, each X-value is associated with a particular dependent-variable value: $y_1, y_2, \ldots, y_n$.

If we drew another sample, however, these Y-values would generally differ. We therefore regard the dependent-variable value associated with each observation as a random variable, taking on different values in different samples. We denote these random variables $Y_1, Y_2, \ldots, Y_n$.

The probability distribution for the ith observation, $p(Y_i)$, gives the conditional distribution of Y for X equal to its ith value: $p(Y|X = x_i)$ or, more simply, $p(Y|x_i)$. Each of these distributions has (we assume) an expected value $\mu_i \equiv E(Y_i)$. The simple-regression model specifies that the expected value of the dependent variable bears a systematic, linear relationship to the independent variable; that is

$$\mu_i = \alpha + \beta x_i \tag{1.7}$$

This model is graphed in Figure 1.6. The coefficients of the model α and β are unknown population parameters: we can observe specific values y_i of Y_i, but we cannot directly observe μ_i. The central task of regression analysis is to make statistical inferences about α and β on the basis of sample data.

The dispersion of Y_i around its expected value implies that there are factors other than X that affect Y. These factors may include other independent variables that influence Y together with random measurement error in Y. Alternatively, we may regard the determination of Y as intrinsically stochastic. The difference between the random variable Y_i and its expected value is the departure of Y_i from the population regression line relating Y to X. This difference, $\varepsilon_i \equiv Y_i - \mu_i$, is therefore called an *error* or *disturbance*. We use a Greek letter for the error random variable, because ε_i, which depends upon μ_i, is unobservable. From its definition, we can see that $E(\varepsilon_i) = 0$, and that,

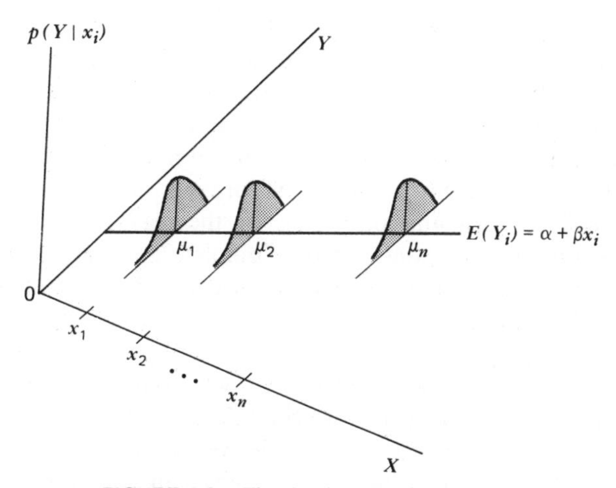

FIGURE 1.6. The simple regression model.

except for the difference in their expectations, the distribution of ε_i is identical to that of Y_i. Using the error random variable, we may rewrite the simple-regression model of equation (1.7) in the following manner:

$$Y_i = \alpha + \beta x_i + \varepsilon_i$$

To complete the specification of the simple-regression model, we make several assumptions in the next paragraph about the distribution of the errors. Since ε_i and Y_i differ only in expectation, these assumptions could also be stated for the dependent variable. The full set of assumptions is not required for all purposes; for example, we shall see that many properties of least-squares regression, hold independently of the assumption of normality (point 3). For completeness, we restate in point 1 what has been established thus far.

1. *Linearity.* Y and X are linearly related: $E(\varepsilon_i) = 0$.
2. *Homoscedasticity* (equality of error variance). Variation around the population regression line is constant across X-values: $V(\varepsilon_i) = \sigma_\varepsilon^2$.
3. *Normality.* Each error random variable follows a normal distribution. Since we have already assumed equal expectations and variances for the ε_i, their distributions are identical: $\varepsilon_i \sim N(0, \sigma_\varepsilon^2)$. Figure 1.6 illustrates the first three assumptions.
4. *Independence.* The observations are sampled independently: Any pair of errors ε_i and ε_j are independent for $i \neq j$. The assumption of independence needs to be justified by the procedure of a study.

1.1.3. Properties of the Least-Squares Estimators

We have now considered two aspects of linear regression: first, least-squares regression for a sample of observations; and second, regression as a statistical model applicable to a population. These two aspects are juxtaposed when we use the least-squares regression coefficients A and B to estimate the regression parameters α and β. As estimators, A and B are random variables, assuming different values in different samples. At present, we shall be content to establish several fundamental properties of the least-squares estimators in simple regression. We postpone a more complete and general discussion to Section 1.2.5. Since the slope coefficient is usually of more interest than the intercept, we shall work directly with B. Analogous results for A are given without proof.

The least-squares slope coefficient B is a linear combination of the observations Y_i. This property of B, which we shall demonstrate presently, is of great importance, because it makes it simple to derive the distribution of B from the distribution of the Y_i. The formula for B, given in equations (1.6), may be rearranged in the following manner:

$$B = \frac{\Sigma(x_i - \bar{x})(Y_i - \bar{Y})}{\Sigma(x_i - \bar{x})^2}$$

$$= \frac{\Sigma(x_i - \bar{x})Y_i - \bar{Y}\Sigma(x_i - \bar{x})}{\Sigma(x_i - \bar{x})^2}$$

$$= \sum_i \left[\frac{x_i - \bar{x}}{\Sigma_j(x_j - \bar{x})^2}\right] Y_i = \sum m_i Y_i \tag{1.8}$$

where $m_i \equiv (x_i - \bar{x})/\Sigma(x_j - \bar{x})^2$ is the weight attached to the ith observation in calculating B. Since the x_i are fixed, and the m_i depend only on the x_i, the weights are fixed as well.

Because B is a linear function of the Y_i, its expected value is easily established:

$$E(B) = \sum m_i E(Y_i) = \sum m_i(\alpha + \beta x_i) = \alpha \sum m_i + \beta \sum m_i x_i$$

$$= \frac{\alpha\Sigma(x_i - \bar{x})}{\Sigma(x_i - \bar{x})^2} + \beta\frac{\Sigma(x_i - \bar{x})x_i}{\Sigma(x_i - \bar{x})^2}$$

$$= 0 + \beta\frac{\Sigma x_i^2 - \bar{x}\Sigma x_i}{\Sigma(x_i - \bar{x})^2} = \beta\frac{\Sigma x_i^2 - n\bar{x}^2}{\Sigma(x_i - \bar{x})^2} = \beta \tag{1.9}$$

B therefore is an unbiased estimator of β. Notice that to establish the unbias of the least-squares estimator we only require the assumption of linearity; that is, $E(\varepsilon_i) = 0$, or equivalently, $E(Y_i) = \alpha + \beta x_i$. It may be shown similarly that $E(A) = \alpha$ (see Problem 1.6).

The sampling variance of B follows from equation (1.8) and from the assumptions of homoscedasticity $[V(Y_i) = V(\varepsilon_i) = \sigma_\varepsilon^2]$ and independence:

$$V(B) = \sum m_i^2 V(Y_i) = \sum m_i^2 \sigma_\varepsilon^2$$

$$= \frac{\sigma_\varepsilon^2 \Sigma(x_i - \bar{x})^2}{\left[\Sigma(x_i - \bar{x})^2\right]^2} = \frac{\sigma_\varepsilon^2}{\Sigma(x_i - \bar{x})^2} \tag{1.10}$$

Likewise, $V(A) = \sigma_\varepsilon^2 \Sigma x_i^2 / n\Sigma(x_i - \bar{x})^2$ (Problem 1.6).

Since B and A are linear functions of the Y_i, if the Y_i are normally distributed, then so are B and A. To summarize,

$$B \sim N\left(\beta, \frac{\sigma_\varepsilon^2}{\Sigma(x_i - \bar{x})^2}\right)$$
$$A \sim N\left(\alpha, \frac{\sigma_\varepsilon^2 \Sigma x_i^2}{n\Sigma(x_i - \bar{x})^2}\right) \tag{1.11}$$

1.1.4. Statistical Inference Concerning Regression Coefficients

Having derived the sampling distributions of the least-squares regression coefficients, we will make inferences about α and β on the basis of A and B. Once more, we postpone a comprehensive discussion to Section 1.2. In the current section we present some basic results for simple regression analysis, without proof, but with some intuitive justification. The theory in this section closely parallels statistical inference for a population mean and, therefore, the form of the argument should be generally familiar.

We already have seen that A and B provide unbiased point estimators of the population regression coefficients. Normal-distribution hypothesis tests and confidence intervals proceed from (1.11), but this result cannot be applied in practice because we generally do not know the population variance of the errors, σ_ε^2. Furthermore, we cannot observe the errors directly. We can, however, use the sample residuals to estimate the errors and, on this basis, secure an estimate of σ_ε^2.

It can be shown that the expectation of the sum of squared residuals is $E(\Sigma E_i^2) = (n - 2)\sigma_\varepsilon^2$. (We prove a more general result in Section 1.2.5.) Thus, an unbiased estimator of the error variance is given by

$$S_E^2 = \frac{\Sigma E_i^2}{n - 2}$$

Heuristically, we lose two *degrees of freedom* in calculating S_E^2 when we estimate the parameters α and β on the basis of A and B. We shall take up the concept of degrees of freedom in greater detail in Section 1.1.6. Using S_E^2, we may estimate the sampling variance of B:

$$S_B^2 = \frac{S_E^2}{\Sigma(x_i - \bar{x})^2}$$

Finally, dividing $B - \beta$ by S_B (called the *estimated standard error* of B) produces a t variable with $n - 2$ degrees of freedom: $t = (B - \beta)/S_B$. This result may be used in the usual manner to construct hypothesis tests and confidence intervals for β.

For example, to test the hypothesis H_0: $\beta = \beta_0$ against the one-sided alternative hypothesis H_a: $\beta > \beta_0$, we may calculate the test statistic $t_0 = (B - \beta_0)/S_B$. The p-value for the hypothesis is then the area to the right of t_0 under $t(n - 2)$. Likewise, to construct a two-sided confidence interval for β at the 100(1 − a) percent level of confidence, we take $\beta = B \pm t_{a/2}S_B$, where $t_{a/2}$ is the value of $t(n - 2)$ with a tail probability of $a/2$. Similar procedures may be applied to make inferences about α.

We shall once again use Sahlins's data for purposes of illustration. Note that the independent variable (consumers/gardener) cannot reasonably be construed as fixed, and hence the theory that we have developed to this point is not strictly applicable. Moreover, and more seriously, if Sahlins's characterization of primitive exchange is correct, production in the several households of the community should not be regarded as independent. With these caveats in mind, we proceed with the example.

Recall that Sahlins's argument is unusual in that its primary implication pertains to the intercept, which is expected to be positive. It is therefore appropriate to test the null hypothesis H_0: $\alpha = 0$ against the directional alternative H_a: $\alpha > 0$. A test of H_0: $\beta = 0$ against H_a: $\beta > 0$ is also of interest. Using the data in Table 1.1, calculation of the test statistics proceeds as follows:

$$\sum x_i^2 = 48.68$$

$$\sum (x_i - \bar{x})^2 = 2.290$$

$$\sum E_i^2 = 3.715$$

$$S_E^2 = \frac{3.715}{18} = 0.2064$$

$$S_A^2 = \frac{0.2064 \times 48.68}{20 \times 2.290} = 0.2194$$

$$S_B^2 = \frac{0.2064}{2.290} = 0.09013$$

For α, we have $t_0 = 1.375/\sqrt{0.2194} = 2.94$; and for β, $t_0 = 0.5166/\sqrt{0.09013} = 1.72$. Both t-statistics have 18 degrees of freedom; their respective p-values are 0.004 and 0.051. Assuming the appropriateness of these tests, we can therefore confidently conclude that α is positive. β also appears to be positive, but the evidence is weaker.

1.1.5. Correlation

Having calculated a least-squares regression line, it is of interest to determine how closely the line fits the scatter of points. This is a vague question, which may be answered in a variety of ways. The standard deviation of the residuals S_E, often called the *standard error of the estimate* (or of the regression), provides one sort of answer, for S_E is a type of average residual: the square root of the estimated population mean-squared error. Since S_E is measured in the units of the dependent variable, it provides an *absolute* measure of fit, and one that is meaningful if the dependent variable has a meaningful metric. For Sahlins's regression, for example, $S_E = \sqrt{0.2064} = 0.454$; therefore, the average distance of observations to the regression line is nearly one-half acres/gardener.

In contrast to the standard error of the estimate, the *correlation coefficient* provides a *relative* measure of fit. To what degree do our predictions of Y improve when we take the linear relation between X and Y into account? A relative index of fit requires a baseline—how well Y may be predicted if X is disregarded. To ignore the independent variable X is implicitly to fit the relation $\hat{Y}_i = C$, or equivalently, $Y_i = C + D_i$, where D_i is a residual. Adopting the least-squares criterion, we seek the value of C that minimizes ΣD_i^2. Proceeding as in Section 1.1.1, we have

$$S(C) = \sum D_i^2 = \sum (Y_i - C)^2$$

$$\frac{dS(C)}{dC} = -1(2)\sum (Y_i - C)$$

Setting the derivative to zero and solving for C produces $C = \Sigma Y_i/n = \bar{Y}$; the sum of squared residuals, therefore, is $\Sigma D_i^2 = \Sigma(Y_i - \bar{Y})^2$.

Note that $\Sigma(Y_i - \bar{Y})^2 \geq \Sigma(Y_i - \hat{Y}_i)^2$, where $\hat{Y}_i = A + Bx_i$ is calculated from the least-squares regression of Y on X. This inequality holds because the null model $\hat{Y} = C$ is a special case of $\hat{Y} = A + Bx$ (where $A = C$ and $B = 0$) and, therefore, cannot have a smaller sum of squared residuals. After all, the least-squares coefficients A and B minimize ΣE_i^2.

$\Sigma(Y_i - \bar{Y})^2$ is called the *total sum of squares* for Y, and is abbreviated *TSS*. $\Sigma(Y_i - \hat{Y}_i)^2 = \Sigma E_i^2$ is called the *residual sum of squares*, or *RSS*. The difference between the two, termed the *regression sum of squares*, *RegrSS* $\equiv$ TSS $-$ RSS, gives the reduction in squared error due to the linear regression. The ratio of RegrSS to TSS is the proportional reduction in squared error; this ratio defines the square of the correlation coefficient: $r^2 \equiv \text{RegrSS}/\text{TSS}$. To find r, we take the positive square root of r^2 when the slope B is positive, and the negative square root when B is negative. Thus, if there is a perfect, positive linear relationship between X and Y (i.e., if all residuals are zero, and $B > 0$), then $r = 1$. A perfect negative linear relationship produces $r = -1$. If there is no linear relationship, then RSS = TSS, RegrSS = 0, and $r = 0$. Between these extremes, r gives the direction of the linear relationship between X and Y, while r^2 may be interpreted as the proportion of the total variation of Y captured by its linear relationship to X.

It is instructive to examine the several sums of squares more closely. Starting with a single observation, we have the identity

$$Y_i - \overline{Y} = (Y_i - \hat{Y}_i) + (\hat{Y}_i - \overline{Y}) \tag{1.12}$$

This equation is interpreted geometrically in Figure 1.7. Squaring both sides of equation (1.12) and summing over observations produces

$$\sum (Y_i - \overline{Y})^2 = \sum (Y_i - \hat{Y}_i)^2 + \sum (\hat{Y}_i - \overline{Y})^2$$
$$+2\sum (Y_i - \hat{Y}_i)(\hat{Y}_i - \overline{Y}) \tag{1.13}$$

The last term in equation (1.13) is zero, for (recalling that $\Sigma E_i = \Sigma E_i \hat{Y}_i = 0$)

$$\sum (Y_i - \hat{Y}_i)(\hat{Y}_i - \overline{Y}) = \sum E_i \hat{Y}_i - \overline{Y} \sum E_i = 0$$

Thus, the regression sum of squares, which we previously defined as the difference TSS − RSS, may be written directly as $\Sigma(\hat{Y}_i - \overline{Y})^2$. This sort of decomposition of the total sum of squares into "explained" and "unexplained" components, paralleling the decomposition of each observation into fitted and residual values, is typical of linear models.

Although we have developed the topic of correlation from a regression perspective, it is possible to arrive at the simple correlation coefficient r by analogy with the correlation ρ between two random variables. Defining first the

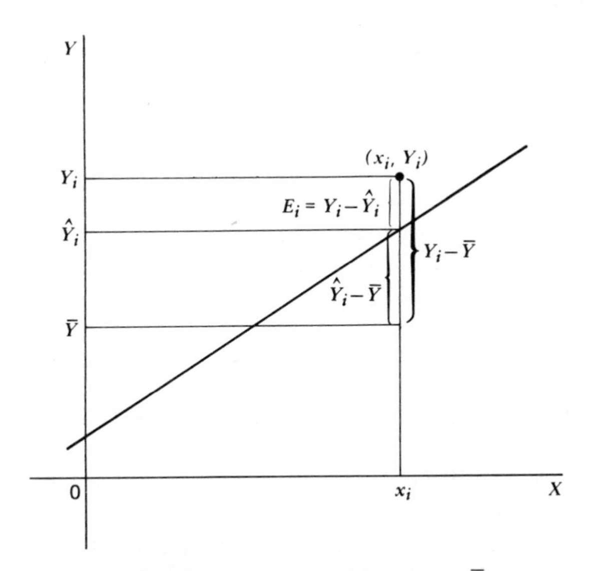

FIGURE 1.7. Decomposition of $Y_i - \overline{Y}$.

sample covariance between X and Y,

$$S_{XY} \equiv \frac{\Sigma(X_i - \bar{X})(Y_i - \bar{Y})}{n - 1}$$

we may then write

$$r = \frac{S_{XY}}{S_X S_Y} = \frac{\Sigma(X_i - \bar{X})(Y_i - \bar{Y})}{\sqrt{\Sigma(X_i - \bar{X})^2 \Sigma(Y_i - \bar{Y})^2}} \tag{1.14}$$

where S_X and S_Y are the sample standard deviations of X and Y, respectively. We shall prove the equivalence of the two formulas for r in the next section. It is immediately apparent from the symmetry of equation (1.14), however, that the correlation does not depend upon which of X and Y is treated as the dependent variable. This property of r is surprising in light of the asymmetry of the regression model used to define the sums of squares.

There is another central property, aside from symmetry, that distinguishes the correlation coefficient r from the regression slope B. B is measured in the units of the dependent variable per unit of the independent variable. r, in contrast, is unitless, as can be seen from either of its definitions. As a consequence, a change in scale of Y or X produces a compensating change in B but leaves r unaffected.

For Sahlins's regression analysis, TSS = 4.326, RSS = 3.715, and RegrSS = 0.6104. Thus $r^2 = 0.6104/4.326 = .1411$, and, since B is positive, $r = +.376$. The linear regression of Y on X, therefore, captures 14 percent of the variation in Y. Equivalently,

$$S_{XY} = \frac{1.182}{19} = 0.06221$$

$$S_X^2 = \frac{2.290}{19} = 0.1205$$

$$S_Y^2 = \frac{4.326}{19} = 0.2277$$

$$r = \frac{0.06221}{\sqrt{0.1205 \times 0.2277}} = .376$$

1.1.6. The Vector Geometry of Simple Regression

It will become clear in this chapter and the next two that linear algebra is the algebra of linear models. Vector geometry provides a spatial representation of linear algebra and therefore furnishes us with a powerful tool for understanding linear models. Few points in this book are developed exclusively in geometric terms. The reader who takes the time to master the geometric perspective, however, will find the effort worthwhile: Certain topics, such as degrees of freedom (discussed in this section), fitting analysis of variance models (Chapter 2), and principal-components analysis (Section 3.1.2), are most simply developed or understood through vector geometry.

We may write the simple regression model in vector form in the following manner:

$$\mathbf{y} = \alpha\mathbf{1} + \beta\mathbf{x} + \boldsymbol{\varepsilon} \tag{1.15}$$

where $\mathbf{y} \equiv (Y_1, Y_2, \ldots, Y_n)'$, $\mathbf{x} \equiv (x_1, x_2, \ldots, x_n)'$, $\boldsymbol{\varepsilon} \equiv (\varepsilon_1, \varepsilon_2, \ldots, \varepsilon_n)'$, and $\mathbf{1} \equiv (1, 1, \ldots, 1)'$; α and β are, as before, the population regression coefficients. The fitted regression equation is similarly

$$\mathbf{y} = A\mathbf{1} + B\mathbf{x} + \mathbf{e} \tag{1.16}$$

where $\mathbf{e} \equiv (E_1, E_2, \ldots, E_n)'$ is the vector of residuals, and A and B are the least-squares regression coefficients. From equation (1.15), we have $E(\mathbf{y}) \equiv \{E(Y_i)\} = \alpha\mathbf{1} + \beta\mathbf{x}$. Analogously, from equation (1.16), $\hat{\mathbf{y}} \equiv \{\hat{Y}_i\} = A\mathbf{1} + B\mathbf{x}$.

We are familiar with a geometric representation of X, Y data—the scatterplot—in which the axes of a two-dimensional coordinate space are defined by the variables X and Y, and the observations are represented as points in the space according to their (X_i, Y_i) coordinates. The scatterplot furnishes us with a practical data-analytic tool as well as with a conceptual device for developing the subject of linear regression.

We now exchange the familiar roles of variables and observations, defining an n-dimensional coordinate space for which the *axes* are given by the *observations* and in which the *variables* are plotted as *vectors*. Of course, since we generally have many more than three observations, it is not possible to draw a graph of the full vector space. Our interest, however, often inheres in two- and three-dimensional subspaces of this larger n-dimensional vector space. In these instances, as we shall see, graphic representation is both possible and illuminating. Moreover, the geometric point of view stands us in good stead even when the subspace of interest has more than three dimensions.

The simple-regression model of equation (1.15) is shown geometrically in Figure 1.8. The subspace depicted in this figure is of dimension three, and is spanned by the vectors $\mathbf{x}$, $\mathbf{y}$, and $\mathbf{1}$. Since $\mathbf{y}$ is a vector random variable which varies from sample to sample, the vector diagram represents a particular sample. The other vectors shown in the diagram clearly lie in the subspace spanned by $\mathbf{x}$, $\mathbf{y}$, and $\mathbf{1}$: $E(\mathbf{y})$ is a linear combination of $\mathbf{x}$ and $\mathbf{1}$; and the error vector $\boldsymbol{\varepsilon}$ is $\mathbf{y} - \alpha\mathbf{1} - \beta\mathbf{x}$. Note that although $\boldsymbol{\varepsilon}$ is nonzero in this sample, on average, over many samples, $E(\boldsymbol{\varepsilon}) = \mathbf{0}$.

Figure 1.9 represents the least-squares simple regression of Y on X, for the same data as shown in Figure 1.8. Figure 1.9 requires some explanation. We know that the fitted values $\hat{\mathbf{y}} = A\mathbf{1} + B\mathbf{x}$ are a linear combination of $\mathbf{1}$ and $\mathbf{x}$,

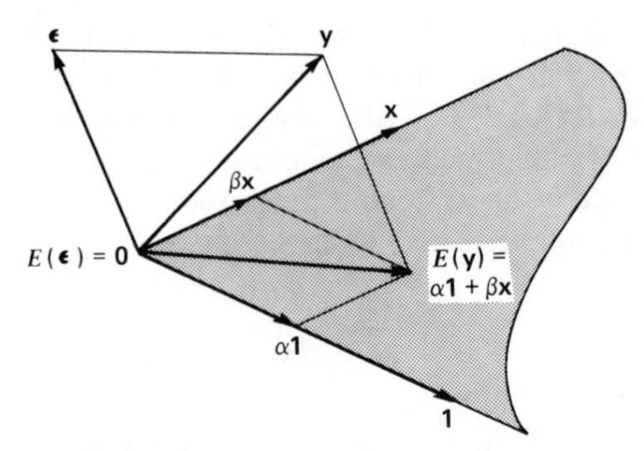

FIGURE 1.8. Vector geometry of simple regression.

and hence lie in the **1**, **x** plane. The residual vector $\mathbf{e} = \mathbf{y} - \hat{\mathbf{y}}$ has length $||\mathbf{e}|| = \sqrt{\Sigma E_i^2}$. The least-squares criterion interpreted geometrically, therefore, specifies that **e** should be as short as possible. Since the length of **e** is the distance between **y** and $\hat{\mathbf{y}}$, this length is minimized by taking $\hat{\mathbf{y}}$ as the orthogonal projection of **y** onto the **1**, **x** plane, as shown in the diagram.

Variables in Mean-Deviation Form We can simplify the vector representation for simple regression by eliminating the *constant regressor* **1** and the intercept coefficient A. This simplification is worthwhile for two reasons: (1) Our diagram becomes two-dimensional rather than three-dimensional; when we turn to multiple regression and introduce a second independent variable, then eliminating the constant leaves us with a three-dimensional rather than a four-dimensional subspace. (2) The various sums of squares that we identified in the previous section appear in the vector diagram when the constant is eliminated.

To delete A, we recall that $\overline{Y} = A + B\overline{x}$. Subtracting this equation from the fitted regression model $Y_i = A + Bx_i + E_i$ produces $(Y_i - \overline{Y}) = B(x_i - \overline{x}) + E_i$. Expressing the variables in *mean-deviation form* eliminates the regression constant. If we define $\mathbf{y}^* \equiv \{Y_i - \overline{Y}\}$ and $\mathbf{x}^* \equiv \{x_i - \overline{x}\}$, the vector form of

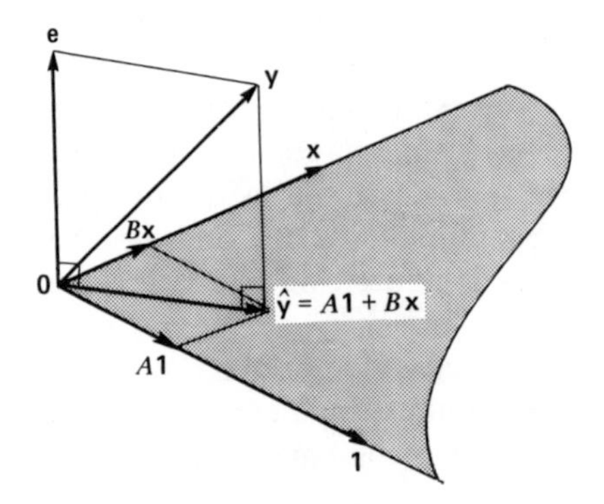

FIGURE 1.9. Vector geometry of least-squares fit.

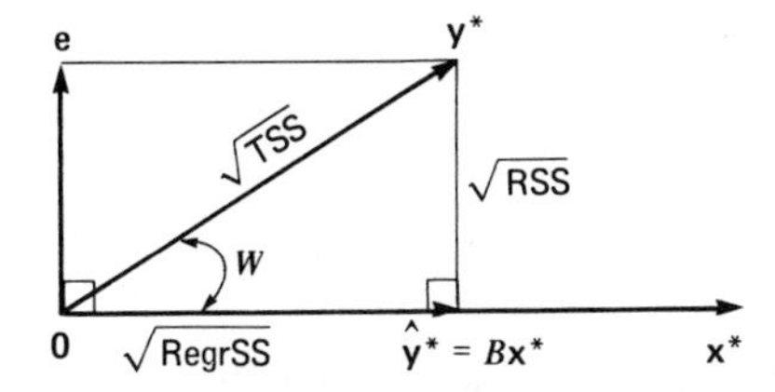

FIGURE 1.10. Vector geometry of least-squares fit for variables in mean-deviation form.

the fitted regression model becomes

$$\mathbf{y}^* = B\mathbf{x}^* + \mathbf{e} \tag{1.17}$$

The vector diagram corresponding to equation (1.17) is shown in Figure 1.10. By the same argument as before,[3] $\hat{\mathbf{y}}^* \equiv \{\hat{Y}_i - \overline{Y}\}$ is a multiple of $\mathbf{x}$; and the length of $\mathbf{e} = \mathbf{y}^* - \hat{\mathbf{y}}^*$ is minimized by taking $\hat{\mathbf{y}}^*$ as the orthogonal projection of $\mathbf{y}^*$ on $\mathbf{x}^*$. Thus,

$$B = \frac{\mathbf{x}^* \cdot \mathbf{y}^*}{\|\mathbf{x}^*\|^2} = \frac{\Sigma(x_i - \bar{x})(Y_i - \overline{Y})}{\Sigma(x_i - \bar{x})^2}$$

which, of course, is the familiar formula for the least-squares slope coefficient (equations (1.6)).

Sums of squares appear on the vector diagram as the squared lengths of vectors. We have already remarked that $\text{RSS} = \Sigma E_i^2 = \|\mathbf{e}\|^2$. Similarly, $\text{TSS} = \Sigma(Y_i - \overline{Y})^2 = \|\mathbf{y}^*\|^2$; and $\text{RegrSS} = \Sigma(\hat{Y}_i - \overline{Y})^2 = \|\hat{\mathbf{y}}^*\|^2$.

The correlation coefficient is therefore $r = \sqrt{\text{RegrSS}/\text{TSS}} = \|\hat{\mathbf{y}}^*\|/\|\mathbf{y}^*\|$. The vectors $\hat{\mathbf{y}}^*$ and $\mathbf{y}^*$ are, respectively, the adjacent side and hypothenuse for the angle W in the right triangle whose vertices are given by the tips of $\mathbf{0}$, $\mathbf{y}^*$, and $\hat{\mathbf{y}}^*$. Thus $r = \cos W$: The correlation between two variables (here, X and Y) is the cosine of the angle separating their mean-deviation vectors. When this angle is zero, one variable is a perfect linear function of the other, and $r = \cos 0 = 1$. When the vectors are orthogonal, $r = \cos 90° = 0$. We shall see shortly that when two variables are negatively correlated, $90° < W \leq 180°$.[4] The correlation $r = \cos W$ may be written directly as

$$r = \frac{\mathbf{x}^* \cdot \mathbf{y}^*}{\|\mathbf{x}^*\|\,\|\mathbf{y}^*\|} = \frac{\Sigma(x_i - \bar{x})(Y_i - \overline{Y})}{\sqrt{\Sigma(x_i - \bar{x})^2 \Sigma(Y_i - \overline{Y})^2}}$$

[3]The mean of the fitted values is the same as the mean of the dependent-variable values: $\overline{Y} = \Sigma Y_i/n = [\Sigma(\hat{Y}_i + E_i)]/n = (\Sigma\hat{Y}_i + \Sigma E_i)/n = \Sigma\hat{Y}_i/n = \overline{\hat{Y}}$.

[4]Recall that we need only consider angles between 0 and 180° for we may always examine the smaller of the two angles separating the $\mathbf{x}^*$ and $\mathbf{y}^*$ vectors. Since $\cos W = \cos(360° - W)$, this convention is of no consequence.

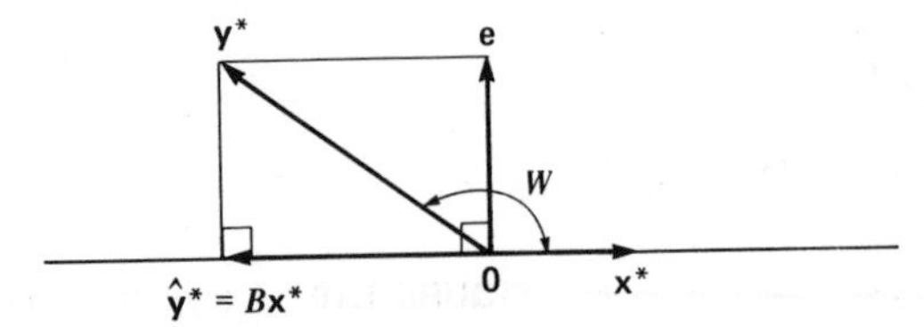

FIGURE 1.11. Vector geometry of least-squares fit for a negative relationship.

This is the alternative formula for the correlation coefficient presented in the previous section (equation (1.14)). The vector representation of simple regression, therefore, demonstrates the equivalence of the two definitions of r.

Figure 1.11 illustrates an inverse relationship between X and Y. All of the conclusions that we based on Figure 1.10 still hold. Since B is negative, $\hat{\mathbf{y}}^* = B\mathbf{x}^*$ is a negative multiple of the $\mathbf{x}^*$ vector. The correlation is still the cosine of W, though if we wish to define r in terms of vector lengths, we need to take the negative root of $\sqrt{\|\hat{\mathbf{y}}^*\|^2/\|\mathbf{y}^*\|^2}$.

Degrees of Freedom The vector representation of simple regression helps to clarify the concept of degrees of freedom. In general, sums of squares for linear models are the squared lengths of variable-vectors. The degrees of freedom associated with a sum of squares is the dimension of the subspace in which its vector must be located.

To begin with, consider the vector $\mathbf{y}$ in Figure 1.9; this vector may be located anywhere in the n-dimensional vector space. The *uncorrected* sum of squares $\Sigma Y_i^2 = \|\mathbf{y}\|^2$, therefore, has n degrees of freedom.

When we convert a variable to mean-deviation form, as for $\mathbf{y}^*$ in Figure 1.10, we confine its vector to an $(n - 1)$-dimensional subspace, losing one degree of freedom. This is easily seen for vectors in two-dimensional space. Let $\mathbf{y} = (Y_1, Y_2)'$, and $\mathbf{y}^* = (Y_1 - \overline{Y}, Y_2 - \overline{Y})'$. Then, since $\overline{Y} = (Y_1 + Y_2)/2$, we may write

$$\mathbf{y}^* = \left(\frac{Y_1 - Y_2}{2}, \frac{Y_2 - Y_1}{2}\right)' = \left(Y_1^*, -Y_1^*\right)'$$

Thus, all vectors $\mathbf{y}^*$ lie on a line through the origin, as shown in Figure 1.12: The subspace of all vectors $\mathbf{y}^*$ is one-dimensional.

Algebraically, by subtracting the mean from each of its coordinates, we have imposed a linear restriction on $\mathbf{y}^*$, insuring that its entries sum to zero: $\Sigma(Y_i - \overline{Y}) = 0$; among the n values of $Y_i - \overline{Y}$, only $n - 1$ are linearly independent. The total sum of squares $\text{TSS} = \Sigma(Y_i - \overline{Y})^2$, therefore, has $n - 1$ degrees of freedom.

We may extend this reasoning to the residual and regression sums of squares. $\hat{\mathbf{y}}^*$ in Figure 1.10 is a multiple of $\mathbf{x}^*$. The vector $\mathbf{x}^*$, in turn, is fixed and spans a one-dimensional subspace. Because $\hat{\mathbf{y}}^*$ necessarily lies somewhere in this one-dimensional subspace, $\text{RegrSS} = \|\hat{\mathbf{y}}^*\|^2$ has one degree of freedom.

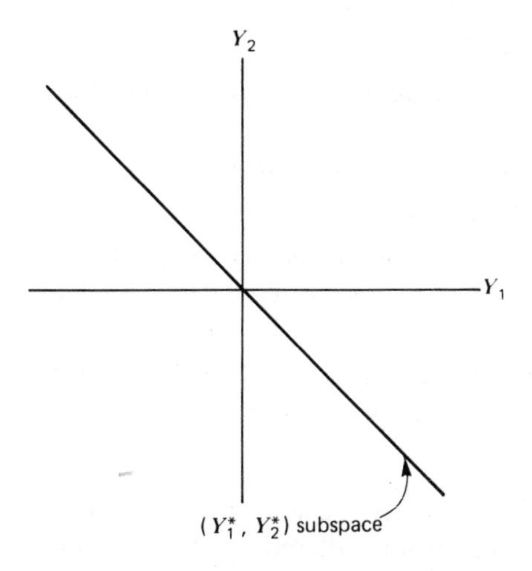

FIGURE 1.12. Subspace of mean-deviation vectors for $n = 2$.

The degrees of freedom for the residual sum of squares may be determined either from Figure 1.9 or Figure 1.10. In Figure 1.9, **y** lies in n-dimensional space. **x** and **1** are fixed and span a subspace of dimension two within the larger vector space. The location of **e** depends upon **y**, but in any event **e** is orthogonal to the plane spanned by **x** and **1**. Consequently, **e** lies in a subspace of dimension $n - 2$, and RSS $= \|\mathbf{e}\|^2$ has $n - 2$ degrees of freedom. In Figure 1.10, $\mathbf{y}^*$ lies in an $n - 1$ dimensional subspace; $\mathbf{x}^*$ spans a subspace of dimension one; **e** is orthogonal to $\mathbf{x}^*$; and, hence, RSS has $(n - 1) - 1 = n - 2$ degrees of freedom. Algebraically, the least-squares residuals **e** satisfy two independent linear restrictions, $\Sigma E_i = 0$, and $\Sigma E_i x_i = 0$ (or, equivalently, one of these and $\Sigma E_i \hat{Y}_i = 0$), accounting for the loss of two degrees of freedom.

PROBLEMS

1.1.# The following five observations were selected from among the 20 households in Mazulu village (Table 1.1):

Household	Consumers/ Gardener X_i	Acres/ Gardener Y_i
1	1.00	1.71
5	1.20	2.21
10	1.46	2.09
15	1.65	2.41
20	2.30	2.36

#Problems marked with a pound sign are intended for "hand" solution.

(a) Construct a scatterplot for Y and X.

(b) Find A and B for the least-squares regression of Y on X, and draw the regression line on the scatterplot.

(c) Calculate the standard error of the estimate S_E and the correlation r between X and Y. Interpret these statistics.

(d) Assuming that the observations were independently sampled, find the standard error of B, S_B, and construct a 90% confidence interval for β.

(e) Construct the geometric vector representation for the regression, showing the $\mathbf{x}^*$, $\mathbf{y}^*$, $\hat{\mathbf{y}}^*$, and $\mathbf{e}$ vectors drawn to scale; find the angle between $\mathbf{x}^*$ and $\mathbf{y}^*$.

1.2. Re-analyze Sahlins's data, given in Table 1.1, deleting the observation for the fourth household. Are the conclusions altered by this deletion?

1.3. Anscombe (1981) presents and analyzes the data given in Table 1.2, showing state per-capita public-school expenditures E (measured in dollars), per-capita personal annual income I (dollars), the proportion of residents under the age of 18 S (per thousand), and the proportion of the population residing in urban areas U (per thousand). Regress E on each of I, S, and U, commenting on the results.

1.4. Linear transformation of X and Y:

(a) Suppose that independent-variable values in Problem 1.2 are transformed according to the equation $X' = 10(X - 1)$, and that Y is regressed on X'. Without redoing the regression calculations in detail, find the following quantities: A', B', S'_E, S'_B, $t'_0 = B'/S'_B$, and r'.

(b) Now suppose that dependent-variable scores are transformed according to the formula $Y'' = 5(Y - 2)$, and that Y'' is regressed on X. Again, find A'', B'', S''_E, S''_B, $t''_0 = B''/S''_B$, and r''.

(c) In general, how are the results of a simple regression analysis affected by linear transformation of X and Y? In what sense, then, is the scaling of these variables arbitrary? Why might one scaling be preferred to another?

1.5. If we wish to estimate β precisely, why is it desirable to have spread-out independent-variable values? (*Hint*: Examine equation (1.10) for the sampling variance of B.) What happens to the least-squares estimator B when there is *no* variation in X?

1.6. Show that, under the assumptions of the regression model, the least-squares coefficient A is an unbiased estimator of α. Derive the sampling variance of A (given in (1.11)). Explain intuitively why the sampling

TABLE 1.2. Data on State Public-School Expenditures

State	E	I	S	U
Maine	189	2824	350.7	508
New Hampshire	169	3259	345.9	564
Vermont	230	3072	348.5	322
Massachusetts	168	3835	335.3	846
Rhode Island	180	3549	327.1	871
Connecticut	193	4256	341.0	774
New York	261	4151	326.2	856
New Jersey	214	3954	333.5	889
Pennsylvania	201	3419	326.2	715
Ohio	172	3509	354.5	753
Indiana	194	3412	359.3	649
Illinois	189	3981	348.9	830
Michigan	233	3675	369.2	738
Wisconsin	209	3363	360.7	659
Minnesota	262	3341	365.4	664
Iowa	234	3265	343.8	572
Missouri	177	3257	336.1	701
North Dakota	177	2730	369.1	443
South Dakota	187	2876	368.7	446
Nebraska	148	3239	349.9	615
Kansas	196	3303	339.9	661
Delaware	248	3795	375.9	722
Maryland	247	3742	364.1	766
Washington D.C.	246	4425	352.1	1000
Virginia	180	3068	353.0	631
West Virginia	149	2470	328.8	390
North Carolina	155	2664	354.1	450
South Carolina	149	2380	376.7	476
Georgia	156	2781	370.6	603
Florida	191	3191	336.0	805
Kentucky	140	2645	349.3	523
Tennessee	137	2579	342.8	588
Alabama	112	2337	362.2	584
Mississippi	130	2081	385.2	445
Arkansas	134	2322	351.9	500
Louisiana	162	2634	389.6	661
Oklahoma	135	2880	329.8	680
Texas	155	3029	369.4	797
Montana	238	2942	368.9	534
Idaho	170	2668	367.7	541
Wyoming	238	3190	365.6	605
Colorado	192	3340	358.1	785
New Mexico	227	2651	421.5	698
Arizona	207	3027	387.5	796
Utah	201	2790	412.4	804
Nevada	225	3957	385.1	809
Washington	215	3688	341.3	726
Oregon	233	3317	332.7	671
California	273	3968	348.4	909
Alaska	372	4146	439.7	484
Hawaii	212	3513	382.9	831

Source: F. J. Anscombe, *Computing in Statistical Science Through APL*, © Springer-Verlag New York Inc., 1981. Reprinted with permission from Anscombe (1981: 228).

variance of A is large when the X-scores are far from zero in one direction or the other.

1.7.* Maximum-likelihood estimation: Under the assumptions of the simple-regression model, the Y_i's are independently and normally distributed random variables with expectations $\alpha + \beta x_i$ and common variance σ_ε^2. Show that if these assumptions hold, then the least-squares coefficients A and B are the maximum-likelihood estimators of α and β, and $\hat{S}_E^2 = \Sigma E_i^2/n$ is the maximum-likelihood estimator of σ_ε^2. (*Hints*: Because of the assumption of independence, the joint probability density for the Y_i's is the product of their marginal probability densities. Find the log of the likelihood function; take the partial derivatives of the log likelihood with respect to the parameters α, β, and σ_ε^2; set these derivatives to zero; and solve for the maximum-likelihood estimators). A more general result is proven in Section 1.2.5.

1.8.* The Gauss–Markov Theorem: Equation (1.8) shows that the least-squares coefficient B is a *linear* function of the Y_i's, and equation (1.9) shows that B is an *unbiased* estimator of β. Prove that of all *linear unbiased* estimators, B has the smallest sampling variance and hence is maximally efficient. (*Hints*: Let $\tilde{B}$ be the minimum-variance linear unbiased estimator. Since $\tilde{B}$ is a linear estimator, $\tilde{B} = \Sigma w_i Y_i$ for some weights w_i. Since $\tilde{B}$ is unbiased,

$$E(\tilde{B}) = \Sigma w_i E(Y_i) = \Sigma w_i(\alpha + \beta x_i)$$

$$= \alpha \Sigma w_i + \beta \Sigma w_i x_i = \beta$$

Thus $\Sigma w_i = 0$ and $\Sigma w_i x_i = 1$. Finally, $V(\tilde{B}) = \Sigma w_i^2 V(Y_i) = \sigma_\varepsilon^2 \Sigma w_i^2$ is a minimum when Σw_i^2 is minimized. Use the *method of Lagrange multipliers* to minimize Σw_i^2 subject to the twin constraints $\Sigma w_i^2 = 0$ and $\Sigma w_i x_i - 1 = 0$. To employ this method, form the function

$$f(w_1, \ldots, w_n, L_1, L_2) = \sum w_i^2 + L_1\left(\sum w_i\right) + L_2\left(\sum w_i x_i - 1\right)$$

where L_1 and L_2 are the Lagrange multipliers; differentiate the function with respect to (each) w_i, L_1, and L_2; set the partial derivatives to zero; and solve for the w_i's (obtaining L_1 and L_2 as by-products). Show that the resulting w_i's are identical to the m_i's of the least-squares coefficient B.) A more general result is proven in Section 1.2.5.

*Starred problems are more difficult and are generally theoretical in nature.

1.2. MULTIPLE REGRESSION

There are several factors limiting the applicability of simple regression analysis: (1) The social world is complex, and social phenomena are likely to have multiple causes. Despite the central role played by abstraction in science, a model that specifies a single cause will not frequently prove satisfactory. In this section on multiple regression we extend the regression model to encompass more than one independent variable. (2) The simple-regression model specifies a linear relationship between X and Y. We deal with the issue of nonlinearity in Chapter 3. (3) Both the independent and dependent variables in regression analysis are quantitative. Qualitative variables, such as sex, social class, and labor-force status, may also be conceived as causes and effects. Chapter 2 introduces qualitative independent variables into linear models, while models for qualitative dependent variables are the subject of Chapter 5. (4) Causation itself may be complex. The effect of one variable may be the cause of another and, indeed, two variables may well affect each other mutually. Structural-equation models, taken up in Chapter 4, permit the specification of detailed causal relations among a set of variables.

In the simple-regression model, we express the dependent variable as a linear function of an independent variable plus error: $Y = f(X_1) + \varepsilon = \alpha + \beta X_1 + \varepsilon$. One of the components of the error is the aggregate effect on Y of independent variables other than X_1. In multiple regression analysis we remove (some of) these additional independent variables from the error and place them in the systematic part of the model: $Y = f(X_1, X_2, \ldots, X_k) + \varepsilon$. In this chapter we assume that f is linear, and thus

$$Y = \beta_0 + \beta_1 X_1 + \beta_2 X_2 + \cdots + \beta_k X_k + \varepsilon$$

In removing X_2 through X_k from the error, we realize two advantages. The first and more obvious gain is that we have reduced the size of the error and are therefore able to predict Y more precisely.

The second, less obvious, but more important advantage is that we may improve our assessment of the effect of X_1 on Y by including the other independent variables in the model, *if* these other variables are correlated with X_1. For if we omit a variable, say X_2, that both affects Y and is correlated with X_1, the expectation of the error is no longer zero for all values of X_1. The assumption that $E(\varepsilon)$ is everywhere zero, incorporated in the simple-regression model, causes us to come to a biased assessment of the effect of X_1 on Y: Part of what we attribute to X_1 is actually due to the correlated but omitted cause X_2. This line of reasoning is more naturally advanced if we conceive of the independent variables as random rather than fixed. We therefore return to the issue in Section 1.4, where we take it up in detail, after considering random independent variables in Section 1.3. Our treatment of multiple regression largely parallels the presentation of simple regression in Section 1.1.

1.2.1. Two Independent Variables

The development of multiple regression is greatly facilitated by casting the model in matrix terms. Matrices and linear algebra provide us not only with a general notation and a powerful system for computation, but also lend conceptual clarity to linear models. Despite (perhaps, because of) the power and compactness of matrix notation, there is a tendency to lose track of details when dealing with matrices. This problem can be partly overcome by remaining aware of what is contained within the matrices. To ease the transition to the matrix treatment of regression analysis, we shall briefly develop two-independent-variable multiple regression employing scalar notation. This preliminary treatment will also be useful to us later on, since many results in multiple regression analysis may be illustrated for two independent variables.

The multiple-regression model for two independent variables is given by the equation

$$Y_i = \beta_0 + \beta_1 x_{1i} + \beta_2 x_{2i} + \varepsilon_i$$

where Y_i is a numerical dependent-variable observation, x_{1i} and x_{2i} are fixed, numerical independent-variable scores, and ε_i is an error random variable with the same properties as the error in simple regression; that is, the errors are assumed to be normally and independently distributed with zero expectations and common variance, σ_ε^2.

Because of the assumption that $E(\varepsilon_i) = 0$, the model specifies that $E(Y)$ is a linear function of X_1 and X_2. The constant β_0 gives the value of $E(Y)$ when both independent variables are zero. β_1 and β_2 are called *partial regression coefficients* or partial slopes. β_1 gives the expected increase in Y for a unit increase in X_1, holding X_2 constant:

$$[\beta_0 + \beta_1(x_{1i} + 1) + \beta_2 x_{2i}] - (\beta_0 + \beta_1 x_{1i} + \beta_2 x_{2i}) = \beta_1$$

Note that because the regression model is linear, β_1 is also the partial derivative of $E(Y)$ with respect to X_1. β_2 has a similar interpretation. Since the effect of simultaneous change in X_1 and X_2 is the sum of their separate (or partial) changes, the linear multiple-regression model is additive:[5] The regression surface $E(Y) = \beta_0 + \beta_1 X_1 + \beta_2 X_2$ describes a plane in the three-dimensional variable space, as shown in Figure 1.13.

The intuitive rationale for least-squares fitting is similar for multiple regression and simple regression. Given a three-dimensional scatter of points, we wish to fit a regression plane that comes as close to the points as possible; since

[5]Though we use shorthand terms such as "increase" and "hold constant," if our data are cross-sectional, the comparison that is implied is a static one, as in the case of the simple-regression model. Moreover, saying that β_1 gives the effect of X_1 with X_2 held constant does not imply that X_1 and X_2 are in fact causally or statistically unrelated.

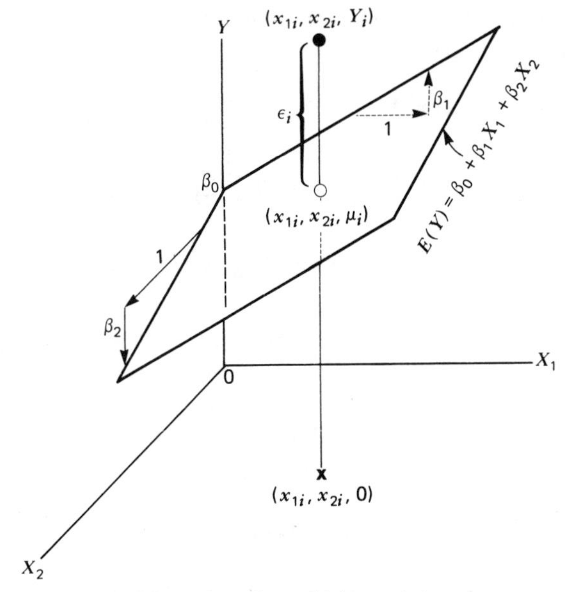

FIGURE 1.13. The multiple-regression plane.

the vertical distances of the points to the plane are the regression residuals, it is reasonable to fit a plane that minimizes the sum of squared vertical distances. The statistical properties of the least-squares estimators in multiple regression will be developed in Section 1.2.5.

Denoting the fitted regression coefficients B_0, B_1, and B_2, the residual for the ith observation may be written

$$E_i = Y_i - \hat{Y}_i = Y_i - B_0 - B_1 x_{1i} - B_2 x_{2i} \tag{1.18}$$

Squaring equation (1.18) and summing over observations, we obtain

$$S(B_0, B_1, B_2) = \sum E_i^2 = \sum (Y_i - B_0 - B_1 x_{1i} - B_2 x_{2i})^2$$

To minimize $S(B_0, B_1, B_2)$ we find its partial derivatives with respect to the coefficients:

$$\frac{\partial S(B_0, B_1, B_2)}{\partial B_0} = \sum (-1)(2)(Y_i - B_0 - B_1 x_{1i} - B_2 x_{2i})$$

$$\frac{\partial S(B_0, B_1, B_2)}{\partial B_1} = \sum (-x_{1i})(2)(Y_i - B_0 - B_1 x_{1i} - B_2 x_{2i})$$

$$\frac{\partial S(B_0, B_1, B_2)}{\partial B_2} = \sum (-x_{2i})(2)(Y_i - B_0 - B_1 x_{1i} - B_2 x_{2i})$$

Setting these partial derivatives to zero and rearranging terms produces the following normal equations:

$$\begin{aligned} B_0 n \quad &+ B_1 \sum x_{1i} \quad + B_2 \sum x_{2i} \quad = \sum Y_i \\ B_0 \sum x_{1i} &+ B_1 \sum x_{1i}^2 \quad + B_2 \sum x_{1i} x_{2i} = \sum x_{1i} Y_i \\ B_0 \sum x_{2i} &+ B_1 \sum x_{1i} x_{2i} + B_2 \sum x_{2i}^2 \quad = \sum x_{2i} Y_i \end{aligned} \tag{1.19}$$

Since (1.19) is a system of three linear equations in three unknowns, it generally provides a unique solution for the least-squares regression coefficients B_0, B_1, and B_2. In fact, we can write out the solution explicitly, if somewhat tediously; dropping the subscript i for observation, and using asterisks to denote variables in mean-deviation form (e.g., $Y^* \equiv Y_i - \overline{Y}$),

$$\begin{aligned} B_0 &= \overline{Y} - B_1 \overline{x}_1 - B_2 \overline{x}_2 \\ B_1 &= \frac{\Sigma x_1^* Y^* \, \Sigma x_2^{*2} - \Sigma x_2^* Y^* \, \Sigma x_1^* x_2^*}{\Sigma x_1^{*2} \Sigma x_2^{*2} - (\Sigma x_1^* x_2^*)^2} \\ B_2 &= \frac{\Sigma x_2^* Y^* \, \Sigma x_1^{*2} - \Sigma x_1^* Y^* \, \Sigma x_1^* x_2^*}{\Sigma x_1^{*2} \Sigma x_2^{*2} - (\Sigma x_1^* x_2^*)^2} \end{aligned} \tag{1.20}$$

The least-squares regression coefficients, therefore, are uniquely defined so long as $\Sigma x_1^{*2} \Sigma x_2^{*2} \neq (\Sigma x_1^* x_2^*)^2$. This condition is satisfied unless X_1 and X_2 are perfectly correlated (as is apparent from the formula for the correlation between two variables, equation (1.14)), or unless one of the independent variables is invariant.

The data in Table 1.3 were employed by Angell (1951)[6] in a multiple regression analysis. Quite apart from their substantive content, these data are of interest because they represent an early use of multiple regression in sociological research. Angell's index of "moral integration" Y combines information on the incidence of crime with data on social-welfare expenditures: Cities that have low crime rates and high levels of welfare effort are assigned high moral-integration scores. The ethnic heterogeneity index X_1 is calculated from the proportions of nonwhites and foreign-born whites residing in the cities, while the mobility index X_2 is constructed from the proportions of residents moving into and out of the cities. Angell argues that the moral integration of cities should decrease with ethnic heterogeneity and geographic

[6]The data reproduced in Table 1.3 are the part of Angell's data pertaining to non-Southern cities. This subset appears in Blalock (1979: 411) as an exercise in regression analysis. The remainder of the dataset is given in Problem 1.9.

TABLE 1.3. Angell's Data on Moral Integration of non-Southern U.S. Cities

City	Moral Integration Y_i	Heterogeneity X_{1i}	Mobility X_{2i}	$\hat{Y}_i$	E_i
Rochester	19.0	20.6	15.0	15.17	3.83
Syracuse	17.0	15.6	20.2	14.89	2.11
Worcester	16.4	22.1	13.6	15.22	1.18
Erie	16.2	14.0	14.8	16.32	-0.12
Milwaukee	15.8	17.4	17.6	15.15	0.65
Bridgeport	15.3	27.9	17.5	13.41	1.89
Buffalo	15.2	22.3	14.7	14.95	0.25
Dayton	14.3	23.7	23.8	12.76	1.54
Reading	14.2	10.6	19.4	15.90	-1.70
Des Moines	14.1	12.7	31.9	12.87	1.23
Cleveland	14.0	39.7	18.6	11.20	2.80
Denver	13.9	13.0	34.5	12.26	1.64
Peoria	13.8	10.7	35.1	12.52	1.28
Wichita	13.6	11.9	42.7	10.68	2.92
Trenton	13.0	32.5	15.8	13.00	0.00
Grand Rapids	12.8	15.7	24.2	14.02	-1.22
Toledo	12.7	19.2	21.6	13.99	-1.29
San Diego	12.5	15.9	49.8	8.49	4.01
Baltimore	12.0	45.8	12.1	11.57	0.43
South Bend	11.8	17.9	27.4	12.96	-1.16
Akron	11.3	20.4	22.1	13.68	-2.38
Detroit	11.1	38.3	19.5	11.24	-0.14
Tacoma	10.9	17.8	31.2	12.16	-1.26
Flint	9.8	19.3	32.2	11.70	-1.90
Spokane	9.6	12.3	38.9	11.43	-1.83
Seattle	9.0	23.9	34.2	10.50	-1.50
Indianapolis	8.8	29.2	23.1	11.99	-3.19
Columbus	8.0	27.4	25.0	11.88	-3.88
Portland Oregon	7.2	16.4	35.8	11.41	-4.21

Source: Angell (1951: Table 9). Reprinted from the *American Journal of Sociology* by permission of the University of Chicago Press. Copyright 1951, The University of Chicago Press.

mobility. Although the method of data collection does not insure independent errors, as would be the case in a randomized experiment or in survey research employing an independent random sample, independence of errors among cities seems a reasonable assumption here. Further, we should probably not conceive of X_1 and X_2 as fixed. We may interpret the least-squares regression descriptively, however.

In analyzing bivariate X, Y data, we begin by drawing a scatterplot. When there are two independent variables, it is possible to construct a three-dimensional scatterplot model, or to draw a two-dimensional representation of the scatterplot employing, for example, a perspective drawing. We shall not pursue these possibilities, however, because they are not easily generalized when the number of independent variables exceeds two.

One possibility is to plot Y separately against each X_j. This is a potentially misleading practice, though, for in multiple regression analysis we are interested in the *partial* relationship between the dependent variable and each independent variable, statistically holding the other independent variable (or variables) constant. Plotting the dependent variable separately against each independent variable displays the simple or *zero-order* relationship between Y and X_j; if the independent variables are correlated with each other, these zero-order relationships may be very different from the partial relationships. The potential discrepancy between zero-order and partial relationships, of course, is one of our primary reasons for employing multiple regression. (See Daniel and Wood (1980: 50–53) for an example of misleading bivariate scatterplots.) In Chapter 3, we shall develop residual-based plotting methods that are appropriate for multiple regression.

The following quantities are calculated from Angell's data given in Table 1.3:

$$\overline{Y} = 12.87$$

$$\bar{x}_1 = 21.18$$

$$\bar{x}_2 = 25.25$$

$$\sum x_1^{*2} = 2268$$

$$\sum x_2^{*2} = 2610$$

$$\sum x_1^* x_2^* = -1248$$

$$\sum x_1^* Y^* = -112.0$$

$$\sum x_2^* Y^* = -350.7$$

Substituting these results into equations (1.20), we obtain $B_1 = -0.1674$, $B_2 = -0.2144$, and $B_0 = 21.83$. Thus, the fitted least-squares regression equation

$$\hat{Y} = 21.8 - 0.167X_1 - 0.214X_2 \tag{1.21}$$

confirms Angell's expectations: Integration has a negative partial relationship both to heterogeneity and to mobility. We conclude, for example, that when ethnic heterogeneity increases one unit, holding geographic mobility constant, moral integration decreases on average 0.167 units.

1.2.2. Multiple Regression in Matrix Form

The general multiple-regression model is given by the equation

$$Y_i = \beta_0 + \beta_1 x_{1i} + \beta_2 x_{2i} + \cdots + \beta_k x_{ki} + \varepsilon_i$$

Collecting the independent variables in a row vector, appending a one for the regression constant, and placing the parameters in a column vector, we may rewrite the regression model as

$$Y_i = (1, x_{1i}, x_{2i}, \ldots, x_{ki}) \begin{pmatrix} \beta_0 \\ \beta_1 \\ \beta_2 \\ \vdots \\ \beta_k \end{pmatrix} + \varepsilon_i$$

$$= \underset{(1\times k+1)}{\mathbf{x}_i'} \; \underset{(k+1\times 1)}{\boldsymbol{\beta}} + \varepsilon_i \qquad (1.22)$$

For a sample of n observations, we have n such equations, which may be combined into a single matrix equation:

$$\begin{pmatrix} Y_1 \\ Y_2 \\ \vdots \\ Y_n \end{pmatrix} = \begin{pmatrix} 1 & x_{11} & \cdots & x_{k1} \\ 1 & x_{12} & \cdots & x_{k2} \\ \vdots & \vdots & & \vdots \\ 1 & x_{1n} & \cdots & x_{kn} \end{pmatrix} \begin{pmatrix} \beta_0 \\ \beta_1 \\ \vdots \\ \beta_k \end{pmatrix} + \begin{pmatrix} \varepsilon_1 \\ \varepsilon_2 \\ \vdots \\ \varepsilon_n \end{pmatrix} \qquad (1.23)$$

$$\underset{(n\times 1)}{\mathbf{y}} = \underset{(n\times k+1)}{\mathbf{X}} \; \underset{(k+1\times 1)}{\boldsymbol{\beta}} + \underset{(n\times 1)}{\boldsymbol{\varepsilon}}$$

With suitable specification of the contents of **X**, called the *design matrix*,[7] equation (1.23) serves not only for multiple regression, but for linear models generally.

[7] The term "design matrix" originates in experimental applications of linear models, where the independent variables and hence the **X** matrix derive from the design of an experiment. Note that to maintain consistency with our scalar notation, we have made the second subscript of **X** the row subscript, violating the usual matrix convention.

Since $\boldsymbol{\varepsilon}$ is a vector random variable, the assumptions of the multiple-regression model may be compactly restated in matrix form. The errors are assumed to be independently and normally distributed with zero expectation and common variance. Thus $\boldsymbol{\varepsilon}$ follows a multivariate-normal distribution with expectation $E(\boldsymbol{\varepsilon}) = \underset{(n \times 1)}{\mathbf{0}}$ and covariance matrix $V(\boldsymbol{\varepsilon}) = E(\boldsymbol{\varepsilon}\boldsymbol{\varepsilon}') = \sigma_\varepsilon^2 \mathbf{I}_n$; in symbols, $\boldsymbol{\varepsilon} \sim N_n(\mathbf{0}, \sigma_\varepsilon^2 \mathbf{I}_n)$. The distribution of $\mathbf{y}$ follows immediately.

$$\boldsymbol{\mu} \equiv E(\mathbf{y}) = E(\mathbf{X}\boldsymbol{\beta} + \boldsymbol{\varepsilon}) = \mathbf{X}\boldsymbol{\beta} + E(\boldsymbol{\varepsilon}) = \mathbf{X}\boldsymbol{\beta}$$

$$V(\mathbf{y}) = E[(\mathbf{y} - \boldsymbol{\mu})(\mathbf{y} - \boldsymbol{\mu})'] = E[(\mathbf{y} - \mathbf{X}\boldsymbol{\beta})(\mathbf{y} - \mathbf{X}\boldsymbol{\beta})']$$

$$= E(\boldsymbol{\varepsilon}\boldsymbol{\varepsilon}') = \sigma_\varepsilon^2 \mathbf{I}_n \tag{1.24}$$

Furthermore, since $\mathbf{y}$ is simply a translation of $\boldsymbol{\varepsilon}$ to a different expectation, $\mathbf{y}$ too is normally distributed: $\mathbf{y} \sim N_n(\mathbf{X}\boldsymbol{\beta}, \sigma_\varepsilon^2 \mathbf{I}_n)$.

To find the least-squares regression coefficients, we write the regression equation

$$\mathbf{y} = \mathbf{X}\mathbf{b} + \mathbf{e}$$

where $\mathbf{b} = (B_0, B_1, \ldots, B_k)'$ is the vector of fitted coefficients, and $\mathbf{e} = (E_1, E_2, \ldots, E_n)'$ is the vector of residuals. We seek the coefficient vector $\mathbf{b}$ that minimizes the residual sum of squares

$$S(\mathbf{b}) = \sum E_i^2 = \mathbf{e}'\mathbf{e} = (\mathbf{y} - \mathbf{X}\mathbf{b})'(\mathbf{y} - \mathbf{X}\mathbf{b})$$

$$= \mathbf{y}'\mathbf{y} - \mathbf{y}'\mathbf{X}\mathbf{b} - \mathbf{b}'\mathbf{X}'\mathbf{y} + \mathbf{b}'\mathbf{X}'\mathbf{X}\mathbf{b}$$

$$= \mathbf{y}'\mathbf{y} - (2\mathbf{y}'\mathbf{X})\mathbf{b} + \mathbf{b}'(\mathbf{X}'\mathbf{X})\mathbf{b} \tag{1.25}$$

Although matrix multiplication is not generally commutative, each product in equation (1.25) is (1×1); thus $\mathbf{y}'\mathbf{X}\mathbf{b} = \mathbf{b}'\mathbf{X}'\mathbf{y}$, which justifies the transition to the last line of the equation.

From the point of view of $\mathbf{b}$, equation (1.25) consists of a constant, a linear form in $\mathbf{b}$, and a quadratic form in $\mathbf{b}$. To minimize $S(\mathbf{b})$, we find its vector derivative with respect to $\mathbf{b}$:

$$\frac{\partial S(\mathbf{b})}{\partial \mathbf{b}} = \mathbf{0} - 2\mathbf{X}'\mathbf{y} + 2\mathbf{X}'\mathbf{X}\mathbf{b}$$

Equating this derivative to zero produces the normal equations in matrix form

$$\mathbf{X}'\mathbf{X}\mathbf{b} = \mathbf{X}'\mathbf{y} \tag{1.26}$$

which will serve us generally for fitting linear models by least squares. There are $k + 1$ linear normal equations in the same number of unknown coefficients. If $\mathbf{X'X}$ is nonsingular, therefore, we may solve uniquely for the least-squares regression coefficients

$$\mathbf{b} = (\mathbf{X'X})^{-1}\mathbf{X'y}$$

The rank of $\mathbf{X'X}$ is equal to the rank of $\mathbf{X}$. (1) Since the rank of $\mathbf{X}$ can be no greater than the smaller of n and $k + 1$, for the least-squares coefficients to be uniquely defined (i.e., for $\mathbf{X'X}$ to be nonsingular), we require at least as many observations (n) as there are coefficients in the model ($k + 1$). This requirement is intuitively sensible: We cannot, for example, fit a unique line to a single data point, nor can we fit a unique plane to two data points. In most applications, n greatly exceeds $k + 1$. (2) The $k + 1$ columns of $\mathbf{X}$ must be linearly independent. This implies that no independent variable may be a perfect linear function of others, and that no independent variable may be invariant. In the latter instance, an invariant independent variable is a perfect multiple of the first column of $\mathbf{X}$, the constant regressor. In applications, these requirements are usually met; $\mathbf{X'X}$, therefore, is generally nonsingular, and the least-squares regression coefficients are uniquely defined. We shall see in Chapter 3, however, that less-than-perfect collinearity of the columns of $\mathbf{X}$ can cause statistical difficulties.

Notice that the second partial derivative of the sum of squared residuals is $\partial^2 S(\mathbf{b})/\partial\mathbf{b}\,\partial\mathbf{b}' = 2\mathbf{X'X}$. Because $\mathbf{X'X}$ is positive definite when $\mathbf{X}$ is of full-column rank, the solution $\mathbf{b} = (\mathbf{X'X})^{-1}\mathbf{X'y}$ represents a minimum of $S(\mathbf{b})$.

Reexamining the normal equations (1.26), we note that the $\mathbf{X'X}$ matrix contains sums of squares and products among the regressors, while $\mathbf{X'y}$ is a vector of sums of cross products between the regressors and the dependent variable. Forming these matrix products, and writing out the normal equations in scalar format, yields a now-familiar pattern (cf. equations (1.5) and (1.19)):

$$\begin{aligned}
B_0 n \quad &+ B_1\sum x_{1i} &+ \cdots + B_k\sum x_{ki} &= \sum Y_i \\
B_0\sum x_{1i} &+ B_1\sum x_{1i}^2 &+ \cdots + B_k\sum x_{1i}x_{ki} &= \sum x_{1i}Y_i \\
&\vdots \\
B_k\sum x_{ki} &+ B_1\sum x_{1i}x_{ki} &+ \cdots + B_k\sum x_{ki}^2 &= \sum x_{ki}Y_i
\end{aligned}$$

To write an explicit scalar solution to these normal equations would be impractical, even for moderate values of k.

For Angell's multiple-regression data, the matrix calculations proceed as follows:

$$\mathbf{X'X} = \begin{pmatrix} 29 & 614.2 & 732.3 \\ 614.2 & 15277. & 14261. \\ 732.3 & 14261. & 21102. \end{pmatrix}$$

$$(\mathbf{X'X})^{-1} = \begin{pmatrix} 0.9405 & -0.01990 & -0.01919 \\ -0.01990 & 0.0005984 & 0.0002862 \\ -0.01919 & 0.0002862 & 0.0005199 \end{pmatrix}$$

$$\mathbf{X'y} = (373.3, 7794., 9076.)'$$

$$\mathbf{b} = (21.8, -0.167, -0.214)' \tag{1.27}$$

1.2.3. Multiple Correlation

In Section 1.1.5 we defined the simple correlation coefficient in terms of sums of squares. That treatment applies, virtually unchanged, to the multiple-regression model. The various sums of squares are defined in the same manner as before:

$$\text{TSS} = \sum (Y_i - \overline{Y})^2 = \mathbf{y'y} - n\overline{Y}^2$$

$$\text{RegrSS} = \sum (\hat{Y}_i - \overline{Y})^2 = \mathbf{b'X'Xb} - n\overline{Y}^2$$

$$\text{RSS} = \sum E_i^2 = \mathbf{e'e}$$

The square of the *multiple correlation coefficient* is defined as the ratio of the regression to the total sum of squares: $R^2 = \text{RegrSS}/\text{TSS}$. The multiple correlation coefficient R is the positive square root of R^2. We shall show in Section 1.2.4: (1) that R is the simple correlation between Y and $\hat{Y}$; and (2) that the total sum of squares in multiple regression may be partitioned in the same manner as in simple regression.

For the Angell multiple-regression data, we calculate

$$\text{TSS} = 226.3$$

$$\text{RegrSS} = 93.97$$

$$\text{RSS} = 132.3$$

$$R^2 = \frac{93.97}{226.3} = .415 \tag{1.28}$$

Thus, heterogeneity and mobility together account for about 40 percent of the variation in moral integration among Angell's cities.

1.2.4. The Vector Geometry of Multiple Regression

To develop the vector geometry of multiple regression, we shall work primarily with two independent variables. By expressing all variables in mean-deviation form and, consequently, eliminating the regression constant, we confine the subspace of interest to three dimensions.

Consider the fitted model

$$\mathbf{y} = B_0\mathbf{1} + B_1\mathbf{x}_1 + B_2\mathbf{x}_2 + \mathbf{e} \tag{1.29}$$

where $\mathbf{y}$ is the vector of dependent-variable observations, $\mathbf{x}_1$ and $\mathbf{x}_2$ are independent-variable vectors, $\mathbf{e}$ is the vector of residuals, and $\mathbf{1}$ is a vector of ones; B_0, B_1, and B_2, of course, are the least-squares regression coefficients. From each observation of equation (1.29) we subtract $\bar{Y} = B_0 + B_1\bar{x}_1 + B_2\bar{x}_2$, obtaining

$$\mathbf{y}^* = B_1\mathbf{x}_1^* + B_2\mathbf{x}_2^* + \mathbf{e} \tag{1.30}$$

Here, $\mathbf{y}^*$, $\mathbf{x}_1^*$, and $\mathbf{x}_2^*$ are vectors of mean deviations.

Figure 1.14(a) shows the three-dimensional vector diagram for the fitted model of equation (1.30), while Figure 1.14(b) depicts the independent-variable plane. Note that $\hat{\mathbf{y}}^* = B_1\mathbf{x}_1^* + B_2\mathbf{x}_2^*$ is a linear combination of the regressors and therefore lies in the $\mathbf{x}_1^*, \mathbf{x}_2^*$ plane. By familiar reasoning, the least-squares criterion implies that $\mathbf{e}$ is perpendicular to the independent-variable plane and, consequently, that $\hat{\mathbf{y}}^*$ is the orthogonal projection of $\mathbf{y}^*$ onto this plane.

The regression coefficients B_1 and B_2 are well defined as long as $\mathbf{x}_1^*$ and $\mathbf{x}_2^*$ are not collinear. This is the geometric version of the requirement that the

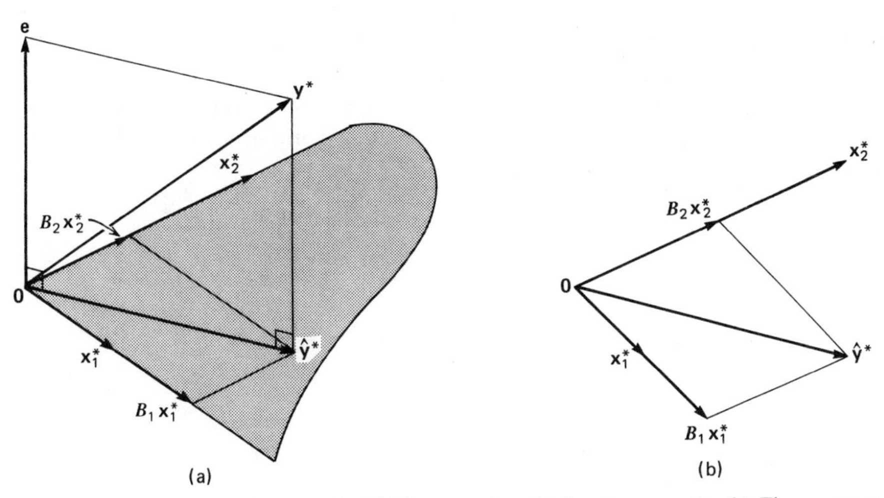

FIGURE 1.14. Vector geometry of multiple regression. (a) Least-squares fit. (b) The regressor plane.

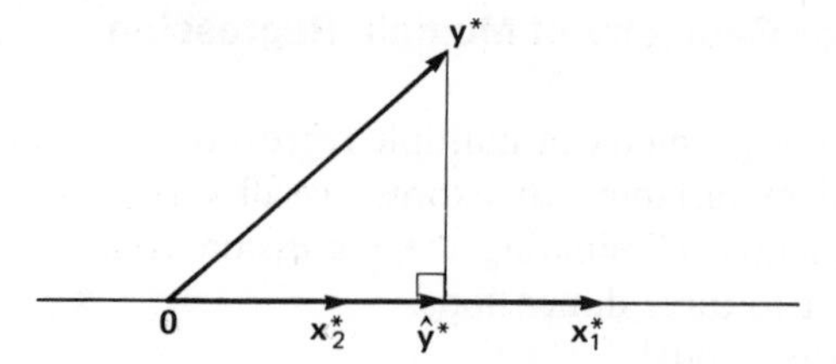

FIGURE 1.15. Two collinear regressors.

independent variables may not be perfectly correlated. If the regressors are collinear, then they span a line rather than a plane; though we can still find $\hat{\mathbf{y}}^*$ by orthogonally projecting $\mathbf{y}^*$ onto this line, as shown in Figure 1.15, we cannot express $\hat{\mathbf{y}}^*$ *uniquely* as a linear combination of $\mathbf{x}_1^*$ and $\mathbf{x}_2^*$.

The sums of squares in multiple regression appear when we examine the plane spanned by $\mathbf{y}^*$ and $\hat{\mathbf{y}}^*$, shown in Figure 1.16. The residual vector also lies in this plane, for $\mathbf{e} = \mathbf{y}^* - \hat{\mathbf{y}}^*$, while the regressor plane is perpendicular to it. As in simple regression analysis, $\text{TSS} = \|\mathbf{y}^*\|^2$, $\text{RegrSS} = \|\hat{\mathbf{y}}^*\|^2$, and $\text{RSS} = \|\mathbf{e}\|^2$. The identity TSS = RegrSS + RSS follows from the Pythagorean theorem. It is clear from Figure 1.16 that $R = \sqrt{\text{RegrSS}/\text{TSS}} = \cos W$. Thus, the multiple correlation is interpretable as the simple correlation between Y and $\hat{Y}$. If there is a perfect linear relation between Y and the independent variables, then $\mathbf{y}^*$ lies in the regressor plane, $\mathbf{y}^* = \hat{\mathbf{y}}^*$, $\mathbf{e} = \mathbf{0}$, $W = 0$, and $R = 1$; if there is no linear relationship between Y and the independent variables, then $\mathbf{y}^*$ is orthogonal to the regressor plane, $\mathbf{y}^* = \mathbf{e}$, $\hat{\mathbf{y}}^* = \mathbf{0}$, $W = 90°$, and $R = 0$.

The vector representation of regression analysis helps clarify the relationship between simple and multiple regression. Figure 1.17(a) is drawn for two positively correlated regressors. The fitted dependent-variable vector is, of course, the orthogonal projection of $\mathbf{y}^*$ onto the $\mathbf{x}_1^*, \mathbf{x}_2^*$ plane. To find the multiple-regression coefficient B_1, we project $\hat{\mathbf{y}}^*$ parallel to $\mathbf{x}_2^*$, locating $B_1\mathbf{x}_1^*$, as shown in Figure 1.17(b), which depicts the regressor plane.

To find the slope coefficient B for the simple regression of Y on X_1, we project the $\mathbf{y}^*$ vector orthogonally onto $\mathbf{x}_1^*$ alone, obtaining $B\mathbf{x}_1^*$; this result also appears in Figure 1.17(a). Since

$$\mathbf{x}_1^* \cdot \mathbf{y}^* = \mathbf{x}_1^* \cdot (\hat{\mathbf{y}}^* + \mathbf{e}) = \mathbf{x}_1^* \cdot \hat{\mathbf{y}}^* + \mathbf{x}_1^* \cdot \mathbf{e} = \mathbf{x}_1^* \cdot \hat{\mathbf{y}}^*$$

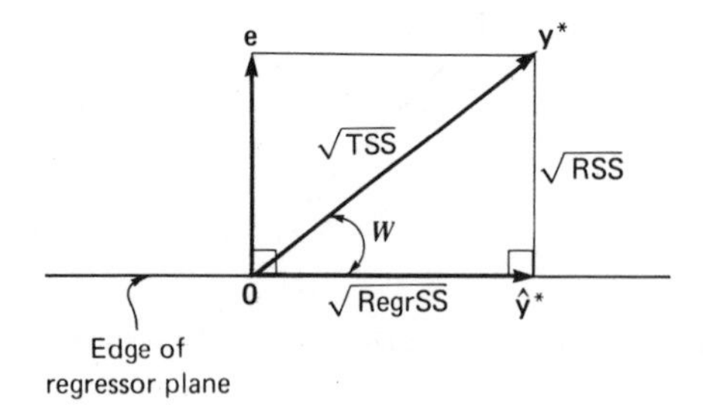

FIGURE 1.16. The $\mathbf{y}^*, \hat{\mathbf{y}}^*$ plane in multiple regression.

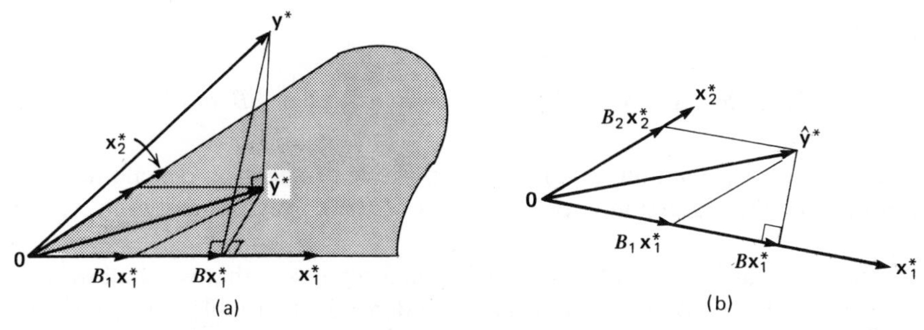

FIGURE 1.17. Simple regression versus multiple regression for correlated regressors. (a) Least-squares fits. (b) The regressor plane.

$B\mathbf{x}_1^*$ is also the orthogonal projection of $\hat{\mathbf{y}}^*$ on $\mathbf{x}_1^*$, as shown in both Figures 1.17(a) and (b). In this instance, projecting $\hat{\mathbf{y}}^*$ perpendicular to $\mathbf{x}_1^*$ rather than parallel to $\mathbf{x}_2^*$ causes B to exceed B_1.

The situation changes fundamentally if the independent variables X_1 and X_2 are uncorrelated, as illustrated in Figures 1.18(a) and (b). Here, $B = B_1$. Another advantage of orthogonal regressors is revealed in Figure 1.18(b): There is a unique partition of the regression sum of squares into components due to each of the two regressors. We have

$$\text{RegrSS} = \hat{\mathbf{y}}^* \cdot \hat{\mathbf{y}}^* = \left(B_1\mathbf{x}_1^* + B_2\mathbf{x}_2^*\right) \cdot \left(B_1\mathbf{x}_1^* + B_2\mathbf{x}_2^*\right)$$

$$= B_1^2\mathbf{x}_1^* \cdot \mathbf{x}_1^* + B_2^2\mathbf{x}_2^* \cdot \mathbf{x}_2^*$$

When the regressors are correlated, as in Figure 1.17(b), no such partition is

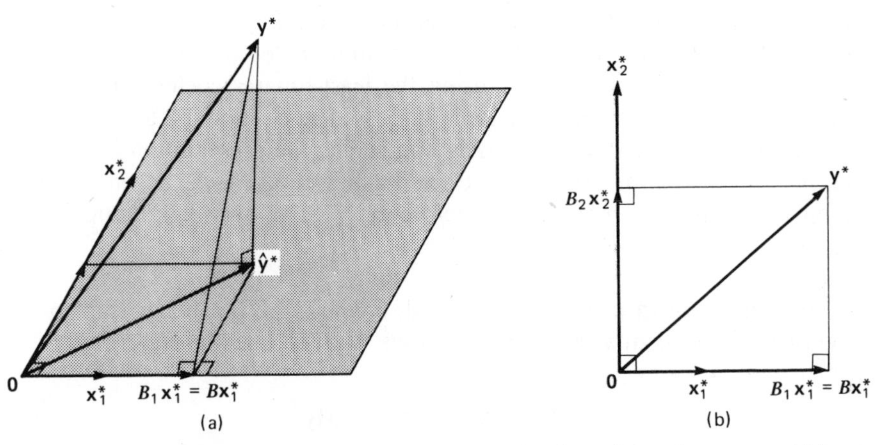

FIGURE 1.18. Simple regression versus multiple regression for orthogonal regressors. (a) Least-squares fits. (b) The regressor plane.

possible, for

$$\text{RegrSS} = \hat{\mathbf{y}}^* \cdot \hat{\mathbf{y}}^* = B_1^2 \mathbf{x}_1^* \cdot \mathbf{x}_1^* + B_2^2 \mathbf{x}_2^* \cdot \mathbf{x}_2^* + 2B_1B_2\mathbf{x}_1^* \cdot \mathbf{x}_2^* \qquad (1.31)$$

The last term in equation (1.31) may be positive or negative, depending upon the signs of the regression coefficients and of the correlation between X_1 and X_2.

There are further advantages to orthogonal regressors that we shall develop later. It would, of course, be unusual for independent variables to be uncorrelated in an observational study. When the independent variables are under the control of the researcher, however, as in experimental work, it is often possible to insure that they are uncorrelated.

As in simple regression, degrees of freedom in multiple regression correspond to the dimension of subspaces of the observation space. Because the $\mathbf{y}^*$ vector, as a vector of mean deviations, is confined to a subspace of dimension $n - 1$, there are $n - 1$ degrees of freedom for TSS. $\hat{\mathbf{y}}^*$ necessarily lies in the fixed $\mathbf{x}_1^*, \mathbf{x}_2^*$ plane, which is a subspace of dimension two; thus RegrSS has two degrees of freedom. $\mathbf{e}$ is orthogonal to the independent-variable plane, and therefore RSS has $(n - 1) - 2 = n - 3$ degrees of freedom.

More generally, k non-collinear regressors in mean-deviation form span a subspace of dimension k. The fitted dependent-variable vector $\hat{\mathbf{y}}^*$ is the orthogonal projection of $\mathbf{y}^*$ onto this subspace, and therefore RegrSS has k degrees of freedom. Similarly, because $\mathbf{e}$ is orthogonal to the regressor subspace, RSS has $(n - 1) - k = n - k - 1$ degrees of freedom.

1.2.5. Properties of the Least-Squares Estimator

In this section we derive a number of results concerning the least-squares estimator $\mathbf{b}$ of the regression parameter vector $\boldsymbol{\beta}$. These results serve two related purposes: (1) They establish certain desirable properties of the least-squares estimator that hold under the assumptions of the regression model; and (2) they furnish a basis for using the least-squares coefficients to make statistical inferences about $\boldsymbol{\beta}$, a topic discussed in Section 1.2.6. Because we wish to derive results that are widely applicable, the presentation in this and the next section is necessarily abstract. Each section, however, ends with a summary of results, and may therefore be passed over lightly upon first reading.

The Distribution of b With $\mathbf{X}$ fixed, $\mathbf{b}$ results from a linear transformation of the dependent-variable observations. That is, $\mathbf{b}$ is a *linear estimator*:

$$\mathbf{b} = (\mathbf{X}'\mathbf{X})^{-1}\mathbf{X}'\mathbf{y} = \mathbf{M}\mathbf{y}$$

defining $\mathbf{M} \equiv (\mathbf{X}'\mathbf{X})^{-1}\mathbf{X}'$. The expected value of $\mathbf{b}$ is established easily from the expectation of **y** (given previously in equations (1.24)):

$$E(\mathbf{b}) = E(\mathbf{My}) = \mathbf{M}E(\mathbf{y}) = (\mathbf{X'X})^{-1}\mathbf{X'}(\mathbf{X}\boldsymbol{\beta}) = \boldsymbol{\beta}$$

b, therefore, is an unbiased estimator of $\boldsymbol{\beta}$. The covariance matrix of the least-squares estimator is similarly derived:

$$V(\mathbf{b}) = \mathbf{M}V(\mathbf{y})\mathbf{M'} = \left[(\mathbf{X'X})^{-1}\mathbf{X'}\right]\sigma_\varepsilon^2\mathbf{I}_n\left[(\mathbf{X'X})^{-1}\mathbf{X'}\right]' \tag{1.32}$$

Moving the scalar σ_ε^2 to the front of this expression, and noting that $(\mathbf{X'X})^{-1}$ is the inverse of a symmetric matrix and is thus itself symmetric, equation (1.32) becomes

$$V(\mathbf{b}) = \sigma_\varepsilon^2(\mathbf{X'X})^{-1}\mathbf{X'X}(\mathbf{X'X})^{-1} = \sigma_\varepsilon^2(\mathbf{X'X})^{-1}$$

The sampling variances and covariances of the regression coefficients, therefore, depend only upon the design matrix and the variance of the errors.

Notice that to derive $E(\mathbf{b})$ and $V(\mathbf{b})$ we do not require the assumption that the dependent-variable observations are normally distributed. If **y** is normally distributed, however, then so is **b** for, as we have seen, **b** results from a linear transformation of **y**. To summarize, then, under the full set of regression-model assumptions, $\mathbf{b} \sim N_{k+1}[\boldsymbol{\beta}, \sigma_\varepsilon^2(\mathbf{X'X})^{-1}]$.

There is a striking parallel, noted by Wonnacott and Wonnacott (1979) and detailed in Table 1.4, between the scalar formulas for simple regression and the matrix formulas for multiple regression. This sort of structural parallel is common in statistical applications where matrix methods are used to generalize a scalar result.

TABLE 1.4. Comparison Between Simple and Multiple Regression

	Simple Regression	Multiple Regression
Model	$Y_i = \alpha + \beta x_i + \varepsilon_i$	$\mathbf{y} = \mathbf{X}\boldsymbol{\beta} + \boldsymbol{\varepsilon}$
Least-Squares Estimator	$B = \dfrac{\Sigma x_i^* Y_i^*}{\Sigma x_i^{*2}} = \left(\Sigma x_i^{*2}\right)^{-1}\Sigma x_i^* Y_i^*$	$\mathbf{b} = (\mathbf{X'X})^{-1}\mathbf{X'y}$
Sampling Variance	$\dfrac{\sigma_\varepsilon^2}{\Sigma x_i^{*2}} = \sigma_\varepsilon^2\left(\Sigma x_i^{*2}\right)^{-1}$	$\sigma_\varepsilon^2(\mathbf{X'X})^{-1}$
Distribution	$B \sim N\left(\beta, \dfrac{\sigma_\varepsilon^2}{\Sigma x_i^{*2}}\right)$	$\mathbf{b} \sim N_{k+1}\left[\boldsymbol{\beta}, \sigma_\varepsilon^2(\mathbf{X'X})^{-1}\right]$

Source: Adapted with permission from Wonnacott and Wonnacott (1979: Table 12-1).

The Gauss–Markov Theorem[8] One of the primary theoretical justifications for least-squares estimation is the Gauss-Markov theorem, which states that if the errors are independently distributed with zero expectation and constant variance, then the least-squares estimator $\mathbf{b}$ is the maximally efficient linear unbiased estimator of $\boldsymbol{\beta}$. That is, of all unbiased estimators that are linear functions of the observations, the least-squares estimator has the smallest sampling variance and, hence, the smallest mean-squared error. For this reason, the least-squares estimator is termed *BLUE*, an acronym for *B*est *L*inear *U*nbiased *E*stimator.

Before proving the Gauss–Markov theorem, it is important to understand its limited scope. First, the theorem holds only if the assumptions of the model hold (although we do not require the assumption of normality). Second, the theorem does not preclude the possibility that there is a linear *biased* estimator more efficient than the least-squares estimator; indeed, we shall pursue this possibility in Section 3.1.4. Third, and finally, the theorem does not rule out the existence of *nonlinear* estimators more efficient than the least-squares estimator. Under the additional assumption of normality, however, it is possible to show that the least-squares estimator is maximally efficient among *all* unbiased estimators (see, e.g., Rao, 1973: 319).

To prove the Gauss–Markov theorem, let $\tilde{\mathbf{b}}$ represent the best linear unbiased estimator of $\boldsymbol{\beta}$. As we know, the least-squares estimator is also a linear estimator, $\mathbf{b} = \mathbf{My}$. It is convenient to write $\tilde{\mathbf{b}} = (\mathbf{M} + \mathbf{A})\mathbf{y}$, where $\mathbf{A}$ gives the difference between the (as yet undetermined) transformation matrix for the BLUE and that for the least-squares estimator. To show that the BLUE and the least-squares estimator coincide, which is the object of the proof, we need to demonstrate that $\mathbf{A} = \mathbf{0}$.

Because $\tilde{\mathbf{b}}$ is by its definition *unbiased*,

$$\begin{aligned}\boldsymbol{\beta} = E(\tilde{\mathbf{b}}) &= E[(\mathbf{M} + \mathbf{A})\mathbf{y}] = E(\mathbf{My}) + E(\mathbf{Ay}) \\ &= E(\mathbf{b}) + \mathbf{A}E(\mathbf{y}) = \boldsymbol{\beta} + \mathbf{AX}\boldsymbol{\beta}\end{aligned}$$

$\mathbf{AX}\boldsymbol{\beta}$, then, is $\mathbf{0}$, regardless of the value of $\boldsymbol{\beta}$, and therefore $\mathbf{AX}$ must be $\mathbf{0}$. To see why this conclusion is warranted, suppose that $\boldsymbol{\beta} = (1, 0, \ldots, 0)'$, which is, of course, one among infinite possibilities. For this $\boldsymbol{\beta}$, $(\mathbf{AX})\boldsymbol{\beta} = \mathbf{0}$ if the first column of $\mathbf{AX}$ consists of zeroes, since the elements of the first column of $\mathbf{AX}$ multiply the first entry in $\boldsymbol{\beta}$ when the matrix product is formed. $\boldsymbol{\beta}$, however, could just as easily have a one in some other position and zeroes elsewhere, and, hence, the only way of insuring that the product $\mathbf{AX}\boldsymbol{\beta}$ is zero for all possible values of $\boldsymbol{\beta}$ is to require that $\mathbf{AX} = \mathbf{0}$.

[8]The theorem is named after the 19th century German mathematician C. F. Gauss and the 20th century Russian mathematician A. A. Markov. Although Gauss worked in the context of measurement errors in the physical sciences, much of the general statistical theory of linear models is due to him. The strategy of proof employed here is adapted from Wonnacott and Wonnacott (1979: 428–430), where it is used in a slightly different context.

We have to this point made use of the linearity and unbias of $\tilde{\mathbf{b}}$. Because $\tilde{\mathbf{b}}$ is the *minimum-variance* linear unbiased estimator, the sampling variances of its elements—that is, the diagonal entries of $V(\tilde{\mathbf{b}})$—are as small as possible.[9] The covariance matrix of $\tilde{\mathbf{b}}$ is given by

$$V(\tilde{\mathbf{b}}) = (\mathbf{M} + \mathbf{A})V(\mathbf{y})(\mathbf{M} + \mathbf{A})' = (\mathbf{M} + \mathbf{A})\sigma_\varepsilon^2\mathbf{I}_n(\mathbf{M} + \mathbf{A})'$$

$$= \sigma_\varepsilon^2(\mathbf{MM}' + \mathbf{MA}' + \mathbf{AM}' + \mathbf{AA}') \tag{1.33}$$

We have shown that $\mathbf{AX} = \mathbf{0}$, and consequently $\mathbf{AM}'$ and its transpose $\mathbf{MA}'$ are zero, for $\mathbf{AM}' = \mathbf{AX}(\mathbf{X}'\mathbf{X})^{-1} = \mathbf{0}(\mathbf{X}'\mathbf{X})^{-1} = \mathbf{0}$.

Equation (1.33) becomes

$$V(\tilde{\mathbf{b}}) = \sigma_\varepsilon^2(\mathbf{MM}' + \mathbf{AA}') \tag{1.34}$$

The sampling variance of the coefficient $\tilde{B}_j$ is the jth diagonal entry of $V(\tilde{\mathbf{b}})$;[10] from equation (1.34)

$$V(\tilde{B}_j) = \sigma_\varepsilon^2\left(\sum_{i=1}^{n} m_{ji}^2 + \sum_{i=1}^{n} a_{ji}^2\right)$$

Both sums in this equation are sums of squares and hence cannot be negative; and since $V(\tilde{B}_j)$ is a minimum, all of the a_{ji} must be zero. Because this argument applies to each coefficient in $\tilde{\mathbf{b}}$, $\mathbf{A} = \mathbf{0}$. Consequently, $\tilde{\mathbf{b}} = (\mathbf{M} + \mathbf{0})\mathbf{y} = \mathbf{My} = \mathbf{b}$, which completes the proof: the BLUE is the least-squares estimator.

The Distribution of Fitted Values and Residuals The vector of fitted values $\hat{\mathbf{y}}$ may be written as a linear function of the observations

$$\hat{\mathbf{y}} \equiv \mathbf{Xb} = \mathbf{X}(\mathbf{X}'\mathbf{X})^{-1}\mathbf{X}'\mathbf{y} = \mathbf{Hy} \tag{1.35}$$

where $\mathbf{H} \equiv \mathbf{X}(\mathbf{X}'\mathbf{X})^{-1}\mathbf{X}'$ is called the *hat matrix* because it transforms $\mathbf{y}$ into $\hat{\mathbf{y}}$ (Hoaglin and Welsch, 1978; Belsley Kuh, and Welsch, 1980). The hat matrix, together with the vector of residuals, will figure prominently in Chapter 3, where we shall develop a variety of procedures for detecting problems in regression analysis. The expectation and covariance matrix of $\hat{\mathbf{y}}$ follow from

[9] It is, in fact, possible to prove a more general result: The best linear unbiased estimator of $\mathbf{a}'\boldsymbol{\beta}$ (an arbitrary linear combination of regression coefficients) is $\mathbf{a}'\mathbf{b}$, where $\mathbf{b}$ is the least-squares estimator (see, e.g., Seber, 1977: 49).

[10] Actually, the variance of $\tilde{B}_0$ is the *first* diagonal entry of $V(\tilde{\mathbf{b}})$. The variance of $\tilde{B}_j$, therefore, is given by the $(j + 1)$th entry. To avoid this awkwardness, we shall index the covariance matrix of $\tilde{\mathbf{b}}$ (and, later, that of $\mathbf{b}$) from zero rather than from one.

equation (1.35):

$$E(\hat{\mathbf{y}}) = \mathbf{X}E(\mathbf{b}) = \mathbf{X}\boldsymbol{\beta} = E(\mathbf{y})$$

$$V(\hat{\mathbf{y}}) = \mathbf{H}V(\mathbf{y})\mathbf{H}' = \left[\mathbf{X}(\mathbf{X}'\mathbf{X})^{-1}\mathbf{X}'\right]\sigma_\varepsilon^2\mathbf{I}_n\left[\mathbf{X}(\mathbf{X}'\mathbf{X})^{-1}\mathbf{X}'\right]'$$

$$= \sigma_\varepsilon^2\mathbf{X}(\mathbf{X}'\mathbf{X})^{-1}\mathbf{X}' = \sigma_\varepsilon^2\mathbf{H}$$

Notice that the hat matrix $\mathbf{H}$ is symmetric and idempotent. Since $E(\hat{\mathbf{y}}) = E(\mathbf{y})$, the fitted values are unbiased estimators of the dependent-variable expectations. If $\mathbf{y}$ is normally distributed, as it is under the assumptions of the regression model, then so is $\hat{\mathbf{y}}$: $\hat{\mathbf{y}} \sim N_n(\mathbf{X}\boldsymbol{\beta}, \sigma_\varepsilon^2\mathbf{H})$.

The least-squares residuals are defined by

$$\mathbf{e} \equiv \mathbf{y} - \hat{\mathbf{y}} = \mathbf{y} - \mathbf{H}\mathbf{y} = (\mathbf{I}_n - \mathbf{H})\mathbf{y} = \mathbf{Q}\mathbf{y} \tag{1.36}$$

where $\mathbf{Q} \equiv \mathbf{I}_n - \mathbf{H}$. Alternatively,

$$\mathbf{e} = \mathbf{y} - \mathbf{H}\mathbf{y} = (\mathbf{X}\boldsymbol{\beta} + \boldsymbol{\varepsilon}) - \mathbf{X}(\mathbf{X}'\mathbf{X})^{-1}\mathbf{X}'(\mathbf{X}\boldsymbol{\beta} + \boldsymbol{\varepsilon})$$

$$= \boldsymbol{\varepsilon} - \mathbf{X}(\mathbf{X}'\mathbf{X})^{-1}\mathbf{X}'\boldsymbol{\varepsilon} = (\mathbf{I}_n - \mathbf{H})\boldsymbol{\varepsilon} = \mathbf{Q}\boldsymbol{\varepsilon}$$

The sample residuals are, then, the *same* linear function of the (unobservable) disturbances and of the dependent-variable observations.

Since the residuals result from linear transformation of the disturbances, their distribution is simply derived. $\mathbf{Q}$, like $\mathbf{H}$, is symmetric and idempotent, and consequently

$$E(\mathbf{e}) = \mathbf{Q}E(\boldsymbol{\varepsilon}) = \mathbf{0}$$

$$V(\mathbf{e}) = \mathbf{Q}V(\boldsymbol{\varepsilon})\mathbf{Q}' = \mathbf{Q}(\sigma_\varepsilon^2\mathbf{I}_n)\mathbf{Q}' = \sigma_\varepsilon^2\mathbf{Q}$$

Assuming normally distributed errors, $\mathbf{e} \sim N_n(\mathbf{0}, \sigma_\varepsilon^2\mathbf{Q})$.

It is apparent that although the errors $\boldsymbol{\varepsilon}$ are (by assumption) independent and identically distributed, the residuals $\mathbf{e}$ are in general correlated, for the off-diagonal entries of $\mathbf{Q}$ are not generally zero (and, in fact, cannot *all* be zero); furthermore, the E_i generally have different variances, since the h_{ii} and hence the q_{ii} generally differ. Finally, $\mathbf{e}$ follows a *singular* normal distribution whose covariance matrix $\sigma_\varepsilon^2\mathbf{Q}$ is of rank $n - k - 1$. To understand this property, recall that there are $n - k - 1$ residual degrees of freedom; the least-squares fit places $k + 1$ linear restrictions on the residual vector $\mathbf{e}$. Since $\mathbf{X}$ and $\mathbf{e}$ are orthogonal,

$$\mathbf{X}'\mathbf{e} = \underset{(k+1\times 1)}{\mathbf{0}}$$

Following Putter (1967), we can transform the least-squares residuals into an independently and identically distributed set by selecting an orthonormal basis for the error subspace, defining transformed residuals in the following manner:

$$\underset{(n-k-1\times 1)}{\mathbf{z}} \equiv \underset{(n-k-1\times n)}{\mathbf{G}}\ \underset{(n\times 1)}{\mathbf{e}} \tag{1.37}$$

In equation (1.37) the transformation matrix $\mathbf{G}$ is selected so that it is orthonormal and orthogonal to $\mathbf{X}$:

$$\mathbf{G}\mathbf{G}' = \mathbf{I}_{n-k-1}$$

$$\mathbf{G}\mathbf{X} = \underset{(n-k-1\times k+1)}{\mathbf{0}}$$

The transformed residuals $\mathbf{z}$ then have the following properties:

$$\mathbf{z} = \mathbf{G}(\mathbf{y} - \hat{\mathbf{y}}) = \mathbf{G}\left[\mathbf{y} - \mathbf{X}(\mathbf{X}'\mathbf{X})^{-1}\mathbf{X}'\mathbf{y}\right] = \mathbf{G}\mathbf{y}$$

$$E(\mathbf{z}) = \mathbf{G}E(\mathbf{y}) = \mathbf{G}\mathbf{X}\boldsymbol{\beta} = \underset{(n-k-1\times 1)}{\mathbf{0}}$$

$$V(\mathbf{z}) = \mathbf{G}\left(\sigma_\varepsilon^2\mathbf{I}_n\right)\mathbf{G}' = \sigma_\varepsilon^2\mathbf{I}_{n-k-1}$$

If the elements of $\boldsymbol{\varepsilon}$ are normally and independently distributed with equal variances, then so are the elements of $\mathbf{z}$. There are, however, n of the former and $n - k - 1$ of the latter; furthermore, $\mathbf{G}$ (and hence $\mathbf{z}$) is not unique, for any rigid rotation of an orthonormal basis is itself orthonormal. The transformed residuals are useful not only for exploring the properties of least-squares estimation, but also in diagnosing certain linear-model problems (see, e.g., Theil, 1971: Ch. 5; and Putter, 1967).

Estimating the Variance of the Errors Transforming $\mathbf{e}$ to $\mathbf{z}$ suggests a simple method for deriving an estimator of the error variance σ_ε^2. The entries of $\mathbf{z}$ have zero expectation and common variance σ_ε^2, so

$$E(\mathbf{z}'\mathbf{z}) = \sum_{i=1}^{n-k-1} E\left(Z_i^2\right) = (n - k - 1)\sigma_\varepsilon^2$$

Thus, an unbiased estimator of the error variance is given by $S_E^2 \equiv \mathbf{z}'\mathbf{z}/(n - k - 1)$. Moreover, because the Z_i are normally and independently distributed, $\mathbf{z}'\mathbf{z}/\sigma_\varepsilon^2 = (n - k - 1)S_E^2/\sigma_\varepsilon^2$ has a χ^2 distribution with $n - k - 1$ degrees of freedom.

The estimator S_E^2 may be computed without finding transformed residuals, for the length of the least-squares residual vector $\mathbf{e}$ is the same as the length of the vector of transformed residuals $\mathbf{z}$: $\sqrt{\mathbf{z}'\mathbf{z}} = \sqrt{\mathbf{e}'\mathbf{e}}$. This result follows from the observation that $\mathbf{z}$ and $\mathbf{e}$ are the same vector represented according to alternative bases: (1) $\mathbf{e}$ gives the coordinates of the residuals relative to the natural basis of the n-dimensional observation space; (2) $\mathbf{z}$ gives the coordinates of the residuals relative to an arbitrary orthonormal basis for the $(n - k - 1)$-dimensional error subspace. A vector does not change its length when the basis changes, and therefore

$$S_E^2 = \frac{\mathbf{e}'\mathbf{e}}{n - k - 1}$$

Heuristically, though $\mathbf{e}$ contains n elements, there are, as we have pointed out, $k + 1$ linear dependencies among them. In calculating an unbiased estimator of the error variance, we divide by the residual degrees of freedom rather than by the number of observations.

Maximum-Likelihood Estimation We shall show that under the assumptions of the regression model, the least-squares estimator $\mathbf{b}$ is the maximum-likelihood estimator of $\boldsymbol{\beta}$. This is an important result, because it establishes an additional justification for least-squares estimation when the assumptions of the model are reasonable, and because it enables us to employ the powerful maximum-likelihood approach to statistical inference.

As we are aware, under the assumptions of the regression model, $\mathbf{y} \sim N_n(\mathbf{X}\boldsymbol{\beta}, \sigma_\varepsilon^2\mathbf{I}_n)$. Thus, for the ith observation, $Y_i \sim N(\mathbf{x}_i'\boldsymbol{\beta}, \sigma_\varepsilon^2)$, where (recall from equation (1.22)) $\mathbf{x}_i'$ is the ith row of the design matrix $\mathbf{X}$. In equation form, the probability density for observation i is given by

$$p(y_i) = \frac{1}{\sigma_\varepsilon\sqrt{2\pi}} \exp\left[-\frac{(y_i - \mathbf{x}_i'\boldsymbol{\beta})^2}{2\sigma_\varepsilon^2}\right]$$

Because the n observations are independent, their joint probability density is the product of their marginal densities,

$$p(\mathbf{y}) = \prod_{i=1}^{n} p(y_i) = \frac{1}{\left(\sigma_\varepsilon\sqrt{2\pi}\right)^n} \exp\left[-\frac{\Sigma(y_i - \mathbf{x}_i'\boldsymbol{\beta})^2}{2\sigma_\varepsilon^2}\right]$$

$$= \frac{1}{\left(2\pi\sigma_\varepsilon^2\right)^{n/2}} \exp\left[-\frac{(\mathbf{y} - \mathbf{X}\boldsymbol{\beta})'(\mathbf{y} - \mathbf{X}\boldsymbol{\beta})}{2\sigma_\varepsilon^2}\right] \quad (1.38)$$

Although equation (1.38) also follows directly from the multivariate-normal distribution of $\mathbf{y}$, the development from $p(y_i)$ to $p(\mathbf{y})$ will prove helpful when we deal with random regressors in Section 1.3.

In light of equation (1.38), the log likelihood is

$$\log L(\boldsymbol{\beta}, \sigma_\varepsilon^2) = -\frac{n}{2}\log(2\pi) - \frac{n}{2}\log \sigma_\varepsilon^2 - \frac{1}{2\sigma_\varepsilon^2}(\mathbf{y} - \mathbf{X}\boldsymbol{\beta})'(\mathbf{y} - \mathbf{X}\boldsymbol{\beta}) \tag{1.39}$$

To maximize the likelihood, we require the partial derivatives of equation (1.39) with respect to the model parameters $\boldsymbol{\beta}$ and σ_ε^2. Differentiation is simplified when we recognize that $(\mathbf{y} - \mathbf{X}\boldsymbol{\beta})'(\mathbf{y} - \mathbf{X}\boldsymbol{\beta})$ is the sum of squared errors.

$$\frac{\partial \log L(\boldsymbol{\beta}, \sigma_\varepsilon^2)}{\partial \boldsymbol{\beta}} = -\frac{1}{2\sigma_\varepsilon^2}(2\mathbf{X}'\mathbf{X}\boldsymbol{\beta} - 2\mathbf{X}'\mathbf{y})$$

$$\frac{\partial \log L(\boldsymbol{\beta}, \sigma_\varepsilon^2)}{\partial \sigma_\varepsilon^2} = -\frac{n}{2}\left(\frac{1}{\sigma_\varepsilon^2}\right) + \frac{1}{2\sigma_\varepsilon^4}(\mathbf{y} - \mathbf{X}\boldsymbol{\beta})'(\mathbf{y} - \mathbf{X}\boldsymbol{\beta})$$

Setting these partial derivatives to zero, and solving for the maximum-likelihood estimators $\hat{\mathbf{b}}$ and $\hat{S}_E^2$ produces

$$\hat{\mathbf{b}} = (\mathbf{X}'\mathbf{X})^{-1}\mathbf{X}'\mathbf{y}$$

$$\hat{S}_E^2 = \frac{(\mathbf{y} - \mathbf{X}\hat{\mathbf{b}})'(\mathbf{y} - \mathbf{X}\hat{\mathbf{b}})}{n} = \frac{\hat{\mathbf{e}}'\hat{\mathbf{e}}}{n}$$

It is apparent that the maximum-likelihood estimator $\hat{\mathbf{b}}$ is the same as the least-squares estimator $\mathbf{b}$. Actually, this identity is clear directly from equation (1.38), without formal maximization of the likelihood: The likelihood is large when the negative exponent is small, and the numerator of this exponent contains the sum of squared errors; minimizing the sum of squared residuals, therefore, maximizes the likelihood.

Notice that the maximum-likelihood estimator $\hat{S}_E^2$ of the error variance is biased, for as we established earlier in this section, $E(\mathbf{e}'\mathbf{e}) = (n - k - 1)\sigma_\varepsilon^2$. As n increases, however, the bias of $\hat{S}_E^2$ decreases: As a maximum-likelihood estimator, $\hat{S}_E^2$ is consistent. We generally employ the unbiased estimator $S_E^2 = \mathbf{e}'\mathbf{e}/(n - k - 1)$ in preference to $\hat{S}_E^2$.

Summary Under the assumptions of the regression model:

1. The least-squares coefficients $\mathbf{b}$ are unbiased estimators of the population regression coefficients $\boldsymbol{\beta}$. The covariance matrix of $\mathbf{b}$ is given by $V(\mathbf{b}) = \sigma_\varepsilon^2(\mathbf{X}'\mathbf{X})^{-1}$.
2. Of all unbiased estimators of $\boldsymbol{\beta}$ that are linear functions of the observations (and hence are mathematically tractable), the least-squares estimator $\mathbf{b}$ has the smallest sampling variance (i.e., is maximally efficient).

3. The fitted values $\hat{\mathbf{y}}$ are normally distributed with expectation $\mathbf{X}\boldsymbol{\beta}$ and covariance matrix $V(\hat{\mathbf{y}}) = \sigma_\varepsilon^2\mathbf{X}(\mathbf{X}'\mathbf{X})^{-1}\mathbf{X}' = \sigma_\varepsilon^2\mathbf{H}$. The residuals $\mathbf{e}$ are normally distributed with expectation $\mathbf{0}$ and covariance matrix $V(\mathbf{e}) = \sigma_\varepsilon^2(\mathbf{I}_n - \mathbf{H}) = \sigma_\varepsilon^2\mathbf{Q}$.
4. An unbiased estimator of the error variance σ_ε^2 is given by $S_E^2 = \mathbf{e}'\mathbf{e}/(n - k - 1)$.
5. The least-squares coefficients $\mathbf{b}$ are the maximum-likelihood estimators of $\boldsymbol{\beta}$.

1.2.6. Statistical Inference for Multiple Regression

The results derived in the previous section provide a basis for statistical inference in multiple regression analysis. We already have determined that $\mathbf{b}$ has certain desirable properties as a point estimator of $\boldsymbol{\beta}$. In this section, we begin by constructing tests and confidence intervals for individual regression coefficients, proceed to develop simultaneous tests for several coefficients, and conclude by considering joint confidence regions for several regression coefficients. Taken as a whole, the methods evolved here provide powerful inferential tools for multiple regression and, by extension, for linear models generally.

Testing Individual Regression Coefficients We are frequently interested in testing the null hypothesis that a population slope is zero, or, alternatively, in constructing a confidence interval for a slope. Less frequently, we may wish to test a hypothesis about the regression constant.

We showed in Section 1.2.5 that the least-squares estimator $\mathbf{b}$ follows a normal distribution with expectation $\boldsymbol{\beta}$ and covariance matrix $\sigma_\varepsilon^2(\mathbf{X}'\mathbf{X})^{-1}$. Consequently, an individual regression coefficient B_j is normally distributed with expectation β_j and sampling variance $\sigma_\varepsilon^2 v_{jj}$, where v_{jj} is the jth diagonal entry of $(\mathbf{X}'\mathbf{X})^{-1}$ (recall footnote 10). The ratio $(B_j - \beta_j)/\sigma_\varepsilon\sqrt{v_{jj}}$, therefore, follows the unit-normal distribution $N(0, 1)$; and to test the hypothesis H_0: $\beta_j = \beta_{j0}$, we may calculate the test statistic $Z_0 = (B_j - \beta_{j0})/\sigma_\varepsilon\sqrt{v_{jj}}$, comparing the obtained value of the statistic to tabled values of the unit-normal distribution. This result is not of practical use, however, because in applications of regression analysis we do not know σ_ε.

Although the error variance is unknown, we have available the unbiased estimator $S_E^2 = \mathbf{e}'\mathbf{e}/(n - k - 1)$. Employing this estimator, we may estimate the covariance matrix of the least-squares regression coefficients:

$$\widehat{V(\mathbf{b})} = S_E^2(\mathbf{X}'\mathbf{X})^{-1} = \frac{\mathbf{e}'\mathbf{e}}{n - k - 1}(\mathbf{X}'\mathbf{X})^{-1}$$

An estimator of the standard error of the regression coefficient B_j is therefore given by $S_E\sqrt{v_{jj}}$, the square root of the diagonal entry of $\widehat{V(\mathbf{b})}$.

We discovered previously that $(n - k - 1)S_E^2/\sigma_\varepsilon^2 = \mathbf{e}'\mathbf{e}/\sigma_\varepsilon^2$ is distributed as $\chi^2(n - k - 1)$. We have recently shown that $(B_j - \beta_j)/\sigma_\varepsilon\sqrt{v_{jj}}$ is distributed as $N(0, 1)$. If we can further prove that the estimators B_j and S_E^2 are independent, then the ratio

$$t_j = \frac{(B_j - \beta_j)/\sigma_\varepsilon\sqrt{v_{jj}}}{\sqrt{\dfrac{\mathbf{e}'\mathbf{e}/\sigma_\varepsilon^2}{n - k - 1}}} = \frac{B_j - \beta_j}{S_E\sqrt{v_{jj}}}$$

follows a t-distribution with $n - k - 1$ degrees of freedom. Heuristically, in estimating σ_ε with S_E, we replace the normal distribution with the t-distribution to reflect the additional source of variability.

We shall demonstrate the independence of B_j and S_E^2 by showing that the residual vector $\mathbf{e}$ (from which S_E^2 is calculated) is independent of the vector $\mathbf{b}$ of least-squares regression coefficients (of which B_j is a member). Since $\mathbf{e}$ and $\mathbf{b}$ are both normally distributed, it suffices to prove that their covariances are zero. The covariance matrix for these two vector random variables is, by definition,

$$\underset{(n \times k+1)}{C(\mathbf{e}, \mathbf{b})} = E[\mathbf{e}(\mathbf{b} - \boldsymbol{\beta})'] \tag{1.40}$$

Now

$$\begin{aligned} \mathbf{b} - \boldsymbol{\beta} &= (\mathbf{X}'\mathbf{X})^{-1}\mathbf{X}'\mathbf{y} - \boldsymbol{\beta} = (\mathbf{X}'\mathbf{X})^{-1}\mathbf{X}'(\mathbf{X}\boldsymbol{\beta} + \boldsymbol{\varepsilon}) - \boldsymbol{\beta} \\ &= (\mathbf{X}'\mathbf{X})^{-1}\mathbf{X}'\boldsymbol{\varepsilon} \end{aligned}$$

It is intuitively sensible, incidentally, that the error in estimation, $\mathbf{b} - \boldsymbol{\beta}$, should be a function of the errors $\boldsymbol{\varepsilon}$. Substituting this result into equation (1.40) and noting that $\mathbf{e} = \mathbf{Q}\boldsymbol{\varepsilon}$ [from equation (1.36)], we obtain

$$\begin{aligned} C(\mathbf{e}, \mathbf{b}) &= E\{[\mathbf{Q}\boldsymbol{\varepsilon}][(\mathbf{X}'\mathbf{X})^{-1}\mathbf{X}'\boldsymbol{\varepsilon}]'\} \\ &= E\{[\mathbf{I}_n - \mathbf{X}(\mathbf{X}'\mathbf{X})^{-1}\mathbf{X}']\boldsymbol{\varepsilon}\boldsymbol{\varepsilon}'\mathbf{X}(\mathbf{X}'\mathbf{X})^{-1}\} \\ &= \sigma_\varepsilon^2\mathbf{X}(\mathbf{X}'\mathbf{X})^{-1} - \sigma_\varepsilon^2\mathbf{X}(\mathbf{X}'\mathbf{X})^{-1} = \mathbf{0} \end{aligned}$$

since $E(\boldsymbol{\varepsilon}\boldsymbol{\varepsilon}') = V(\boldsymbol{\varepsilon}) = \sigma_\varepsilon^2\mathbf{I}_n$.

To test the hypothesis H_0: $\beta_j = \beta_{j0}$, therefore, we calculate the test statistic

$$t_{j0} = \frac{B_j - \beta_{j0}}{S_E\sqrt{v_{jj}}} \tag{1.41}$$

comparing the obtained value of t_{j0} with the tabled percentage points of $t(n - k - 1)$. To test the null hypothesis H_0: $\beta_j = 0$, which is generally of interest, we simply divide the least-squares regression coefficient by its standard error. A useful rule of thumb, valid unless the residual degrees of freedom are very small (say, fewer than ten), is that $|t_0| \simeq 2.0$ is associated with a one-tail p-value of 2.5 percent.

To construct a two-sided confidence interval for β_j, at the $100(1 - a)$ percent level of confidence, we take

$$\beta_j = B_j \pm t_{a/2} S_E \sqrt{v_{jj}}$$

We illustrate these results by applying them to Angell's data on the moral integration of U.S. cities. Using quantities previously calculated in equations (1.27) and (1.28), we get

$$S_E^2 = \frac{132.3}{29 - 3} = 5.088$$

$$\widehat{V(\mathbf{b})} = 5.088(\mathbf{X}'\mathbf{X})^{-1}$$

$$= \begin{pmatrix} 4.785 & -0.1012 & -0.09764 \\ -0.1012 & 0.003044 & 0.001456 \\ -0.09764 & 0.001456 & 0.002645 \end{pmatrix}$$

Equation (1.42) shows the fitted regression, with coefficient standard errors in parentheses and t-values in brackets.

$$\begin{array}{llll} \hat{Y} = 21.8 & -0.167X_1 & -0.214X_2 & \\ \quad (2.19) & (0.0552) & (0.0514) & \qquad (1.42) \\ & [-3.03] & [-4.16] & \end{array}$$

We have reported no t-value for B_0 because the hypothesis H_0: $\beta_0 = 0$ is not of substantive interest here. Since the t-values for the slope coefficients exceed two in magnitude and are negative as predicted, both coefficients are statistically significant beyond the .025 level.

Hypothesis Tests Concerning Several Regression Coefficients Although we usually test regression coefficients individually, these tests may not be sufficient, for in general the least-squares estimators of different regression parameters are correlated: The off-diagonal entries of $V(\mathbf{b}) = \sigma_\varepsilon^2(\mathbf{X}'\mathbf{X})^{-1}$, giving the sampling covariances of the least-squares coefficients, are zero only when the regressors themselves are uncorrelated.[11] Simultaneous tests for sets

[11] Our concern here pertains to sampling correlations among the k *slope* coefficients, induced by correlations among the independent variables. The constant regressor may be made orthogonal to the other regressors by expressing the independent variables in mean-deviation form. This approach, called *centering*, has certain computational advantages (it tends to reduce rounding errors in regression calculations), but it does not affect the slope coefficients or the sampling covariances among them.

of regression coefficients, taking their intercorrelation into account, may be constructed by the likelihood-ratio principle.

Suppose that we fit the model

$$Y = \beta_0 + \beta_1 x_1 + \cdots + \beta_k x_k + \varepsilon \tag{1.43}$$

obtaining the least-squares estimator $\mathbf{b} = (B_0, B_1, \ldots, B_k)'$ and the maximum-likelihood estimator of the error variance $\hat{S}_E^2$. We wish to test the null hypothesis that a subset of regression parameters is zero; for convenience, let these coefficients be the first $p \leq k$, so that we have the null hypothesis H_0: $\beta_1 = \cdots = \beta_p = 0$. This null hypothesis corresponds to the model

$$\begin{aligned} Y &= \beta_0 + 0x_1 + \cdots + 0x_p + \beta_{p+1}x_{p+1} + \cdots + \beta_k x_k + \varepsilon \\ &= \beta_0 + \beta_{p+1}x_{p+1} + \cdots + \beta_k x_k + \varepsilon \end{aligned} \tag{1.44}$$

which is a *specialization* (or restriction) of model (1.43). Fitting model (1.44) by least squares—that is, regressing Y on X_{p+1} through X_k—we obtain $\mathbf{b}_0 = (B_0', 0, \ldots, 0, B_{p+1}', \ldots, B_k')'$, and $\hat{S}_{E0}^2$. Note that the coefficients in $\mathbf{b}_0$ generally will differ from those in $\mathbf{b}$ (hence the primes), and that $\hat{S}_E^2 \leq \hat{S}_{E0}^2$, since both models are fit by least squares.

The likelihood for the full model (1.43), evaluated at the maximum-likelihood estimators, may be obtained from equation (1.38):

$$\begin{aligned} L &= \left(2\pi\hat{S}_E^2\right)^{-n/2} \exp\left[-\frac{(\mathbf{y} - \mathbf{X}\mathbf{b})'(\mathbf{y} - \mathbf{X}\mathbf{b})}{2\hat{S}_E^2}\right] \\ &= \left(2\pi\frac{\mathbf{e}'\mathbf{e}}{n}\right)^{-n/2} \exp\left(-\frac{\mathbf{e}'\mathbf{e}}{2\dfrac{\mathbf{e}'\mathbf{e}}{n}}\right) \\ &= \left(2\pi\frac{\mathbf{e}'\mathbf{e}}{n}\right)^{-n/2} e^{-n/2} \end{aligned}$$

Likewise, for the restricted model (1.44) with estimators $\mathbf{b}_0$ and $\hat{S}_{E0}^2 = \mathbf{e}_0'\mathbf{e}_0/n$, we have likelihood

$$L_0 = \left(2\pi\frac{\mathbf{e}_0'\mathbf{e}_0}{n}\right)^{-n/2} e^{-n/2}$$

The likelihood ratio for testing H_0 is therefore

$$\frac{L_0}{L} = \left(\frac{\mathbf{e}_0'\mathbf{e}_0}{\mathbf{e}'\mathbf{e}}\right)^{-n/2} = \left(\frac{\mathbf{e}'\mathbf{e}}{\mathbf{e}_0'\mathbf{e}_0}\right)^{2/n} \tag{1.45}$$

Equation (1.45) is the root of a ratio of residual sums of squares. $\mathbf{e}_0'\mathbf{e}_0 \geq \mathbf{e}'\mathbf{e}$, and thus the likelihood ratio is small when the residual sum of squares from the restricted model is appreciably larger than that from the general model—circumstances under which we should doubt the truth of the null hypothesis. A test of H_0 is provided by the generalized likelihood-ratio test statistic, $G_0^2 = -2\log(L_0/L)$, which is asymptotically distributed as $\chi^2(p)$ under the null hypothesis.

It is unnecessary to use this asymptotic result, however, for an exact test may be obtained. In Section 1.2.5 we determined that $\text{RSS}/\sigma_\varepsilon^2 = \mathbf{e}'\mathbf{e}/\sigma_\varepsilon^2$ is distributed as $\chi^2(n-k-1)$. If the null hypothesis is true, then $\text{RSS}_0/\sigma_\varepsilon^2 = \mathbf{e}_0'\mathbf{e}_0/\sigma_\varepsilon^2$ is also distributed as χ^2 with $n-(k-p)-1 = n-k+p-1$ degrees of freedom. Consequently, the difference

$$\frac{\text{RSS}_0}{\sigma_\varepsilon^2} - \frac{\text{RSS}}{\sigma_\varepsilon^2} = \frac{\text{RSS}_0 - \text{RSS}}{\sigma_\varepsilon^2} = \frac{\mathbf{e}_0'\mathbf{e}_0 - \mathbf{e}'\mathbf{e}}{\sigma_\varepsilon^2}$$

has a χ^2 distribution with $(n-k+p-1)-(n-k-1) = p$ degrees of freedom, equal to the number of parameters set to zero in the restricted model. Furthermore, the difference $\text{RSS}_0 - \text{RSS}$ is independent of RSS; we shall demonstrate their independence by proving that $\mathbf{e}_0 - \mathbf{e}$ and $\mathbf{e}$ are orthogonal.

$$\begin{aligned}\mathbf{e}'(\mathbf{e}_0 - \mathbf{e}) &= \mathbf{e}'\mathbf{e}_0 - \mathbf{e}'\mathbf{e} = \mathbf{e}'\mathbf{e}_0 - \mathbf{e}'(\mathbf{y} - \hat{\mathbf{y}}) \\ &= \mathbf{e}'\mathbf{e}_0 - \mathbf{e}'\mathbf{y} \\ &= \mathbf{e}'\mathbf{e}_0 - \mathbf{e}'(\hat{\mathbf{y}}_0 + \mathbf{e}_0) \\ &= \mathbf{e}'\mathbf{e}_0 - \mathbf{e}'\mathbf{e}_0 = 0\end{aligned}$$

The crucial steps in this series of rearrangements, the elimination of $\mathbf{e}'\hat{\mathbf{y}}$ and $\mathbf{e}'\hat{\mathbf{y}}_0$, may be justified by appealing to the vector geometry of regression analysis: Both $\hat{\mathbf{y}}$ and $\hat{\mathbf{y}}_0$ lie in the subspace spanned by the columns of $\mathbf{X}$; since the residual vector $\mathbf{e}$ is orthogonal to the regressor subspace, it is orthogonal to $\hat{\mathbf{y}}$ and $\hat{\mathbf{y}}_0$.

$(\text{RSS}_0 - \text{RSS})$ and RSS, each divided by σ_ε^2, are independent χ^2 random variables with p and $n-k-1$ degrees of freedom, respectively; the ratio

$$F_0 = \frac{(\text{RSS}_0 - \text{RSS})/p}{\text{RSS}/(n-k-1)} \tag{1.46}$$

therefore, is distributed as $F(p, n-k-1)$. In light of the identity TSS = RegrSS + RSS, the numerator of the F statistic in equation (1.46) may also be expressed as an increment in the regression sum of squares or in the multiple correlation R^2:

$$F_0 = \frac{n-k-1}{p} \times \frac{\text{RegrSS} - \text{RegrSS}_0}{\text{RSS}}$$

$$= \frac{n-k-1}{p} \times \frac{\dfrac{\text{RegrSS}}{\text{TSS}} - \dfrac{\text{RegrSS}_0}{\text{TSS}}}{\dfrac{\text{RSS}}{\text{TSS}}}$$

$$= \frac{n-k-1}{p} \times \frac{R^2 - R_0^2}{1 - R^2} \tag{1.47}$$

F_0 is large and leads to rejection of the null hypothesis, therefore, when the full model is a statistically significant improvement over the restricted model. This improvement is reflected variously and equivalently in an increase in the regression sum of squares, an increase in the squared multiple correlation, and a decrease in the residual sum of squares. We shall call the change in the sum of squares, $\text{RSS}_0 - \text{RSS} = \text{RegrSS} - \text{RegrSS}_0$, the *incremental sum of squares* for hypothesis H_0. (Draper and Smith, 1981, use the term "extra sum of squares.")

Although it is often convenient to find an incremental sum of squares by fitting alternative regression models, it is also possible to calculate this quantity directly from the least-squares coefficient vector $\mathbf{b}$ and the $(\mathbf{X}'\mathbf{X})^{-1}$ matrix for the full model. Let $\mathbf{b}_1$ represent the coefficients of interest selected from among the entries of $\mathbf{b}$: $\mathbf{b}_1 = (B_1, \ldots, B_p)'$;[12] and let $\mathbf{V}_{11}$ represent the square submatrix consisting of entries in the p rows and columns of $(\mathbf{X}'\mathbf{X})^{-1}$ that pertain to the coefficients in $\mathbf{b}_1$. Then it may be shown that the incremental sum of squares $\text{RegrSS} - \text{RegrSS}_0$ is equal to $\mathbf{b}_1'\mathbf{V}_{11}^{-1}\mathbf{b}_1$, and thus the F-statistic for the hypothesis H_0: $\beta_1 = \cdots = \beta_p = 0$ may be written $F_0 = \mathbf{b}_1'\mathbf{V}_{11}^{-1}\mathbf{b}_1 / pS_E^2$. More generally, for the hypothesis H_0: $\boldsymbol{\beta}_1 = \boldsymbol{\beta}_0$, we have the test statistic

$$F_0 = \frac{(\mathbf{b}_1 - \boldsymbol{\beta}_0)'\mathbf{V}_{11}^{-1}(\mathbf{b}_1 - \boldsymbol{\beta}_0)}{pS_E^2} \tag{1.48}$$

We may apply the incremental sum of squares approach to devise a test for a single regression coefficient. Suppose, for example, that we wish to test the hypothesis H_0: $\beta_1 = 0$. Then the restricted model under the hypothesis is $Y = \beta_0 + \beta_2 x_2 + \cdots + \beta_k x_k + \varepsilon$. The vector geometry of the test is shown in Figure 1.19. Applying equation (1.47), we have the test statistic $F_0 = (n-k-1)(\text{RegrSS} - \text{RegrSS}_0)/\text{RSS}$, which is distributed as $F(1, n-k-1)$.

[12]Note the difference between the $\mathbf{b}_1$ vector (here) and $\mathbf{b}_0$ (used previously): $\mathbf{b}_1$ consists of coefficients from $\mathbf{b}$, which results from fitting the full model; $\mathbf{b}_0$ consists of the coefficients that result from fitting the restricted model.

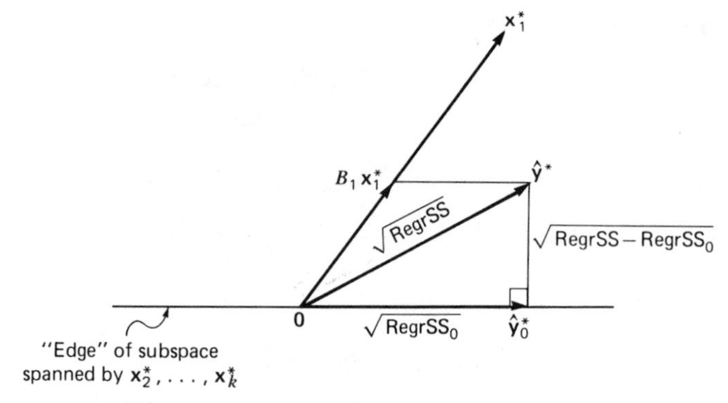

FIGURE 1.19. Incremental sum of squares for H_0: $\beta_1 = 0$.

Note the partition of $\hat{\mathbf{y}}^*$ into two orthogonal components: (1) $\hat{\mathbf{y}}_0^*$, due to $\mathbf{x}_2^*,\ldots,\mathbf{x}_k^*$ in the absence of $\mathbf{x}_1^*$; and (2) $\hat{\mathbf{y}}^* - \hat{\mathbf{y}}_0^*$, giving the incremental contribution of $\mathbf{x}_1^*$.

Using equation (1.48), and letting v_{11} denote the first diagonal entry of $(\mathbf{X}'\mathbf{X})^{-1}$, the incremental F-statistic for H_0: $\beta_1 = 0$ may be written

$$F_0 = \frac{B_1^2/v_{11}}{S_E^2} = \frac{B_1^2}{S_E^2 v_{11}} = \left(\frac{B_1}{S_E\sqrt{v_{11}}}\right)^2 = t_0^2$$

where t_0 is the t-statistic introduced in equation (1.41) to test this hypothesis. This is a sensible result, for $F(1, n-k-1) = t^2(n-k-1)$. To test the contribution of a single regressor, therefore, we may either divide the coefficient for that regressor by its standard error or, equivalently, find the incremental sum of squares due to the regressor.

Another hypothesis of broad applicability is the global or omnibus null hypothesis that all slopes are zero, H_0: $\beta_1 = \cdots = \beta_k = 0$. In this instance, the null model includes only the constant regressor: $Y = \beta_0 + \varepsilon$. Then, from equation (1.47) we have the test statistic

$$F_0 = \frac{n-k-1}{k} \times \frac{\text{RegrSS}}{\text{RSS}} = \frac{n-k-1}{k} \times \frac{R^2}{1-R^2}$$

which, if the null hypothesis is correct, is distributed as F with k and $n-k-1$ degrees of freedom.

A heuristic justification for the omnibus F-test follows from the expectations of the regression and residual sums of squares. We already know that $E(\text{RSS}) = (n-k-1)\sigma_\varepsilon^2$. Letting $\boldsymbol{\beta}_1 = (\beta_1,\ldots,\beta_k)'$ represent the vector of slope coefficients (i.e., excluding the constant β_0), and letting $\underset{(n\times k)}{\mathbf{X}^*} = \{x_{ij} - \bar{x}_j\}$ denote the matrix of independent-variable mean deviations, with the constant

TABLE 1.5. Analysis of Variance Table for Multiple Regression

Source of Variation	Sum of Squares	df	Mean Square	Expected Mean Square	F	H_0
Regression	$\text{RegrSS} = \hat{\mathbf{y}}^{*\prime}\hat{\mathbf{y}}^{*}$ $= \hat{\mathbf{y}}'\hat{\mathbf{y}} - n\bar{Y}^2$	k	RegrMS $= \dfrac{\text{RegrSS}}{k}$	$\sigma_\varepsilon^2 + \dfrac{\boldsymbol{\beta}_1'(\mathbf{X}^{*\prime}\mathbf{X}^{*})\boldsymbol{\beta}_1}{k}$	$\dfrac{\text{RegrMS}}{\text{RMS}}$	$\boldsymbol{\beta}_1 = \mathbf{0}$
Residuals	$\text{RSS} = \mathbf{e}'\mathbf{e}$	$n - k - 1$	$\text{RMS} = \dfrac{\text{RSS}}{n-k-1}$ $= S_E^2$	σ_ε^2		
Total	$\text{TSS} = \mathbf{y}^{*\prime}\mathbf{y}^{*}$ $= \mathbf{y}'\mathbf{y} - n\bar{Y}^2$	$n - 1$				

TABLE 1.6. ANOVA Table for Angell's Multiple Regression

Source	Sum of Squares	df	Mean Square	F	p
Mobility and Heterogeneity	93.97	2	46.99	9.24	$< .001$
Residuals	132.3	26	5.088		
Total	226.3	28			

regressor deleted, the expectation of the regression sum of squares may be written (see Seber, 1977: Ch. 4):

$$E(\text{RegrSS}) = \boldsymbol{\beta}_1'(\mathbf{X}^{*\prime}\mathbf{X}^{*})\boldsymbol{\beta}_1 + k\sigma_\varepsilon^2$$

When the null hypothesis is true (and, indeed, $\boldsymbol{\beta}_1 = \mathbf{0}$), RegrSS$/k$ and RSS$/(n-k-1)$, which are termed *mean squares*, both are unbiased estimators of the error variance σ_ε^2. We therefore expect the F-ratio to be close to one.[13] When H_0 is false, however, $E(\text{RegrSS}/k) > \sigma_\varepsilon^2$, since $\mathbf{X}_1^{*\prime}\mathbf{X}_1^{*}$ is positive definite, and thus $\boldsymbol{\beta}_1'(\mathbf{X}^{*\prime}\mathbf{X}^{*})\boldsymbol{\beta}_1 > 0$ for $\boldsymbol{\beta}_1 \neq \mathbf{0}$; we tend then to observe values of F_0 that are in excess of one.

This information may be summarized in an *analysis of variance* (*ANOVA*) *table*, which shows the partition of the total sum of squares into explained and unexplained components. A general ANOVA for multiple regression is given in Table 1.5. An illustrative ANOVA for Angell's data appears in Table 1.6. It is clear from this table that we may reject the null hypothesis H_0: $\beta_1 = \beta_2 = 0$. In this instance, the separate tests for the slope coefficients, reported in equation (1.42), showed that each coefficient is statistically significant.

[13]The expectation of F_0 is not precisely one, because the expectation of the ratio of independent random variables is not necessarily the ratio of their expectations.

Joint Confidence Regions The test statistic in equation (1.48) is distributed as $F(p, n-k-1)$ under the hypothesis H_0: $\boldsymbol{\beta}_1 = \boldsymbol{\beta}_0$; consequently

$$Pr\left[\frac{(\mathbf{b}_1 - \boldsymbol{\beta}_0)'\mathbf{V}_{11}^{-1}(\mathbf{b}_1 - \boldsymbol{\beta}_0)}{pS_E^2} \leq F_a\right] = 1 - a \tag{1.49}$$

where F_a is the value of $F(p, n-k-1)$ corresponding to a right-tail probability of a. From equation (1.49) we may derive the $100(1-a)$ percent *joint confidence region* for $\boldsymbol{\beta}_1$:

$$(\mathbf{b}_1 - \boldsymbol{\beta}_1)'\mathbf{V}_{11}^{-1}(\mathbf{b}_1 - \boldsymbol{\beta}_1) \leq pS_E^2 F_a \tag{1.50}$$

Any parameter vector $\boldsymbol{\beta}_1$ that satisfies this inequality is within the confidence region and thus is acceptable; any parameter vector that does not satisfy the inequality is unacceptable. Inequality (1.50) describes a region in p-dimensional parameter space whose boundary is an ellipsoid.

Like a confidence interval, a joint confidence region is a portion of the parameter space constructed in such a manner that, with repeated sampling, a preselected percentage of regions will contain the true parameter values. Unlike a confidence interval, which pertains to a single coefficient β_j, a joint confidence region encompasses all *combinations* of values for parameters $\beta_1, \beta_2, \ldots, \beta_p$ that are *simultaneously* acceptable at the specified level of confidence. Indeed, the familiar confidence interval may be regarded as a one-dimensional confidence region.

Although joint confidence regions do not appear frequently in applications, they help to shed light on the problems for statistical inference that are encountered when regressor variables are highly correlated with one another. To illustrate this point, we shall work with the two-independent-variable model and shall examine the confidence region for the slope coefficients $\boldsymbol{\beta}_1 = (\beta_1, \beta_2)'$. In this instance, the joint confidence region of (1.50) becomes

$$(B_1 - \beta_1, B_2 - \beta_2)\begin{pmatrix} \sum x_{1i}^{*2} & \sum x_{1i}^* x_{2i}^* \\ \sum x_{1i}^* x_{2i}^* & \sum x_{2i}^{*2} \end{pmatrix}\begin{pmatrix} B_1 - \beta_1 \\ B_2 - \beta_2 \end{pmatrix} \leq 2S_E^2 F_a \tag{1.51}$$

(Recall that $x_{ji}^* = x_{ji} - \bar{x}_{j\cdot}$.) Notice that $\mathbf{V}_{11}^{-1}$ is the mean-deviation sum of squares and products matrix for the independent variables. The boundary of the confidence region, obtained when the equality holds, is an ellipse centered at (B_1, B_2) in the β_1, β_2 parameter space.

If the independent variables are uncorrelated, then the sum of cross products $\Sigma x_{1i}^* x_{2i}^*$ is zero, and the axes of the ellipse in (1.51) are parallel to the axes of the parameter space. This situation is illustrated in Figure 1.20(a). Individual confidence intervals for the two parameters are also shown in this figure. Note that the joint confidence region is very nearly the intersection of these separate confidence intervals.

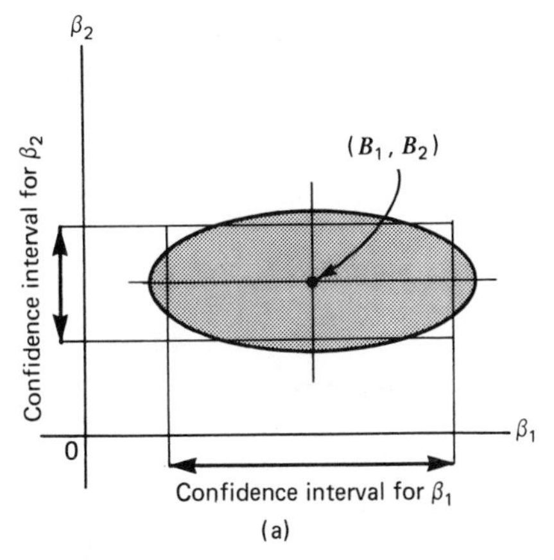

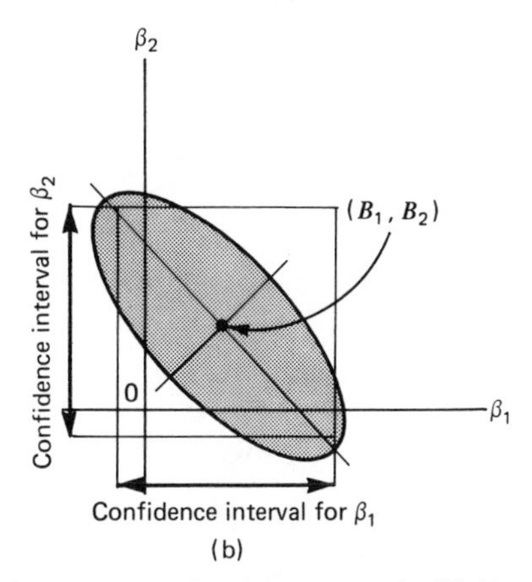

FIGURE 1.20. Joint confidence region for β_1 and β_2. (a) Uncorrelated regressors. (b) Highly correlated regressors.

In contrast, when X_1 and X_2 are correlated, the orientation of the confidence ellipse changes, as shown in Figure 1.20(b). There is, consequently, no longer a close correspondence between the separate confidence intervals for the coefficients and the joint confidence region. In the hypothetical (but not uncommon) case illustrated in Figure 1.20(b), for example, we should conclude that *either* β_1 *or* β_2 may be zero (since there are points in the confidence region where one or the other coefficient is zero), but that they are *not both* zero (since the point $(0, 0)$ is not in the confidence region). Here the correlation of X_1 and X_2 makes it difficult for us to draw conclusions about their separate effects. We shall expand upon the issue of correlated regressors in Section 3.1.

Summary Although our consideration of statistical inference in regression analysis has been conceptually fairly complex, the procedures that we have derived are simple to apply:

1. The estimated covariance matrix of $\mathbf{b}$ is $\widehat{V(\mathbf{b})} = S_E^2(\mathbf{X}'\mathbf{X})^{-1}$. The estimated standard error of each B_j is therefore given by the square root of the corresponding diagonal entry of $\widehat{V(\mathbf{b})}$. To test the null hypothesis that an individual β_j is zero, divide B_j by its estimated standard error, producing a t-statistic with $n - k - 1$ degrees of freedom.
2. To test a hypothesis that a specific subset of p regression coefficients are zero, fit the model with and without the p independent variables in question. The difference in the regression sums of squares for the two models is called the incremental sum of squares for the hypothesis. Dividing the incremental sum of squares by pS_E^2 produces an F-statistic for the hypothesis with p and $n - k - 1$ degrees of freedom (where S_E^2 is the estimated error variance for the full model).

3. To test the null hypothesis that all k population slopes are zero, calculate $F_0 = \text{RegrSS}/kS_E^2$, with k and $n - k - 1$ degrees of freedom.

PROBLEMS

1.9.# The following 14 Southern U.S. cities were omitted from Table 1.3:

City	Moral Integration Y_i	Heterogeneity X_{1i}	Mobility X_{2i}
Richmond*	10.4	65.3	24.9
Houston	10.2	49.0	36.1
Fort Worth	10.2	30.5	36.8
Oklahoma City	9.7	20.7	47.2
Chattanooga*	9.3	57.7	27.2
Nashville*	8.6	57.4	25.4
Birmingham*	8.2	83.1	25.9
Dallas	8.0	36.8	37.8
Louisville*	7.7	31.5	19.4
Jacksonville*	6.0	73.7	27.7
Memphis*	5.4	84.5	26.7
Tulsa	5.3	23.8	44.9
Miami*	5.1	50.2	41.8
Atlanta*	4.2	70.6	32.6

Employing data for the nine Southeastern cities (marked with asterisks):

(a) Find B_0, B_1, and B_2 for the least-squares regression of Y on X_1 and X_2. (*Hint*: The calculation of the slope coefficients $\mathbf{b}_1 = (B_1, B_2)'$ may be simplified by expressing Y, X_1, and X_2 in mean-deviation form, thus eliminating B_0 from the regression equation. Then $\mathbf{b}_1 = (\mathbf{X}^{*\prime}\mathbf{X}^*)^{-1}\mathbf{X}^{*\prime}\mathbf{y}^*$, where $\mathbf{X}^*$ is the matrix of mean-deviation scores for the independent variables and $\mathbf{y}^*$ is the mean-deviation vector for the dependent variable. Finally, you may recover the constant from the formula $B_0 = \overline{Y} - B_1\overline{X}_1 - B_2\overline{X}_2$.)

(b) Calculate the standard error of the estimate for the regression, S_E, and the multiple correlation coefficient R.

(c) Assuming independently distributed errors, estimate the covariance matrix $V(\mathbf{b})$ of the regression coefficients. (Alternatively, if you found $\mathbf{b}_1 = (B_1, B_2)'$ in part (a) by expressing X_1 and X_2 in mean-deviation form, you may calculate $\widehat{V(\mathbf{b}_1)} = S_E^2(\mathbf{X}^{*\prime}\mathbf{X}^*)^{-1}$.) Separately test the null hypotheses H_0: $\beta_1 = 0$ and H_0: $\beta_2 = 0$.

(d) Using the incremental sum of squares approach, compute an F-test for the hypothesis H_0: $\beta_1 = 0$. Confirm that this test is equivalent to the t-test in part (c).

(e) Construct an ANOVA table for the regression, testing the omnibus null hypothesis H_0: $\beta_1 = \beta_2 = 0$.

(f)* Draw a graph of the 90 percent simultaneous confidence region for β_1 and β_2.

(g) Find the correlation r_{12} between X_1 and X_2. Then construct the geometric vector representation for the regression, drawing separate graphs for (i) the $\mathbf{x}_1^*, \mathbf{x}_2^*$ plane, showing the $\hat{\mathbf{y}}^*$ vector, B_1, and B_2; and (ii) the $\mathbf{y}^*, \hat{\mathbf{y}}^*$ plane, showing $\mathbf{e}$. Draw all vectors to scale, but suppose that X_1 and X_2 scores are divided by 10 prior to the regression calculations; that is, let $\mathbf{x}_1^* = \{(X_{1i} - \bar{X}_1)/10\}$ and $\mathbf{x}_2^* = \{(X_{2i} - \bar{X}_2)/10\}$. Note that as a consequence of these transformations, the previously computed values of B_1 and B_2 are each multiplied by a factor of 10. Find the angle between $\mathbf{y}^*$ and $\hat{\mathbf{y}}^*$.

1.10. Re-analyze Angell's data on the moral integration of cities (from Table 1.3), including the data for 14 Southern cities given in Problem 1.9. Does the inclusion of these Southern cities affect the results? If so, is it desirable to report the results of the analysis without taking region explicitly into account? How might region be taken into account in the data analysis? (This topic is pursued in Section 2.1.)

1.11. Using Anscombe's data on state education expenditures (given in Problem 1.3, Table 1.2), regress expenditures (E) on income (I), proportion of population under 18 (S), and proportion urban (U). Compare the results of this multiple regression with those of the simple regressions fit in Problem 1.3. Are there any important differences in the regression coefficients? Which analysis do you prefer?

1.12. Nearly collinear regressors: Consider the geometric-vector representation for the two-independent-variable multiple-regression equation $\hat{\mathbf{y}}^* = B_1\mathbf{x}_1^* + B_2\mathbf{x}_2^*$, distinguishing between two cases: (i) when $\mathbf{x}_1^*$ and $\mathbf{x}_2^*$ are highly correlated; (ii) when $\mathbf{x}_1^*$ and $\mathbf{x}_2^*$ are orthogonal. By examining the regressor plane, show that slight changes in the position of the $\hat{\mathbf{y}}^*$ vector (due, for example, to sampling fluctuations) can cause dramatic changes in the regression coefficients B_1 and B_2 in the first case but not in the second. (The problem of collinearity is discussed further in Section 3.1.)

1.13. Show that the sampling variances of A and B in simple-regression analysis (given in (1.11)) may be obtained from the general matrix result $V(\mathbf{b}) = \sigma_\varepsilon^2(\mathbf{X}'\mathbf{X})^{-1}$.

1.14. Consider the regression model $Y = \beta_0 + \beta_1 X_1 + \beta_2 X_2 + \varepsilon$. How can the incremental sum of squares approach be used to test the hypothesis that the slopes are equal to each other, H_0: $\beta_1 = \beta_2 \equiv \beta$? (*Hint*: Under H_0, the model becomes $Y = \beta_0 + \beta X_1 + \beta X_2 + \varepsilon = \beta_0 + \beta(X_1 + X_2) + \varepsilon$.) Under what circumstances would a hypothesis of this form be substantively meaningful?

1.15. Consider the general regression equation

$$Y = B_0 + B_1 X_1 + B_2 X_2 + \cdots + B_k X_k + E$$

An alternative procedure for calculating the least-squares coefficient B_1 is as follows: (i) regress Y on X_2 through X_k, obtaining residuals $E_{Y:2\ldots k}$; (ii) regress X_1 on X_2 through X_k, obtaining residuals $E_{1:2\ldots k}$; then (iii) regress $E_{Y:2\ldots k}$ on $E_{1:2\ldots k}$ producing B_1.

(a) Confirm that the coefficient B_1 in Problem 1.9(a) may be obtained in this manner.

(b)* Show generally for the two-independent-variable model that B_1 calculated by this procedure is the same as B_1 computed in the usual manner. (*Hint*: work with variables in mean-deviation form; vector notation may be helpful.)

(c) Is it reasonable to describe B_1 as the effect of X_1 on Y when the influence of $X_2, \ldots, X_k$ is removed from both X_1 and Y?

1.16.† Partial correlation: The *partial correlation* between Y and X_1 "holding constant" $X_2, \ldots, X_k$, denoted $r_{Y1:2\ldots k}$, is defined as the simple correlation between $E_{Y:2\ldots k}$ and $E_{1:2\ldots k}$ (defined in Problem 1.15).

(a) Calculate the partial correlation $r_{Y1:2}$ for the data analyzed in Problem 1.9.

(b) In light of the interpretation of partial regression coefficients developed in Problem 1.15, why is $r_{Y1:2\ldots k}$ zero if and only if B_1 (from the regression of Y on $X_1, X_2, \ldots, X_k$) is zero?

(c) Show how the partial correlation $r_{Y1:2}$ may be represented using geometric vectors; draw the vectors $\mathbf{y}^*$, $\mathbf{x}_1^*$, and $\mathbf{x}_2^*$, and define $\mathbf{e}_1 \equiv \{E_{1:2;\,i}\}$ and $\mathbf{e}_Y \equiv \{E_{Y:2;\,i}\}$ (where i is the subscript for observation).

(d)* Starting with the vector representation of partial correlation in part (c), show that the incremental sum of squares F-test for the hypothesis H_0: $\beta_1 = 0$ can be written

$$F_0 = \frac{(n-k-1)r_{Y1:2}^2}{1 - r_{Y1:2}^2}$$

Recalling part (b), why is this result intuitively plausible?

†Problems marked with a dagger substantially extend the material covered in the text. The reader may wish to examine these problems even if they are not worked out in detail.

1.17.† Prediction: One use of a fitted regression equation is to *predict* dependent-variable values for particular combinations of independent-variable scores. Suppose that we fit the model $\mathbf{y} = \mathbf{X}\boldsymbol{\beta} + \boldsymbol{\varepsilon}$, obtaining the least-squares estimate $\mathbf{b}$ of $\boldsymbol{\beta}$. Let $\mathbf{x}_0' = (1, x_{10}, \ldots, x_{k0})$ represent a set of independent-variable scores for which a prediction is desired, and let Y_0 be the corresponding value of Y. $\mathbf{x}_0'$ does not necessarily correspond to an observation in the sample for which the model was fit.

(a) If we use $\hat{Y}_0 = \mathbf{x}_0'\mathbf{b}$ to estimate $E(Y_0)$, then the error in estimation is $\delta \equiv \hat{Y}_0 - E(Y_0)$. Show that $E(\delta) = 0$ (i.e., that $\hat{Y}_0$ is an unbiased estimator of $E(Y_0)$) and that $V(\delta) = \sigma_\varepsilon^2 \mathbf{x}_0'(\mathbf{X}'\mathbf{X})^{-1}\mathbf{x}_0$.

(b) At times, interest may inhere not in estimating the *expected* value of Y_0 but in predicting or forecasting an *actual* value $Y_0 = \mathbf{x}_0'\boldsymbol{\beta} + \varepsilon_0$ that will be observed. The error in the forecast is then

$$D \equiv \hat{Y}_0 - Y_0 = \mathbf{x}_0'\mathbf{b} - (\mathbf{x}_0'\boldsymbol{\beta} + \varepsilon_0) = \mathbf{x}_0'(\mathbf{b} - \boldsymbol{\beta}) - \varepsilon_0$$

Show that $E(D) = 0$ and that $V(D) = \sigma_\varepsilon^2[1 + \mathbf{x}_0'(\mathbf{X}'\mathbf{X})^{-1}\mathbf{x}_0]$. Why is the variance of the forecast error greater than the variance of δ found in part (a)?

(c) Use these results and the data for Problem 1.10 to calculate a predicted moral-integration score for a city with a heterogeneity index of 50 and a mobility index of 25. Place a 90 percent confidence interval around the prediction assuming (i) that you wish to estimate $E(Y_0)$, and (ii) that you wish to forecast an actual Y_0 score. (Note that since σ_ε^2 is not known, you will need to use S_E^2 and the t-distribution.)

1.18.† Adjusted R^2: The squared multiple correlation *adjusted for degrees of freedom* is defined as

$$\tilde{R}^2 \equiv 1 - \frac{S_E^2}{S_Y^2} = 1 - \frac{\dfrac{\text{RSS}}{n-k-1}}{\dfrac{\text{TSS}}{n-1}}$$

(a) Show that $\tilde{R}^2 < R^2$, but that when n is large $\tilde{R}^2 \simeq R^2$.

(b) Calculate $\tilde{R}^2$ for Angell's data analyzed in Problem 1.10 and compare its value to that of R^2.

1.3. RANDOM REGRESSORS

The theory of regression analysis developed in this chapter has proceeded from the premise that $\mathbf{X}$ is fixed. If we repeat a study, we expect the dependent-variable observations $\mathbf{y}$ to change, but if $\mathbf{X}$ is fixed, the independent-variable values are constant across replications. We mentioned near the beginning of the chapter that this situation is realistically descriptive of an experiment, where the independent variables are manipulated by the researcher. Most research in the social sciences, however, is observational rather than experimental; and in an observational study (survey research, for example), we would in general obtain different independent-variable values upon replication of the study. In these cases, therefore, $\mathbf{X}$ is *random* rather than fixed.

It is remarkable that our theory of regression analysis applies even when $\mathbf{X}$ is random, so long as certain assumptions are met. For fixed independent variables, the assumptions underlying the model took the form: $\boldsymbol{\varepsilon} \sim N_n(\mathbf{0}, \sigma_\varepsilon^2 \mathbf{I}_n)$. That is, the distribution of the error random variable is the same for all combinations of independent-variable values represented by the rows of the design matrix. When $\mathbf{X}$ is random, we assume that this property holds for *all possible* combinations of independent-variable values in the population that we are sampling. That is, $\mathbf{X}$ and $\boldsymbol{\varepsilon}$ are assumed to be independent, and thus the *conditional* distribution of the error for a sample of independent-variable values $(\boldsymbol{\varepsilon}|\mathbf{X}_0)$ is $N_n(\mathbf{0}, \sigma_\varepsilon^2 \mathbf{I}_n)$, regardless of the particular sample $\mathbf{X}_0 = \{x_{ji}\}$ that is chosen.

Because $\mathbf{X}$ is random, it has some (multivariate) probability distribution. We need to make no assumptions about this distribution, however, beyond (1) requiring that $\mathbf{X}$ and $\boldsymbol{\varepsilon}$ are independent, and (2) assuming that the distribution of $\mathbf{X}$ does not depend upon the parameters, $\boldsymbol{\beta}$ and σ_ε^2, of the regression model. In particular, we need not assume that the *independent variables* (as opposed to the error) are normally distributed. This is fortunate for, in many instances, regressor variables are highly discrete, taking on as few as two different values. We shall not, of course, repeat the entire argument of this chapter, but we shall show that some key results hold, under the new assumptions, when the independent variables are random.

For a particular sample of $\mathbf{X}$ values, $\mathbf{X}_0$, the conditional expectation of $\mathbf{y}$ is

$$\begin{aligned} E(\mathbf{y}|\mathbf{X}_0) &= E[(\mathbf{X}\boldsymbol{\beta} + \boldsymbol{\varepsilon})|\mathbf{X}_0)] = \mathbf{X}_0\boldsymbol{\beta} + E(\boldsymbol{\varepsilon}|\mathbf{X}_0) \\ &= \mathbf{X}_0\boldsymbol{\beta} \end{aligned}$$

Consequently, the conditional expectation of the least-squares estimator is

$$\begin{aligned} E(\mathbf{b}|\mathbf{X}_0) &= E\left[(\mathbf{X}'\mathbf{X})^{-1}\mathbf{X}'\mathbf{y}|\mathbf{X}_0)\right] = \left(\mathbf{X}_0'\mathbf{X}_0\right)^{-1}\mathbf{X}_0'E(\mathbf{y}|\mathbf{X}_0) \\ &= \left(\mathbf{X}_0'\mathbf{X}_0\right)^{-1}\mathbf{X}_0'\mathbf{X}_0\boldsymbol{\beta} = \boldsymbol{\beta} \end{aligned}$$

Since we can repeat this argument for any value of $\mathbf{X}$, $\mathbf{b}$ is conditionally unbiased for any and every such value; it is therefore unconditionally unbiased as well: $E(\mathbf{b}) = \boldsymbol{\beta}$.

Suppose that we use the procedures developed in Section 1.2.6 to perform statistical inference for $\boldsymbol{\beta}$. For concreteness, imagine that we calculate a p-value for the null hypothesis H_0: $\beta_1 = \cdots = \beta_k = 0$. Because $(\boldsymbol{\varepsilon}|\mathbf{X}_0) \sim N_n(\mathbf{0}, \sigma_\varepsilon^2\mathbf{I}_n)$, as was required of $\boldsymbol{\varepsilon}$ when we treated $\mathbf{X}$ as fixed, the p-value obtained is correct for $\mathbf{X} = \mathbf{X}_0$. There is, however, nothing special about a particular $\mathbf{X}_0$: Since $\boldsymbol{\varepsilon}$ is independent of $\mathbf{X}$, the distribution of $\boldsymbol{\varepsilon}$ is $N_n(\mathbf{0}, \sigma_\varepsilon^2\mathbf{I}_n)$ for any and every value of $\mathbf{X}$. The p-value, therefore, is unconditionally valid.

Finally, we show that the maximum-likelihood estimators of $\boldsymbol{\beta}$ and σ_ε^2 are unchanged when $\mathbf{X}$ is random, as long as the new assumptions are met. When $\mathbf{X}$ is random, sampled observations consist not just of dependent-variable values but also of independent-variable values; we denote the former $Y_1, \ldots, Y_n$, and the latter $\mathbf{x}_1', \ldots, \mathbf{x}_n'$. Since the observations are sampled independently, their joint probability density is the product of their marginal densities:

$$p(y_1, \mathbf{x}_1'; \ldots; y_n, \mathbf{x}_n') = p(y_1, \mathbf{x}_1') \ldots p(y_n, \mathbf{x}_n') \tag{1.52}$$

Now, the probability density $p(y_i, \mathbf{x}_i')$ for observation i may be written as $p(y_i|\mathbf{x}_i')p(\mathbf{x}_i')$. According to the regression model, the conditional distribution of Y_i given $\mathbf{x}_i'$ is normal:

$$p(y_i|\mathbf{x}_i') = \frac{1}{\sigma_\varepsilon\sqrt{2\pi}} \exp\left[-\frac{(y_i - \mathbf{x}_i'\boldsymbol{\beta})^2}{2\sigma_\varepsilon^2}\right]$$

Thus, the joint probability density for all observations, given in equation (1.52), becomes

$$\begin{aligned} p(\mathbf{y}, \mathbf{X}) &= \prod_{i=1}^{n} p(\mathbf{x}_i') \frac{1}{\sigma_\varepsilon\sqrt{2\pi}} \exp\left[-\frac{(y_i - \mathbf{x}_i'\boldsymbol{\beta})^2}{2\sigma_\varepsilon^2}\right] \\ &= \left[\prod_{i=1}^{n} p(\mathbf{x}_i')\right] \frac{1}{(2\sigma_\varepsilon^2\pi)^{n/2}} \exp\left[-\frac{(\mathbf{y} - \mathbf{X}\boldsymbol{\beta})'(\mathbf{y} - \mathbf{X}\boldsymbol{\beta})}{2\sigma_\varepsilon^2}\right] \end{aligned} \tag{1.53}$$

So long as $p(\mathbf{x}_i')$ does not depend upon the parameters $\boldsymbol{\beta}$ and σ_ε^2, we may ignore the joint density of the independent-variable observations in maximizing equation (1.53) with respect to these parameters. Consequently, the maximum-likelihood estimator of $\boldsymbol{\beta}$ is the least-squares estimator, as was the case for fixed $\mathbf{X}$ (cf. equation (1.38)).

1.4. SPECIFICATION ERROR, EMPIRICAL RELATIONS, AND STRUCTURAL RELATIONS

We have developed the theory of regression analysis without addressing in detail the meaning of regression coefficients. There are, in fact, two fundamentally different interpretations of regression coefficients, and failure to distinguish clearly between them is the source of much confusion. Borrowing Goldberger's (1973) terminology, we may interpret a regression descriptively, as an *empirical association* among variables, or causally, as a *structural relation* among variables.

Let us deal first with empirical relationships. Suppose that, in a population of interest, the relationship between two variables, X_1 and Y, is well described by the simple-regression model $Y = \beta_0 + \beta_1 X_1 + \varepsilon$. That is to say, the conditional expectation of Y is linearly related to X_1—i.e., $E(Y|X_1 = x_1) = \beta_0 + \beta_1 x_1$—and the variation of Y around this regression line is normal with constant variance. We do not assume that X_1 necessarily causes Y or, if it does, that residual causes of Y are independent of X_1: There is, quite simply, a linear empirical relationship between X_1 and Y in the population. If we draw a random sample from this population, the least-squares sample slope B_1 is an unbiased estimator of the population slope β_1, and we may validly make statistical inferences from the sample to the population.

Suppose, now, that we introduce a second independent variable X_2, and that, in the same sense as before, the relationship between Y and the two X's is linear. That is, $E(Y) = \beta_0' + \beta_1' X_1 + \beta_2' X_2$, and the distribution of Y about the regression plane is normal and homoscedastic. The slope β_1' of the population regression plane for the multiple regression can, and generally will, differ from β_1, the simple-regression slope. The sample least-squares coefficients B_1' and B_2' are unbiased estimators of β_1' and β_2'. That β_1' differs from β_1, and that therefore B_1 is a biased estimator of β_1', is not problematic, for these are simply empirical relationships, and we do not in this context interpret a regression coefficient as the *effect* of an independent variable on the dependent variable. The issue of *specification error*—fitting a false model to the data—does not arise, as long as the linear form of the regression model adequately describes the empirical population relationship between the dependent variable and the independent variables.

The situation is drastically altered, however, if we view the regression equation as representing a structural relationship—that is, a model of how dependent-variable scores are determined. Imagine now that dependent-variable scores are *constructed* according to the model $Y = \beta_0 + \beta_1 X_1 + \beta_2 X_2 + \varepsilon$, where the error ε satisfies the usual regression assumptions; in particular, $E(\varepsilon) = 0$, and ε is independent of X_1 and X_2. If we use least squares to fit this model to sample data, we obtain unbiased estimators of β_1 and β_2. Suppose, however, that instead we fit the model $Y = \beta_0 + \beta_1 X_1 + \varepsilon'$ where, implicitly, the effect of X_2 on Y is absorbed by the error $\varepsilon' = \varepsilon + \beta_2 X_2$. In the event that X_1 and X_2 are correlated, then X_1 becomes correlated with ε'. If we assume that X_1 and ε' are uncorrelated, as we do if we proceed to fit the model by least squares, we make an error of specification. The consequence of this error is that our estimator of β_1 is biased: Because X_1 and X_2 are correlated, and because X_2 is deleted from the model, part of the effect of X_2 is wrongly attributed to X_1. This point is elaborated in Section 4.5.

In generalizing our treatment of misspecified structural relationships, it is convenient to work with probability limits. Suppose that the dependent variable Y is determined by the model

$$\mathbf{y}^* = \mathbf{X}^*\boldsymbol{\beta} + \boldsymbol{\varepsilon} = \mathbf{X}_1^*\boldsymbol{\beta}_1 + \mathbf{X}_2^*\boldsymbol{\beta}_2 + \boldsymbol{\varepsilon}$$

where the error $\boldsymbol{\varepsilon}$ behaves according to the usual assumptions. Here we have expressed each variable as deviations from its expectation (e.g., $\mathbf{y}^* = \{Y_i - E(Y)\}$), and have partitioned the independent-variable matrix into two sets of regressors; the parameter vector is partitioned in the same way.

Imagine that we ignore $\mathbf{X}_2^*$, so that $\mathbf{y}^* = \mathbf{X}_1^*\boldsymbol{\beta}_1 + \tilde{\boldsymbol{\varepsilon}}$, where $\tilde{\boldsymbol{\varepsilon}} = \mathbf{X}_2^*\boldsymbol{\beta}_2 + \boldsymbol{\varepsilon}$. If we fit this model by least squares, then we employ the estimator

$$\begin{aligned}
\mathbf{b}_1 &= \left(\mathbf{X}_1^{*\prime}\mathbf{X}_1^*\right)^{-1}\mathbf{X}_1^{*\prime}\mathbf{y}^* \\
&= \left(\frac{1}{n}\mathbf{X}_1^{*\prime}\mathbf{X}_1^*\right)^{-1}\frac{1}{n}\mathbf{X}_1^{*\prime}\mathbf{y}^* \\
&= \left(\frac{1}{n}\mathbf{X}_1^{*\prime}\mathbf{X}_1^*\right)^{-1}\frac{1}{n}\mathbf{X}_1^{*\prime}\left(\mathbf{X}_1^*\boldsymbol{\beta}_1 + \mathbf{X}_2^*\boldsymbol{\beta}_2 + \boldsymbol{\varepsilon}\right) \\
&= \boldsymbol{\beta}_1 + \left(\frac{1}{n}\mathbf{X}_1^{*\prime}\mathbf{X}_1^*\right)^{-1}\left(\frac{1}{n}\mathbf{X}_1^{*\prime}\mathbf{X}_2^*\right)\boldsymbol{\beta}_2 + \left(\frac{1}{n}\mathbf{X}_1^{*\prime}\mathbf{X}_1^*\right)^{-1}\frac{1}{n}\mathbf{X}_1^{*\prime}\boldsymbol{\varepsilon} \qquad (1.54)
\end{aligned}$$

Taking probability limits in equation (1.54) produces

$$\begin{aligned}
\operatorname{plim}\mathbf{b}_1 &= \boldsymbol{\beta}_1 + \boldsymbol{\Sigma}_{11}^{-1}\boldsymbol{\Sigma}_{12}\boldsymbol{\beta}_2 + \boldsymbol{\Sigma}_{11}^{-1}\boldsymbol{\sigma}_{1\varepsilon} \\
&= \boldsymbol{\beta}_1 + \boldsymbol{\Sigma}_{11}^{-1}\boldsymbol{\Sigma}_{12}\boldsymbol{\beta}_2
\end{aligned}$$

$\boldsymbol{\Sigma}_{11} \equiv \operatorname{plim}(1/n)\mathbf{X}_1^{*\prime}\mathbf{X}_1^*$ is the population variance-covariance matrix for $\mathbf{X}_1$; $\boldsymbol{\Sigma}_{12} \equiv \operatorname{plim}(1/n)\mathbf{X}_1^{*\prime}\mathbf{X}_2^*$ is the population covariance matrix for $\mathbf{X}_1$ and $\mathbf{X}_2$; and $\boldsymbol{\sigma}_{1\varepsilon} = \operatorname{plim}(1/n)\mathbf{X}_1^{*\prime}\boldsymbol{\varepsilon}$ is the vector of population covariances between $\mathbf{X}_1$ and $\boldsymbol{\varepsilon}$, which is zero by the assumed independence of the error and the independent variables. The asymptotic (or population) covariance of $\mathbf{X}_1$ and $\tilde{\boldsymbol{\varepsilon}}$ is not generally zero, however, as is readily established:

$$\begin{aligned}
\operatorname{plim}\frac{1}{n}\mathbf{X}_1^{*\prime}\tilde{\boldsymbol{\varepsilon}} &= \operatorname{plim}\frac{1}{n}\mathbf{X}_1^{*\prime}\left(\mathbf{X}_2^*\boldsymbol{\beta}_2 + \boldsymbol{\varepsilon}\right) \\
&= \boldsymbol{\Sigma}_{12}\boldsymbol{\beta}_2 + \boldsymbol{\sigma}_{1\varepsilon} = \boldsymbol{\Sigma}_{12}\boldsymbol{\beta}_2
\end{aligned}$$

The estimator $\mathbf{b}_1$, therefore, is consistent if $\boldsymbol{\Sigma}_{12}$ is zero; that is, if the independent variables in the two sets are uncorrelated. In this case, incorporating $\mathbf{X}_2^*$ in the error does not cause $\mathbf{X}_1^*$ to become correlated with $\tilde{\boldsymbol{\varepsilon}}$. Note that $\mathbf{b}_1$ is also consistent if $\boldsymbol{\beta}_2 = \mathbf{0}$: Excluding irrelevant regressors does not have negative consequences. Incidentally, *including* irrelevant regressors does not cause the least-squares estimator to become inconsistent; after all, if the assumptions of the model hold, then $\mathbf{b}$ is an unbiased estimator regardless of the value of $\boldsymbol{\beta}$, even when some of the entries of $\boldsymbol{\beta}$ are zero (but see Problem 1.20).

To regard a regression model as a structural model, we must be prepared to argue that the aggregated omitted causes of the dependent variable, which comprise the error, are uncorrelated with the independent variables included in the model. For observational data, this argument must be made on substantive, theoretical grounds. We can, of course, avoid the issue of specification error by construing the regression empirically, but for most social research this approach is not intellectually satisfying.

Causal inference is more straightforward in experimental research, where *randomization* may be practiced. By randomly assigning independent-variable values to experimental units (say, subjects), we attempt to insure that the residual causes of the dependent variable are statistically independent of the independent variables. The implicit assumption here is that the experimental manipulation affects only the independent variables in the experimental design, and has no effect on the residual causes of the dependent variable. Practices such as standardization of procedures across experimental conditions and double-blind assignment of subjects to conditions are meant to remove possible sources of correlation between independent variables and the error.

PROBLEMS

1.19. Examples of specification error:

(a) Describe a non-experimental research situation—real or contrived—in which failure to control statistically for an omitted variable induces a correlation between the error and an independent variable, producing erroneous conclusions.

(b) Describe an experiment—real or contrived—in which faulty experimental practice causes an independent variable to become correlated with the error, compromising the validity of the results produced by the experiment.

(c) Is it fair to conclude that a researcher is *never* able to absolutely rule out the possibility that an independent variable of interest is correlated with the error? Explain your answer.

1.20. Suppose that the "true" model generating a set of data is $Y = \beta_0 + \beta_1 X_1 + \varepsilon$, where the error ε follows the usual regression-model assumptions. A researcher fits the model $Y = \beta_0 + \beta_1 X_1 + \beta_2 X_2 + \varepsilon$, which includes the irrelevant independent variable X_2; that is, $\beta_2 = 0$. Were we to fit the (correct) simple-regression model, the variance of B_1 would be $V(B_1) = \sigma_\varepsilon^2 / \Sigma(X_{1i} - \bar{X}_1)^2$.

(a) Show that the variance of B_1 for the multiple-regression model is

$$V(B_1) = \frac{\sigma_\varepsilon^2}{\Sigma(X_{1i} - \bar{X}_1)^2} \times \frac{1}{1 - r_{12}^2}$$

where r_{12} is the correlation between X_1 and X_2. (*Hint*: Write all variables in mean-deviation form to eliminate the constant B_0 from the regression equation, and invert $\mathbf{X}^{*\prime}\mathbf{X}^*$, where $\mathbf{X}^*$ is the mean-deviation matrix for X_1 and X_2.)

(b) What, then, is the cost of including an irrelevant regressor? How does this compare to the cost of failing to include a relevant regressor?

1.21. Assess the relative merits of the following opposed arguments:

(a) Angell's 43 American cities do not constitute a sample drawn from some larger population of cities. Because the cities in fact comprise a population, any observed nonzero regression coefficient is necessarily "statistically significant" regardless of its size. Indeed, tests of statistical hypotheses are meaningless for these and similar data.

(b) Angell's 43 American cities do not constitute a sample drawn randomly from some larger population of cities. It is nevertheless meaningful to test for the statistical significance of, say, the effect of mobility on moral integration, because our interest inheres not in this specific group of cities and their moral-integration scores, but in the presumed process that generated these scores. If a hypothetical model in which the effect of mobility is nil (corresponding to H_0: $\beta_2 = 0$) could easily have given rise to the observed data, then we can have little confidence in the importance of mobility as a determinant of moral integration. Of course, the statistical assumptions of the model, including the assumption of independent errors, must be judged reasonable in order for the test to be valid.

1.5. STANDARDIZED REGRESSION COEFFICIENTS

Social researchers often wish to compare the effects of different independent variables in a regression analysis. When the independent variables are commensurable (that is, measured in the same units) or when they can be reduced to a common standard, comparison is straightforward. In most instances, however, independent variables in a regression equation are not commensurable. Standardized regression coefficients permit a limited assessment of the relative effects of incommensurable independent variables. The limitations of the comparison afforded by standardized coefficients often are not appreciated by researchers, and it is fair to say that these coefficients are overused and misused in social research. Aside from their admittedly restricted utility in data analysis, however, standardized coefficients have a role in regression computations and are useful in exploring certain properties of regression analysis (e.g., collinearity, discussed in Chapter 3).

To place standardized coefficients in perspective, let us first consider an example in which the regressors are commensurable. Imagine that the annual dollar income of wage workers is regressed on their years of education and years of labor-force experience, according to the model

$$\text{Income} = \beta_0 + \beta_1\text{Education} + \beta_2\text{Experience} + \varepsilon$$

Since the two independent variables in this equation are measured in years, the coefficients β_1 and β_2 are both expressed in dollars/year, and consequently may be directly compared. If, for example, β_1 is larger than β_2, then a year's increment in education yields a greater expected income return than a year's increment in labor-force experience.

Now let us consider incommensurable independent variables, such as those in Angell's regression of the moral integration of cities on their residents' ethnic heterogeneity and geographic mobility. This example was first discussed in Section 1.2.1. Heterogeneity and mobility are measured in arbitrary, and presumably different, units. The estimated regression coefficients computed from Angell's data are -0.167 for heterogeneity and -0.214 for mobility [equation (1.21)]. We cannot conclude from these coefficients that mobility has a "larger" effect on integration than heterogeneity has, for the relative magnitudes of the coefficients are dependent upon the scaling of the independent variables; and the independent variables, we recall, are not measured in the same units. If, for instance, we were to divide heterogeneity scores by 10, the coefficient for that independent variable would become -1.67.

By the very meaning of the term, incommensurable quantities cannot be directly compared. Still, in certain instances, incommensurables can be reduced to a common (e.g., monetary) standard. For Angell's data, however, there does not seem to be an obvious basis for this sort of reduction.

In the absence of a theoretically meaningful basis for comparison, an empirical comparison may be made by scaling regression coefficients according to a measure of independent-variable dispersion. We may, for example, multiply each regression coefficient by the observed range of the corresponding independent variable. For Angell's data, the range of heterogeneity scores is $45.8 - 10.6 = 35.2$, and the range of mobility scores is $49.8 - 12.1 = 37.7$. When heterogeneity is manipulated over this range, the estimated expected change in moral integration is $-0.167 \times 35.2 = -5.88$; similarly, for mobility the estimated expected change is $-0.214 \times 37.7 = -8.07$. Thus, mobility has a somewhat larger effect than heterogeneity over the range of scores observed in the data. Note that this conclusion is distinctly limited in scope. For other data, where the variation in mobility and heterogeneity may be different, the relative impact of the two variables may also differ, even if the regression coefficients are unchanged.

It is more common to standardize regression coefficients by the standard deviations of independent variables than by their ranges. The usual practice is to standardize the dependent variable as well, although this procedure does not change the *relative* sizes of different slope coefficients.

Beginning with the multiple regression model

$$Y_i = \beta_0 + \beta_1 X_{1i} + \cdots + \beta_k X_{ki} + \varepsilon_i$$

we express all variables in mean-deviation form by subtracting

$$E(Y) = \beta_0 + \beta_1 E(X_1) + \cdots + \beta_k E(X_k)$$

$$\mu_y = \beta_0 + \beta_1 \mu_1 + \cdots + \beta_k \mu_k$$

which produces

$$Y_i - \mu_y = \beta_1(X_{1i} - \mu_1) + \cdots + \beta_k(X_{ki} - \mu_k) + \varepsilon_i \tag{1.55}$$

Then we divide both sides of equation (1.55) by the standard deviation of the dependent variable, σ_y; we simultaneously multiply and divide the jth term on the right-hand side of the equation by the standard deviation σ_j of X_j. These operations effectively standardize each variable in the regression equation:

$$\frac{Y_i - \mu_y}{\sigma_y} = \left(\beta_1 \frac{\sigma_1}{\sigma_y}\right)\frac{X_{1i} - \mu_1}{\sigma_1} + \cdots + \left(\beta_k \frac{\sigma_k}{\sigma_y}\right)\frac{X_{ki} - \mu_k}{\sigma_k} + \frac{\varepsilon_i}{\sigma_y} \tag{1.56}$$

$$Z_{yi} = \beta_1^* Z_{1i} + \cdots + \beta_k^* Z_{ki} + \varepsilon_i^*$$

In equation (1.56), $Z_y \equiv (Y - \mu_y)/\sigma_y$ is the dependent variable linearly transformed to an expectation of zero and a standard deviation of one; $Z_1, \ldots, Z_k$ are the independent variables similarly standardized; $\varepsilon^* \equiv \varepsilon/\sigma_y$ is the transformed error which, we note, *does not* have a standard deviation of one; and $\beta_j^* \equiv \beta_j(\sigma_j/\sigma_y)$ is the *standardized partial regression coefficient* for the jth independent variable. β_j^* is interpretable as the expected change in Y, in standard-deviation units, for a one-standard-deviation increment in X_j, holding constant the other independent variables. In the sample, where population expectations and variances are unknown, we take[14]

$$Z_{yi} = \frac{Y_i - \overline{Y}}{S_y}$$

$$Z_{ji} = \frac{X_{ji} - \overline{X}_j}{S_j}$$

$$B_j^* = B_j \frac{S_j}{S_y}$$

[14]The sample standardized regression coefficient B_j^* is sometimes called a "beta-weight." We shall avoid this unfortunate term, however, since it frequently leads to the confusion of sample and population coefficients.

As we shall show presently, the least-squares estimates B_j^* of the standardized regression parameters may be computed directly from the correlations among variables in the model. This approach is often computationally advantageous, for it tends to reduce rounding errors. With the standardized coefficients in hand, we may recover the unstandardized coefficients from the relations

$$B_j = B_j^* \frac{S_y}{S_j}$$

$$B_0 = \bar{Y} - B_1\bar{X}_1 - \cdots - B_k\bar{X}_k$$

In matrix format, the fitted standardized regression model may be written

$$\underset{(n\times 1)}{\mathbf{z}_y} = \underset{(n\times k)}{\mathbf{Z}_X}\ \underset{(k\times 1)}{\mathbf{b}^*} + \underset{(n\times 1)}{\mathbf{e}^*}$$

where $\mathbf{z}_y$ is the vector of standardized dependent-variable scores, $\mathbf{Z}_X$ is the matrix of standardized independent variables, $\mathbf{b}^*$ is the vector of standardized regression coefficients, and $\mathbf{e}^*$ is a vector of residuals. The least-squares regression coefficients are given by

$$\mathbf{b}^* = \left(\mathbf{Z}_X'\mathbf{Z}_X\right)^{-1}\mathbf{Z}_X'\mathbf{z}_y \tag{1.57}$$

Because the sum of products for two standardized variables is $n - 1$ times their correlation, equation (1.57) may be rewritten

$$\begin{aligned}\mathbf{b}^* &= \left[(n-1)\mathbf{R}_{XX}\right]^{-1}\left[(n-1)\mathbf{r}_{Xy}\right]\\ &= (n-1)^{-1}(n-1)\mathbf{R}_{XX}^{-1}\mathbf{r}_{Xy} = \underset{(k\times k)}{\mathbf{R}_{XX}^{-1}}\ \underset{(k\times 1)}{\mathbf{r}_{Xy}}\end{aligned}$$

Here, $\mathbf{R}_{XX}$ is the matrix of correlations among the independent variables, and $\mathbf{r}_{Xy}$ is the vector of correlations between the independent variables and the dependent variable. The estimated covariance matrix for $\mathbf{b}^*$ is

$$\widehat{V(\mathbf{b}^*)} = S_E^{*2}\left(\mathbf{Z}_X'\mathbf{Z}_X\right)^{-1} = \frac{S_E^{*2}}{n-1}\mathbf{R}_{XX}^{-1}$$

where $S_E^{*2} = \mathbf{e}^{*\prime}\mathbf{e}^*/(n-k-1)$. Note that when n is large relative to k

$$S_E^{*2} = \frac{\text{RSS}^*}{n-k-1} \simeq \frac{\text{RSS}^*}{n-1} = \frac{\text{RSS}^*}{\text{TSS}^*} = 1 - R^2$$

Since t-statistics (and other statistical tests) are independent of the scales of the variables in a regression, the same t-values are obtained for standardized variables as for unstandardized ones.

Applying these results to Angell's data, we compute

$$\begin{pmatrix} B_1^* \\ B_2^* \end{pmatrix} = \begin{pmatrix} 1 & -.5130 \\ -.5130 & 1 \end{pmatrix}^{-1} \begin{pmatrix} -.1564 \\ -.4564 \end{pmatrix}$$

$$= \begin{pmatrix} -0.5301 \\ -0.7283 \end{pmatrix}$$

$$S_E^{*2} = \frac{16.37}{29-3} = 0.6296$$

$$\widehat{V(\mathbf{b}^*)} = \frac{0.6296}{29-1} \begin{pmatrix} 1 & -.5130 \\ -.5130 & 1 \end{pmatrix}^{-1}$$

$$= \begin{pmatrix} 0.03051 & 0.01565 \\ 0.01565 & 0.03051 \end{pmatrix}$$

$$S_{B_1^*} = S_{B_2^*} = \sqrt{0.03051} = 0.1747$$

We have stressed the restricted extent to which standardized regression coefficients permit the comparison of effects for incommensurable regressors. A common misuse of these coefficients is to employ them to make comparisons of the effects of the *same* independent variable in two or more samples drawn from different populations. If the independent variable in question has different standard deviations in these samples, spurious differences between coefficients can be produced even when unstandardized coefficients are similar; alternatively, differences in unstandardized coefficients can be masked by compensating differences in dispersion.

PROBLEMS

1.22. Standardized regression coefficients and correlations:

(a) Show that in simple regression analysis, the standardized slope coefficient B^* is equal to the correlation r between the independent and dependent variables.

(b) Use the geometric vector representation to illustrate how it is possible to obtain standardized slope coefficients in multiple regression that are greater than 1.0 or smaller than -1.0. (*Hint*: Work in the regressor plane, noting that all standardized variables have vectors of length $\sqrt{n-1}$ and that $\|\hat{\mathbf{y}}^*\| \leq \|\mathbf{y}^*\|$.)

1.23. Calculate standardized coefficients for Angell's regression of moral integration on ethnic heterogeneity and geographic mobility, using data for all 43 cities. (The data appear in Table 1.3 and Problem 1.9.) Compare the standardized coefficients to the unstandardized coefficients found in Problem 1.10.

TABLE 1.7. Blau and Duncan's Stratification Data

	X_1	X_2	X_3	X_4	Y
X_1	1.000				
X_2	.516	1.000			
X_3	.453	.438	1.000		
X_4	.332	.417	.538	1.000	
Y	.322	.405	.596	.541	1.000

X_1 = Father's Education

X_2 = Father's Occupational Status

X_3 = Respondent's Education

X_4 = Status of Respondent's First Job

Y = Respondent's Current Occupational Status

Source: Blau and Duncan (1967: 169).

1.24. Calculate standardized coefficients for Anscombe's regression of state education expenditures on per-capita personal income, proportion of the population under 18, and proportion urban. (The data are given in Problem 1.3, Table 1.2.) Compare the standardized coefficients to the unstandardized coefficients found in Problem 1.11.

1.25. The correlation matrix in Table 1.7 is taken from Blau and Duncan's (1967) work on social stratification (also discussed in Chapter 4). Using these correlations, find the standardized coefficients for the regression of Y on X_1, X_2, X_3, and X_4. Why is the slope for father's education (X_1) so small? Is it legitimate to conclude that father's education is unimportant as a cause of respondent's occupational status? (See Section 4.5).

1.26. Prove that the squared multiple correlation for the regression of Y on $X_1, \ldots, X_k$ may be written as

$$R^2 = B_1^* r_{Y1} + \cdots + B_k^* r_{Yk} = \mathbf{r}_{yX}' \mathbf{b}^*$$

(*Hint*: Multiply the standardized fitted regression model $\mathbf{z}_y = \mathbf{Z}_X \mathbf{b}^* + \mathbf{e}^*$ through by $\mathbf{z}_y'$, and then divide both sides by $n - 1$ to convert to correlations.) Use this result to calculate the multiple correlation for the Blau and Duncan data analyzed in the previous problem.

2

Dummy-Variable Regression and Analysis of Variance

We mentioned in Chapter 1 that one of the serious limitations of linear regression analysis is its restriction to quantitative independent variables. The first section of the present chapter demonstrates how the regression model may be extended through the use of dummy-variable regressors to include qualitative independent variables. Section 2.2 develops linear models for single qualitative independent variables (called one-way analysis of variance). These models are broadened in Section 2.3 to two qualitative independent variables (two-way analysis of variance), and in Section 2.4 to any number of qualitative independent variables. Section 2.5 presents an alternative linear model incorporating both qualitative and quantitative independent variables (analysis of covariance).

2.1. QUALITATIVE INDEPENDENT VARIABLES IN REGRESSION ANALYSIS

In this section we develop a technique called *dummy-variable regression*, which permits us to enter qualitative independent variables into a regression equation. We begin with a *dichotomous* (i.e., two-category) qualitative variable, proceed to show that the dummy-variable approach may be adapted to *polychotomous* (many-category) variables, and finally explain how to model interactions between qualitative and quantitative independent variables in a regression.

2.1.1. Representing a Dichotomous Independent Variable by a Dummy Regressor

We shall initially consider the simplest case: one dichotomous and one quantitative independent variable. As in Chapter 1, we assume that relationships are additive. Our motivations for including qualitative independent variables in a linear model are the usual ones (cf. Section 1.2): to explain more fully how dependent-variable values are determined; and to avoid a biased assessment of the impact of an independent variable, as a consequence of omitting another independent variable related to it.

For concreteness, suppose that we are interested in investigating the relationship between education and income among blacks and whites. Figures 2.1(a) and (b) represent two possible (idealized) situations. In both cases, the within-race regressions of income on education are parallel; we shall consider nonparallel regressions in Section 2.1.3. In Figure 2.1(a), the independent variables race and education are unrelated to each other—blacks and whites have similar distributions of education scores. Here, if we ignore race and regress income on education alone, we obtain the same slope as would be produced by separate within-race regressions; since blacks have lower incomes than whites of equal educational level, however, by ignoring race we have inflated the size of the residuals.

The situation depicted in Figure 2.1(b) is importantly different. Here, race and education are related, and therefore if we regress income on education alone, we come to a biased assessment of the effect of education on income: Since blacks tend to have both lower education and, for a given educational level, lower income, part of the effect of race (i.e., what should be interpreted as income discrimination) is wrongly attributed to education.[1]

In light of these considerations, we might proceed to partition our sample by race and perform separate regressions for blacks and whites. This approach, though reasonable, is not ideal. Fitting separate regressions makes it difficult to estimate and test for racial differences in income. Furthermore, if we may reasonably assume parallel regressions for blacks and whites, we can more efficiently estimate the common education slope by pooling data from both subsamples. In particular, if the usual assumptions of the regression model (Section 1.1.2) hold, then it is desirable to fit the common-slope model by least squares.

One way of writing the common-slope model is

$$Y_i = \alpha + \gamma D_i + \beta X_i + \varepsilon_i \qquad (2.1)$$

[1] This discussion has proceeded as if we had access to population data. In a sample, of course, we cannot expect within-group regressions to be perfectly parallel, if only because of sampling error. Moreover, we cannot demonstrate "bias" without at least implicit reference to population parameters. The treatment here has heuristic value, however.

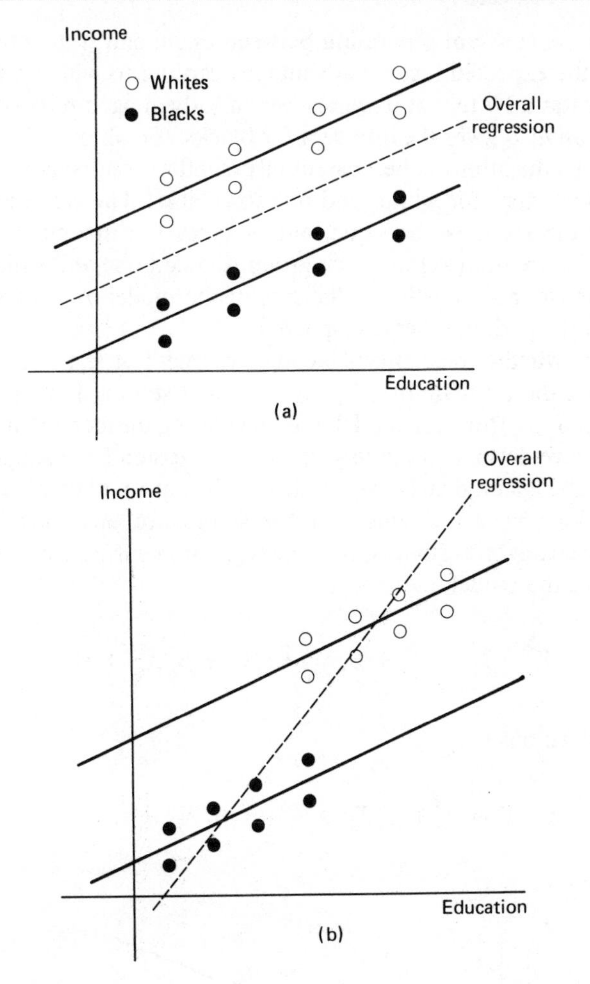

FIGURE 2.1. Income by race and education (hypothetical data). (a) Race and education unrelated. (b) Race and education related.

where D, called an *indicator variable* or a *dummy-variable regressor*, is coded zero for blacks and one for whites. Thus, for blacks the model becomes

$$Y_i = \alpha + \gamma(0) + \beta X_i + \varepsilon_i = \alpha + \beta X_i + \varepsilon_i$$

and for whites

$$Y_i = \alpha + \gamma(1) + \beta X_i + \varepsilon_i = (\alpha + \gamma) + \beta X_i + \varepsilon_i$$

These regression equations are graphed in Figure 2.2.

The coefficient γ for the dummy regressor gives the difference in the intercepts of the two regression lines. Since the regression lines are parallel, γ

also represents the constant separation between them, and it may therefore be interpreted as the expected income advantage accruing to whites when education is held constant. Note that if whites were *dis*advantaged relative to blacks, γ would be negative. α gives the intercept for blacks, for whom D is zero, and β is the common education slope. Essentially similar results are obtained if instead we code D zero for whites and one for blacks: The sign of γ changes, but its magnitude remains the same, and α gives the income intercept for whites. It is therefore immaterial which group is coded one and which is coded zero, so long as we interpret the coefficients of the model in a manner that is consistent with the coding scheme employed.

To determine whether race affects income, we may test H_0: $\gamma = 0$, either by a t-test, dividing the estimate of γ by its estimated standard error, or equivalently by dropping D from the model and calculating an incremental F-test.

Though we have developed dummy-variable regression for a single quantitative regressor, the method may be applied with any number of quantitative regressors, so long as we assume that the slopes are the same in the two dummy-variable categories (that is, that the regression surfaces are parallel). In general, if we fit the model

$$Y_i = \beta_0 + \gamma D_i + \beta_1 X_{1i} + \cdots + \beta_k X_{ki} + \varepsilon_i$$

then for $D = 0$ we have

$$Y_i = \beta_0 + \beta_1 X_{1i} + \cdots + \beta_k X_{ki} + \varepsilon_i$$

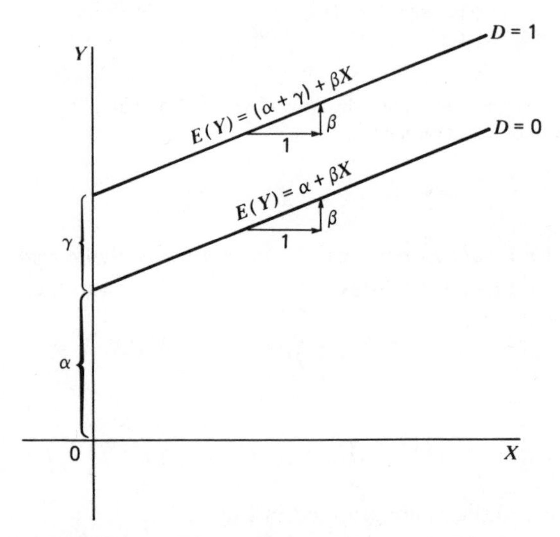

FIGURE 2.2. The dummy-variable regression model.

and for $D = 1$

$$Y_i = (\beta_0 + \gamma) + \beta_1 X_{1i} + \cdots + \beta_k X_{ki} + \varepsilon_i$$

2.1.2. Extension to Polychotomous Independent Variables: Sets of Dummy Regressors

The method of the previous section generalizes straightforwardly to polychotomous variables. By way of illustration, recall Angell's regression of moral integration on ethnic heterogeneity and geographic mobility for 29 U.S. cities (Section 1.2.1). The data for this illustration were given in Table 1.3. We have divided Angell's cities into three regional groups: East, Midwest, and West, as shown in Table 2.1. The three-category regional classification may be entered into the regression equation by coding *two* dummy variables; we may, for instance, employ the coding

	D_1	D_2
East	1	0
Midwest	0	1
West	0	0

(2.2)

and fit the model

$$Y_i = \beta_0 + \gamma_1 D_{1i} + \gamma_2 D_{2i} + \beta_1 X_{1i} + \beta_2 X_{2i} + \varepsilon_i$$

TABLE 2.1. Angell's 29 U.S. Cities Classified by Region

East	Midwest	West
Rochester	Milwaukee	Denver
Syracuse	Dayton	San Diego
Worcester	Des Moines	Tacoma
Erie	Cleveland	Spokane
Bridgeport	Peoria	Seattle
Buffalo	Wichita	Portland
Reading	Grand Rapids	
Trenton	Toledo	
Baltimore	South Bend	
	Akron	
	Detroit	
	Flint	
	Indianapolis	
	Columbus	

This model describes three parallel regression planes that may differ in their intercepts:

$$\text{East:} \quad Y_i = (\beta_0 + \gamma_1) + \beta_1 X_{1i} + \beta_2 X_{2i} + \varepsilon_i$$

$$\text{Midwest:} \quad Y_i = (\beta_0 + \gamma_2) + \beta_1 X_{1i} + \beta_2 X_{2i} + \varepsilon_i$$

$$\text{West:} \quad Y_i = \beta_0 \qquad + \beta_1 X_{1i} + \beta_2 X_{2i} + \varepsilon_i$$

β_0, therefore, gives the intercept for cities in the West, γ_1 represents the constant vertical difference between the regression plane for the East and that for the West, and γ_2 represents the difference between the Midwest and the West.

Since western cities are coded zero for both dummy regressors, the West implicitly serves as a *baseline* category with which the other regions are compared. The choice of a baseline category is essentially arbitrary, for we would fit precisely the same three regression planes regardless of which region is selected for this role. The meaning of the individual dummy-variable coefficients γ_1 and γ_2 depends, however, upon which category is chosen as the baseline.

It is sometimes natural to select a particular category as a basis for comparison—an experiment that includes a control group is a case in point. In these instances, the individual dummy-variable coefficients will be of interest. In most applications, however, a baseline category is chosen arbitrarily, as it was for Angell's data. We are therefore interested in testing the null hypothesis of no regional effects, H_0: $\gamma_1 = \gamma_2 = 0$, but the individual hypotheses H_0: $\gamma_1 = 0$ and H_0: $\gamma_2 = 0$, which test differences between the East and West and between the Midwest and West, are of less intrinsic interest. The hypothesis H_0: $\gamma_1 = \gamma_2 = 0$ may be tested by the incremental sum of squares approach.

We have shown how to model the effects of a three-category qualitative variable by coding two dummy regressors. It may seem more natural to code three dummy regressors according to the following scheme:

	D_1	D_2	D_3
East	1	0	0
Midwest	0	1	0
West	0	0	1

(2.3)

Then, for the jth regional category we would have

$$Y_i = (\beta_0 + \gamma_j) + \beta_1 X_{1i} + \beta_2 X_{2i} + \varepsilon_i$$

The difficulty with this procedure is that the set of three dummy variables is perfectly collinear (for in (2.3), $D_3 = 1 - D_1 - D_2$), leading to a design matrix of deficient rank and hence to underdetermined normal equations. We shall pursue this issue more fully in Section 2.2.

In general, then, for a polychotomous independent variable with m categories, we need to code $m - 1$ dummy regressors, so that $D_j = 1$ when an observation falls in category j, and $D_j = 0$ otherwise. Consequently, all $D_j = 0$ for an observation in the mth category. When there is more than one qualitative independent variable, and if we assume that these variables have additive effects, we simply code a set of dummy regressors for each. To test the hypothesis that the effects of a qualitative variable are nil, we delete its dummy regressors from the model and compute the incremental sum of squares.

In Chapter 1, we used Angell's data to regress moral integration on heterogeneity and mobility, obtaining the fitted equation

$$\underset{(2.19)}{\hat{Y} = 21.8} \; - \underset{(0.0552)}{0.167X_1} \; - \underset{(0.0514)}{0.214X_2} \quad R^2 = .415$$

As is our usual practice, standard errors are shown in parentheses beneath the regression coefficients. Inserting regional dummy variables in the regression equation (employing the coding scheme shown in (2.2)) produces the following results:

$$\begin{gathered}\hat{Y} = \underset{(3.65)}{14.9} + \underset{(1.92)}{4.47D_1} + \underset{(1.30)}{1.75D_2} - \underset{(0.0536)}{0.128X_1} - \underset{(0.0804)}{0.0593X_2} \\ R^2 = .534\end{gathered} \tag{2.4}$$

The fitted regressions for the three regions are therefore:

$$\text{East:} \quad \hat{Y} = 19.4 - 0.128X_1 - 0.0593X_2$$

$$\text{Midwest:} \quad \hat{Y} = 16.7 - 0.128X_1 - 0.0593X_2$$

$$\text{West:} \quad \hat{Y} = 14.9 - 0.128X_1 - 0.0593X_2$$

Note that the coefficients for both heterogeneity and mobility become smaller in magnitude when region is controlled and, indeed, the coefficient for mobility is no longer statistically significant. This outcome reflects the fact that eastern cities tend to be low in mobility and high in moral integration; when region is ignored, then, part of its effect is absorbed by mobility. Even when mobility and heterogeneity are held constant, the level of integration for eastern cities is greater than that for cities in the Midwest, which in turn have a higher level of integration than western cities.

To test the null hypothesis of no regional differences, H_0: $\gamma_1 = \gamma_2 = 0$, we calculate

$$F_0 = \frac{n - k - 1}{p} \times \frac{R_1^2 - R_0^2}{1 - R_1^2}$$

$$= \frac{29 - 5}{2} \times \frac{.534 - .415}{1 - .534} = 3.06$$

with 2 and 24 degrees of freedom. The p-value for the test statistic is .06, indicating that regional differences in moral integration are marginally statistically significant.

2.1.3. Modeling Interactions

Two independent variables are said to *interact* in determining a dependent variable when the effect of one depends upon the value of the other. The linear additive models we have considered thus far therefore specify the absence of interactions. In this section we shall show how the dummy-variable regression model may be modified to accommodate interactions between quantitative and qualitative independent variables. Since the two concepts are frequently confused, we must state at the outset that *interaction and correlation of independent variables are logically and empirically distinct*: *two independent variables may interact whether or not they are related to one another statistically.*

Our treatment of dummy-variable regression has assumed parallel regressions across the several categories of a qualitative variable. If these regressions are not parallel, then the qualitative variable interacts with one or more quantitative regressors. The dummy-regression model may be modified to reflect these interactions.

For simplicity, we return to the example of Section 2.1.1, examining the effects of race and education on income. Consider the hypothetical data shown in Figure 2.3. In Figure 2.3(a), race and education are related, while in Figure 2.3(b) they are independent. It is apparent in both instances that the within-race regressions of income on education are not parallel: The slope for whites is larger than that for blacks. Because the effect of education varies by race, race and education interact in determining income. It is also the case, incidentally, that the effect of race varies by level of education; since the regressions are not parallel, the relative income advantage of whites (i.e., income discrimination) increases with education. Interaction, then, is a symmetric concept—the effect of education varies by race, and the effect of race varies by education.

We might model the data in Figure 2.3 by fitting separate regressions for blacks and whites. As before, however, it would be more convenient to fit a combined model. Among other advantages, a combined model facilitates a test of the race-by-education interaction. Moreover, a unified model permitting

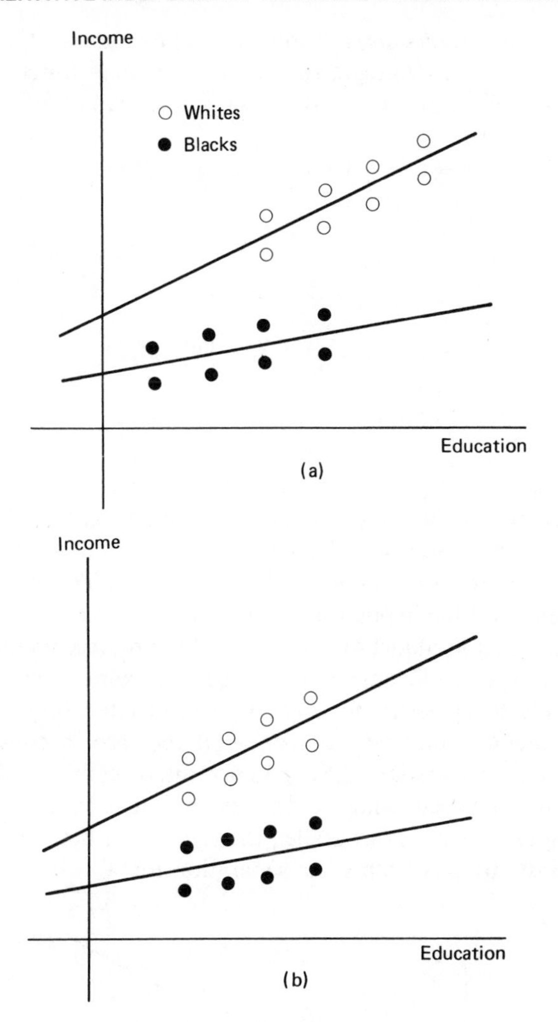

FIGURE 2.3. Income by race and education (hypothetical data). (a) Race and education related. (b) Race and education unrelated.

different intercepts and slopes in the two groups produces the same fit as separate regressions, because the residual sum of squares for the full sample is minimized when the residual sum of squares is minimized in each group.

The model given in equation (2.5) permits different intercepts and slopes for blacks and whites:

$$Y_i = \alpha + \gamma D_i + \beta X_i + \zeta(D_i X_i) + \varepsilon_i \tag{2.5}$$

Along with the dummy regressor D and the quantitative regressor X, we have entered an *interaction regressor* DX into the regression equation. Since DX is the product of the other two regressors, it is not a *linear* function of D and X, and thus perfect collinearity is avoided. For blacks, equation (2.5) becomes

$$\begin{aligned} Y_i &= \alpha + \gamma(0) + \beta X_i + \zeta(0 \cdot X_i) + \varepsilon_i \\ &= \alpha + \beta X_i + \varepsilon_i \end{aligned}$$

and for whites

$$\begin{aligned} Y_i &= \alpha + \gamma(1) + \beta X_i + \zeta(1 \cdot X_i) + \varepsilon_i \\ &= (\alpha + \gamma) + (\beta + \zeta) X_i + \varepsilon_i \end{aligned}$$

These equations are graphed in Figure 2.4: α and β are respectively the intercept and slope for the regression of income on education among blacks; γ gives the difference in intercepts between the white and black groups; and ζ gives the difference in slopes between the two groups. To test for interaction, therefore, we may test the hypothesis H_0: $\zeta = 0$.

In the no-interaction model of equation (2.1), γ represents the partial effect of race (i.e., the expected income difference between whites and blacks of equal education), while β represents the partial effect of education (the within-race expected increment in income for a one-unit increase in education). In the interaction model of equation (2.5), γ is not interpretable as the unqualified income difference between whites and blacks of equal education. Because the within-race regressions are not parallel, the separation between the regression lines is not constant; γ is simply the separation for $X = 0$. It is generally no

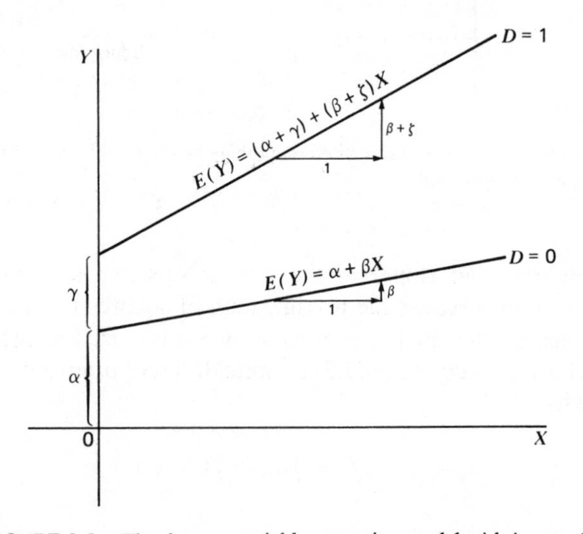

FIGURE 2.4. The dummy-variable regression model with interaction.

more important to assess the expected income difference between whites and blacks of zero education than at other educational levels, and therefore the difference-in-intercepts parameter γ is not of intrinsic interest when interaction is present.

Likewise, in the interaction model, β is not the unqualified effect of education, but rather the effect of education among blacks. Though this coefficient is of some substantive interest, it is not necessarily more important than the effect of education among whites ($\beta + \zeta$), which does not appear directly in the model.

Following Nelder (1974, 1976, 1977) we say that the separate partial effects, or *main effects*, of education and race are *marginal* to the education-by-race interaction. In general, we do not test for or seek to interpret main effects of independent variables that interact. If, however, we can rule out interaction either on theoretical or on empirical grounds, we may proceed to test, estimate, and interpret main effects.

Furthermore, it does not generally make sense to specify and fit models that include interaction regressors but delete main effects marginal to them. This is not to say that such models are uninterpretable—they are, rather, not broadly applicable. Suppose, for example, that we fit the model

$$Y_i = \alpha + \beta X_i + \zeta D_i X_i + \varepsilon_i$$

As shown in Figure 2.5(a), this model describes regression lines for blacks and whites that have the same intercept but may differ in slope, a specification of little substantive interest. Similarly, the model

$$Y_i = \alpha + \gamma D_i + \zeta D_i X_i + \varepsilon_i$$

graphed in Figure 2.5(b), constrains the slope for blacks to zero, which is needlessly restrictive.

The method presented in this section is easily generalized to polychotomous variables, to several qualitative independent variables, and to several quantitative regressors. We shall use Angell's multiple-regression data to illustrate the application of the method. We entertain the possibility that region interacts both with heterogeneity and with mobility, yielding the model

$$\begin{aligned} Y_i = \beta_0 + \gamma_1 D_{1i} + \gamma_2 D_{2i} + \beta_1 X_{1i} + \beta_2 X_{2i} + \zeta_{11}(D_{1i} X_{1i}) + \zeta_{12}(D_{1i} X_{2i}) \\ + \zeta_{21}(D_{2i} X_{1i}) + \zeta_{22}(D_{2i} X_{2i}) + \varepsilon_i \end{aligned} \tag{2.6}$$

Notice that we have constructed one interaction regressor for each product of a dummy regressor with a quantitative regressor. This model permits different intercepts and slopes in the three regions:

$$\begin{aligned} \text{East:} \quad & Y_i = (\beta_0 + \gamma_1) + (\beta_1 + \zeta_{11}) X_{1i} + (\beta_2 + \zeta_{12}) X_{2i} + \varepsilon_i \\ \text{Midwest:} \quad & Y_i = (\beta_0 + \gamma_2) + (\beta_1 + \zeta_{21}) X_{1i} + (\beta_2 + \zeta_{22}) X_{2i} + \varepsilon_i \\ \text{West:} \quad & Y_i = \beta_0 + \beta_1 X_{1i} + \beta_2 X_{2i} + \varepsilon_i \end{aligned}$$

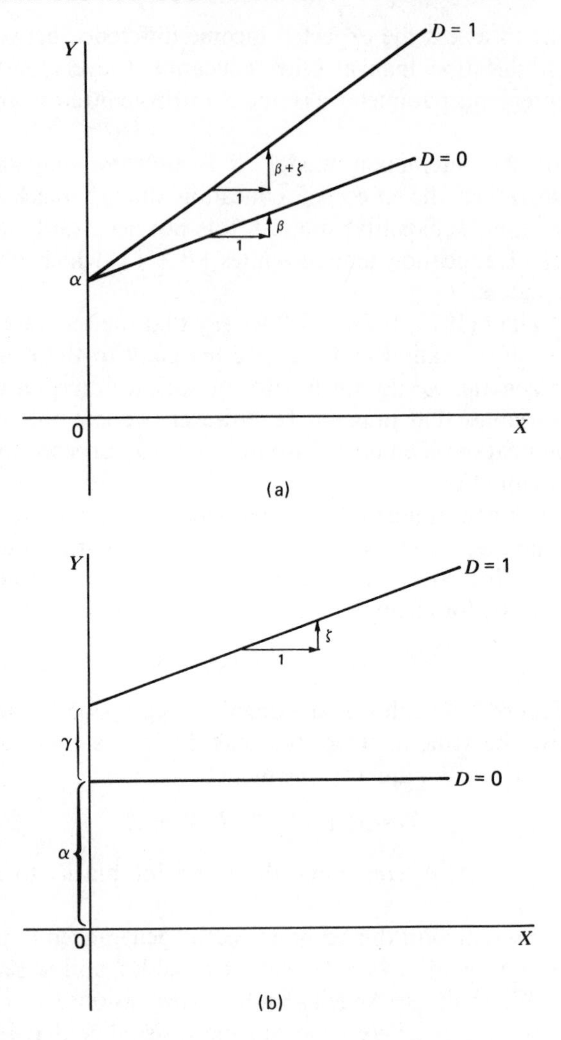

FIGURE 2.5. The principle of marginality in dummy-variable regression. (a) $Y = \alpha + \beta X + \zeta(XD) + \varepsilon$. (b) $Y = \alpha + \gamma D + \zeta(XD) + \varepsilon$.

The West, which is coded zero for both dummy regressors, therefore serves as a baseline for the intercepts and slopes of the other regions. As in the no-interaction model, the choice of a baseline category is generally arbitrary, as it is here, and of no consequence.

Fitting model (2.6) to Angell's data produces the following results:

$$\begin{aligned}\hat{Y} = 11.4 + 11.1D_1 + 6.67D_2 - 0.199X_1 + 0.0658X_2 \\ +0.0432D_1X_1 - 0.284D_1X_2 \\ +0.0604D_2X_1 - 0.171D_2X_2 \qquad R^2 = .562\end{aligned} \tag{2.7}$$

We have not reported standard errors in equation (2.7) because the regression coefficients are not of individual interest. Normally, we should test the region × heterogeneity and region × mobility interactions separately, using two degrees of freedom for each test. Contrasting R^2 for the interaction model of equation (2.7) with R^2 for the no-interaction model of equation (2.4), however, it is apparent that both sources of interaction are negligible. To test the combined interactions, we compute

$$F_0 = \frac{29 - 9}{4} \times \frac{.562 - .534}{1 - .562} = 0.32$$

with 4 and 20 degrees of freedom. The p-value for the null hypothesis of no interactions, H_0: $\zeta_{11} = \zeta_{12} = \zeta_{21} = \zeta_{22} = 0$, is .86. Having ruled out interaction between region and heterogeneity and between region and mobility, it is reasonable to proceed to test for separate effects of these independent variables. These tests were reported in Section 2.1.2.[2]

2.1.4. Standardized Coefficients in Dummy-Variable Regression

A cautionary note is appropriate here: Inexperienced researchers sometimes report standardized coefficients for dummy-variable regressors. As we have explained, an unstandardized dummy-variable coefficient is interpretable as the expected dependent-variable difference between a particular independent-variable category and the baseline category for the dummy-variable set. If a dummy-variable coefficient is standardized, this interpretation is lost. Furthermore, because a zero/one dummy variable cannot be increased by one standard deviation, the usual interpretation of a standardized regression coefficient is not meaningful.

These difficulties may be avoided by standardizing only the dependent variable and quantitative independent variables in a regression, leaving dummy variables in zero/one form. There is, of course, no harm in using standardized dummy variables for computations, as long as coefficients are translated back into unstandardized form (as explained in Section 1.5) prior to interpretation.

An analogous point may be made with respect to interaction regressors. We may legitimately standardize a numerical independent variable prior to taking its product with a dummy-variable regressor, but to standardize the interaction regressor itself is not sensible, since this regressor is not interpreted separately from the main effects marginal to it.

[2] The test reported earlier (implicitly) used RSS from the no-interaction model to estimate the error variance. Our general practice is to estimate σ_ε^2 using RSS from the largest model fit even when, as is frequently the case, this model includes effects that are not statistically significant. The largest model necessarily has the smallest residual sum of squares, but it also has the fewest residual degrees of freedom. These two factors tend to offset one another, and in the present instance it makes little difference which of the two estimates of the error variance is used.

PROBLEMS

2.1. Employing Angell's data for non-Southern U.S. cities given in Table 1.3 together with the data for Southern cities given in Problem 1.9, use dummy-variable regression to analyze the influence of ethnic heterogeneity, geographic mobility, and region on moral integration. Use four regional categories in your analysis: East, Midwest, West, and South (see Table 2.1).

(a) Fit a model that assumes no interaction between region and the other two independent variables. Write out the fitted regression equation for each region. Test the statistical significance of the heterogeneity, mobility, and region effects. Interpret the results of the analysis.

(b) Include interactions between region and heterogeneity and between region and mobility. Test the statistical significance of these interactions. Write out the fitted equation for each region.

(c) Fit a separate regression of moral integration on heterogeneity and mobility for the 14 Southern cities. Confirm that the fitted equation for the South is the same as that obtained in part (b).

(d) Refit the full model after standardizing integration, heterogeneity, and mobility. Compare the results to those obtained in part (b).

2.2. The data in Table 2.2 are taken from a study by Ornstein (1976) of interlocking directorates among major Canadian firms. Construct dummy variables to represent the 10 industrial sectors and four nations of control. Then regress number of interlocks on assets, sector, and nation of control. Test the statistical significance of each of these effects, and summarize the results of your analysis.

2.3. The geometry of dummy-variable regression for the no-interaction model $E(Y) = \alpha + \beta X + \gamma D$ was illustrated in Figure 2.2. Note that this graph is two-dimensional even though there are three variables (Y, X, and D) in the model. Draw a three-dimensional representation of the dummy-variable regression model, employing a separate axis for each of Y, X, and D. (*Hint*: D takes on only the values zero and one.) How are the three-dimensional and two-dimensional graphs related? Why is the two-dimensional graph adequate?

2.4.† Adjusted Means (also see Problems 2.11 and 2.14): Let $\overline{Y}_1$ represent the ("unadjusted") mean moral-integration score of Angell's Eastern cities, $\overline{Y}_2$ that of Midwestern cities, $\overline{Y}_3$ that of Western cities, and $\overline{Y}_4$ that of Southern cities. Differences among the $\overline{Y}_j$ may partly reflect regional differences in heterogeneity and mobility. In the dummy-variable regression of Problem 2.1(a), regional differences are "controlled" for heterogeneity and mobility, producing the fitted regression equation

$$\hat{Y} = B_0 + B_1 X_1 + B_2 X_2 + C_1 D_1 + C_2 D_2 + C_3 D_3$$

TABLE 2.2. Ornstein's Interlocking-Directorate Data

Firm	Assets[a]	Industry[b]	Nation of Control[c]	Interlocks[d]
1	147670	8	1	87
2	133000	8	1	107
3	113230	8	1	94
4	85418	8	1	48
5	75477	8	1	66
6	40742	9	1	69
7	40140	6	1	46
8	26866	8	1	16
9	24500	6	1	77
10	23700	2	2	6
11	23482	9	1	18
12	21512	9	1	29
13	20780	2	2	36
14	18688	9	1	20
15	18286	8	1	13
16	17910	2	2	6
17	17784	9	1	31
18	16631	9	1	27
19	16458	9	1	23
20	15280	2	2	13
21	15140	9	1	32
22	14362	9	1	28
23	14163	1	1	4
24	13820	10	1	42
25	13787	9	1	17
26	12810	6	1	40
27	12080	2	2	29
28	11250	2	1	29
29	11090	5	2	21
30	10580	2	4	3
31	10570	5	1	32
32	10568	9	1	29
33	10320	2	1	40
34	10110	6	2	5
35	9044	3	1	33
36	8395	9	1	21
37	8182	9	2	18
38	7994	6	2	18
39	7930	9	1	13
40	7877	5	2	2
41	7564	9	2	22
42	7510	9	2	13
43	7287	9	2	2
44	7018	8	1	0
45	6629	2	2	18
46	6571	9	1	9
47	6498	6	1	31
48	6407	1	1	16
49	6286	2	1	28
50	5932	9	1	34
51	5704	2	1	33
52	5479	7	1	7
53	5437	6	2	12
54	5429	7	2	15
55	5366	6	2	13
56	5035	2	2	16
57	5021	2	4	27
58	4980	1	1	12
59	4838	6	1	11
60	4634	2	2	0
61	4592	6	2	16
62	4390	6	1	17
63	4304	3	1	55
64	4298	1	4	15
65	4227	9	1	44
66	4210	2	2	18
67	4154	9	4	20
68	4100	5	2	19

TABLE 2.2. *(Continued)*

Firm	Assets[a]	Industry[b]	Nation of Control[c]	Interlocks[d]
69	4099	5	2	9
70	4088	9	1	12
71	3960	4	4	17
72	3896	1	1	15
73	3879	3	1	27
74	3673	7	1	30
75	3654	2	4	27
76	3631	6	2	12
77	3606	2	3	11
78	3570	7	1	28
79	3561	2	2	8
80	3274	6	1	12
81	3152	5	2	18
82	3058	3	3	23
83	2958	3	1	51
84	2927	2	4	35
85	2878	7	1	8
86	2814	1	1	43
87	2807	2	1	4
88	2801	3	1	18
89	2786	1	4	20
90	2757	2	3	13
91	2667	2	3	22
92	2625	1	3	10
93	2566	7	2	1
94	2549	10	2	6
95	2488	1	1	30
96	2285	2	2	6
97	2281	3	2	11
98	2182	1	2	20
99	2165	2	2	7
100	2164	2	4	13
101	2141	2	2	0
102	2108	5	3	14
103	2086	2	2	19
104	2025	1	1	4
105	1881	1	1	2
106	1876	7	1	2
107	1841	3	2	7
108	1656	2	4	25
109	1655	7	1	29
110	1612	7	1	5
111	1603	2	2	12
112	1601	3	1	25
113	1591	1	2	2
114	1583	10	1	25
115	1561	2	2	2
116	1520	5	2	16
117	1511	3	2	3
118	1487	2	4	9
119	1482	5	2	1
120	1477	5	2	0
121	1469	2	2	1
122	1434	2	2	1
123	1427	10	1	1
124	1416	7	1	6
125	1378	2	2	12
126	1372	2	2	5
127	1343	3	3	5
128	1337	2	2	0
129	1335	5	1	4
130	1315	2	4	5
131	1235	5	1	33
132	1172	1	1	11
133	1154	5	2	3
134	1154	10	1	3
135	1112	1	2	5
136	1060	1	1	25
137	1027	2	4	14

TABLE 2.2. (*Continued*)

Firm	Assets[a]	Industry[b]	Nation of Control[c]	Interlocks[d]
138	984	7	2	1
139	978	5	2	0
140	953	2	2	12
141	950	3	1	18
142	943	7	2	11
143	904	2	1	39
144	898	1	3	3
145	888	3	1	2
146	848	3	1	8
147	844	5	2	0
148	839	6	1	11
149	832	1	1	13
150	830	2	3	1
151	816	2	1	10
152	809	5	3	0
153	802	5	1	0
154	798	1	1	11
155	789	2	3	9
156	789	5	2	6
157	782	7	1	11
158	780	5	2	1
159	779	7	1	14
160	761	3	2	1
161	751	1	1	8
162	742	1	1	7
163	727	1	1	1
164	707	1	2	9
165	704	10	1	10
166	702	5	1	3
167	690	3	4	0
168	677	5	1	12
169	638	5	2	6
170	637	1	2	1
171	636	2	2.	0
172	614	4	1	2
173	590	5	2	2
174	589	2	4	23
175	586	6	1	10
176	575	1	1	1
177	566	1	2	0
178	559	5	2	7
179	558	1	1	14
180	552	1	1	7
181	548	2	2	5
182	540	5	1	6
183	539	1	1	9
184	533	6	1	5
185	523	5	2	8
186	519	5	2	8
187	519	1	1	0
188	516	6	1	5
189	511	10	1	0
190	510	5	1	11
191	508	5	4	1
192	497	1	1	4
193	495	5	1	0
194	494	7	1	8
195	488	5	1	1
196	487	5	1	8
197	471	7	2	0
198	456	2	2	5
199	456	5	2	0
200	444	1	2	1
201	438	5	3	18
202	432	5	2	1
203	432	5	1	3
204	422	3	2	11
205	407	2	2	6
206	402	5	2	0

TABLE 2.2. (*Continued*)

Firm	Assets[a]	Industry[b]	Nation of Control[c]	Interlocks[d]
207	391	1	1	28
208	387	7	1	11
209	386	4	4	2
210	379	5	2	16
211	376	7	1	5
212	375	5	2	8
213	372	1	2	8
214	370	1	3	3
215	364	6	2	5
216	361	2	2	2
217	359	1	2	0
218	358	1	2	0
219	352	3	1	21
220	350	5	2	1
221	345	7	2	8
222	332	5	2	0
223	326	1	1	3
224	326	1	2	0
225	326	5	1	28
226	325	5	4	0
227	318	1	1	2
228	312	5	2	2
229	305	1	3	4
230	304	7	2	4
231	303	3	3	3
232	297	4	1	2
233	276	2	1	9
234	270	5	1	4
235	261	4	3	1
236	256	5	2	1
237	245	2	1	11
238	241	2	2	3
239	225	1	2	6
240	225	2	3	8
241	220	1	2	0
242	201	1	2	5
243	200	5	2	0
244	188	5	2	0
245	160	1	1	0
246	158	1	1	5
247	119	1	1	6
248	62	2	2	0

Source: Personal communication from M. Ornstein.

[a]Assets in millions of dollars.

[b]Industrial Sector

1. Agriculture, Food, Light Industry
2. Mining, Metals, etc.
3. Wood and Paper
4. Construction
5. Heavy Manufacturing
6. Transport
7. Merchandizing
8. Banks
9. Other Financials
10. Holding Companies

[c]Nation of Control

1. Canada
2. United States
3. Britain
4. Other Foreign

[d]Number of interlocking director and executive positions.

Consequently, if we fix heterogeneity and mobility at particular values, say $X_1 = x_1$ and $X_2 = x_2$, then the fitted integration scores for the several regions are given by

$$\hat{Y}_1 = B_0 + C_1 + B_1 x_1 + B_2 x_2$$

$$\hat{Y}_2 = B_0 + C_2 + B_1 x_1 + B_2 x_2$$

$$\hat{Y}_3 = B_0 + C_3 + B_1 x_1 + B_2 x_2$$

$$\hat{Y}_4 = B_0 \qquad + B_1 x_1 + B_2 x_2$$

(a) Note that the *differences* among the $\hat{Y}_j$ depend only upon C_1, C_2, and C_3, and not upon the values of x_1 and x_2. Why is this so?
(b) When $x_1 = \bar{X}_1$ and $x_2 = \bar{X}_2$, the $\hat{Y}_j$ are called *adjusted means*, and are denoted $\tilde{Y}_j$. How may $\tilde{Y}_j$ be interpreted? In what sense is $\tilde{Y}_j$ an "adjusted" mean?
(c) Locate the "unadjusted" and adjusted mean incomes of blacks and whites in each of Figures 2.1(a) and (b). Construct a similar example in which the difference between adjusted means is *larger* than the corresponding difference in unadjusted means.
(d) Using the results of the regression from Problem 2.1(a), compute adjusted mean moral-integration scores for each of the four regions. Find the unadjusted mean integration for each region, and comment on the differences, if any, between the unadjusted and adjusted means.
(e) Can the concept of an adjusted mean be extended to a model that includes interactions?

2.2. ONE-WAY ANALYSIS OF VARIANCE

The term *analysis of variance* was introduced in Chapter 1 to describe a partition of the dependent-variable sum of squares into explained and unexplained components; we noted then that this partition applies generally to linear models. For historical reasons,[3] *analysis of variance* also refers to procedures for fitting and testing linear models in which the independent variables are qualitative. When there is a single qualitative independent variable (termed a *classification* or *factor*), these procedures are called *one-way* analysis of variance. One-way analysis of variance, like simple regression analysis, is a method of limited applicability; also like simple regression,

[3] The methods and terminology of analysis of variance were introduced by the British statistician R. A. Fisher (1925). Fisher's many other contributions to statistics include the fundamental technique of randomization in experimental design and the method of maximum-likelihood estimation.

however, one-way analysis of variance provides an elementary context within which to introduce more broadly useful techniques.

2.2.1. Fitting and Testing the One-Way ANOVA Model

In Sections 2.1.1 and 2.1.2, we learned how to employ dummy regressors to represent the effects of a qualitative variable alongside those of quantitative independent variables. The quantitative regressors may be removed, leaving a simpler model. For example, for an m-category classification, we have the model

$$Y_i = \alpha + \gamma_1 D_{1i} + \cdots + \gamma_{m-1} D_{m-1,i} + \varepsilon_i$$

where the dummy variables D_j are coded according to the usual scheme. Then, for an observation in the jth category (or *group*), $j = 1,\ldots,m-1$, the model becomes $\mu_j \equiv E(Y_i|D_{ji} = 1) = \alpha + \gamma_j$; and for an observation in the last or baseline group, $\mu_m \equiv E(Y_i|\text{all } D_{ji} = 0) = \alpha$. To test for differences in the population means $\mu_1, \mu_2, \ldots, \mu_m$, we therefore need merely to test the omnibus null hypothesis H_0: $\gamma_1 = \gamma_2 = \cdots = \gamma_{m-1} = 0$.

Our consideration of one-way analysis of variance might well end here, but for a desire to develop methods that generalize easily to more complex situations in which there are several potentially interacting factors. Our first innovation is notational: Since observations may be partitioned according to independent-variable categories, it is convenient to let Y_{ij} represent the jth observation in the ith group; n_i is the number of observation in the ith group, and therefore $n = \sum_{i=1}^{m} n_i$. As above, $\mu_i \equiv E(Y_{ij})$ represents the population mean in category i.

The one-way analysis-of-variance model is written in the following manner:

$$Y_{ij} = \mu + \alpha_i + \varepsilon_{ij} \tag{2.8}$$

where μ represents the general level of the dependent variable, α_i is the effect on the dependent variable of membership in the ith group, and ε_{ij} is an error random variable that follows the usual linear-model assumptions; that is, the ε_{ij} are normally and independently distributed with zero expectations and common variance. Upon taking expectations, equation (2.8) becomes

$$\mu_i = \mu + \alpha_i$$

The parameters of the model, therefore, are underdetermined, for there are but m population means and $m + 1$ parameters in the model: Even if we knew the population means, we could not uniquely determine μ and the α_i.

Because the parameters of model (2.8) are themselves underdetermined, they cannot be uniquely estimated. Suppose that we try to use least squares to fit the model to data. In matrix form, the linear model is

$$
\begin{array}{l}
\\
\text{group 1}\\ \\ \\
\text{group 2}\\ \\ \\ \\
\text{group } m
\end{array}
\begin{pmatrix} Y_{11} \\ \vdots \\ Y_{1n_1} \\ \hline Y_{21} \\ \vdots \\ Y_{2n_2} \\ \hline \vdots \\ \hline Y_{m1} \\ \vdots \\ Y_{mn_m} \end{pmatrix}
=
\overset{\begin{matrix}(\mu) & (\alpha_1) & (\alpha_2) & & (\alpha_m)\end{matrix}}{
\begin{pmatrix}
1 & 1 & 0 & \cdots & 0 \\
\vdots & \vdots & \vdots & & \vdots \\
1 & 1 & 0 & \cdots & 0 \\ \hline
1 & 0 & 1 & \cdots & 0 \\
\vdots & \vdots & \vdots & & \vdots \\
1 & 0 & 1 & \cdots & 0 \\ \hline
\vdots & \vdots & \vdots & & \vdots \\ \hline
1 & 0 & 0 & \cdots & 1 \\
\vdots & \vdots & \vdots & & \vdots \\
1 & 0 & 0 & \cdots & 1
\end{pmatrix}}
\begin{pmatrix} \mu \\ \alpha_1 \\ \alpha_2 \\ \vdots \\ \alpha_m \end{pmatrix}
+
\begin{pmatrix} \varepsilon_{11} \\ \vdots \\ \varepsilon_{1n_1} \\ \hline \varepsilon_{21} \\ \vdots \\ \varepsilon_{2n_2} \\ \hline \vdots \\ \hline \varepsilon_{m1} \\ \vdots \\ \varepsilon_{mn_m} \end{pmatrix}
$$

$$\underset{(n \times 1)}{\mathbf{y}} = \underset{(n \times m+1)}{\mathbf{X}} \quad \underset{(m+1 \times 1)}{\boldsymbol{\beta}} + \underset{(n \times 1)}{\boldsymbol{\varepsilon}} \qquad (2.9)$$

For ease of reference, the parameters are shown above the columns of the design matrix. It is clear that there is one linear dependency among the columns of $\mathbf{X}$: The first column, for example, is the sum of the rest. Consequently $\mathbf{X}$ is of rank m, $\mathbf{X}'\mathbf{X}$ is singular, and the normal equations cannot be solved uniquely for estimates of the parameters. It is important to understand that our inability to estimate the ANOVA model proceeds directly from the indeterminacy of the model itself. As we mentioned, even if we knew the population means, we could not uniquely calculate the parameters of the model. Our goal, we recall, is to test for group differences. It is perhaps surprising that we can accomplish this purpose without modifying the underdetermined ANOVA model. The null hypothesis of no group differences H_0: $\mu_1 = \cdots = \mu_m$ is equivalent to H_0: $\alpha_1 = \cdots = \alpha_m$.[4] To find the sum of squares attributable to this hypothesis, it is unnecessary to estimate the parameters of the model; we only need to find the fitted dependent-variable vector.

The $m + 1$ columns of the design matrix in equation (2.9) span a subspace of dimension m. We may project $\mathbf{y}$ onto this subspace, locating $\hat{\mathbf{y}}$, even though the individual parameter estimates are not uniquely determined.[5] This situation is illustrated in Figure 2.6 for $m = 2$.

[4] If all the α_i are equal (say to α), then they may be "absorbed" in the constant μ. The null model is therefore $Y_{ij} = \mu + \alpha + \varepsilon_{ij} = \mu' + \varepsilon_{ij}$.

[5] Since the $\hat{\mathbf{y}}$ vector in this treatment is not in mean-deviation form, to find the sum of squares for the hypothesis, we must subtract $\overline{Y}$ from each of its entries. ($\overline{Y}$ is the least-squares estimator of μ' in the null model given in footnote 4.)

One convenient way of locating $\hat{\mathbf{y}}$ is arbitrarily to delete one of the columns of $\mathbf{X}$; the remaining columns are linearly independent and thus provide a basis for the $\mathbf{X}$ subspace. To delete one of the columns of $\mathbf{X}$ is implicitly to set one parameter to zero. If, for example, we delete the last column of the design matrix, we set $\alpha_m = 0$. Then $\mu = \mu_m$, and $\alpha_i = \mu_i - \mu_m$ for $i = 1,\ldots,m-1$. This is simply the dummy-variable coding scheme of the previous section. Alternatively, if we set $\mu = 0$, then $\alpha_i = \mu_i$.

Indeed, any linear restriction of the form

$$w_0\mu + \sum_{i=1}^{m} w_i\alpha_i = 0$$

(where the w_i are constants, not all zero) placed on the parameters of the model permits us to solve for the parameters in terms of the population means. Equivalently, a linear restriction applied to the parameter estimates yields a solution to the normal equations. For purposes of testing the null hypothesis of no group differences, it is immaterial which restriction is employed, since we shall arrive at the same regression sum of squares regardless of the basis selected for the $\mathbf{X}$ subspace.

There is, however, an advantage in selecting a restriction that produces easily interpretable parameters and estimates, and that generalizes usefully to more complex models. For these reasons, we shall impose the constraint $\Sigma_{i=1}^{m}\alpha_i = 0$. Employing this restriction to solve for the parameters produces

$$\mu = \frac{\Sigma\mu_i}{m} \equiv \mu.$$

$$\alpha_i = \mu_i - \mu. \tag{2.10}$$

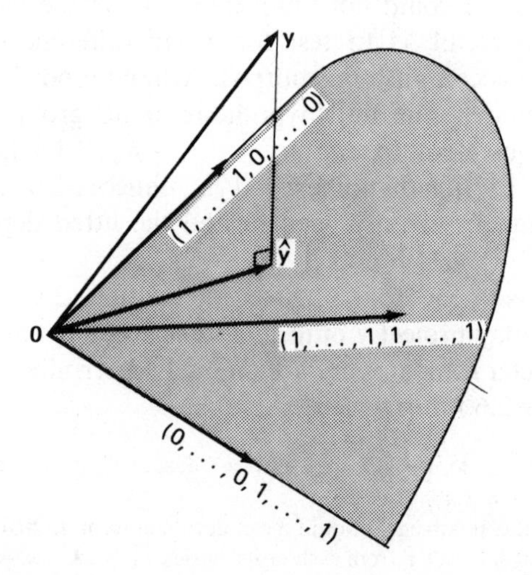

FIGURE 2.6. Vector representation of the one-way ANOVA model.

We use a dot to indicate averaging over the range of a subscript. The *grand* or *general mean* μ, then, is the average of the population group means, while α_i gives the difference between the ith group mean and the grand mean. Thus, under this restriction, the hypothesis of no group differences H_0: $\mu_1 = \cdots = \mu_m$ is equivalent to H_0: $\alpha_1 = \cdots = \alpha_m = 0$.

Imposing a constraint of the form $\mu = 0$ or $\alpha_m = 0$ produces a full-rank design matrix by deleting a column. Under the restriction $\Sigma\alpha_i = 0$, any α_i—say the last—is the negative of the sum of the others; that is, $\alpha_m = -\Sigma_{i=1}^{m-1}\alpha_i$. Thus, we may delete α_m from the vector of parameters and substitute $-\alpha_1 - \alpha_2 - \cdots - \alpha_{m-1}$, producing the following basis for the column space of $\mathbf{X}$:

$$\underset{(n\times m)}{\mathbf{X}_R} = \begin{array}{c} (\mu)\ (\alpha_1)\ (\alpha_2)\ \quad (\alpha_{m-1}) \\ \left(\begin{array}{ccccc} 1 & 1 & 0 & \cdots & 0 \\ \vdots & \vdots & \vdots & & \vdots \\ 1 & 1 & 0 & \cdots & 0 \\ \hline 1 & 0 & 1 & \cdots & 0 \\ \vdots & \vdots & \vdots & & \vdots \\ 1 & 0 & 1 & \cdots & 0 \\ \hline \vdots & \vdots & \vdots & & \vdots \\ \hline 1 & -1 & -1 & \cdots & -1 \\ \vdots & \vdots & \vdots & & \vdots \\ 1 & -1 & -1 & \cdots & -1 \end{array}\right) \end{array} \begin{array}{l} \text{group 1} \\ \\ \text{group 2} \\ \vdots \\ \text{group } m \end{array} \tag{2.11}$$

We call $\mathbf{X}_R$ the *reduced* or *full-rank* design matrix under the constraint $\Sigma\alpha_i = 0$.

There is, then, the following relation between the group means and the parameters of the constrained model:

$$\begin{array}{c} \\ \begin{pmatrix} \mu_1 \\ \mu_2 \\ \vdots \\ \mu_m \end{pmatrix} \end{array} = \begin{array}{c} (\mu)\ \ (\alpha_1)\ \ (\alpha_2)\ \quad (\alpha_{m-1}) \\ \begin{pmatrix} 1 & 1 & 0 & \cdots & 0 \\ 1 & 0 & 1 & \cdots & 0 \\ \vdots & \vdots & \vdots & & \vdots \\ 1 & -1 & -1 & \cdots & -1 \end{pmatrix} \end{array} \begin{pmatrix} \mu \\ \alpha_1 \\ \alpha_2 \\ \vdots \\ \alpha_{m-1} \end{pmatrix}$$

$$\underset{(m\times 1)}{\boldsymbol{\mu}} = \underset{(m\times m)}{\mathbf{X}_B} \quad \underset{(m\times 1)}{\boldsymbol{\beta}_R} \tag{2.12}$$

In this *parametric equation*, $\mathbf{X}_B$ is called the *row basis* of the reduced design matrix. Solving equation (2.12) for $\boldsymbol{\beta}_R = \mathbf{X}_B^{-1}\boldsymbol{\mu}$ produces the results shown previously in equations (2.10).

TABLE 2.3. One-Way ANOVA Table

Source	Sum of Squares	*df*	Mean Square	*F*	H_0
Groups	$\Sigma n_i(\overline{Y}_i - \overline{Y})^2$	$m - 1$	$\frac{\text{RegrSS}}{m-1}$	$\frac{\text{RegrMS}}{\text{RMS}}$	$\alpha_1 = \cdots = \alpha_m = 0$ $(\mu_1 = \cdots = \mu_m)$
Residuals	$\Sigma\Sigma(Y_{ij} - \overline{Y}_i)^2$	$n - m$	$\frac{\text{RSS}}{n-m}$		
Total	$\Sigma\Sigma(Y_{ij} - \overline{Y})^2$	$n - 1$			

To fit the one-way ANOVA model to data, we may employ the reduced design matrix given in equation (2.11). $\mathbf{X}_R$ has been constructed to be of full-column rank, and thus we may calculate $\mathbf{b}_R = (\mathbf{X}_R'\mathbf{X}_R)^{-1}\mathbf{X}_R'\mathbf{y}$. Because the sample mean in the ith group, $\overline{Y}_i$, is the least-squares estimator of μ_i, it is often easier to compute

$$M = \frac{\Sigma\overline{Y}_i}{m} = \overline{Y}.$$

$$A_i = \overline{Y}_i - \overline{Y}.$$

Furthermore, since $\hat{Y}_{ij} = M + A_i = \overline{Y}_i$, the regression and residual sums of squares take particularly simple forms in one-way analysis of variance:[6]

$$\begin{aligned} \text{RegrSS} &= \sum_{i=1}^{m}\sum_{j=1}^{n_i}\left(\hat{Y}_{ij} - \overline{Y}\right)^2 = \sum_{i=1}^{m} n_i\left(\overline{Y}_i - \overline{Y}\right)^2 \\ \text{RSS} &= \sum_{i=1}^{m}\sum_{j=1}^{n_i}\left(Y_{ij} - \hat{Y}_{ij}\right)^2 = \Sigma\Sigma\left(Y_{ij} - \overline{Y}_i\right)^2 \end{aligned} \tag{2.13}$$

This information may be presented in the form of an ANOVA table, as shown in Table 2.3.

We shall use Angell's data (from Table 1.3) to illustrate one-way analysis of variance, examining the relationship of mobility to region. A plot of mobility by region is shown in Figure 2.7; regional frequencies, means, and standard deviations are given in Table 2.4. Mobility appears to increase from East to West, and there is some indication that city mobility scores are less variable in the East than in the other two regions. The one-way ANOVA for these data appears in Table 2.5. The relationship between region and mobility is strong ($R^2 = 1682/2610 = .644$) and statistically significant.

[6]If the n_i are unequal, as is usually the case in observational research, then the mean of group means $\overline{Y}.$ generally differs from the overall mean $\overline{Y}$ of the dependent-variable observations; for $\overline{Y} = \Sigma n_i\overline{Y}_i/n$, while $\overline{Y}. = \Sigma\overline{Y}_i/m$.

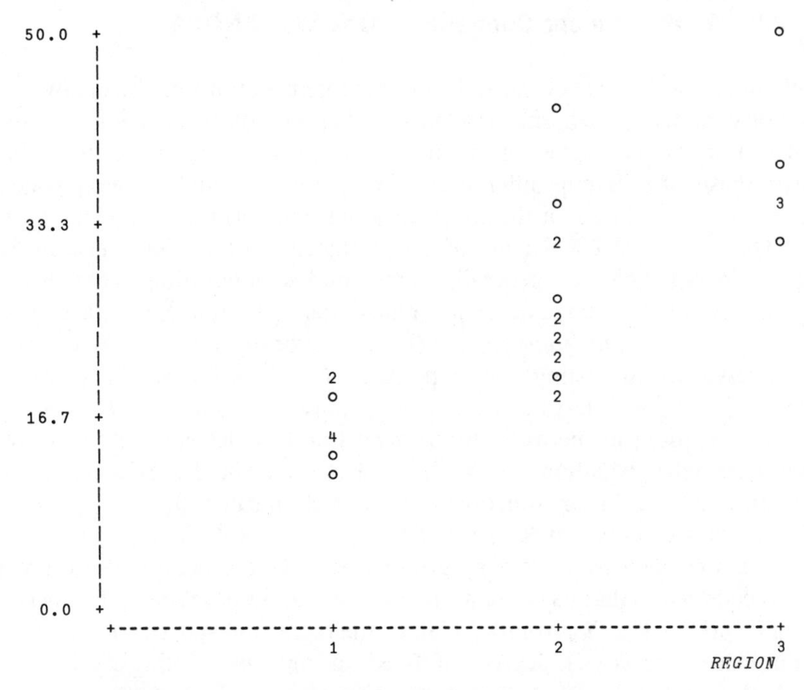

FIGURE 2.7. Geographic mobility by region for 29 U.S. cities.

TABLE 2.4. Geographic Mobility by Region for 29 U.S. Cities

	East	Midwest	West
mean	15.90	26.06	37.40
standard deviation	2.658	7.106	6.567
frequency	9	14	6

2.5. ANOVA for Mobility by Region

Source	Sum of Squares	*df*	Mean Square	*F*	*p*
Region	1681.7	2	840.85	23.54	< .001
Residuals	928.65	26	35.717		
Total	2610.3	28			

2.2.2. Testing Linear Contrasts in One-Way ANOVA

We have considered several full-rank parameterizations of the one-way analysis-of-variance model, showing how each proceeds from a linear restriction placed on the parameters of the model. For purposes of testing the global null hypothesis of no group differences, there parameterizations are equivalent, for each provides a basis for the m-dimensional column space of the design matrix.

The individual coefficients of the parameterizations developed in Section 2.2.1 do not, however, generally correspond to interesting hypotheses about group means. For example, if we impose the restriction $\Sigma\alpha_i = 0$, then to test H_0: $\alpha_i = 0$ for a *particular* group i (by a t-test or one-degree-of-freedom F-test) is equivalent to testing the hypothesis H_0: $\mu_i = \mu$.. Similarly, under the restriction $\alpha_m = 0$, testing H_0: $\alpha_i = 0$ is equivalent to testing H_0: $\mu_i = \mu_m$.

The relationship between group means and model parameters is given by the parametric equation $\boldsymbol{\mu} = \mathbf{X}_B\boldsymbol{\beta}_R$. Thus, as we noted previously, the model parameters are linear functions of the cell means: $\boldsymbol{\beta}_R = \mathbf{X}_B^{-1}\boldsymbol{\mu}$. In many instances, we can select $\mathbf{X}_B$ so that the elements of $\boldsymbol{\beta}_R$ incorporate interesting *contrasts* or *comparisons* among group means. One generally useful procedure is to code $\mathbf{X}_B$ so that its columns are orthogonal. In developing this method, we shall work back to $\mathbf{X}_B$ from $\mathbf{X}_B^{-1}$. Since the regression sum of squares for group differences has $m - 1$ degrees of freedom, only $m - 1$ linearly independent columns of the design matrix may be devoted to comparisons among group means; the remaining column is used for the grand mean.

The data given in Table 2.6 are drawn from an experimental study by Friendly and Franklin (1980) of the effects of presentation format on learning

TABLE 2.6. Friendly and Franklin's Data on the Effects of Presentation on Recall: Number of Words Recalled by Experimental Condition

Experimental Condition		
SFR	*B*	*M*
39	40	40
25	38	39
37	39	34
25	37	37
29	39	40
39	24	36
21	30	36
39	39	38
24	40	36
25	40	30

Source: Personal communication from M. Friendly.

and memory. Subjects participating in the experiment read a list of 40 words. Then, after performing a brief distracting task, the subjects were asked to recall as many of the words as possible. This procedure was repeated for five trials. Thirty subjects were randomly assigned to three conditions. In the control or "standard free recall" (*SFR*) condition, the order of presentation of words on the list was randomized for each of the five trials of the experiment. In the two experimental conditions, recalled words were presented in the order in which they were listed by the subject on the previous trial. In one of these conditions (labeled *B*), the recalled words were presented as a group *before* the forgotten ones, while in the other condition (labeled *M* for *meshed*), the recalled and forgotten words were interspersed. The data recorded in the table are the number of words correctly recalled by each subject for the final trial of the experiment. Friendly and Franklin expected that making the order of presentation contingent upon the subject's previous performance would increase recall.

From Friendly and Franklin's study we wish to determine: (1) whether the experimental groups differ from the control; and (2) whether the experimental groups differ from each other. Letting μ_1, μ_2, and μ_3 represent the dependent-variable expectations in the three conditions, we have the null hypotheses: (1) H_0: $\mu_1 = (\mu_2 + \mu_3)/2$; and (2) H_0: $\mu_2 = \mu_3$.

Each hypothesis may be coded in a model parameter, employing the following relationship between parameters and group means:

$$\begin{pmatrix} \mu \\ \gamma_1 \\ \gamma_2 \end{pmatrix} = \begin{pmatrix} 1/3 & 1/3 & 1/3 \\ 1 & -1/2 & -1/2 \\ 0 & 1 & -1 \end{pmatrix} \begin{pmatrix} \mu_1 \\ \mu_2 \\ \mu_3 \end{pmatrix}$$

$$\boldsymbol{\beta}_R = \mathbf{X}_B^{-1} \boldsymbol{\mu} \tag{2.14}$$

Thus, H_0: $\gamma_1 = 0$ is equivalent to the first hypothesis and H_0: $\gamma_2 = 0$ is equivalent to the second hypothesis.

The *rows* of $\mathbf{X}_B^{-1}$ in equation (2.14) are orthogonal and, consequently, the *columns* of $\mathbf{X}_B$ are orthogonal as well. Indeed, each column of $\mathbf{X}_B$ is equal to the corresponding row of $\mathbf{X}_B^{-1}$ divided by the sum of squared entries in that row; thus

$$\begin{pmatrix} \mu_1 \\ \mu_2 \\ \mu_3 \end{pmatrix} = \begin{pmatrix} 1 & 2/3 & 0 \\ 1 & -1/3 & 1/2 \\ 1 & -1/3 & -1/2 \end{pmatrix} \begin{pmatrix} \mu \\ \gamma_1 \\ \gamma_2 \end{pmatrix}$$

Because of the one-to-one correspondence between rows of $\mathbf{X}_B^{-1}$ and columns of $\mathbf{X}_B$, we may specify the latter matrix directly. In fact, we can rescale the columns of the row basis for more convenient coding, as shown in equation (2.15), without altering the hypotheses incorporated in the contrast coefficients

γ_1 and γ_2: If, for example, $\gamma_1 = 0$, then any multiple of γ_1 is zero as well.

$$\mathbf{X}_B = \begin{pmatrix} 1 & 2 & 0 \\ 1 & -1 & 1 \\ 1 & -1 & -1 \end{pmatrix} \tag{2.15}$$

To code comparisons directly in the design matrix, we therefore need merely follow these rules:

1. The first column of $\mathbf{X}_B$ consists of ones (for the general mean), while the remaining columns each code a linear comparison among group means.
2. The columns after the first each sum to zero, making them orthogonal to the first column.
3. All pairs of columns of $\mathbf{X}_B$ are orthogonal.

When comparisons of interest cannot be constructed by this procedure, it is always possible to work backward from $\mathbf{X}_B^{-1}$ (which expresses the model parameters as linear functions of the population group means) to $\mathbf{X}_B$, as long as the comparisons specified by $\mathbf{X}_B^{-1}$ are linearly independent. Linear independence is required to insure that $\mathbf{X}_B^{-1}$ is nonsingular. A simpler method for calculating the sum of squares for a contrast is developed in Problem 2.7.

If there are equal numbers of observations in the several groups, then an orthogonal design-matrix basis $\mathbf{X}_B$ produces an orthogonal reduced design matrix $\mathbf{X}_R$. It is, of course, unlikely that equal group frequencies would occur by chance in an observational study, but they may be produced by design in experimental research (such as the Friendly and Franklin study). The columns of an orthogonal design matrix represent independent sources of variation in the dependent variable, and therefore a set of orthogonal comparisons partitions the regression sum of squares into one-degree-of-freedom components, each testing a hypothesis of interest. When it is applicable, this is an elegant approach to linear-model analysis. We should not forget, however, that linear comparisons among means may be of interest even if group frequencies are unequal, causing contrasts to be correlated.

Means and standard deviations for the Friendly and Franklin memory data appear in Table 2.7; the data are graphed in Figure 2.8. The mean number of

TABLE 2.7. Recall by Condition

	Experimental Condition		
	SFR	*B*	*M*
mean	30.30	36.60	36.60
standard deviation	7.334	5.337	3.026

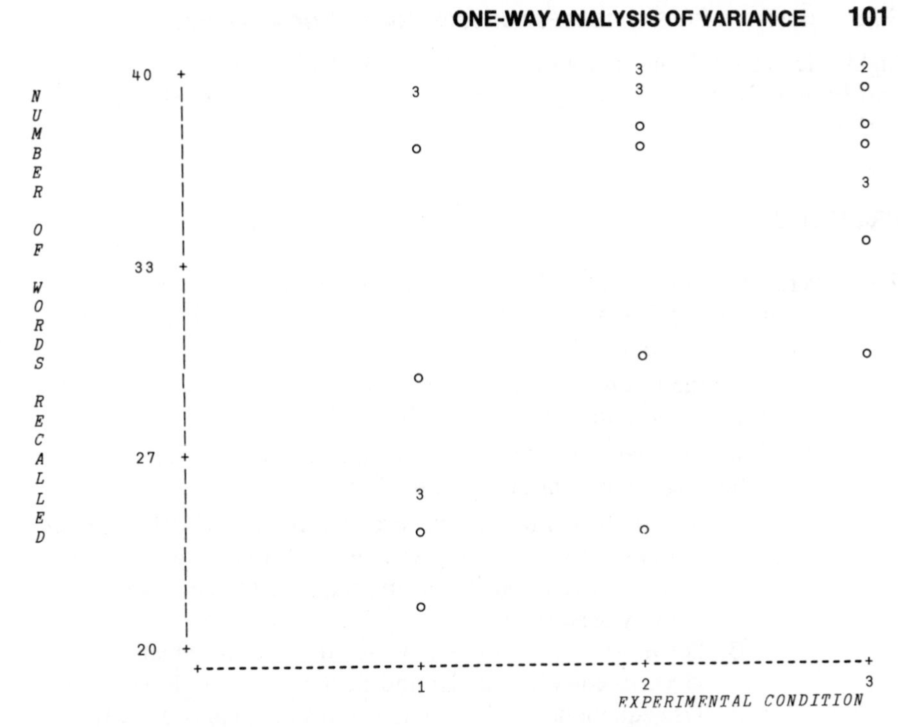

FIGURE 2.8. Number of words recalled by experimental condition.

words recalled is higher in the experimental conditions than in the control, as the researchers predicted. Notice that an apparent "ceiling effect" (grouping of scores near the top of the scale) has resulted in negatively skewed distributions within each condition and in a somewhat larger dispersion of scores in the control condition.

An analysis of variance for these data, employing orthogonal comparisons, is shown in Table 2.8. In this case, the sum of squares explained by conditions is due wholly to the contrast between the control group and the experimental groups, since the *B* and *M* conditions have identical means. In interpreting tests for contrasts—indeed, in analysis of variance generally—it is important to examine the group means. The *SFR* versus *B* and *M* contrast is statistically

TABLE 2.8. ANOVA for Recall by Condition

Source	SS	*df*	MS	*F*	*p*
Condition	264.60	2	132.30	4.34	.02
SFR vs. *B* & *M*	264.60	1	264.60	8.68	.005
B vs. *M*	0.0	1	0.0	0.0	1.0
Residuals	822.90	27	30.478		
Total	1087.5	29			

highly significant. Note that a one-tail test is appropriate here, since Friendly and Franklin's alternative to the first null hypothesis is directional: $t = \sqrt{F} = 2.95$, $p = .003$.

PROBLEMS

2.5. Perform a one-way analysis of variance of geographic mobility by region, employing Angell's data on 43 U.S. cities, given in Tables 1.3 and 2.1 and Problem 1.9.

(a) Find the mean and standard deviation for mobility in each of the four regions, and comment on the results.

(b) Confirm that identical sums of squares are produced by the following three computational methods:

(i) Set $\alpha_4 = 0$ in the analysis-of-variance model; use the dummy-variable coding scheme employed in Section 2.1 to produce a reduced design matrix; and fit the model to the data by least squares.

(ii) Set $\alpha_4 = -(\alpha_1 + \alpha_2 + \alpha_3)$; use the coding scheme for $\mathbf{X}_R$ given in equation (2.11); and fit the model by least squares.

(iii) Use equations (2.13) to obtain sums of squares directly.

(c) Having obtained the regression and residual sums of squares, summarize this information in an ANOVA table, testing the null hypothesis of no regional differences in mobility.

2.6. The usual t-statistic for testing a difference between two means is given by $t_0 = (\overline{Y}_1 - \overline{Y}_2)/S_{\overline{Y}_1 - \overline{Y}_2}$, where

$$S^2_{\overline{Y}_1 - \overline{Y}_2} = \frac{\sum_{j=1}^{n_1} \left(Y_{1j} - \overline{Y}_1\right)^2 + \sum_{j=1}^{n_2} \left(Y_{2j} - \overline{Y}_2\right)^2}{n_1 + n_2 - 2} \times \left(\frac{1}{n_1} + \frac{1}{n_2}\right)$$

Here $\overline{Y}_1$ and $\overline{Y}_2$ are the means in the two groups; n_1 and n_2 are the numbers of observations in the groups; and Y_{1j} and Y_{2j} are the observations themselves. Let F_0 be the one-way ANOVA F-statistic for testing the null hypothesis H_0: $\mu_1 = \mu_2$. Prove that $t_0^2 = F_0$ and that, consequently, the two tests are equivalent.

2.7.†# Testing Contrasts Using Group Means: Suppose that we wish to test a hypothesis concerning a contrast of group means

$$H_0: c_1\mu_1 + c_2\mu_2 + \cdots + c_m\mu_m = 0$$

where $c_1 + c_2 + \cdots + c_m = 0$. Define the sample *value of the contrast* as

$$C \equiv c_1\overline{Y}_1 + c_2\overline{Y}_2 + \cdots + c_m\overline{Y}_m$$

and let

$$C'^2 \equiv \frac{C^2}{\dfrac{c_1^2}{n_1} + \dfrac{c_2^2}{n_2} + \cdots + \dfrac{c_m^2}{n_m}}$$

(a) Show that under H_0: (i) $E(C) = 0$ and

$$V(C) = \sigma_\varepsilon^2 \left(\frac{c_1^2}{n_1} + \frac{c_2^2}{n_2} + \cdots + \frac{c_m^2}{n_m} \right)$$

(ii) $t_0 \equiv C'/S_E$ follows a t-distribution with $n - m$ degrees of freedom. (*Hint*: The $\bar{Y}_i$ are independent, and each is distributed as $N(\mu_i, \sigma_\varepsilon^2/n_i)$.) C'^2 is the sum of squares for the contrast.

(b) Using Friendly and Franklin's data (Table 2.6), verify that the test statistics obtained by the method of part (a) are the same as those produced by the incremental sum of squares approach.

(c) Pecknold et al. (1982) describe an experiment in which psychiatric patients suffering from anxiety were assigned randomly to one of three groups: (i) a treatment group that received a standard anti-anxiety drug (diazepam); (ii) a treatment group that received a new anti-anxiety drug (fenobam); and (iii) a control group that received a placebo. At several points in the study, subjects were administered standard tests for anxiety, including the Hamilton Rating Scale; high scores on the scale are indicative of severe anxiety. After three weeks of treatment, the experiment produced the following results:

	Group		
	Diazepam	Fenobam	Placebo
Mean Hamilton Anxiety Score	16.00	15.11	23.54
Standard Deviation	8.33	7.14	12.30
Number of Subjects	9	9	11

Source: Reprinted with permission from Pecknold et al. (1982: 130), © 1982 The Williams and Wilkins Co., Baltimore.

Use the method derived in part (a) to test the null hypotheses that (i) the average anxiety score of patients receiving the drugs is no different from that of patients receiving the placebo; and (ii) the average anxiety score of patients receiving the new drug is the same as that of patients receiving diazepam. Note that the residual mean square can be calculated according to the formula $S_E^2 = [1/(n-m)]\Sigma_{i=1}^m S_i^2(n_i - 1)$. Why?

2.3. TWO-WAY ANALYSIS OF VARIANCE

Two-way analysis of variance examines the relationship of a quantitative dependent variable to two qualitative independent variables. The inclusion of a second factor permits us to model and test partial relationships, as well as to introduce interactions. In fact, most issues pertaining to analysis of variance may be developed for the two-factor design.

2.3.1. Patterns of Relationship in the Two-Way Classification

Before immersing ourselves in the details of model specification and hypothesis testing, we shall imagine that we have direct access to population means. In this manner, we may establish simply the patterns of relationship that occur when a quantitative dependent variable is classified by two qualitative variables.

Notation for the two-way classification is shown in Table 2.9. The two independent variables, R and C, have r and c categories, respectively. The factor categories are denoted R_i and C_j. Within each *cell* of the design—that is, for each combination of categories (R_i, C_j) of the two factors—there is a population cell mean μ_{ij}. Extending the dot notation introduced in the previous section, $\mu_{i.} \equiv \sum_{j=1}^{c} \mu_{ij}/c$ is the *marginal mean* in row i; $\mu_{.j} \equiv \sum_{i=1}^{r} \mu_{ij}/r$ is the marginal mean in column j; and

$$\mu_{..} \equiv \sum_i \sum_j \mu_{ij}/rc = \sum_i \mu_{i.}/r = \sum_j \mu_{.j}/c$$

is the grand mean.

If R and C do not interact in determining the dependent variable, then the partial relationship between each factor and the dependent variable does not depend upon the level of the other factor. For R, therefore, the difference $\mu_{ij} - \mu_{i'j}$ (where i and i' denote *different* categories of R) is constant across categories of C—that is, this difference is the same for all $j = 1, \ldots, c$; consequently, $\mu_{ij} - \mu_{i'j} = \mu_{i.} - \mu_{i'.}$. Likewise, for C, the difference $\mu_{ij} - \mu_{ij'}$ is constant across all categories of R, and is equal to the marginal difference

TABLE 2.9. Population Means in the Two-Way Classification

	C_1	C_2	$\cdots$	C_c	
R_1	μ_{11}	μ_{12}	$\cdots$	μ_{1c}	$\mu_{1.}$
R_2	μ_{21}	μ_{22}	$\cdots$	μ_{2c}	$\mu_{2.}$
$\vdots$	$\vdots$	$\vdots$		$\vdots$	$\vdots$
R_r	μ_{r1}	μ_{r2}	$\cdots$	μ_{rc}	$\mu_{r.}$
	$\mu_{.1}$	$\mu_{.2}$	$\cdots$	$\mu_{.c}$	$\mu_{..}$

$\mu_{.j} - \mu_{.j'}$. When interaction is absent, then, the partial effect of a factor—the factor's main effect—is given by differences in the population marginal means.

Several patterns of relationship are graphed in Figure 2.9. Plots of means, incidentally, not only serve to clarify the ideas underlying analysis of variance, but also provide a useful tool for examining and presenting data. In the illustrations, factor R has three levels, marked off along the horizontal axis; since R is qualitative, the order of its categories and the spacing between them

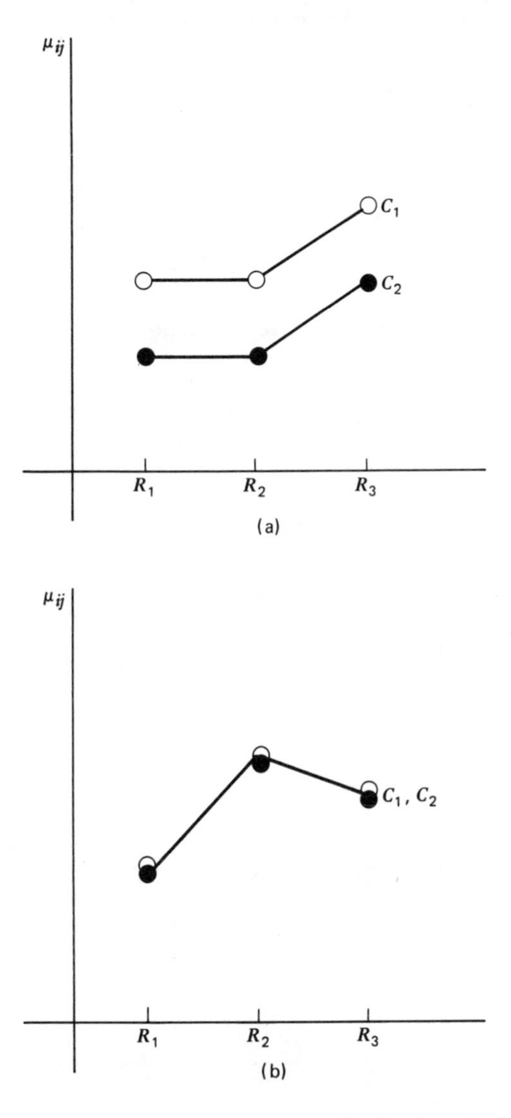

FIGURE 2.9. Patterns of relationship in the two-way classification. (a) R and C main effects. (b) R main effects. (c) C main effects. (d) No effects. (e) RC interactions. (f) RC interactions.

are arbitrary. Factor C has two categories. The six cell means are plotted as points, connected by lines according to the levels of factor C.

The separation between the lines at level R_i represents the difference $\mu_{i1} - \mu_{i2}$. If there is no interaction, therefore, the separation between the lines is constant and the lines themselves are parallel. Note that we may arrive at the same conclusion by examining factor C. The rise (or fall) in the line for C_j from R_i to R_{i+1} is $\mu_{i+1,j} - \mu_{ij}$; if there is no interaction, then this difference is the same for $j = 1$ and $j = 2$, and thus the lines from R_i to R_{i+1} are parallel. As we mentioned in Section 2.1.3, interaction is a symmetric concept: If R interacts with C, then C interacts with R.

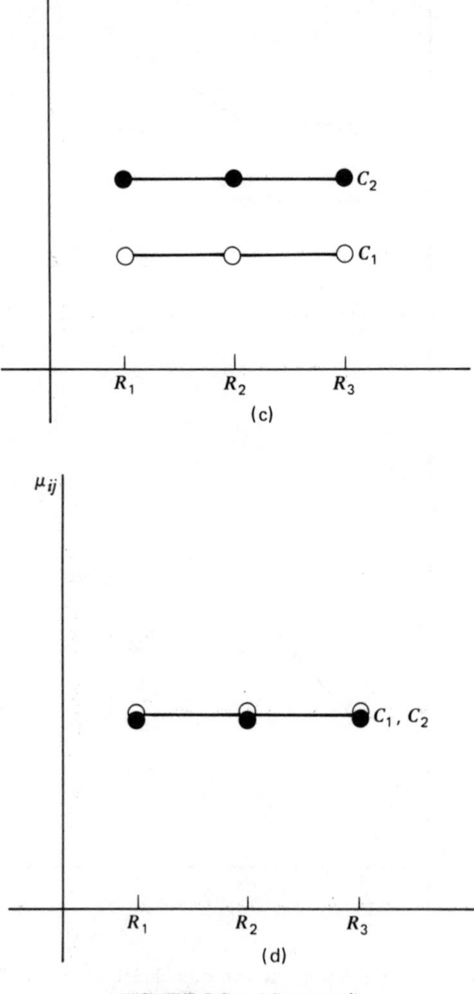

FIGURE 2.9. (*Continued*).

In Figure 2.9(a) both R and C have nonzero main effects. In Figure 2.9(b), the differences $\mu_{i1} - \mu_{i2}$ are zero, and consequently the C main effects are nil. In Figure 2.9(c), the R main effects are nil, since the differences $\mu_{ij} - \mu_{i'j}$ are zero. Finally, in Figure 2.9(d) both sets of main effects are nil.

From the previous discussion, it is clear that R and C interact when the differences $\mu_{ij} - \mu_{i'j}$ are not constant across all values of $j = 1,\ldots,c$; equivalently, in the presence of interaction the differences $\mu_{ij} - \mu_{ij'}$ are not constant across $i = 1,\ldots,r$. In a plot of cell means, interaction is reflected in a lack of parallelism between the profiles of means appearing in the plot. Two illustrative patterns of interaction are shown in Figures 2.9(e) and (f).

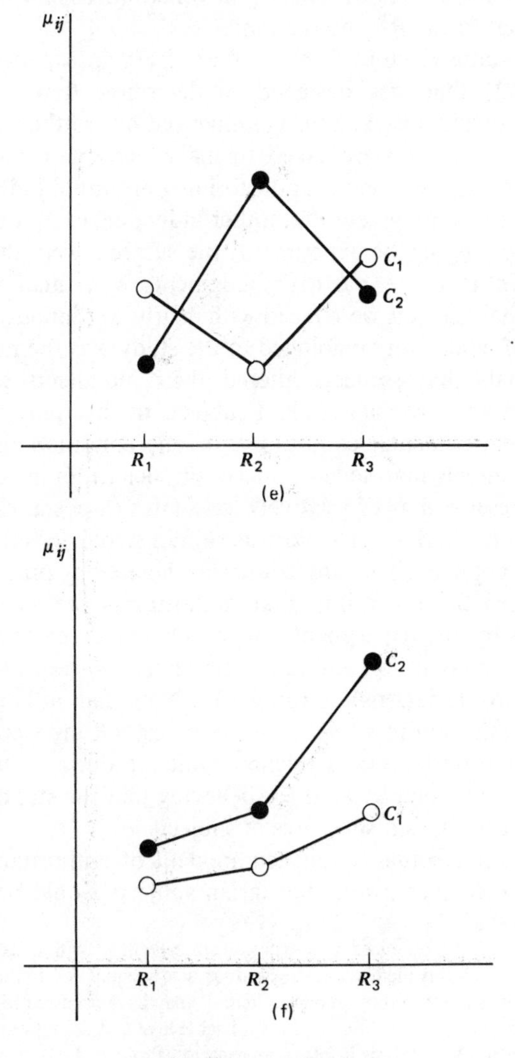

FIGURE 2.9. (*Continued*).

Even when interactions are absent in the population, we cannot expect perfectly parallel plots of sample means. There is, of course, sampling error in sampled data. We have to determine whether departures from parallelism observed in a sample are sufficiently large to be statistically significant, or whether they could easily be the product of chance. Moreover, in large samples we need to determine whether interactions are of sufficient magnitude to be substantively important. We may well decide to ignore interactions that are statistically significant but trivially small. In general, however, if we conclude that interactions are present and nonnegligible, then we do not interpret the main effects of the factors—after all, to conclude that two variables interact is to deny that they have *separate* effects. In other words, the R and C main effects are marginal to the RC interaction.

The illustrative data given in Table 2.10 are from an experiment by Moore and Krupat (1971) that was designed to determine how the relationship between conformity and social status is influenced by "authoritarianism." The subjects in the experiment were asked to make perceptual judgments of an intrinsically ambiguous character. Upon forming an initial judgment, subjects were presented with the judgment of another individual (a "partner") who was ostensibly participating in the experiment; the subjects were then asked for a final judgment. In fact, the partner's judgments were manipulated by the experimenter so that subjects were faced with nearly continuous disagreement.

The measure of conformity employed in the study was the number of times in 40 critical trials that subjects altered their judgments in response to disagreement. The 45 university-student subjects in the study were randomly assigned to two experimental conditions: In one condition the partner was described as of relatively high social status (a physician); in the other condition the partner was described as of relatively low status (a postal clerk).

A standard authoritarianism scale was administered to subjects after the experiment was completed. This procedure was dictated by practical considerations, but it raises the possibility that authoritarianism scores were inadvertently affected by the experimental manipulation of partner's status. The relationship between condition and authoritarianism is shown (along with cell means and standard deviations) in Table 2.11. Note that authoritarianism was trichotomized by the authors.[7] A chi-square test of independence for the condition by authoritarianism frequency table produces a p-value of .08, indicating that there is some ground for believing that the status manipulation has affected the authoritarianism scores of subjects.

Because of the conceptual-rigidity component of authoritarianism, Moore and Krupat predicted that low-authoritarian subjects would be more respon-

[7]Moore and Krupat trichotomized authoritarianism *separately* within each condition. This approach, while not strictly justified, serves to produce nearly equal cell frequencies (see Section 2.3.5) and yields results similar to those reported here. It may have occurred to the reader that the dummy-regression procedures of Section 2.1 are applicable here and do not require the categorization of authoritarianism. This analysis is left as an exercise (Problem 2.13). Moore and Krupat do in fact report the difference between slopes for the within-condition regressions of conformity on authoritarianism.

sive than high-authoritarian subjects to the social status of their partner. In other words, authoritarianism and partner's status are expected to interact in determining conformity. The cell means, graphed in Figure 2.10, appear to confirm the experimenters' expectations.

We note, for future reference, that the standard deviation of conformity in one cell (high-authoritarian, low-status partner) is appreciably larger than in the others. Upon inspection of the data, it is clear that the relatively large

TABLE 2.10. Moore and Krupat Conformity Data

	Partner's Status			
	Low		High	
Authoritarianism	Conformity	F^a	Conformity	F
Low	8	37	21	16
	7	20	23	15
	10	36	15	30
	6	18	12	22
	13	31	16	35
	12	36		
	7	28		
	6	28		
	8	17		
	12	35		
Medium	12	51	12	41
	4	44	17	48
	4	42	14	42
	9	43	16	45
			20	44
			8	42
			14	42
			17	41
			7	50
			17	39
			15	44
High	4	57	19	68
	8	65	9	63
	13	56	7	54
	9	65	11	59
	24	57	14	52
	7	61	13	57
	23	57	10	52
	13	55		

Source: Personal communication from J. Moore.
[a]Authoritarianism-scale score.

TABLE 2.11. Conformity by Authoritarianism and Partner's Status

Authoritarianism	Partner's Status Low	Partner's Status High
Low	8.900[a]	17.40
	2.644[b]	4.506
	10[c]	5
Medium	7.250	14.27
	3.948	3.952
	4	11
High	12.63	11.86
	7.347	3.934
	8	7

[a] Cell mean.
[b] Standard deviation.
[c] Frequency.

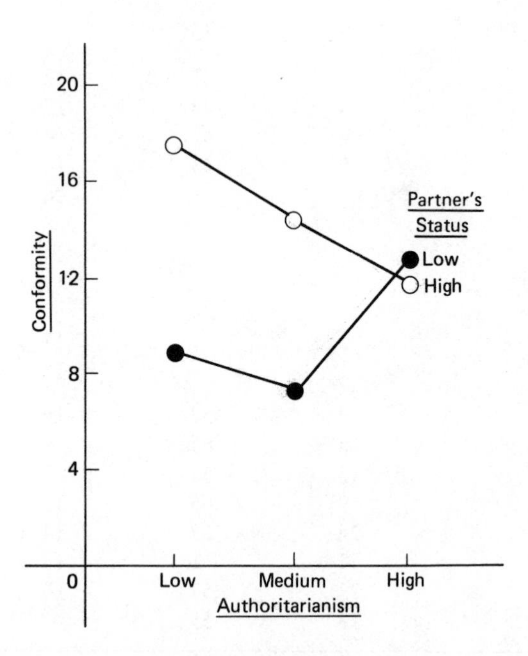

FIGURE 2.10. Conformity by authoritarianism and partner's status.

dispersion in this cell is due to two subjects who have atypically high conformity scores of 23 and 24.

2.3.2. The Two-Way ANOVA Model

Because interpretation of results in two-way analysis of variance depends centrally on the presence or absence of interaction, our first concern is to test the null hypothesis of no interaction. Based on the discussion in Section 2.3.1, this hypothesis may be expressed in terms of the cell means:

$$H_0: \mu_{ij} - \mu_{i'j} = \mu_{ij'} - \mu_{i'j'} \quad \text{for all } i, i' \text{ and } j, j' \tag{2.16}$$

Note that by rearranging the terms in equation (2.16), we may write the null hypothesis as

$$H_0: \mu_{ij} - \mu_{ij'} = \mu_{i'j} - \mu_{i'j'} \quad \text{for all } i, i' \text{ and } j, j' \tag{2.17}$$

once more demonstrating the symmetry of the concept of interaction.

If we accept the hypothesis of zero interactions, we may proceed to test for main effects of the two factors. It is convenient to express hypotheses concerning main effects in terms of the marginal means. Thus, for the row classification R we have the null hypothesis

$$H_0: \mu_{1.} = \mu_{2.} = \cdots = \mu_{r.} \tag{2.18}$$

and for the column classification C

$$H_0: \mu_{.1} = \mu_{.2} = \cdots = \mu_{.c} \tag{2.19}$$

Hypotheses (2.18) and (2.19) are testable whether interactions are present or absent, but these hypotheses generally are of substantive interest only when the interactions are nil.

The two-way analysis-of-variance model, suitably defined, provides a convenient means for testing the hypotheses (2.16), (2.18) and (2.19). The model is given by the equation

$$Y_{ijk} = \mu + \alpha_i + \beta_j + \gamma_{ij} + \varepsilon_{ijk} \tag{2.20}$$

where μ is the general mean of Y, α_i and β_j are main-effect parameters, γ_{ij} represents the RC interaction, and ε_{ijk} is an error random variable satisfying the usual linear-model assumptions. Taking expectations, model (2.20) becomes

$$\mu_{ij} = E(Y_{ijk}) = \mu + \alpha_i + \beta_j + \gamma_{ij} \tag{2.21}$$

Since there are $r \times c$ population cell means and $1 + r + c + (r \times c)$ parameters in model (2.21), the parameters of the model are not uniquely determined by the cell means. By reasoning that is familiar from our treatment of one-way ANOVA in Section 2.2.1, the indeterminacy of model (2.21) may be overcome by imposing $1 + r + c$ linearly independent restrictions on its parameters. It is desirable to select restrictions so that the resulting model makes it simple to test the hypotheses of interest.

With this purpose in mind, we specify the following constraints on the model parameters:

$$
\begin{aligned}
&\sum_{i=1}^{r} \alpha_i = 0 \\
&\sum_{j=1}^{c} \beta_j = 0 \\
&\sum_{i=1}^{r} \gamma_{ij} = 0 \quad \text{for all } j = 1,\ldots,c \\
&\sum_{j=1}^{c} \gamma_{ij} = 0 \quad \text{for all } i = 1,\ldots,r
\end{aligned}
\tag{2.22}
$$

At first glance, it seems that we have specified too many constraints, for the equations in (2.22) define $1 + 1 + c + r$ restrictions. The $c + r$ restrictions on the interactions include one linear dependency, however: One of these restrictions is redundant. If all r rows of interaction parameters sum to zero, then all of the interaction parameters sum to zero: $\Sigma_{i=1}^{r}\Sigma_{j=1}^{c}\gamma_{ij} = 0$. Thus, if the first $c - 1$ columns sum to zero, the last column sums to zero as well; for

$$
\sum_{i=1}^{r}\sum_{j=1}^{c} \gamma_{ij} = \sum_{j=1}^{c-1}\left(\sum_{i=1}^{r}\gamma_{ij}\right) + \sum_{i=1}^{r}\gamma_{ic}
$$

$$
0 = 0 + \sum_{i=1}^{r}\gamma_{ic}
$$

In shorthand form, we say that the constraints (2.22) specify that every set of effects sums to zero over each of its coordinates.

The constraints produce the following solution for model parameters in terms of population means:

$$
\begin{aligned}
\mu &= \mu_{..} \\
\alpha_i &= \mu_{i.} - \mu_{..} \\
\beta_j &= \mu_{.j} - \mu_{..} \\
\gamma_{ij} &= \mu_{ij} - \mu - \alpha_i - \beta_j \\
&= \mu_{ij} - \mu_{i.} - \mu_{.j} + \mu_{..}
\end{aligned}
\tag{2.23}
$$

The hypothesis of no row main effects (2.18) is therefore equivalent to H_0: all $\alpha_i = 0$, for under this hypothesis $\mu_{1.} = \mu_{2.} = \cdots = \mu_{r.} = \mu_{..}$. Likewise, the hypothesis of no column main effects (2.19) is equivalent to H_0: all $\beta_j = 0$, since then $\mu_{.1} = \mu_{.2} = \cdots = \mu_{.c} = \mu_{..}$. Finally, we shall demonstrate that the hypothesis of no interactions (given in (2.16) or (2.17)) is equivalent to H_0: all

$\gamma_{ij} = 0$. If all interaction parameters are zero, then

$$\gamma_{ij} = \mu_{ij} - \mu_{i.} - \mu_{.j} + \mu_{..} = 0$$

$$\gamma_{i'j} = \mu_{i'j} - \mu_{i'.} - \mu_{.j} + \mu_{..} = 0$$

$$\gamma_{ij'} = \mu_{ij'} - \mu_{i.} - \mu_{.j'} + \mu_{..} = 0$$

$$\gamma_{i'j'} = \mu_{i'j'} - \mu_{i'.} - \mu_{.j'} + \mu_{..} = 0$$

and thus

$$\gamma_{ij} - \gamma_{i'j} = \gamma_{ij'} - \gamma_{i'j'} \qquad \text{(i.e., } 0 - 0 = 0 - 0\text{)}$$

$$\mu_{ij} - \mu_{i'j} - \mu_{i.} + \mu_{i'.} = \mu_{ij'} - \mu_{i'j'} - \mu_{i.} + \mu_{i'.}$$

$$\mu_{ij} - \mu_{i'j} = \mu_{ij'} - \mu_{i'j'}$$

which is the hypothesis of no interaction (2.16).

2.3.3. Fitting the Two-Way ANOVA Model to Data

Since the least-squares estimator of μ_{ij} is the sample cell mean $\overline{Y}_{ij} = \sum_{k=1}^{n_{ij}} Y_{ijk}/n_{ij}$, least-squares estimators of the constrained ANOVA-model parameters follow immediately from equations (2.23):

$$M = \overline{Y}_{..} = \frac{\sum\sum \overline{Y}_{ij}}{rc}$$

$$A_i = \overline{Y}_{i.} - \overline{Y}_{..} = \frac{\sum_j \overline{Y}_{ij}}{c} - \overline{Y}_{..}$$

$$B_j = \overline{Y}_{.j} - \overline{Y}_{..} = \frac{\sum_i \overline{Y}_{ij}}{r} - \overline{Y}_{..}$$

$$C_{ij} = \overline{Y}_{ij} - \overline{Y}_{i.} - \overline{Y}_{.j} + \overline{Y}_{..}$$

The residuals are given by the deviations of the observations from their cell means,

$$E_{ijk} = Y_{ijk} - (M + A_i + B_j + C_{ij})$$
$$= Y_{ijk} - \overline{Y}_{ij}$$

In testing hypotheses about model parameters, however, we require incremental sums of squares, and it is therefore advantageous to fit the model by employing a full-rank design matrix $\mathbf{X}_R$ corresponding to the restricted model. As in one-way analysis of variance, the restrictions on the two-way ANOVA model may be used to reduce the over-parameterized design matrix to full column rank. For illustrative purposes, we take $r = 3$ and $c = 2$; the principles underlying this example extend simply to the general case, as we shall explain.

In light of the restriction $\alpha_1 + \alpha_2 + \alpha_3 = 0$, α_3 may be deleted from the model, substituting $-\alpha_1 - \alpha_2$. Likewise, β_2 may be replaced by $-\beta_1$. More generally, $-\Sigma_{i=1}^{r-1}\alpha_i$ replaces α_r, and $-\Sigma_{j=1}^{c-1}\beta_j$ replaces β_c. Since there are, then, $r-1$ linearly independent α_i parameters and $c-1$ linearly independent β_j parameters, the degrees of freedom for the two sets of main effects are $r-1$ and $c-1$, respectively.

The interactions in the 3×2 classification satisfy the following linear constraints:

$$\begin{aligned}
\gamma_{11} + \gamma_{12} &= 0 \\
\gamma_{21} + \gamma_{22} &= 0 \\
\gamma_{31} + \gamma_{32} &= 0 \\
\gamma_{11} + \gamma_{21} + \gamma_{31} &= 0 \\
\gamma_{12} + \gamma_{22} + \gamma_{32} &= 0
\end{aligned}$$

(Recall that although there are five such constraints, the fifth follows from the first four—there are but four linearly independent constraints.) We may, therefore delete all interactions but γ_{11} and γ_{21}, substituting

$$\begin{aligned}
\gamma_{12} &= -\gamma_{11} \\
\gamma_{22} &= -\gamma_{21} \\
\gamma_{31} &= -\gamma_{11} - \gamma_{21} \\
\gamma_{32} &= -\gamma_{31} = \gamma_{11} + \gamma_{21}
\end{aligned}$$

More generally, we can write all $r \times c$ of the γ_{ij}'s in terms of $(r-1)(c-1)$ linearly independent interaction parameters, and there are, consequently, $(r-1)(c-1)$ degrees of freedom for interaction.

The row basis of the reduced design matrix is given in the parametric equation

$$\begin{matrix} & & (\mu) & (\alpha_1) & (\alpha_2) & (\beta_1) & (\gamma_{11}) & (\gamma_{21}) & \end{matrix}$$

$$\begin{pmatrix} \mu_{11} \\ \mu_{12} \\ \mu_{21} \\ \mu_{22} \\ \mu_{31} \\ \mu_{32} \end{pmatrix} = \begin{pmatrix} 1 & 1 & 0 & 1 & 1 & 0 \\ 1 & 1 & 0 & -1 & -1 & 0 \\ 1 & 0 & 1 & 1 & 0 & 1 \\ 1 & 0 & 1 & -1 & 0 & -1 \\ 1 & -1 & -1 & 1 & -1 & -1 \\ 1 & -1 & -1 & -1 & 1 & 1 \end{pmatrix} \begin{pmatrix} \mu \\ \alpha_1 \\ \alpha_2 \\ \beta_1 \\ \gamma_{11} \\ \gamma_{21} \end{pmatrix}$$

$$\boldsymbol{\mu} = \mathbf{X}_B \boldsymbol{\beta}_R \tag{2.24}$$

We have constructed $\mathbf{X}_B$ according to the model constraints, but the row basis may also be coded mechanically by applying these rules:

1. The first column of $\mathbf{X}_B$ consists of ones, and represents the grand mean μ.

2. There are $r-1$ columns for the R main effects. The ith such column contains ones for cells i, j ($j = 1,\ldots,c$), minus-ones for cells r, j ($j = 1,\ldots,c$), and zeroes elsewhere (i.e., for cells i', j: $i' \neq i, r$; $j = 1,\ldots,c$).
3. There are $c-1$ columns for the C main effects. The jth such column contains ones for cells i, j ($i = 1,\ldots,r$), minus-ones for cells i, c ($i = 1,\ldots,r$), and zeroes elsewhere.
4. There are $(r-1)(c-1)$ columns for the RC interactions. These columns consist of all pairwise products of the $(r-1)$ columns for R main effects and the $(c-1)$ columns for C main effects.

The reduced design matrix $\mathbf{X}_R$ is produced by repeating the i, jth row of $\mathbf{X}_B$ n_{ij} times, where (we recall) n_{ij} is the number of observations in the i, jth cell of the design. The matrix form of the full-rank two-way analysis-of-variance model is shown in equation (2.25). In practice, it is of course unnecessary to order observations by cells, so long as the proper row of $\mathbf{X}_R$ corresponds to each entry of $\mathbf{y}$.

$$
\begin{array}{c}
\text{cell}\\ 11 \\ \\ \\ 12 \\ \\ \\ 21 \\ \\ \\ 22 \\ \\ \\ 31 \\ \\ \\ 32
\end{array}
\begin{bmatrix}
Y_{111} \\ \vdots \\ Y_{11n_{11}} \\ \hline Y_{121} \\ \vdots \\ Y_{12n_{12}} \\ \hline Y_{211} \\ \vdots \\ Y_{21n_{21}} \\ \hline Y_{221} \\ \vdots \\ Y_{22n_{22}} \\ \hline Y_{311} \\ \vdots \\ Y_{31n_{31}} \\ \hline Y_{321} \\ \vdots \\ Y_{32n_{32}}
\end{bmatrix}
=
\begin{bmatrix}
1 & 1 & 0 & 1 & 1 & 0 \\
\vdots & \vdots & \vdots & \vdots & \vdots & \vdots \\
1 & 1 & 0 & 1 & 1 & 0 \\ \hline
1 & 1 & 0 & -1 & -1 & 0 \\
\vdots & \vdots & \vdots & \vdots & \vdots & \vdots \\
1 & 1 & 0 & -1 & -1 & 0 \\ \hline
1 & 0 & 1 & 1 & 0 & 1 \\
\vdots & \vdots & \vdots & \vdots & \vdots & \vdots \\
1 & 0 & 1 & 1 & 0 & 1 \\ \hline
1 & 0 & 1 & -1 & 0 & -1 \\
\vdots & \vdots & \vdots & \vdots & \vdots & \vdots \\
1 & 0 & 1 & -1 & 0 & -1 \\ \hline
1 & -1 & -1 & 1 & -1 & -1 \\
\vdots & \vdots & \vdots & \vdots & \vdots & \vdots \\
1 & -1 & -1 & 1 & -1 & -1 \\ \hline
1 & -1 & -1 & -1 & 1 & 1 \\
\vdots & \vdots & \vdots & \vdots & \vdots & \vdots \\
1 & -1 & -1 & -1 & 1 & 1
\end{bmatrix}
\begin{pmatrix}
\mu \\ \alpha_1 \\ \alpha_2 \\ \beta_1 \\ \gamma_{11} \\ \gamma_{21}
\end{pmatrix}
+
\begin{bmatrix}
\varepsilon_{111} \\ \vdots \\ \varepsilon_{11n_{11}} \\ \hline \varepsilon_{121} \\ \vdots \\ \varepsilon_{12n_{12}} \\ \hline \varepsilon_{211} \\ \vdots \\ \varepsilon_{21n_{21}} \\ \hline \varepsilon_{221} \\ \vdots \\ \varepsilon_{22n_{22}} \\ \hline \varepsilon_{311} \\ \vdots \\ \varepsilon_{31n_{31}} \\ \hline \varepsilon_{321} \\ \vdots \\ \varepsilon_{32n_{32}}
\end{bmatrix}
$$

$$
\mathbf{y} \quad = \quad \mathbf{X}_R \quad \boldsymbol{\beta}_R \quad + \quad \boldsymbol{\varepsilon} \qquad (2.25)
$$

Deleting the last two columns of $\mathbf{X}_R$ produces the design matrix for a full-rank no-interaction model. This model may be fit to the data by the usual least-squares procedure: $\mathbf{b}_R = (\mathbf{X}_R'\mathbf{X}_R)^{-1}\mathbf{X}_R'\mathbf{y}$. It is also possible to fit models that retain interactions but delete main-effect regressors. Although these models violate marginality and therefore are not generally meaningful, they do have a role in producing incremental sums of squares for hypothesis tests, as we shall see presently.

2.3.4. Testing Hypotheses in Two-Way ANOVA

We have specified constraints on the two-way ANOVA model in such a manner that testing hypotheses about the parameters of the model is equivalent to testing hypotheses of interest concerning interactions and main effects of the factors. Tests for sets of model parameters may be constructed by the incremental sum of squares approach.

For ease of reference, we write $\text{SS}(\boldsymbol{\alpha}, \boldsymbol{\beta}, \boldsymbol{\gamma})$ to denote the regression sum of squares for the full model. The regression sums of squares for other models are similarly represented. For example, for the no-interaction model, we have $\text{SS}(\boldsymbol{\alpha}, \boldsymbol{\beta})$; and for the model that deletes the column main-effect regressors, we have $\text{SS}(\boldsymbol{\alpha}, \boldsymbol{\gamma})$. Incremental sums of squares, as usual, are given by differences between regression sums of squares for alternative models. We shall use the following notation for incremental sums of squares in ANOVA:[8]

$$\text{SS}(\boldsymbol{\gamma}|\boldsymbol{\alpha}, \boldsymbol{\beta}) = \text{SS}(\boldsymbol{\alpha}, \boldsymbol{\beta}, \boldsymbol{\gamma}) - \text{SS}(\boldsymbol{\alpha}, \boldsymbol{\beta})$$

$$\text{SS}(\boldsymbol{\alpha}|\boldsymbol{\beta}, \boldsymbol{\gamma}) = \text{SS}(\boldsymbol{\alpha}, \boldsymbol{\beta}, \boldsymbol{\gamma}) - \text{SS}(\boldsymbol{\beta}, \boldsymbol{\gamma})$$

$$\text{SS}(\boldsymbol{\beta}|\boldsymbol{\alpha}, \boldsymbol{\gamma}) = \text{SS}(\boldsymbol{\alpha}, \boldsymbol{\beta}, \boldsymbol{\gamma}) - \text{SS}(\boldsymbol{\alpha}, \boldsymbol{\gamma})$$

$$\text{SS}(\boldsymbol{\alpha}|\boldsymbol{\beta}) = \text{SS}(\boldsymbol{\alpha}, \boldsymbol{\beta}) - \text{SS}(\boldsymbol{\beta})$$

$$\text{SS}(\boldsymbol{\beta}|\boldsymbol{\alpha}) = \text{SS}(\boldsymbol{\alpha}, \boldsymbol{\beta}) - \text{SS}(\boldsymbol{\alpha})$$

The residual sum of squares is $\text{RSS} = \Sigma\Sigma\Sigma E_{ijk}^2 = \Sigma\Sigma\Sigma(Y_{ijk} - \overline{Y}_{ij})^2 = \text{TSS} - \text{SS}(\boldsymbol{\alpha}, \boldsymbol{\beta}, \boldsymbol{\gamma})$.

[8]The reader may encounter variations of the SS notation. One common approach (used, for example, in Searle, 1971) is to include μ in the argument to the sum-of-squares function, and to let $R(\ )$ denote the raw (rather than mean-deviation) sum of squares. Thus, in this scheme, $R(\mu, \boldsymbol{\alpha}, \boldsymbol{\beta}) = \Sigma\Sigma\Sigma\hat{Y}_{ijk}^2$ for the no-interaction model, while

$$R(\boldsymbol{\alpha}, \boldsymbol{\beta}|\mu) = R(\mu, \boldsymbol{\alpha}, \boldsymbol{\beta}) - R(\mu) = \sum\sum\sum\left(\hat{Y}_{ijk} - \overline{Y}\right)^2$$
$$= \text{SS}(\boldsymbol{\alpha}, \boldsymbol{\beta})$$

is the mean-deviation explained sum of squares for the same model.

SS($\gamma|\alpha, \beta$), the incremental sum of squares due to interaction, is appropriate for testing the null hypothesis of no interaction, H_0: all $\gamma_{ij} = 0$. In the presence of interactions, we may use SS($\alpha|\beta, \gamma$) and SS($\beta|\alpha, \gamma$) to test null hypotheses concerning main effects, though as we argued previously, these hypotheses are not of general substantive interest when interactions are important. In the absence of interactions, SS($\alpha|\beta$) and SS($\beta|\alpha$) may be used to test for main effects, but the use of SS($\alpha|\beta, \gamma$) and SS($\beta|\alpha, \gamma$) would be appropriate as well. If interactions are present, tests based on SS($\alpha|\beta$) and SS($\beta|\alpha$) do not test the null hypotheses H_0: all $\alpha_i = 0$ and H_0: all $\beta_j = 0$; instead, interaction parameters become implicated in the tests that ostensibly pertain to main effects. These comments are summarized in Table 2.12, which shows a general analysis-of-variance table for the two-way classification. It is important to note that in general, SS(α) and SS(β) are inappropriate for testing hypotheses that the R and C main effects are nil, since each of these sums of squares depends upon the other set of main effects and upon the interactions (if they are present).

Certain authors (e.g., Nelder, 1976, 1977) prefer SS($\alpha|\beta$) and SS($\beta|\alpha$) for tests of main effects because, if interactions are absent, tests based on these sums of squares are more powerful than those based on SS($\alpha|\beta, \gamma$) and SS($\beta|\alpha, \gamma$). Other authors (e.g., Hocking and Speed, 1975) prefer SS($\alpha|\beta, \gamma$) and SS($\beta|\alpha, \gamma$) because, in the presence of interactions, tests based on these sums of squares have a straightforward (if substantively uninteresting) interpretation.

TABLE 2.12. Two-Way ANOVA Table

Source	df	Sum of Squares	MS[a]	F[b]	H_0
R	$r-1$	SS($\alpha\|\beta, \gamma$)			$\boldsymbol{\alpha} = \mathbf{0}$ $(\mu_{i.} = \mu_{i'.})$
		SS($\alpha\|\beta$)			$\boldsymbol{\alpha} = \mathbf{0}\|\boldsymbol{\gamma} = \mathbf{0}$ $(\mu_{i.} = \mu_{i'.}\|$no interaction$)$
C	$c-1$	SS($\beta\|\alpha, \gamma$)			$\boldsymbol{\beta} = \mathbf{0}$ $(\mu_{.j} = \mu_{.j'})$
		SS($\beta\|\alpha$)			$\boldsymbol{\beta} = \mathbf{0}\|\boldsymbol{\gamma} = \mathbf{0}$ $(\mu_{.j} = \mu_{.j'}\|$no interaction$)$
RC	$(r-1)(c-1)$	SS($\gamma\|\alpha, \beta$)			$\boldsymbol{\gamma} = \mathbf{0}$ $(\mu_{ij} - \mu_{i'j} = \mu_{ij'} - \mu_{i'j'})$
Residuals	$n - rc$	TSS − SS(α, β, γ)	$\dfrac{\text{RSS}}{n - rc}$		
Total	$n-1$	TSS			

[a]MS = SS/df.
[b]F = MS/RMS.

For the Moore and Krupat data, given earlier in Table 2.10, factor R is authoritarianism and factor C is partner's status. Sums of squares for these data are

$$\begin{aligned} SS(\alpha, \beta, \gamma) &= 391.44 \\ SS(\alpha, \beta) &= 215.95 \\ SS(\alpha, \gamma) &= 151.87 \\ SS(\beta, \gamma) &= 355.42 \\ SS(\alpha) &= 3.7333 \\ SS(\beta) &= 204.33 \\ TSS &= 1209.2 \end{aligned}$$

The ANOVA for the experiment is shown in Table 2.13. The predicted status × authoritarianism interaction proves to be statistically significant. A researcher would not normally report both sets of main-effect sums of squares; in the present instance, where the interactions are not negligible, $SS(\alpha|\beta)$ and $SS(\beta|\alpha)$ do not test hypotheses about main effects, as we explained. The tests based on $SS(\alpha|\beta, \gamma)$ and $SS(\beta|\alpha, \gamma)$, while appropriate in the presence of interactions, are not really of interest here.

2.3.5. Equal Cell Frequencies

Equal cell frequencies simplify—but do not change fundamentally—the analysis of the preceding section. When all cell frequencies are the same, the

TABLE 2.13. ANOVA for Conformity by Authoritarianism and Partner's Status

Source	SS	*df*	MS	*F*	*p*
Authoritarianism		2			
$\alpha\|\beta, \gamma$	36.02		18.01	0.86	.43
$\alpha\|\beta$	11.62		5.81	0.28	.76
Partner's Status		1			
$\beta\|\alpha, \gamma$	239.57		239.57	11.43	.001
$\beta\|\alpha$	212.22		212.22	10.12	.002
Authoritarianism × Status	175.49	2	87.745	4.18	.02
Residuals	817.76	39	20.968		
Total	1209.2	44			

regressors for *different* sets of effects in the reduced design matrix are uncorrelated with one another. This is readily verified by examining the row basis of the reduced design matrix for the 3×2 classification (equation (2.24)): Each effect regressor is a contrast in cell means, and regressors in different sets (the two sets of main effects and the interactions) are orthogonal. For this reason, equal-cell-frequencies data are termed *balanced* or *orthogonal*.

Orthogonality of the main-effect and interaction subspaces permits a unique decomposition of the explained sum of squares $\text{SS}(\alpha, \beta, \gamma)$ into components due to the three sets of effects. Indeed, for balanced data,

$$\text{SS}(\alpha|\beta, \gamma) = \text{SS}(\alpha|\beta) = \text{SS}(\alpha)$$

$$\text{SS}(\beta|\alpha, \gamma) = \text{SS}(\beta|\alpha) = \text{SS}(\beta)$$

$$\text{SS}(\gamma|\alpha, \beta) = \text{SS}(\gamma)$$

and hence $\text{SS}(\alpha, \beta, \gamma) = \text{SS}(\alpha) + \text{SS}(\beta) + \text{SS}(\gamma)$. These results lead to particularly simple formulas for the several sums of squares:

$$\text{SS}(\alpha) = n^*c \sum_{i=1}^{r} (\bar{Y}_{i.} - \bar{Y}_{..})^2$$

$$\text{SS}(\beta) = n^*r \sum_{j=1}^{c} (\bar{Y}_{.j} - \bar{Y}_{..})^2 \qquad (2.26)$$

$$\text{SS}(\gamma) = n^* \sum_{i=1}^{r} \sum_{j=1}^{c} (\bar{Y}_{ij} - \bar{Y}_{i.} - \bar{Y}_{.j} + \bar{Y}_{..})^2$$

where $n^* = n/rc$ is the number of observations in each cell of the design.

2.3.6. Linear Contrasts in Two-Way ANOVA

In Section 2.2.2, we showed how linear comparisons among group means may be coded into the one-way analysis-of-variance design matrix. In the present section, we shall briefly indicate how this technique may be applied to two-way ANOVA.

A straightforward approach to comparisons in the two-way classification is to specify contrasts separately for each set of main effects, obtaining interaction contrasts by finding all pairwise products of the main-effect contrasts. As long as we take care to specify main-effect contrasts that are orthogonal in the row basis of the design matrix, the interaction contrasts will be orthogonal as well.

Imagine, for example, a 3×2 classification arising from an experiment in which the first factor consists of a control group (R_1) and two experimental groups subjected to different treatments (R_2 and R_3). The second factor is, say, sex, with categories male (C_1) and female (C_2). A possible set of main-effect contrasts for this experiment is

	(δ_1)	(δ_2)
R_1	2	0
R_2	−1	1
R_3	−1	−1

	(β_1)
C_1	1
C_2	−1

The parametric equation generated from these contrasts is

$$\begin{matrix} & & (\mu) & (\delta_1) & (\delta_2) & (\beta_1) & (\zeta_{11}) & (\zeta_{21}) & \end{matrix}$$

$$\begin{pmatrix} \mu_{11} \\ \mu_{12} \\ \mu_{21} \\ \mu_{22} \\ \mu_{31} \\ \mu_{32} \end{pmatrix} = \begin{pmatrix} 1 & 2 & 0 & 1 & 2 & 0 \\ 1 & 2 & 0 & -1 & -2 & 0 \\ 1 & -1 & 1 & 1 & -1 & 1 \\ 1 & -1 & 1 & -1 & 1 & -1 \\ 1 & -1 & -1 & 1 & -1 & -1 \\ 1 & -1 & -1 & -1 & 1 & 1 \end{pmatrix} \begin{pmatrix} \mu \\ \delta_1 \\ \delta_2 \\ \beta_1 \\ \zeta_{11} \\ \zeta_{21} \end{pmatrix} \tag{2.27}$$

Note that we use two degrees of freedom for condition main effects, one degree of freedom for the sex main effect, and two degrees of freedom for interaction.

This coding permits us to test the following hypotheses:

1. H_0: $\zeta_{11} = 0$. The difference (if any) between the control group and the experimental groups is identical for males and females.
2. H_0: $\zeta_{21} = 0$. The difference (if any) between the two experimental groups is identical for males and females.
3. H_0: $\delta_1 = 0$. There is no marginal difference between the control group and the experimental groups.

4. H_0: $\delta_2 = 0$. There is no marginal difference between the two experimental groups.
5. H_0: $\beta_1 = 0$. There is no marginal difference between males and females.

Hypothesis (3), though testable in any event, is generally of interest only when hypothesis (1) is acceptable. Hypotheses (4) and (2) are similarly related. The null hypothesis of no sex main effect, hypothesis (5), would normally be examined only when the interaction null hypotheses (1) and (2) are not rejected.

If the data are balanced, then the full-rank design matrix produced from equation (2.27) is orthogonal. Thus, these contrasts provide a decomposition into one-degree-of-freedom components of the explained sum of squares for the two-way ANOVA model.

2.3.7. Some Cautionary Remarks

R. A. Fisher (1925) originally formulated analysis of variance for orthogonal data. Yet, as early as 1934, Fisher's colleague at the Rothamsted Experimental Station in England, F. Yates, indicated how the analysis of variance could be extended to unbalanced data. Apart from approximate methods motivated by the desire to reduce the effort of calculation, Yates suggested two approaches to the two-way classification, naming both approaches for the computational techniques that he developed. The first approach, which he called "the method of weighted squares of means," calculates the sums of squares $SS(\alpha|\beta, \gamma)$, $SS(\beta|\alpha, \gamma)$, and $SS(\gamma|\alpha, \beta)$ (using our notation). Yates's second approach, which he called "the method of fitting constants," assumes that interactions are absent and calculates $SS(\alpha|\beta)$ and $SS(\beta|\alpha)$.

Considering the apparent simplicity of the two-way classification and the clarity of Yates's treatment of it, it is ironic that the analysis of unbalanced data has become the subject of controversy and confusion. While it is not our purpose to present a complete account of the debate regarding the proper handling of unbalanced data, and while it is tempting to ignore this debate altogether, there are two reasons for addressing the topic briefly here: (1) The reader may encounter confused applications of ANOVA or may have occasion to consult other accounts of the method; and (2) computer programs for analysis of variance are often misleading in their documentation and output or even incorrect in their calculations (see Francis, 1973). With respect to the second point, it is good practice to test a computer program with known data before trusting it to analyze new data properly. Indeed, this advice applies not just to ANOVA calculations, but generally.

Much of the confusion about the analysis of unbalanced data has its source in the relationship between the overparameterized (deficient-rank) model and the constrained (full-rank) model. Equation (2.28) shows the relation between cell means and the overparameterized-model parameters for the 3 × 2 classification. As we remarked in Section 2.3.2, the parameters of equation (2.28) are underdetermined in the absence of further constraints.

$$\begin{pmatrix} \mu_{11} \\ \mu_{12} \\ \mu_{21} \\ \mu_{22} \\ \mu_{31} \\ \mu_{32} \end{pmatrix} = \overset{(\mu)\ (\alpha_1)\ (\alpha_2)\ (\alpha_3)\ (\beta_1)\ (\beta_2)\ (\gamma_{11})\ (\gamma_{12})\ (\gamma_{21})\ (\gamma_{22})\ (\gamma_{31})\ (\gamma_{32})}{\begin{pmatrix} 1 & 1 & 0 & 0 & 1 & 0 & 1 & 0 & 0 & 0 & 0 & 0 \\ 1 & 1 & 0 & 0 & 0 & 1 & 0 & 1 & 0 & 0 & 0 & 0 \\ 1 & 0 & 1 & 0 & 1 & 0 & 0 & 0 & 1 & 0 & 0 & 0 \\ 1 & 0 & 1 & 0 & 0 & 1 & 0 & 0 & 0 & 1 & 0 & 0 \\ 1 & 0 & 0 & 1 & 1 & 0 & 0 & 0 & 0 & 0 & 1 & 0 \\ 1 & 0 & 0 & 1 & 0 & 1 & 0 & 0 & 0 & 0 & 0 & 1 \end{pmatrix}} \begin{bmatrix} \mu \\ \alpha_1 \\ \alpha_2 \\ \alpha_3 \\ \beta_1 \\ \beta_2 \\ \gamma_{11} \\ \gamma_{12} \\ \gamma_{21} \\ \gamma_{22} \\ \gamma_{31} \\ \gamma_{32} \end{bmatrix} \tag{2.28}$$

By employing an arbitrary basis for the subspace spanned by the design matrix, we may nevertheless find the explained sums of squares for the unconstrained model, which we denote $SS^*(\boldsymbol{\alpha}, \boldsymbol{\beta}, \boldsymbol{\gamma})$. Since the usual restrictions that we employ (given in (2.22)) provide a basis for the design matrix, $SS^*(\boldsymbol{\alpha}, \boldsymbol{\beta}, \boldsymbol{\gamma}) = SS(\boldsymbol{\alpha}, \boldsymbol{\beta}, \boldsymbol{\gamma})$.

Searle (1971: 300–301) has pointed out, however, that *any* unconstrained model that includes the interactions spans the same subspace as the full model. Consider, for example, the model

$$Y_{ijk} = \gamma_{ij} + \varepsilon_{ijk}$$

from which the general mean and main effects have been deleted. The parametric equation for this model is

$$\begin{pmatrix} \mu_{11} \\ \mu_{12} \\ \mu_{21} \\ \mu_{22} \\ \mu_{31} \\ \mu_{32} \end{pmatrix} = \begin{pmatrix} 1 & 0 & 0 & 0 & 0 & 0 \\ 0 & 1 & 0 & 0 & 0 & 0 \\ 0 & 0 & 1 & 0 & 0 & 0 \\ 0 & 0 & 0 & 1 & 0 & 0 \\ 0 & 0 & 0 & 0 & 1 & 0 \\ 0 & 0 & 0 & 0 & 0 & 1 \end{pmatrix} \begin{pmatrix} \gamma_{11} \\ \gamma_{12} \\ \gamma_{21} \\ \gamma_{22} \\ \gamma_{31} \\ \gamma_{32} \end{pmatrix}$$

Clearly, the design matrix for the "interactions-only" model is of rank six (indeed, $\gamma_{ij} = \mu_{ij}$), producing the same explained sums of squares as the full model: $SS^*(\boldsymbol{\gamma}) = SS^*(\boldsymbol{\alpha}, \boldsymbol{\beta}, \boldsymbol{\gamma})$. In the unconstrained model, the main-effect subspaces are (quite literally) marginal to the interaction subspace. Thus, $SS^*(\boldsymbol{\alpha}|\boldsymbol{\beta}, \boldsymbol{\gamma}) = SS^*(\boldsymbol{\beta}|\boldsymbol{\alpha}, \boldsymbol{\gamma}) = 0$.

Working with the constrained model, we found in Section 2.3.4 that $SS(\alpha|\beta,\gamma)$ and $SS(\beta|\alpha,\gamma)$ are appropriate for testing hypotheses about main effects. The values of these incremental sums of squares, however, *depend upon* the constraints used to insure that the reduced design matrix is of full column rank. In particular, these constraints should be selected so that meaningful hypotheses about cell means are tested. This is precisely the approach we adopted. The SS notation is frequently used carelessly without attention to the constraints that are employed and to the hypotheses that follow from them.

Further discussion of the points raised in this section may be found in a variety of sources, including Hocking and Speed (1975); Speed and Hocking (1976); Speed, Hocking, and Hackney (1978); Speed and Monlezun (1979); Searle, Speed, and Henderson (1981); and Steinhorst (1982).

PROBLEMS

2.8. The following balanced data are from an experiment reported by Fox and Guyer (1978). Twenty four-person groups of subjects played 30 trials of a "prisoner's dilemma" game. On every trial of the experiment, each subject selected either a competitive or a cooperative choice. The value reported below for each group is the number of cooperative choices (of 120 choices) made by subjects in that group. Ten of the groups recorded their choices anonymously, while the remaining groups made public choices (i.e., subjects' choices were made known to other group members); half of the groups in each experimental condition were composed of males and half of females. The experimenters expected to observe a higher level of cooperation in the public-choice condition, but did not make predictions about sex effects or sex-by-condition interaction.

Number of Cooperative Choices

	Sex	
Condition	Male	Female
Public-Choice	49	54
	64	61
	37	79
	52	64
	68	29
Anonymous	27	40
	58	39
	52	44
	41	34
	30	44

(a) Calculate the mean and standard deviation for cooperation in each cell; graph the cell means, and comment on the results of the experiment.

(b) Form the reduced design matrix $\mathbf{X}_R$ for a two-way ANOVA of cooperation by condition and sex. Verify that the columns of $\mathbf{X}_R$ are orthogonal.

(c) Confirm that the sums of squares for the ANOVA may be obtained in the following four ways:
   (i) as $SS(\boldsymbol{\alpha}|\boldsymbol{\beta})$, $SS(\boldsymbol{\beta}|\boldsymbol{\alpha})$, and $SS(\boldsymbol{\gamma}|\boldsymbol{\alpha}, \boldsymbol{\beta})$;
   (ii) as $SS(\boldsymbol{\alpha}|\boldsymbol{\beta}, \boldsymbol{\gamma})$, $SS(\boldsymbol{\beta}|\boldsymbol{\alpha}, \boldsymbol{\gamma})$, and $SS(\boldsymbol{\gamma}|\boldsymbol{\alpha}, \boldsymbol{\beta})$;
   (iii) as $SS(\boldsymbol{\alpha})$, $SS(\boldsymbol{\beta})$, and $SS(\boldsymbol{\gamma})$;
   (iv) using equations (2.26).

(d) Construct the ANOVA table for the two-way analysis of variance.

2.9.* Suppose that the two-way ANOVA model

$$Y_{ijk} = \mu + \alpha_i + \beta_j + \gamma_{ij} + \varepsilon_{ijk}$$

is reduced to full rank by imposing the following constraints (for $r = 3$ and $c = 2$): $\alpha_3 = 0$; $\beta_2 = 0$; $\gamma_{31} = \gamma_{12} = \gamma_{22} = \gamma_{32} = 0$. These constraints lead to a dummy-variable coding of the reduced design matrix.

(a) Write out the row basis of the reduced design matrix.

(b) Solve for the parameters of the constrained model in terms of the cell means. What is the nature of the hypotheses H_0: all $\alpha_i = 0$, H_0: all $\beta_j = 0$, and H_0: all $\gamma_{ij} = 0$ for this parameterization of the model?

(c) Let $SS^{**}(\boldsymbol{\alpha}, \boldsymbol{\beta}, \boldsymbol{\gamma})$ represent the regression sum of squares for the full model, calculated under the constraints defined above; let $SS^{**}(\boldsymbol{\alpha}, \boldsymbol{\beta})$ represent the regression sum of squares for the model that deletes the interaction regressors; and so on. Using the Moore and Krupat data recorded in Table 2.10, confirm that

$$SS^{**}(\boldsymbol{\alpha}|\boldsymbol{\beta}) = SS(\boldsymbol{\alpha}|\boldsymbol{\beta})$$

$$SS^{**}(\boldsymbol{\beta}|\boldsymbol{\alpha}) = SS(\boldsymbol{\beta}|\boldsymbol{\alpha})$$

$$SS^{**}(\boldsymbol{\gamma}|\boldsymbol{\alpha}, \boldsymbol{\beta}) = SS(\boldsymbol{\gamma}|\boldsymbol{\alpha}, \boldsymbol{\beta})$$

$$SS^{**}(\boldsymbol{\alpha}|\boldsymbol{\beta}, \boldsymbol{\gamma}) \neq SS(\boldsymbol{\alpha}|\boldsymbol{\beta}, \boldsymbol{\gamma})$$

$$SS^{**}(\boldsymbol{\beta}|\boldsymbol{\alpha}, \boldsymbol{\gamma}) \neq SS(\boldsymbol{\beta}|\boldsymbol{\alpha}, \boldsymbol{\gamma})$$

(Recall that the unstarred SS notation gives regression sums of squares under the constraints (2.22).)

(d) Explain the equalities in part (c) by showing that, in general,

$$SS^*(\alpha, \beta, \gamma) = SS(\alpha, \beta, \gamma) = SS^{**}(\alpha, \beta, \gamma)$$

$$SS^*(\alpha, \beta) = SS(\alpha, \beta) = SS^{**}(\alpha, \beta)$$

$$SS^*(\alpha) = SS(\alpha) = SS^{**}(\alpha)$$

$$SS^*(\beta) = SS(\beta) = SS^{**}(\beta)$$

(*Hint*: In each instance, examine the subspace spanned by the appropriate columns of the design matrix—either full or reduced. Recall that the single-star SS* notation gives regression sums of squares for the unconstrained model.)

(e) Analyze the Moore and Krupat data using one or more ANOVA computer programs available at your computing center. How do the programs compute sums of squares?

2.4. HIGHER-WAY ANALYSIS OF VARIANCE

The methods of Section 2.3 may be extended to any number of factors. We shall examine the three-way classification in some detail before commenting briefly on the general case.

2.4.1. The Three-Way Classification

We label the factors in the three-way classification A, B, and C; the factors have a, b, and c categories, consecutively. A dependent-variable observation is represented by Y_{ijkl}, where the last subscript gives the index of the observation within its cell. The number of observations sampled in cell i, j, k is n_{ijk}; and μ_{ijk} is the population mean in this cell. Quantities such as $\mu_{\cdots}$, $\mu_{i..}$, and $\mu_{ij.}$ denote marginal means formed by averaging over the dotted subscripts.

The three-way ANOVA model is

$$\begin{aligned} Y_{ijkl} &= \mu_{ijk} + \varepsilon_{ijkl} \\ &= \mu + \alpha_{A(i)} + \alpha_{B(j)} + \alpha_{C(k)} + \alpha_{AB(ij)} + \alpha_{AC(ik)} \\ &\quad + \alpha_{BC(jk)} + \alpha_{ABC(ijk)} + \varepsilon_{ijkl} \end{aligned}$$

We make the usual linear-model assumptions about the errors ε_{ijkl}, and constrain all sets of parameters to sum to zero over every coordinate; for example

$$\sum_{i=1}^{a} \alpha_{A(i)} = 0$$

$$\sum_{i=1}^{a} \alpha_{AB(ij)} = \sum_{j=1}^{b} \alpha_{AB(ij)} = 0 \qquad \text{for all } i, j$$

$$\sum_{i=1}^{a} \alpha_{ABC(ijk)} = \sum_{j=1}^{b} \alpha_{ABC(ijk)} = \sum_{k=1}^{c} \alpha_{ABC(ijk)} = 0 \qquad \text{for all } i, j, k$$

Notice that to avoid the proliferation of symbols we have introduced a new and easily extended notation for model parameters: The first set of subscripts indicates the factors to which a parameter pertains, while the parenthetical subscripts index factor categories.

The three-way ANOVA model includes parameters for main effects, for *two-way* (or *first-order*) *interactions* between each pair of factors, and for *three-way* (or *second-order*) *interactions* among all three factors. The two-way interactions have the same interpretation as in two-way ANOVA. If, for instance, A and B interact, then the effect of either factor on the dependent variable varies across the categories of the other factor. Similarly, if the ABC interaction is nonzero, then the joint effect of any pair of factors (say, A and B) varies across the categories of the third factor (C).

In interpreting effects in three-way ANOVA, we once more appeal to the principle of marginality. Thus main effects (e.g., of A) are not interpreted if they are marginal to non-null interactions (AB, AC, or ABC). Likewise, *lower-order* interactions (e.g., AB) are not interpreted if they are marginal to non-null *higher-order* interactions (ABC): If the joint effects of A and B are different in different categories of factor C, then it is not generally sensible to speak of the unconditional AB effects.

The parametric equation for the three-way ANOVA model, showing the row basis of the reduced design matrix, is given in equation (2.29) (page 127), for $a = 2$, $b = 2$, and $c = 3$. For ease of reference, each column of $\mathbf{X}_B$ is labeled with the subscript of the parameter to which it corresponds. The following points are noteworthy:

1. The 12 cell means are expressed in terms of an equal number of linearly independent model parameters, thus underscoring the point that three-way interactions may be required to account for the pattern of cell means. More generally in the three-way classification, there are abc cells and $1 + (a-1) + (b-1) + (c-1) + (a-1)(b-1) + (a-1)(c-1) + (b-1)(c-1) + (a-1)(b-1)(c-1) = abc$ independent parameters.
2. Degrees of freedom for a set of effects correspond, as usual, to the number of linearly independent parameters in the set. There are, for example, $a - 1$ degrees of freedom for the A main effects, $(a-1)(b-1)$ degrees of freedom for the AB interactions, and $(a-1)(b-1)(c-1)$ degrees of freedom for the ABC interactions.

$$
\begin{array}{c}
\quad (\mu) \quad A(1) \quad B(1) \quad C(1) \quad C(2) \quad AB(11) \quad AC(11) \quad AC(12) \quad BC(11) \quad BC(12) \quad ABC(111) \quad ABC(112)
\end{array}
$$

$$
\begin{bmatrix} \mu_{111} \\ \mu_{112} \\ \mu_{113} \\ \mu_{121} \\ \mu_{122} \\ \mu_{123} \\ \mu_{211} \\ \mu_{212} \\ \mu_{213} \\ \mu_{221} \\ \mu_{222} \\ \mu_{223} \end{bmatrix}
=
\begin{bmatrix}
1 & 1 & 1 & 1 & 0 & 1 & 1 & 0 & 1 & 0 & 1 & 0 \\
1 & 1 & 1 & 0 & 1 & 1 & 0 & 1 & 0 & 1 & 0 & 1 \\
1 & 1 & 1 & -1 & -1 & 1 & -1 & -1 & -1 & -1 & -1 & -1 \\
1 & 1 & -1 & 1 & 0 & -1 & 1 & 0 & -1 & 0 & -1 & 0 \\
1 & 1 & -1 & 0 & 1 & -1 & 0 & 1 & 0 & -1 & 0 & -1 \\
1 & 1 & -1 & -1 & -1 & -1 & -1 & -1 & 1 & 1 & 1 & 1 \\
1 & -1 & 1 & 1 & 0 & -1 & -1 & 0 & 1 & 0 & -1 & 0 \\
1 & -1 & 1 & 0 & 1 & -1 & 0 & -1 & 0 & 1 & 0 & -1 \\
1 & -1 & 1 & -1 & -1 & -1 & 1 & 1 & -1 & -1 & 1 & 1 \\
1 & -1 & -1 & 1 & 0 & 1 & -1 & 0 & -1 & 0 & 1 & 0 \\
1 & -1 & -1 & 0 & 1 & 1 & 0 & -1 & 0 & -1 & 0 & 1 \\
1 & -1 & -1 & -1 & -1 & 1 & 1 & 1 & 1 & 1 & -1 & -1
\end{bmatrix}
\begin{bmatrix} \mu \\ \alpha_{A(1)} \\ \alpha_{B(1)} \\ \alpha_{C(1)} \\ \alpha_{C(2)} \\ \alpha_{AB(11)} \\ \alpha_{AC(911)} \\ \alpha_{AC(12)} \\ \alpha_{BC(11)} \\ \alpha_{BC(12)} \\ \alpha_{ABC(111)} \\ \alpha_{ABC(112)} \end{bmatrix}
$$

$$
\boldsymbol{\mu} = \mathbf{X}_B \boldsymbol{\beta}_R \tag{2.29}
$$

3. Main-effect regressors in the reduced design matrix are determined according to the rules employed in two-way ANOVA (Section 2.3.3).
4. Interaction regressors may be simply constructed by taking all products of main effects marginal to the interaction in question. Thus, the AC columns in equation (2.29) consist of the products $A(1) \times C(1)$ and $A(1) \times C(2)$. Likewise, the ABC columns consist of the products $A(1) \times B(1) \times C(1)$ and $A(1) \times B(1) \times C(2)$.

Solving for the constrained parameters in terms of population means produces the following results:

$$\mu = \mu_{...}$$

$$\alpha_{A(i)} = \mu_{i..} - \mu_{...}$$

$$\alpha_{AB(ij)} = \mu_{ij.} - \mu - \alpha_{A(i)} - \alpha_{B(j)}$$

$$= \mu_{ij.} - \mu_{i..} - \mu_{.j.} + \mu_{...}$$

$$\alpha_{ABC(ijk)} = \mu_{ijk} - \mu - \alpha_{A(i)} - \alpha_{B(j)} - \alpha_{C(k)} - \alpha_{AB(ij)} - \alpha_{AC(ik)} - \alpha_{BC(jk)}$$

$$= \mu_{ijk} - \mu_{ij.} - \mu_{i.k} - \mu_{.jk} + \mu_{i..} + \mu_{.j.} + \mu_{..k} - \mu_{...}$$

(The patterns for $\alpha_{B(j)}$, $\alpha_{C(k)}$, $\alpha_{AC(ik)}$, and $\alpha_{BC(jk)}$ are similar, and are omitted for brevity.) As in two-way analysis of variance, therefore, the null hypothesis H_0: all $\alpha_{A(i)} = 0$ is equivalent to H_0: $\mu_{1..} = \mu_{2..} = \cdots = \mu_{a..}$; and the hypothesis H_0: all $\alpha_{AB(ij)} = 0$ is equivalent to H_0: $\mu_{ij.} - \mu_{i'j.} = \mu_{ij'.} - \mu_{i'j'.}$ for all i, i' and j, j'. Likewise, some algebraic manipulation shows that the null hypothesis H_0: all $\alpha_{ABC(ijk)} = 0$ is equivalent to

$$H_0: (\mu_{ijk} - \mu_{i'jk}) - (\mu_{ij'k} - \mu_{i'j'k}) = (\mu_{ijk'} - \mu_{i'jk'}) - (\mu_{ij'k'} - \mu_{i'j'k'})$$

$$\text{for all } i, i'; j, j'; \text{ and } k, k' \quad (2.30)$$

Notice that the second-order differences in hypothesis (2.30) are equal when the pattern of AB interactions is invariant across categories of factor C—a reasonable extension of the notion of no interaction to three factors. Rearranging the terms in (2.30) produces similar results for AC and BC: Interaction, as always, is a symmetric concept. As in two-way ANOVA, this simple relationship between model parameters and population means is dependent upon the restrictions imposed on the over-parameterized model.

Tests for parameters in the three-way ANOVA model may be constructed according to the incremental sum of squares approach. A general ANOVA table, using the SS notation of Section 2.3.4 and showing alternative tests for main effects and lower-order interactions, is given in Table 2.14. Once more, for compactness, only tests involving factor A are shown. Note that a main-effect

TABLE 2.14. Three-Way ANOVA Table

Source	df	Sum of Squares	MS[a]	F[b]	H_0
A	$a-1$	$SS(\alpha_A \mid \alpha_B, \alpha_C, \alpha_{AB}\, \alpha_{AC}, \alpha_{BC}, \alpha_{ABC})$			$\alpha_A = 0$
		$SS(\alpha_A \mid \alpha_B, \alpha_C, \alpha_{BC})$			$\alpha_A = 0 \mid \alpha_{AB} = \alpha_{AC} = \alpha_{ABC} = 0$
AB	$(a-1)(b-1)$	$SS(\alpha_{AB} \mid \alpha_A, \alpha_B, \alpha_C, \alpha_{AC}, \alpha_{BC}, \alpha_{ABC})$			$\alpha_{AB} = 0$
		$SS(\alpha_{AB} \mid \alpha_A, \alpha_B, \alpha_C, \alpha_{AC}, \alpha_{BC})$			$\alpha_{AB} = 0 \mid \alpha_{ABC} = 0$
ABC	$(a-1)(b-1)(c-1)$	$SS(\alpha_{ABC} \mid \alpha_A, \alpha_B, \alpha_C, \alpha_{AB}, \alpha_{AC}, \alpha_{BC})$			$\alpha_{ABC} = 0$
Residuals	$n - abc$	$TSS - SS(\alpha_A, \alpha_B, \alpha_C, \alpha_{AB}, \alpha_{AC}, \alpha_{BC}, \alpha_{ABC})$	RMS		
Total	$n-1$	TSS			

[a] MS = SS/df.
[b] F = MS/RMS.

hypothesis such as H_0: all $\alpha_{A(i)} = 0$ is interesting even when the BC interactions are present, since A is not marginal to BC; if these interactions are absent, however, we may then legitimately use $SS(\alpha_A \mid \alpha_B, \alpha_C)$ for testing the A main effects, so long as the AB, AC, and ABC interactions are also negligible. Data for an illustrative three-way ANOVA are given in Problem 2.12.

2.4.2. Higher-Order Classifications

Extension of the analysis of variance to more than three factors is algebraically and computationally straightforward. The general k-way classification may be described by a model containing terms for every combination of factors; the highest-order term, therefore, is for the k-way interactions. If the k-way interactions are nonzero, then the joint effects of any $k-1$ factors vary across the categories of the remaining factor. In general, therefore, we may be guided by the principle of marginality in interpreting effects.

Our consideration of the three-way classification revealed that second-order interactions are conceptually fairly complex. The even greater complexity of higher-order interactions can make their substantive interpretation exceedingly difficult. Yet, at times we may expect to observe a high-order interaction of a particular sort, as when a specific *combination* of traits predisposes individuals to act in a certain manner. In any event, it is common to find that high-order interactions are not statistically significant or that they are negligibly small relative to other effects.

There is, moreover, no rule of data analysis that requires us to fit and test all possible interactions. In working with higher-way classifications, we may limit consideration to effects that are of theoretical interest, or at least to effects that are substantively interpretable. It is therefore not uncommon for researchers to fit models containing only main effects:

$$Y_{ij\ldots lm} = \mu + \alpha_{A(i)} + \alpha_{B(j)} + \cdots + \alpha_{K(l)} + \varepsilon_{ij\ldots lm} \tag{2.31}$$

This approach has been called *multiple-classification analysis* or *MCA*, though the term is equally descriptive of any ANOVA model fit to the k-way classification. In a similar spirit a researcher might entertain models including only main effects and two-way interactions.

2.4.3. Empty Cells in ANOVA

As the number of factors increases, the number of cells grows at a much faster rate: For k dichotomous factors, for example, the number of cells is 2^k. One consequence of this proliferation is that some combinations of independent-variable categories may not be observed; that is, certain cells may be empty.

A general rule of thumb is that we may safely use our usual approach to estimation and testing in the presence of empty cells, as long as the *marginal* frequency tables corresponding to the effects that we examine contain no empty cells, and as long as other effects are nil. This rule, therefore, never covers the k-way interactions when empty cells are present in a k-way classification. It may in fact be possible to estimate and test effects not covered by the rule, but determining whether tests are possible and specifying meaningful hypotheses to be tested become more complex in this instance. For details, see, for example, Searle (1971: 318–324), Hocking and Speed (1975: 711–712), and Speed, Hocking, and Hackney (1978: 110–111).

For purposes of illustration, we develop a very simple example, employing a 2×2 classification. Suppose that $n_{22} = 0$, but that all other cells contain data. Then, deleting the population mean μ_{22}, and imposing the usual constraints on the model parameters, the parametric equation for the complete two-way ANOVA model becomes

$$\begin{pmatrix} \mu_{11} \\ \mu_{12} \\ \mu_{21} \end{pmatrix} = \begin{pmatrix} 1 & 1 & 1 & 1 \\ 1 & 1 & -1 & -1 \\ 1 & -1 & 1 & -1 \end{pmatrix} \begin{pmatrix} \mu \\ \alpha_1 \\ \beta_1 \\ \gamma_{11} \end{pmatrix}$$

Since there are four parameters in the constrained model and just three observed cell means, the parameters of the model are not uniquely determined.

Now imagine that we may reasonably specify zero two-way interactions for these data. Then, according to our rule of thumb, we should be able to test the R and C main effects, since there are observations at all levels of each factor;

for example, there are $n_{11} + n_{12}$ observations at R_1 and n_{21} observations at R_2. The parametric equation for the main-effects model is

$$\begin{pmatrix} \mu_{11} \\ \mu_{12} \\ \mu_{21} \end{pmatrix} = \begin{pmatrix} 1 & 1 & 1 \\ 1 & 1 & -1 \\ 1 & -1 & 1 \end{pmatrix} \begin{pmatrix} \mu \\ \alpha_1 \\ \beta_1 \end{pmatrix}$$

Solving for the parameters of this model produces[9]

$$\mu = \frac{\mu_{12} + \mu_{21}}{2}$$

$$\alpha_1 = \frac{\mu_{11} - \mu_{21}}{2}$$

$$\beta_1 = \frac{\mu_{11} - \mu_{12}}{2}$$

These results make sense, for in the *absence* of interaction: (1) μ_{12} and μ_{21} are balanced with respect to both sets of main effects and therefore their average serves as a suitable definition of the grand mean; (2) the difference $\mu_{11} - \mu_{21}$ gives the effect of changing R while C is held constant; and (3) $\mu_{11} - \mu_{12}$ gives the partial effect of C holding R constant.

The advice given in Section 2.3.7 regarding care in the use of computer programs for the analysis of variance of unbalanced data applies even more urgently when there are empty cells.

PROBLEMS

2.10.† The Geometry of Effects in Three-Way ANOVA: Contrived parameters for a three-way ANOVA model are given in Table 2.15. Let $\mu = 0$. Use these values to construct population cell means for each of the following models (simply sum the parameters that pertain to each of the 12 cells of the design):

(i) $\mu_{ijk} = \mu + \alpha_{A(i)} + \alpha_{B(j)} + \alpha_{C(k)}$
(ii) $\mu_{ijk} = \mu + \alpha_{A(i)} + \alpha_{B(j)} + \alpha_{C(k)} + \alpha_{AC(ik)}$
(iii) $\mu_{ijk} = \mu + \alpha_{A(i)} + \alpha_{B(j)} + \alpha_{C(k)} + \alpha_{AB(ij)} + \alpha_{AC(ik)} + \alpha_{BC(jk)}$
(iv) the full model

(a) Draw a graph of each set of means, placing factor C on the horizontal axis. Use different lines (solid and dashed) for the

[9] The 2×2 classification with one empty cell is particularly simple because the number of parameters in the main-effects model is equal to (i.e., no fewer than) the number of observed cell means. This is not generally the case, making a general analysis considerably more complex.

levels of factor A, and different symbols (circle and square) for the levels of factor B. Note that there will be four connected profiles of means on each plot, one profile for each combination of categories of A and B across the levels of C. Attempt to interpret the graphs in terms of the effects included in the several models.

(b)* Using the two sets of means generated from models (iii) and (iv), for each set plot the six differences $\mu_{i1k} - \mu_{i2k}$ by the categories of factors A and C. Can you account for the different patterns of these two graphs?

2.11.† More on Adjusted Means (also see Problems 2.4 and 2.14): Consider the main-effects model for the k-way classification, given in equation

TABLE 2.15. Three-Way ANOVA Parameters for Illustrations

		$\alpha_{A(i)}$	
	A_1		A_2
	2		−2
		$\alpha_{B(j)}$	
	B_1		B_2
	−3		3
		$\alpha_{C(k)}$	
	C_1	C_2	C_3
	1	−3	2
		$\alpha_{AB(ij)}$	
	B_1		B_2
A_1	−2		2
A_2	2		−2
		$\alpha_{AC(ik)}$	
	C_1	C_2	C_3
A_1	1	1	−2
A_2	−1	−1	2
		$\alpha_{BC(jk)}$	
	C_1	C_2	C_3
B_1	0	3	−3
B_2	0	−3	3
		$\alpha_{ABC(ijk)}$	
	C_1	C_2	C_3
A_1B_1	1	−2	1
A_1B_2	−1	2	−1
A_2B_1	−1	2	−1
A_2B_2	1	−2	1

(2.31) and reproduced here in expectation form:

$$\mu_{ij\ldots l} \equiv E\left(Y_{ij\ldots lm}\right) = \mu + \alpha_{A(i)} + \alpha_{B(j)} + \cdots + \alpha_{K(l)}$$

(a) Show that if we constrain each set of effects to sum to zero, the population marginal mean for category i of factor A is given by $\mu_{i\ldots\ldots} = \mu + \alpha_{A(i)}$.

(b) We define $\tilde{Y}_{i\ldots\ldots} = M + A_{A(i)}$ to be the *adjusted mean* in category i of factor A. How is this quantity to be interpreted?

(c) Does the definition of adjusted means in part (b) depend fundamentally upon the constraint that each set of effects sums to zero?

(d) Can the idea of an adjusted mean be extended to ANOVA models that include interactions?

2.12.†# ANOVA With Equal Cell Frequencies: In higher-way ANOVA, as in two-way ANOVA, when cell frequencies are equal, the sum of squares for each set of effects may be calculated directly from the parameter estimates for the full model, or equivalently in terms of the cell and marginal means. To get the sum of squares for a particular set of effects, we merely square the parameter estimate associated with each cell, sum over all cells, and multiply by the common cell frequency, n^*. For example, for a balanced three-way ANOVA,

$$\text{SS}(\boldsymbol{\alpha}_{AB}) = n^* \sum_{i=1}^{a} \sum_{j=1}^{b} \sum_{k=1}^{c} A_{AB(ij)}^2$$

$$= n^* c \sum_i \sum_j A_{AB(ij)}^2$$

$$= n^* c \sum_i \sum_j \left(\overline{Y}_{ij.} - \overline{Y}_{i..} - \overline{Y}_{.j.} + \overline{Y}_{...}\right)^2$$

(a) Write out similar expression for $\text{SS}(\boldsymbol{\alpha}_A)$ and $\text{SS}(\boldsymbol{\alpha}_{ABC})$ in three-way ANOVA. Show that $\text{RSS} = (n^* - 1)\Sigma_i\Sigma_j\Sigma_k S_{ijk}^2$, where S_{ijk}^2 is the variance in cell i, j, k of the design.

(b) Table 2.16 shows the results of an experiment on interpersonal attraction reported by Riordan, Quigley-Fernandez, and Tedeschi (1982). Subjects in the study interacted with an experimenter's confederate whose attitudes were manipulated to be either similar or dissimilar to those of the subjects. In the course of the study, it was arranged for some subjects to request the confederate's help in completing a task, while other subjects did not ask for help. Finally, the confederate provided help to some subjects but not to others. These three factors combine to produce eight experimental conditions. Nine subjects were assigned to each condition. (Actually, two conditions contained eight subjects, but we shall disregard this complication.) At the beginning and again at the end of the study, subjects rated their attraction to the confederate on a two-item scale, with possible composite scores ranging from two through 14. Table 2.16 reports means and standard deviations for changes in attraction over the course of the study. Riordan, Quigley-Fernandez, and Tedeschi (1982: 364) make the following predictions with regard to these changes:

> Changes in attraction should be associated with an interaction of all three factors. Subjects should show an increment in attraction when help is provided by a dissimilar other and the increment should be greater when the help was not requested than when it was requested. A decrement in attraction should occur when help is not given by a similar other and the decrement should be greater when help was requested. No change should occur when a dissimilar other does not provide help and it was not requested, and very little change in a negative direction when the dissimilar other does not provide help and it was requested. No change is also predicted when a similar other provides help and it was requested, and very little change in a positive direction when it was not requested.

(i) Graph the cell means from Table 2.16. Do the results of the study appear to confirm the authors' predictions?

TABLE 2.16. Means and Standard Deviations of Changes in Attraction by Experimental Conditions

			Attraction Change	
Attitude Similarity	Help Requested	Help Provided	Mean	Standard Deviation
Similar	Yes	Yes	0.22	1.2
		No	−4.62	1.3
	No	Yes	−0.12	1.2
		No	−2.56	1.4
Dissimilar	Yes	Yes	1.44	1.9
		No	−2.22	2.3
	No	Yes	2.10	2.1
		No	−1.98	1.3

Source: Reprinted with permission from Riordan, Quigley-Fernandez, and Tedeschi (1982: Table 1), and personal communication from C. Riordan.

(ii) Using the results of part (a), compute an analysis of variance for the data in Table 2.16. What conclusions would you draw?

2.5. ANALYSIS OF COVARIANCE

Analysis of covariance (*ANOCOVA*) is a term used to describe linear models with both qualitative and quantitative independent variables. The method therefore is equivalent to dummy-variable regression, covered in Section 2.1, although the ANOCOVA model is parameterized differently from the dummy-regression model. In analysis of covariance, an ANOVA formulation is used for the main effects and interactions of the qualitative independent variables (or *factors*), and quantitative independent variables (or *covariates*) are generally expressed in mean-deviation form. Although it is possible to construct ANOCOVA models that incorporate interactions between factors and covariates (as we shall show presently), these models typically include only additive effects of covariates.

Since analysis of covariance is a simple extension of the material covered previously in this chapter, we shall present a single, brief example including two factors and two covariates. In this instance, the ANOCOVA model is written in the following manner:

$$Y_{ijk} = \mu + \alpha_i + \beta_j + \gamma_{ij} + \delta_1\left(X_{1ijk} - \overline{X}_1\right) + \delta_2\left(X_{2ijk} - \overline{X}_2\right) + \varepsilon_{ijk} \tag{2.32}$$

where the factor main-effect and interaction parameters α_i, β_j, and γ_{ij} follow the usual ANOVA constraints, and δ_1 and δ_2 are the slopes for the covariates X_1 and X_2. An analysis-of-variance table for testing effects in this model is shown in Table 2.17.

TABLE 2.17. ANOVA Table for Analysis of Covariance

Source	Sum of Squares	*df*	MS[a]	F[b]	H_0
R	$SS(\boldsymbol{\alpha} \mid \boldsymbol{\beta}, \boldsymbol{\gamma}, \delta_1, \delta_2)$	$r-1$			$\boldsymbol{\alpha} = \mathbf{0}$
C	$SS(\boldsymbol{\beta} \mid \boldsymbol{\alpha}, \boldsymbol{\gamma}, \delta_1, \delta_2)$	$c-1$			$\boldsymbol{\beta} = \mathbf{0}$
RC	$SS(\boldsymbol{\gamma} \mid \boldsymbol{\alpha}, \boldsymbol{\beta}, \delta_1, \delta_2)$	$(r-1)(c-1)$			$\boldsymbol{\gamma} = \mathbf{0}$
X_1	$SS(\delta_1 \mid \boldsymbol{\alpha}, \boldsymbol{\beta}, \boldsymbol{\gamma}, \delta_2)$	1			$\delta_1 = 0$
X_2	$SS(\delta_2 \mid \boldsymbol{\alpha}, \boldsymbol{\beta}, \boldsymbol{\gamma}, \delta_1)$	1			$\delta_2 = 0$
Residuals	TSS $- SS(\boldsymbol{\alpha}, \boldsymbol{\beta}, \boldsymbol{\gamma}, \delta_1, \delta_2)$	$n - rc - 2$	RMS		
Total	TSS	$n-1$			

[a] MS = SS/*df*.
[b] F = MS/RMS.

To incorporate two-way interactions between factors and covariates, we may write the model

$$Y_{ijk} = \mu + \alpha_i + \beta_j + \gamma_{ij} + (\delta_{1i} + \zeta_{1j})(X_{1ijk} - \overline{X}_1)$$

$$+ (\delta_{2i} + \zeta_{2j})(X_{2ijk} - \overline{X}_2) + \varepsilon_{ijk} \qquad (2.33)$$

Note that model (2.32) is a restricted version of model (2.33), obtained when interactions between factors and covariates are nil; that is, when $\delta_{1i} = \delta_{1i'}$, $\delta_{2i} = \delta_{2i'}$, $\zeta_{1j} = \zeta_{1j'}$, and $\zeta_{2j} = \zeta_{2j'}$, for all i, i' and j, j'. Tests for these interactions may therefore be constructed by the incremental sum of squares approach.

Finally, the following model incorporates three-way interactions among the two factors and each covariate:

$$Y_{ijk} = \mu + \alpha_i + \beta_j + \gamma_{ij} + \delta_{1ij}(X_{1ijk} - X_1) + \delta_{2ij}(X_{2ijk} - \overline{X}) + \varepsilon_{ijk} \qquad (2.34)$$

δ_{1ij} and δ_{2ij} represent the slopes for the regression of Y on X_1 and X_2 *within* cell i, j. A test of the null hypothesis of no three-way interaction follows from the observation that model (2.33) is a restricted version of model (2.34).

PROBLEMS

2.13. Reanalyze the Moore and Krupat data given in Table 2.10, treating authoritarianism (i.e., the F-scale scores) as a covariate. Be sure to test for interaction between authoritarianism and partner's status.

2.14.† Adjusted Means, Concluded: Recall the discussions of *adjusted means* in Problems 2.4 and 2.11.

(a) How can the analysis-of-covariance model given in equation (2.32) be used to compute adjusted cell means for the factors R and C?

(b) In computing adjusted means, is anything gained by expressing the covariates as deviations from their means rather than as raw scores?

(c) If the interactions γ_{ij} are deleted from the model, how may we calculate adjusted means for the categories of R and C?

Further information on adjusted means may be found in Searle, Speed, and Milliken (1980).

3

Diagnosing and Treating Linear-Model Problems

Chapters 1 and 2 introduced powerful and general methods for specifying, fitting, and testing linear models. The mechanical application of these methods is inadvisable, however, since in practice linear models and least-squares fitting are subject to a variety of difficulties. A careful investigator must, therefore, remain sensitive to the data as well as to the substantive concerns—such as theory and hypothesis—that motivate research.

This chapter develops methods for detecting linear-model problems and describes techniques for dealing with difficulties that are revealed. Section 3.1 takes up the problem of highly collinear regressors, showing how to assess the presence of collinearity and the harm that collinearity produces, and describing strategies for estimating linear models when collinearity is present. Section 3.2 explains how residuals can be used to diagnose model deficiencies such as nonlinearity and heteroscedastic errors. The detection of outliers and influential data is also dealt with here. The use of data transformations for correcting linear-model difficulties is discussed in Section 3.3. Section 3.4 takes up the related topic of nonlinear models and develops the method of nonlinear least squares. The final section of the chapter deals briefly with two difficulties often encountered in the analysis of social-survey data: missing data and statistical inference for complex sampling designs.

3.1. COLLINEARITY

We touched upon the issue of collinear regressors at several points in Chapter 1. Recall from that discussion that when regressors are perfectly linearly related, the least-squares regression coefficients are not uniquely defined. Less-than-perfect collinearity causes regression estimates to be unstable: Regression coefficients have large standard errors; coefficients very different from the least-squares estimates produce residual sums of squares near the (least-squares) minimum; and small changes in the data can effect large changes in the least-squares estimates.

We shall first examine the problem of collinearity in the simplified context of the two-independent-variable multiple-regression model. Then we shall develop the method of principal components, which permits us to describe the correlational structure of a number of variables—here, the independent variables in a regression. Using principal components and related statistics, we shall learn how to detect the presence of collinearity in regression analysis; how to assess which regression coefficients are adversely affected by collinearity and to what degree; and how to determine the specific collinear relations that are present in the data. Finally, we shall briefly describe methods that may yield improved estimates in the presence of serious collinearity.

The term *multicollinearity* is often used to denote collinear relations among the regressors in a linear model. Although we shall not employ this terminology here, it does serve to emphasize the fact that collinearity is not restricted to pairwise relationships between regressors.

3.1.1. Two Independent Variables

Figure 3.1 illustrates the effect of collinearity on estimation when there are two independent variables in a regression. (The black dots in this figure represent data points, while the white dots represent fitted values lying in the regression plane; the **X**'s show the projections of the data points onto the X_1, X_2 plane.)

In Figure 3.1(a), the correlation between the independent variables X_1 and X_2 is slight, as indicated by the broad scatter of observations in the X_1, X_2 plane. The least-squares regression plane, also shown in this figure, therefore has a firm base of support: Small changes in the regression plane cause relatively large increases in the residual sum of squares. In Figure 3.1(b), X_1 and X_2 are perfectly collinear. Because the independent-variable observations are not spread out in the X_1, X_2 plane, the regression plane is supported only by a line, and the plane can tip about this line without changing the residual sum of squares; thus, the least-squares regression plane is not uniquely defined. Finally, in Figure 3.1(c), the linear relationship between X_1 and X_2 is strong, though not perfect: The support afforded to the regression plane is tenuous, so that the plane can be tipped without causing large increases in the residual sum of squares.

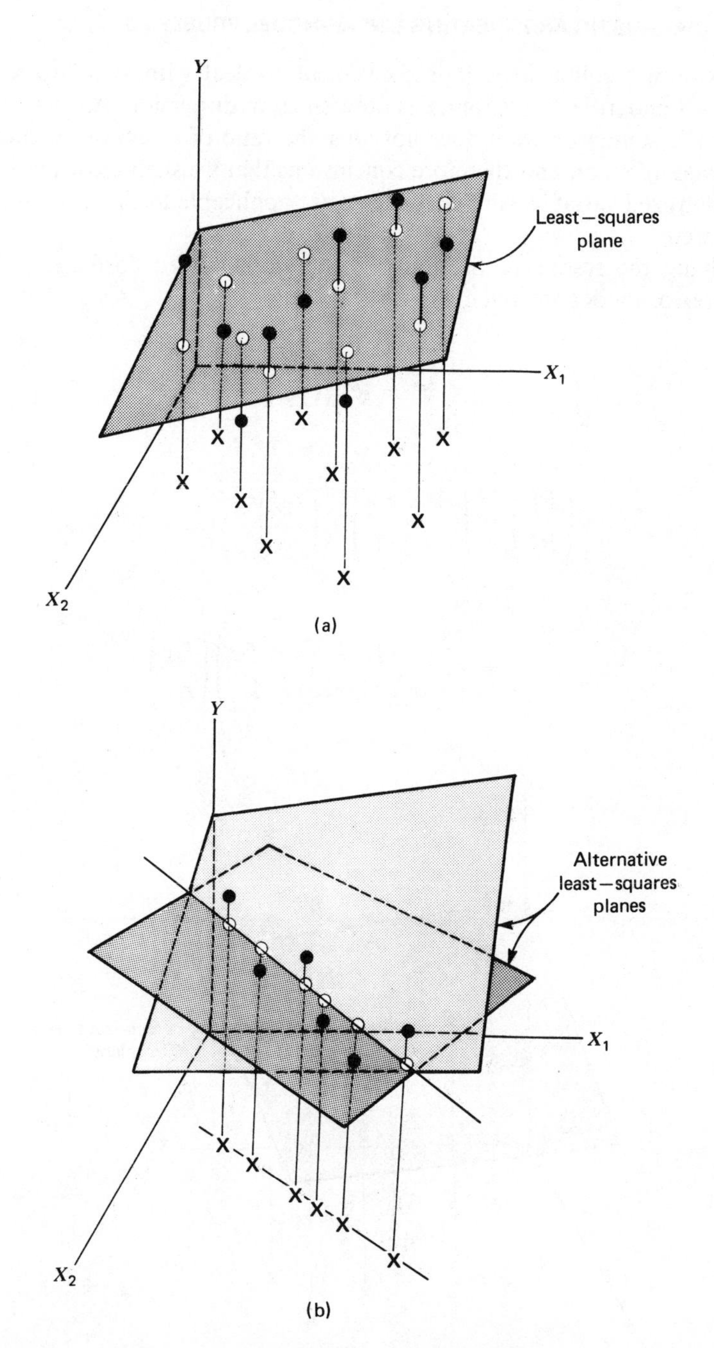

FIGURE 3.1. The effect of collinearity on estimation. (a) Small correlation between X_1 and X_2: regression plane well supported. (b) Perfect correlation between X_1 and X_2: regression plane not uniquely defined. (c) Strong correlation between X_1 and X_2: regression plane defined but not well supported. (Adapted with permission from Belsley, Kuh, and Welsch, 1980: 87–88.)

In exploring collinearity, it is convenient to deal with standardized variables, for standardization eliminates differences in dispersion. As mentioned in Section 1.5, standardization does not alter the ratio of a regression coefficient to its standard error, and therefore conclusions that we shall draw on the basis of standardized variables and coefficients are applicable to the unstandardized case as well.

Applying the results of Section 1.5, the standardized coefficients for the two-regressor model are given by

$$\mathbf{b}^* = \mathbf{R}_{XX}^{-1}\mathbf{r}_{Xy}$$

$$\begin{pmatrix} B_1^* \\ B_2^* \end{pmatrix} = \begin{pmatrix} 1 & r_{12} \\ r_{12} & 1 \end{pmatrix}^{-1} \begin{pmatrix} r_{1Y} \\ r_{2Y} \end{pmatrix}$$

$$= \frac{1}{1 - r_{12}^2} \begin{pmatrix} 1 & -r_{12} \\ -r_{12} & 1 \end{pmatrix} \begin{pmatrix} r_{1Y} \\ r_{2Y} \end{pmatrix}$$

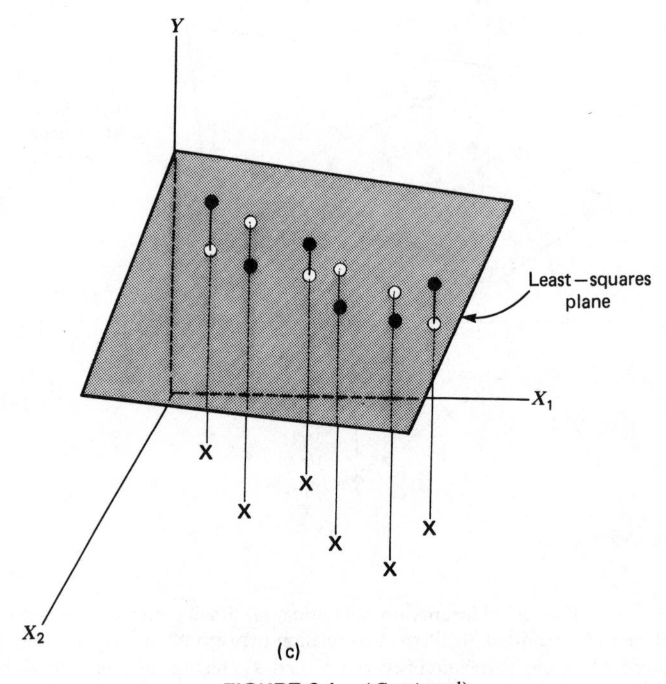

(c)

FIGURE 3.1. *(Continued).*

with covariance matrix

$$V(\mathbf{b}^*) = \frac{\sigma_\varepsilon^{*2}}{n-1}\mathbf{R}_{XX}^{-1}$$

$$V\begin{pmatrix} B_1^* \\ B_2^* \end{pmatrix} = \frac{\sigma_\varepsilon^{*2}}{n-1}\begin{pmatrix} 1 & r_{12} \\ r_{12} & 1 \end{pmatrix}^{-1}$$

$$= \frac{\sigma_\varepsilon^{*2}}{n-1} \times \frac{1}{1-r_{12}^2}\begin{pmatrix} 1 & -r_{12} \\ -r_{12} & 1 \end{pmatrix}$$

Thus the sampling variance for either standardized regression coefficient is

$$V\left(B_j^*\right) = \frac{\sigma_\varepsilon^{*2}}{n-1} \times \frac{1}{1-r_{12}^2} \tag{3.1}$$

For orthogonal data, the second factor on the right of equation (3.1) is unity, and as the squared correlation between X_1 and X_2 increases, this term grows larger. When $r_{12}^2 = 1$, the sampling variance becomes infinite, reflecting the singularity of $\mathbf{R}_{XX}$.

The quantity $1/(1 - r_{12}^2)$, called the *variance-inflation factor* (*VIF*), shows the degree to which the precision of estimation is degraded by collinearity. This result is intuitively reasonable, since r_{12}^2 expresses the strength of linear relationship between the independent variables. In the two-regressor model, the variance-inflation factors for the independent variables are identical. We shall show in Section 3.1.3 that when there are more regressors, the variance-inflation factors, though still functions of the correlations among the independent variables, generally differ from one another.

3.1.2. Principal Components

The method of principal components, due originally to K. Pearson and H. Hotelling, provides a useful representation of the correlational structure of a set of variables. We develop the method briefly here; more complete accounts may be obtained from texts on multivariate statistics (such as Morrison, 1976: Ch. 8). Because the material in this section is relatively complex, the section ends with a summary; the first-time reader may wish to pass lightly over most of the section and refer primarily to the summary and to the two-variable case, which is treated near the end.

We begin with the vectors of standardized regressors $\mathbf{z}_1, \mathbf{z}_2, \ldots, \mathbf{z}_k$. As we shall see, the *principal components* $\mathbf{w}_1, \mathbf{w}_2, \ldots, \mathbf{w}_p$ provide an orthogonal basis for the regressor subspace. The first principal component, $\mathbf{w}_1$, is oriented so as to account for maximum collective variation in the $\mathbf{z}_j$; the second principal component, $\mathbf{w}_2$, is orthogonal to $\mathbf{w}_1$ and, under the restriction of orthogonality, is oriented to account for maximum remaining variation in the $\mathbf{z}_j$; $\mathbf{w}_3$ is orthogonal to both $\mathbf{w}_1$ and $\mathbf{w}_2$; and so on. There are therefore as many principal components as there are linearly independent regressors; that is, $p = \text{rank}(\mathbf{Z}_X)$. Although the method of principal components is general, we shall assume through most of our discussion that the regressors are not perfectly collinear and, consequently, that $p = k$. Each principal component is scaled so that its variance is equal to the combined regressor variance for which it accounts.

Since the principal components lie in the regressor subspace, each is a linear combination of the regressors. Thus, the first principal component may be written

$$\underset{(n\times 1)}{\mathbf{w}_1} = A_{11}\mathbf{z}_1 + A_{21}\mathbf{z}_2 + \cdots + A_{k1}\mathbf{z}_k$$
$$= \underset{(n\times k)}{\mathbf{Z}_X}\ \underset{(k\times 1)}{\mathbf{a}_1}$$

The variance of the first component is

$$S_{W_1}^2 = \frac{1}{n-1}\mathbf{w}_1'\mathbf{w}_1 = \frac{1}{n-1}\mathbf{a}_1'\mathbf{Z}_X'\mathbf{Z}_X\mathbf{a}_1 = \mathbf{a}_1'\mathbf{R}_{XX}\mathbf{a}_1$$

where, we recall, $\mathbf{R}_{XX} = [1/(n-1)]\mathbf{Z}_X'\mathbf{Z}_X$ is the correlation matrix of the regressors. We wish to maximize $S_{W_1}^2$, but to make maximization meaningful it is necessary to constrain the coefficients $\mathbf{a}_1$. In the absence of a constraint, $S_{W_1}^2$ may be made as large as we please simply by picking large coefficients. The normalizing constraint

$$\mathbf{a}_1'\mathbf{a}_1 = 1 \tag{3.2}$$

proves convenient, but any constraint of this form would do.[1]

We may maximize $S_{W_1}^2$ subject to the restriction (3.2) by employing a Lagrange multiplier[2] L_1, defining

$$F_1 \equiv \mathbf{a}_1'\mathbf{R}_{XX}\mathbf{a}_1 - L_1(\mathbf{a}_1'\mathbf{a}_1 - 1) \tag{3.3}$$

[1]Normalizing the coefficients so that $\mathbf{a}_1'\mathbf{a}_1 = 1$ causes the variance of the first principal component to be equal to the combined variance of the standardized regressors accounted for by this component, as will become clear presently.

[2]The method of *Lagrange multipliers*, named after the 18th-century French mathematician J. L. Lagrange, permits us to maximize or minimize a function of the form $f(\mathbf{x})$ under a constraint of the form $g(\mathbf{x}) = 0$. We define the function $h(\mathbf{x}, L) \equiv f(\mathbf{x}) - Lg(\mathbf{x})$, introducing the Lagrange multiplier L. Differentiating $h(\mathbf{x})$ with respect to $\mathbf{x}$ and L produces

$$\frac{\partial h(\mathbf{x}, L)}{\partial \mathbf{x}} = \frac{\partial f(\mathbf{x})}{\partial \mathbf{x}} - L\frac{\partial g(\mathbf{x})}{\partial \mathbf{x}}$$
$$\frac{\partial h(\mathbf{x}, L)}{\partial L} = -g(\mathbf{x})$$

Setting these partial derivatives to zero produces a system of equations which may be solved for $\mathbf{x}$ (and L). Since the equation system includes $g(\mathbf{x}) = 0$, a solution vector $\mathbf{x}$ satisfies the constraint. (This method was used in Problem 1.8.) We may introduce additional Lagrange multipliers to handle more than one constraint.

Differentiating equation (3.3) with respect to $\mathbf{a}_1$ we obtain

$$\frac{\partial F_1}{\partial \mathbf{a}_1} = 2\mathbf{R}_{XX}\mathbf{a}_1 - 2L_1\mathbf{a}_1$$

$$-\frac{\partial F_1}{\partial L_1} = \mathbf{a}_1'\mathbf{a}_1 - 1$$

Setting these partial derivatives to zero produces the equations

$$\begin{aligned} (\mathbf{R}_{XX} - L_1\mathbf{I}_k)\mathbf{a}_1 &= \mathbf{0} \\ \mathbf{a}_1'\mathbf{a}_1 &= 1 \end{aligned} \tag{3.4}$$

The first equation in (3.4) has nontrivial solutions for $\mathbf{a}_1$ only when $(\mathbf{R}_{XX} - L_1\mathbf{I}_k)$ is singular; that is, when $|\mathbf{R}_{XX} - L_1\mathbf{I}_k| = 0$. L_1, therefore, is an eigenvalue of $\mathbf{R}_{XX}$, and $\mathbf{a}_1$ is the corresponding eigenvector, scaled so that $\mathbf{a}_1'\mathbf{a}_1 = 1$.

There are, however, k solutions to equations (3.4) corresponding to the k eigenvalues and eigenvectors of $\mathbf{R}_{XX}$, so we must decide which solution to choose. From (3.4), we have $\mathbf{R}_{XX}\mathbf{a}_1 = L_1\mathbf{a}_1$. Consequently,

$$S_{W_1}^2 = \mathbf{a}_1'\mathbf{R}_{XX}\mathbf{a}_1 = L_1\mathbf{a}_1'\mathbf{a}_1 = L_1$$

Since our purpose is to maximize $S_{W_1}^2$ (subject to the constraint on $\mathbf{a}_1$), we must select the *largest* eigenvalue of $\mathbf{R}_{XX}$ to define the first principal component.

The second principal component is derived similarly, under the further restriction that it be orthogonal to the first. We have

$$\mathbf{w}_2 = A_{12}\mathbf{z}_1 + A_{22}\mathbf{z}_2 + \cdots + A_{k2}\mathbf{z}_k = \mathbf{Z}_X\mathbf{a}_2$$

with variance

$$S_{W_2}^2 = \mathbf{a}_2'\mathbf{R}_{XX}\mathbf{a}_2$$

We wish to maximize this variance subject to the normalizing constraint

$$\mathbf{a}_2'\mathbf{a}_2 = 1$$

and the orthogonality constraint

$$\mathbf{w}_1'\mathbf{w}_2 = 0 \tag{3.5}$$

But

$$\begin{aligned} \mathbf{w}_1'\mathbf{w}_2 &= \mathbf{a}_1'\mathbf{Z}_X'\mathbf{Z}_X\mathbf{a}_2 = (n-1)\mathbf{a}_1'\mathbf{R}_{XX}\mathbf{a}_2 \\ &= (n-1)L_1\mathbf{a}_1'\mathbf{a}_2 \end{aligned} \tag{3.6}$$

and therefore the orthogonality constraint (3.5) is equivalent to $\mathbf{a}_1'\mathbf{a}_2 = 0$ (since $L_1 \neq 0$). The constraints may be represented by Lagrange multipliers, L_2 and M_{21}; let

$$F_2 \equiv \mathbf{a}_2'\mathbf{R}_{XX}\mathbf{a}_2 - L_2(\mathbf{a}_2'\mathbf{a}_2 - 1) - M_{21}(\mathbf{a}_1'\mathbf{a}_2)$$

Differentiating this expression, we get

$$\frac{\partial F_2}{\partial \mathbf{a}_2} = 2\mathbf{R}_{XX}\mathbf{a}_2 - 2L_2\mathbf{a}_2 - M_{21}\mathbf{a}_1$$

$$-\frac{\partial F_2}{\partial L_2} = \mathbf{a}_2'\mathbf{a}_2 - 1$$

$$-\frac{\partial F_2}{\partial M_{21}} = \mathbf{a}_1'\mathbf{a}_2$$

Setting the partial derivatives to zero produces

$$\mathbf{R}_{XX}\mathbf{a}_2 - L_2\mathbf{a}_2 = \frac{M_{21}}{2}\mathbf{a}_1 \tag{3.7}$$

(together with the normalizing and orthogonality constraints). Then, multiplying equation (3.7) through on the left by $\mathbf{a}_1'$, we obtain

$$\mathbf{a}_1'\mathbf{R}_{XX}\mathbf{a}_2 - L_2\mathbf{a}_1'\mathbf{a}_2 = \frac{M_{21}}{2}\mathbf{a}_1'\mathbf{a}_1 \tag{3.8}$$

Both terms on the left-hand side of equation (3.8) are zero (from equation (3.6)), and thus M_{21} must be zero, since $\mathbf{a}_1'\mathbf{a}_1$ is one (i.e., nonzero). Equation (3.7) consequently reduces to

$$(\mathbf{R}_{XX} - L_2\mathbf{I}_k)\mathbf{a}_2 = \mathbf{0} \tag{3.9}$$

As in the case of the first principal component, therefore, L_2 is an eigenvalue of $\mathbf{R}_{XX}$, and $\mathbf{a}_2$ is the corresponding eigenvector scaled so that $\mathbf{a}_2'\mathbf{a}_2 = 1$. Because $S_{W_2}^2 = L_2$, and because we wish to make $S_{W_2}^2$ as large as possible, L_2 must be the *second-largest* eigenvalue of $\mathbf{R}_{XX}$.[3]

The remaining principal components follow in a similar manner. We order the eigenvalues of $\mathbf{R}_{XX}$ so that $L_1 \geq L_2 \geq \cdots \geq L_k > 0$.[4] The matrix of principal-component coefficients

$$\underset{(k \times k)}{\mathbf{A}} = [\mathbf{a}_1, \mathbf{a}_2, \ldots, \mathbf{a}_k]$$

[3]Although the largest eigenvalue (L_1) is also a root of equation (3.9), its use would violate the condition that $\mathbf{a}_2$ be orthogonal to $\mathbf{a}_1$.

[4]It is possible, but unlikely, that two or more eigenvalues of $\mathbf{R}_{XX}$ are equal. In this event, the orientation of the principal components corresponding to the equal eigenvalues is not unique, though the subspace spanned by these components is still well defined.

contains normalized eigenvectors of $\mathbf{R}_{XX}$. $\mathbf{A}$ is therefore orthonormal: $\mathbf{A}'\mathbf{A} = \mathbf{A}\mathbf{A}' = \mathbf{I}_k$. The principal components

$$\underset{(n \times k)}{\mathbf{W}} = \underset{(n \times k)}{\mathbf{Z}_X} \underset{(k \times k)}{\mathbf{A}} \tag{3.10}$$

have covariance matrix

$$\begin{aligned} \frac{1}{n-1}\mathbf{W}'\mathbf{W} &= \frac{1}{n-1}\mathbf{A}'\mathbf{Z}_X'\mathbf{Z}_X\mathbf{A} = \mathbf{A}'\mathbf{R}_{XX}\mathbf{A} \\ &= \mathbf{A}'\mathbf{A}\mathbf{L} = \mathbf{L} \end{aligned}$$

where $\mathbf{L} \equiv \text{diag}(L_1, L_2, \ldots, L_k)$ is the diagonal matrix of eigenvalues of $\mathbf{R}_{XX}$. $\mathbf{W}$, therefore, is orthogonal, as required. Furthermore,

$$\text{trace}(\mathbf{L}) = \sum_{j=1}^{k} L_j = k = \text{trace}(\mathbf{R}_{XX})$$

and thus the principal components partition the combined variance of the standardized variables $Z_1, Z_2, \ldots, Z_k$.

Solving equation (3.10) for $\mathbf{Z}_X$ produces

$$\mathbf{Z}_X = \mathbf{W}\mathbf{A}^{-1} = \mathbf{W}\mathbf{A}'$$

and, consequently,

$$\mathbf{R}_{XX} = \frac{1}{n-1}\mathbf{Z}_X'\mathbf{Z}_X = \frac{1}{n-1}\mathbf{A}\mathbf{W}'\mathbf{W}\mathbf{A}' = \mathbf{A}\mathbf{L}\mathbf{A}'$$

Thus

$$\mathbf{R}_{XX}^{-1} = (\mathbf{A}')^{-1}\mathbf{L}^{-1}\mathbf{A}^{-1} = \mathbf{A}\mathbf{L}^{-1}\mathbf{A}' \tag{3.11}$$

We shall use this result in Section 3.1.3 in our investigation of collinearity.

The vector geometry of principal components is illustrated for two variables in Figure 3.2. The symmetry of this figure is peculiar to the two-dimensional case. The length of each principal-component vector is the square root of the sum of squared orthogonal projections of $\mathbf{z}_1$ and $\mathbf{z}_2$ on the component. The direction of $\mathbf{w}_1$ is chosen to maximize the combined length of these projections and, hence, to maximize the length of $\mathbf{w}_1$. The direction of $\mathbf{w}_2$ is chosen to be orthogonal to $\mathbf{w}_1$. Note that $\|\mathbf{w}_j\|^2 = L_j(n-1)$. It is clear from the figure that as the correlation between Z_1 and Z_2 increases, the first principal component is lengthened at the expense of the second; thus L_1 grows and L_2 becomes smaller. If, alternatively, $\mathbf{z}_1$ and $\mathbf{z}_2$ are orthogonal, then $\|\mathbf{w}_1\| = \|\mathbf{w}_2\| = \sqrt{n-1}$, and $L_1 = L_2 = 1$.

The algebra of the two-variable case is also particularly simple. The eigenvalues of $\mathbf{R}_{XX}$ are the solutions of the determinantal equation

$$\begin{vmatrix} 1 - L & r_{12} \\ r_{12} & 1 - L \end{vmatrix} = 0$$

$$(1 - L)^2 - r_{12}^2 = L^2 - 2L + 1 - r_{12}^2 = 0 \tag{3.12}$$

Using the quadratic formula to find the roots of equation (3.12) yields

$$L_1 = 1 + \sqrt{r_{12}^2}$$

$$L_2 = 1 - \sqrt{r_{12}^2} \tag{3.13}$$

Thus, consistent with the geometry of Figure 3.2, as the magnitude of the correlation between the two variables increases, the variation attributed to the first principal component also grows. If r_{12} is positive, then solving for $\mathbf{A}$ from the relation $\mathbf{R}_{XX}\mathbf{A} = \mathbf{L}\mathbf{A}$ under the restriction $\mathbf{A}'\mathbf{A} = \mathbf{I}_2$ gives

$$\mathbf{A} = \begin{pmatrix} \frac{\sqrt{2}}{2} & \frac{\sqrt{2}}{2} \\ \frac{\sqrt{2}}{2} & -\frac{\sqrt{2}}{2} \end{pmatrix}$$

More generally, for k orthogonal standardized regressors, all $L_j = 1$ and all $\|\mathbf{w}_j\| = \sqrt{n-1}$. As collinearities among the variables increase, some eigenvalues become large while others grow small. Small eigenvalues and the corresponding short principal components represent dimensions along which the regressor subspace has collapsed. Perfect collinearities are associated with zero eigenvalues.

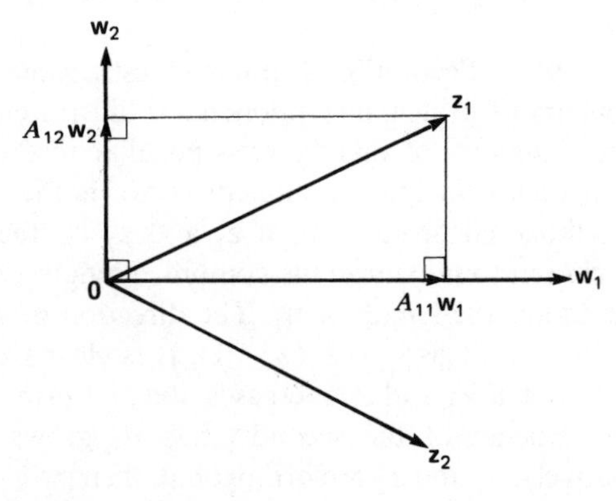

FIGURE 3.2. Vector geometry of principal components.

Summary

1. The principal components of the k standardized regressors $\mathbf{Z}_X$ are a new set of k variables derived from $\mathbf{Z}_X$ by a linear transformation: $\mathbf{W} = \mathbf{Z}_X\mathbf{A}$, where $\mathbf{A}$ is the $(k \times k)$ transformation matrix.
2. The transformation $\mathbf{A}$ is chosen so that the columns of $\mathbf{W}$ are orthogonal—that is, the principal components are uncorrelated. In addition, $\mathbf{A}$ is selected so that the first component accounts for maximum variance in the Z's; the second for maximum variance under the constraint that it be orthogonal to the first; and so on. Each principal component is scaled so that its variance is equal to the variance in the Z's for which it accounts. The principal components therefore partition the variance of the Z's.
3. The transformation matrix $\mathbf{A}$ contains (by columns) normalized eigenvectors of $\mathbf{R}_{XX}$, the correlation matrix of the regressors. The columns of $\mathbf{A}$ are ordered by their corresponding eigenvalues: The first column corresponds to the largest eigenvalue, and the last column to the smallest. The eigenvalue L_j associated with the jth component represents the variance attributable to that component.
4. If there are perfect collinearities in $\mathbf{Z}_X$, then some eigenvalues of $\mathbf{R}_{XX}$ will be zero, and there will be fewer than k principal components. Near collinearities are associated with small eigenvalues and corresponding short principal components.

3.1.3. Diagnosing Collinearity

In Section 1.5 we showed that the least-squares regression estimator for standardized variables is $\mathbf{b}^* = \mathbf{R}_{XX}^{-1}\mathbf{r}_{Xy}$, with covariance matrix

$$V(\mathbf{b}^*) = \frac{\sigma_\varepsilon^{*2}}{n-1}\mathbf{R}_{XX}^{-1}$$

The sampling variance of B_j^* is given by the jth diagonal entry of $V(\mathbf{b}^*)$. It may be shown that the jth diagonal entry of $\mathbf{R}_{XX}^{-1}$ is $1/(1 - R_j^2)$, where R_j^2 is the square of the multiple correlation for the regression of X_j on the other independent variables (see Theil, 1971: 166). Thus

$$V\left(B_j^*\right) = \frac{\sigma_\varepsilon^{*2}}{n-1} \times \frac{1}{1 - R_j^2}$$

Following the usage introduced in Section 3.1.1, we call $1/(1 - R_j^2)$ the variance-inflation factor for the jth regressor (VIF_j). For orthogonal data, all VIFs attain the minimum value of one; in the face of strong collinearity, at least some of the VIFs are large.

Using equation (3.11), the variance-inflation factors may be expressed as functions of the eigenvalues of $\mathbf{R}_{XX}$ and the principal-component coefficients; specifically,

$$\mathrm{VIF}_j = \sum_{l=1}^{k} \frac{A_{jl}^2}{L_l}$$

Thus, it is the small eigenvalues that contribute to large sampling variances, but only for those regressors that have large coefficients associated with the short principal components. This result is sensible, for small eigenvalues and their short components correspond to collinear relations among the regressors; regressors with large coefficients for these components are the regressors implicated in the collinearities.

The relative size of the eigenvalues serves as an indicator of the degree of collinearity present in the data. The square root of the ratio of the largest to the smallest eigenvalue, $K \equiv \sqrt{L_1/L_k}$, called the *condition number*, is a commonly employed index of the global instability of the least-squares regression coefficients: A very large condition number (say, in the neighborhood of 30),[5] indicates that relatively small changes in the data tend to produce large changes in the least-squares solution. In this event, $\mathbf{R}_{XX}$ is said to be *ill conditioned.*

It is instructive to examine the condition number in the simplified context of the two-regressor model. From equations (3.13), we have

$$K = \sqrt{\frac{L_1}{L_2}} = \sqrt{\frac{1 + \sqrt{r_{12}^2}}{1 - \sqrt{r_{12}^2}}}$$

Thus, $K = 30$ corresponds to $r_{12}^2 = .9956$, which in turn produces a variance-inflation factor of $1/(1 - .9956) = 227$, clearly indicative of degrading collinearity. In contrast, $K = 10$ corresponds to $r_{12}^2 = .9608$ and the much smaller (though still troubling) VIF = 26.

Belsley, Kuh, and Welsch (1980: Ch. 3) define a *condition index* $K_j \equiv \sqrt{L_1/L_j}$ for each principal component of $\mathbf{R}_{XX}$.[6] Then, the number of large condition indices points to the number of different collinear relations among the regres-

[5]This rule of thumb, from Belsley, Kuh, and Welsch (1980: Chapter 3) is supported by the results of random-sampling simulation experiments. The context of their analysis is slightly different from ours, however, since these authors do not express the regressors in mean-deviation form prior to checking for collinearity.

[6]Primarily for reasons of computational accuracy, Belsley Kuh, and Welsch develop diagnostic methods for collinearity in terms of the *singular-value decomposition* of the regressor matrix, scaled so that each variable has a sum of squares of one. We employ an equivalent eigenvalue-eigenvector approach because of its conceptual simplicity and broader familiarity. The eigenvalues of $\mathbf{R}_{XX}$ are the squares of the singular values of $(1/\sqrt{n-1})\mathbf{Z}_X$. Indeed, the condition number K defined here is actually the condition number of $(1/\sqrt{n-1})\mathbf{Z}_X$ (and hence of $\mathbf{Z}_X$). Information on the singular-value decomposition and its role in linear-model analysis may be found in Belsley, Kuh, and Welsch (1980: Chapter 3) and in Mandel (1982).

sors. By examining how the several principal components contribute to the VIF for each regressor, it is possible to identify the variables involved in these collinear relations. Belsley, Kuh, and Welsch suggest that we examine the proportional contributions of the components to the variance-inflation factors, defining

$$P_{jl} = \frac{A_{jl}^2/L_l}{\text{VIF}_j}$$

When P_{jl} is large (say, near 0.5), and K_l is also large, then the jth regressor is implicated in the near dependency represented by the lth principal component.[7]

In most applications, it suffices to determine (1) whether serious collinearity is present; (2) which regression coefficients are affected by collinearity; and perhaps (3) which regressors are involved in each near dependency. Point (1) follows from the condition indices, point (2) from the variance-inflation factors, and point (3) from the component contributions to the VIFs. It is possible, however, to proceed further.

Once the variables associated with each near dependency are identified, the relative weights of these variables in the dependency may be estimated by regressing one variable in the set on the others. The multiple correlation for this regression should be close to one. Alternatively, Chatterjee and Price (1977: Ch. 7) employ the principal-component coefficients to estimate variable weights for collinear relations. A component $\mathbf{w}_l$ associated with a very small eigenvalue $L_l \simeq 0$ is itself approximately equal to the zero vector; consequently

$$A_{1l}\mathbf{z}_1 + A_{2l}\mathbf{z}_2 + \cdots + A_{kl}\mathbf{z}_k \simeq \mathbf{0}$$

and we may use the large A_{jl}'s to specify a function of the $\mathbf{z}_j$'s approximately equal to zero.

Illustrative data on Canadian women's labor-force participation in the post-war period, drawn from B. Fox (1980), are shown in Table 3.1. Note that these are time-series data, with yearly observations.[8] Fox was interested in determining how women's labor-force participation rate (L, measured here as percent of adult women in the work force) responds to a variety of factors indicative of the supply of and demand for women's labor. The independent variables in the analysis include the total fertility rate (F, the expected number

[7]When there are several coexisting collinear relations, it is not always simple to separate the variables involved in each (Belsley, Kuh, and Welsch, 1980: 154–156).

[8]The use of time-series data in regression generally casts doubt upon the assumption that errors from different observations are independent, since the observation for one period is likely to share characteristics with observations from other periods close to it in time. If, in fact, errors are *autocorrelated* in this manner, different estimation techniques than ordinary least-squares regression are called for (see, e.g., Johnston, 1972: Ch. 8). In the present instance, however, examination of the least-squares residuals supports the reasonableness of the assumption of independent errors.

TABLE 3.1. Fox's Canadian Women's Labor-Force Participation Data

L^a	T	F	M	W	D	P
25.3	1946	3748	25.35	14.05	18.18	10.28
24.4	1947	3996	26.14	14.61	28.33	9.28
24.2	1948	3725	25.11	14.23	30.55	9.51
24.2	1949	3750	25.45	14.61	35.81	8.87
23.7	1950	3669	26.79	15.26	38.39	8.54
24.2	1951	3682	26.33	14.58	26.52	8.84
24.1	1952	3845	27.89	15.66	45.65	8.60
23.8	1953	3905	29.15	16.30	52.99	5.49
23.6	1954	4047	29.52	16.57	54.84	6.67
24.3	1955	4043	32.05	17.99	65.53	6.25
25.1	1956	4092	32.98	18.33	72.56	6.32
26.2	1957	4168	32.25	17.64	69.49	7.30
26.6	1958	4073	32.52	18.16	71.71	8.65
26.9	1959	4100	33.95	18.58	78.89	8.80
27.9	1960	4119	34.63	18.95	84.99	9.39
29.1	1961	4159	35.14	18.78	87.71	10.23
29.9	1962	4134	34.49	18.74	95.31	10.77
29.8	1963	4017	35.99	19.71	104.40	10.84
30.9	1964	3886	36.68	20.06	116.80	11.70
32.1	1965	3467	37.96	20.94	130.99	12.33
33.2	1966	3150	38.68	21.20	135.25	12.18
34.5	1967	2879	39.65	21.95	142.93	13.67
35.1	1968	2681	41.20	22.68	155.47	13.82
36.1	1969	2563	42.44	23.75	165.04	14.91
36.9	1970	2571	42.02	25.63	164.53	15.52
37.0	1971	2503	45.32	26.79	169.63	15.47
37.9	1972	2302	45.61	27.51	190.62	15.85
40.1	1973	2931	45.59	27.35	209.60	15.40
40.6	1974	1875	48.06	29.64	216.66	16.23
42.2	1975	1866	46.12	29.33	224.34	16.71

Source: Reprinted with permission from B. Fox (1980: 449).
[a]See text for definition of variables.

of births to a cohort of 1000 women proceeding through their child-bearing years at current age-specific fertility rates); men's (M) and women's (W) average weekly wages (in constant 1935 dollars and adjusted for current tax rates); per-capita consumer debt (D, in constant dollars); and the availability of part-time work (P, measured as the percent of the active work force working 34 hours a week or less). Women's wages, consumer debt, and the availability of part-time work are expected to affect women's labor-force participation positively. Fertility and men's wages are expected to have negative effects. Since all of the series, including that for the dependent variable, manifest strong linear trends over the 30-year period of the study, year (T, coded from 1946 to 1975) is also included in the regression as an independent variable.[9]

The results of a least-squares regression of women's labor-force participation on the several independent variables prove highly disappointing. The fitted regression is shown both in standardized and unstandardized form in equations (3.14); standard errors for the standardized regression coefficients

[9]The specification of time as an independent variable in time-series regression is a common, if crude, strategy for controlling for factors that change regularly with time but are not included explicitly in the model.

are given in parentheses.

$$\hat{L} = 50.4 - 0.0180T + 0.000262F - 0.0415M$$
$$+0.0572W + 0.0669D + 0.6865P$$

$$\hat{Z}_L = \underset{(0.147)}{-\ 0.0267}\ Z_T + \underset{(0.0502)}{0.0321}\ Z_F - \underset{(0.176)}{0.0495}\ Z_M \tag{3.14}$$
$$+ \underset{(0.139)}{0.0452}\ Z_W + \underset{(0.162)}{0.697}\ Z_D + \underset{(0.0460)}{0.386}\ Z_P$$

Despite a very large multiple correlation ($R^2 = .9936$) and a highly statistically significant omnibus F-test ($F(6, 23) = 598$), only two regressors, consumer debt and part-time work, have statistically significant regression coefficients. Moreover, the coefficient for fertility is unexpectedly positive, and that for time is negative, though both of these coefficients are very small. The coefficients for men's and women's wages are of the expected sign, but are small.

A correlation matrix for Fox's timeseries data is shown in Table 3.2. The first row and column of this matrix are for the dependent variable. It is clear from the correlations in the remainder of the table, many of them above .95, that the independent variables are highly collinear. Table 3.3 reports a principal-components analysis for the independent variables. The condition indices reveal at least one, and possibly two, near linear dependencies among the regressors, corresponding to the indices 33.68 and 23.70. The effects of these dependencies, and the variables implicated in them, are shown in Table 3.4, which reports variance-inflation factors and the component contributions to the VIFs. Although all regression coefficients are adversely affected by collinearity, those for fertility (VIF = 9.09) and part-time work (VIF = 7.65) are by far the least-seriously degraded. (Recall, however, that the coefficient for fertility is unexpectedly positive and not statistically significant.) The strongest

TABLE 3.2. Correlations for Fox's Data

	L	T	F	M	W	D	P
L	1.0						
T	.9531	1.0					
F	−.8784	−.7612	1.0				
M	.9595	.9891	−.7942	1.0			
W	.9674	.9637	−.8537	.9830	1.0		
D	.9819	.9805	−.8437	.9861	.9868	1.0	
P	.9504	.8459	−.8904	.8533	.8715	.8875	1.0
Mean	29.997	1960.5	3464.9	35.169	19.986	102.79	10.947
Standard Deviation	5.950	8.8	729.4	7.099	4.700	61.98	3.347

TABLE 3.3. Principal-Component Analysis for Fox's Regressors

	Principal Component					
Regressor	W_1	W_2	W_3	W_4	W_5	W_6
T	0.4121	0.3957	0.1213	0.6095	−0.0885	−0.5288
F	−0.3802	0.6988	0.5277	−0.2785	0.1047	0.0092
M	0.4168	0.3302	−0.0528	−0.0388	−0.5814	0.6123
W	0.4204	0.1529	−0.2706	−0.7231	−0.0141	−0.4512
D	0.4223	0.1757	−0.0642	0.0562	0.8011	0.3765
P	0.3961	−0.4379	0.7916	−0.1530	−0.0349	−0.0079
Eigenvalue	5.504	0.3527	0.1092	0.01944	0.009798	0.004853
Condition Index	1.00	3.95	7.10	16.83	23.70	33.68

collinear relation among the regressors (shown in column six) involves time, men's wages, women's wages, and, perhaps, consumer debt. The next-strongest near dependency (column five) appears to involve men's wages and consumer debt.

The methods described in this section are not fully applicable to models that include polychotomous qualitative variables whose effects are represented by sets of dummy regressors. The reason underlying this qualification is subtle, but may be clarified by appealing to the vector representation of linear models.

The correlations among a set of dummy variables are affected by the choice of a baseline category. It is, indeed, always possible to select an orthogonal basis for the dummy-regressor subspace (though such a basis does not employ dummy-variable coding). What is at issue is the subspace itself and not the arbitrary basis chosen for it (although a particular basis may be a poor choice for purposes of computation if it produces numerically unstable results). We are not concerned, therefore, with the "artificial" collinearity among the

TABLE 3.4. Variance-Inflation Factors and Proportional Principal-Component Contributions to VIFs for Fox's Data

		Principal Component						
Regressor	VIF	W_1	W_2	W_3	W_4	W_5	W_6	Total
T	78.14	0.0004	0.0057	0.0017	0.2445	0.0102	0.7374	0.9999
F	9.09	0.0029	0.1524	0.2806	0.4391	0.1231	0.0019	1.0000
M	112.19	0.0003	0.0028	0.0002	0.0007	0.3075	0.6885	1.0000
W	69.64	0.0005	0.0010	0.0096	0.3862	0.0003	0.6024	1.0000
D	95.03	0.0003	0.0009	0.0004	0.0017	0.6892	0.3074	0.9999
P	7.65	0.0037	0.0711	0.7499	0.1573	0.0162	0.0017	0.9999

dummy regressors in the same set; we are interested instead in the relationship between the subspaces generated to represent the effects of *different* independent variables, be they qualitative or quantitative. As a consequence, we may legitimately employ the methods of this section to examine the impact of collinearity on the coefficients of numerical regressors (or on any single-degree-of-freedom effects) even when sets of dummy regressors are present in the model.

3.1.4. Estimation in the Presence of Collinearity

As is apparent from the preceding discussion, collinear data make it exceedingly difficult to assess the effects of independent variables in a linear model. This is true even when the model is correctly specified—that is, when the model adequately represents the mechanism according to which dependent-variable scores are generated. In the extreme case of perfect collinearity, it becomes impossible to separate the effects of regressors involved in the collinear relations.

Because, as is commonly remarked, collinearity is a data problem rather than a model-specification problem, the ideal solution is to collect more adequate data. Thus, when independent variables are manipulable, we may design an experiment in which regressors are orthogonal. In observational research it is more difficult, and indeed at times impossible, to avoid collinear data, but even here intelligent data collection can often prove helpful. If, for example, we wish to assess the relative contributions of family background and school environment to students' educational achievement, we should avoid collecting data from a community in which the children of poor parents attend one school and those of wealthy parents attend another.

The advice to collect better data is, however, cold comfort to the researcher who has collinear data in hand. Although collinear data do not imply that the model entertained is misspecified, respecification sometimes makes substantive sense. It may be the case, for example, that several highly correlated regressors are alternative indicators of a single construct (i.e., abstract variable). In this instance, the several measures might be combined to form an index, or one measure may be selected to represent the others. A more elaborate approach to multiple indicators is developed in Section 4.8. Further discussion of scale construction may be found in many sources, including Torgerson (1958) and Lord and Novick (1968).

Beyond the collection of more adequate data and model respecification, there are three general approaches that have been taken to the problem of estimation in the presence of collinearity. The first approach introduces theoretically based prior information to resolve the ambiguity produced by collinearity. Suppose, for instance, that theory leads us to expect that two regressors have equal coefficients. We may efficiently estimate the common coefficient even if the two variables are highly correlated by entering the sum of the two variables into the regression equation in place of the individual variables. Notice that we have purchased the possibility of estimating the model at the expense of our ability to test the hypothesis that the coefficients of the two independent variables are equal (see Problem 1.14).

Prior information of this sort is in practice difficult to come by for most social-science applications of linear models. In principle, however, the introduction of prior information regarding regression coefficients is similar in character to the assumption that the error in a linear model is uncorrelated with the regressors. In both of these cases, information external to the regression equation itself, and justifiable only on substantive grounds, is employed to make estimation possible. Belsley, Kuh, and Welsch (1980: 193–204), Theil (1971: 346–352), and others pursue this solution to the problem of collinearity.

A second approach to collinear data is independent-variable selection—reducing collinearity by eliminating some of the offending regressors. When there are k regressors under consideration for inclusion in a model, 2^k subsets of regressors may be selected: the model including the constant regressor only; each regressor appearing individually (along with the constant); each combination of two regressors; and so on, through the model including all k regressors.

Many techniques for variable selection have been proposed. "Stepwise" methods select independent variables in an incremental fashion, either entering these variables one at a time into the model on the basis of some criterion (such as maximum increase in R^2), or successively deleting regressors from the model. The first general strategy is called "forward selection," while the second is termed "backward elimination"; some stepwise-regression procedures combine the two approaches. In any event, selection proceeds until some stopping rule is satisfied.

A general criticism of stepwise methods is that they do not insure optimal results: These methods may not identify the subset of regressors of a given size that maximizes R^2, even if R^2 is used as the criterion for inclusion in the model. More seriously, stepwise regression is frequently abused in practice: Researchers sometimes attempt to interpret the order of entry of independent variables into the regression equation as somehow indicative of the relative "importance" of these variables; and, at times, investigators are led to fit regression equations that do not make substantive sense, such as models that include interaction regressors but that delete the main effects marginal to them. The wide availability of computer programs for stepwise regression appears to invite these abuses. Detailed accounts of stepwise methods may be found in Draper and Smith (1966: Ch. 6) and Chatterjee and Price (1977: Ch. 9).

Some of the deficiencies of stepwise regression are overcome by methods that examine a larger proportion of all subsets of regressors. In fact, the increase in computer power over the past two decades, together with improved computational methods, make it possible to examine all 2^k regressor subsets even when k is fairly large (say, 15). Daniel and Wood (1980) and Chatterjee and Price (1977) contain helpful treatments of subset selection methods.

Hocking's (1976) review article on variable selection includes a comparative discussion of stepwise and subset techniques.

The third approach to collinear data is to employ a method of estimation other than ordinary least-squares regression. The Gauss–Markov theorem (Section 1.2.5) assures that the least-squares method provides maximally efficient linear unbiased estimators under the assumptions of the regression model. In the presence of collinearity, the sampling variance of least-squares estimators is increased, and hence their efficiency is impaired. The Gauss–Markov theorem does not rule out the possibility that we may improve the efficiency of estimation by employing linear biased estimators (or, indeed, by employing nonlinear estimators). To be more efficient than least squares, a biased estimator must realize a sufficient decrease in sampling variance to offset the contribution of its bias to estimator mean-squared error: Recall that the mean-squared error of an estimator is the sum of its sampling variance and squared bias.

Ridge regression, originally proposed in 1962 by Hoerl and elaborated by Hoerl and Kennard (1970a, 1970b), is one of a class of biased estimation methods that seek to improve the performance of least-squares estimation in the presence of collinear data. The ridge-regression estimator is given by the equation

$$\mathbf{b}_h^* = (\mathbf{R}_{XX} + h\mathbf{I}_k)^{-1}\mathbf{r}_{Xy} \tag{3.15}$$

where $h \geq 0$ is a scalar constant. When the data are collinear, some off-diagonal entries of $\mathbf{R}_{XX}$ generally are large, making this matrix ill-conditioned. Heuristically, the ridge-regression method improves the conditioning of $\mathbf{R}_{XX}$ by inflating its diagonal entries.

Although the least-squares estimator $\mathbf{b}^*$ is unbiased, its entries tend to be too large in absolute value, a tendency that is magnified as collinearity increases. In practice, researchers working with collinear data often compute wildly large regression coefficients (reexamine Problem 1.22). The ridge estimator may be thought of as a "shrunken" version of the least-squares estimator, correcting the tendency of the latter to produce coefficients that are too large.

Using the least-squares normal equations $\mathbf{R}_{XX}\mathbf{b}^* = \mathbf{r}_{Xy}$, the ridge estimator of equation (3.15) may be rewritten in the following manner:

$$\begin{aligned}
\mathbf{b}_h^* &= (\mathbf{R}_{XX} + h\mathbf{I}_k)^{-1}\mathbf{R}_{XX}\mathbf{b}^* \\
&= \left[\mathbf{R}_{XX}^{-1}(\mathbf{R}_{XX} + h\mathbf{I}_k)\right]^{-1}\mathbf{R}_{XX}^{-1}\mathbf{R}_{XX}\mathbf{b}^* \\
&= \left(\mathbf{I}_k + h\mathbf{R}_{XX}^{-1}\right)^{-1}\mathbf{b}^* = \mathbf{U}\mathbf{b}^*
\end{aligned} \tag{3.16}$$

where the transformation matrix $\mathbf{U} \equiv (\mathbf{I}_k + h\mathbf{R}_{XX}^{-1})^{-1}$. As h increases, the entries of $\mathbf{U}$ tend to grow smaller, and therefore $\mathbf{b}_h^*$ is driven towards zero. In fact, Hoerl and Kennard (1970a) show that for any value of $h > 0$, $\mathbf{b}_h^{*\prime}\mathbf{b}_h^* < \mathbf{b}^{*\prime}\mathbf{b}^*$. When $h = 0$, of course, the ridge and least-squares estimators coincide.

The expected value of the ridge estimator may be determined from its relation to the least-squares estimator, given in equation (3.16); treating $\mathbf{Z}_X$, and hence $\mathbf{U}$, as fixed,

$$E(\mathbf{b}_h^*) = \mathbf{U}E(\mathbf{b}_h^*) = \mathbf{U}\boldsymbol{\beta}^* = (\mathbf{I}_k + h\mathbf{R}_{xx}^{-1})^{-1}\boldsymbol{\beta}^*$$

The bias of $\mathbf{b}_h^*$ is therefore

$$\text{bias}(\mathbf{b}_h^*) \equiv E(\mathbf{b}_h^*) - \boldsymbol{\beta}^* = (\mathbf{U} - \mathbf{I}_k)\boldsymbol{\beta}^*$$

and since the departure of $\mathbf{U}$ from $\mathbf{I}_k$ increases with h, the bias of the ridge estimator is an increasing function of h.

The variance of the ridge estimator is also simply derived. We may write the ridge estimator as a linear transformation of the standardized dependent-variable observations:

$$\mathbf{b}_h^* = \mathbf{W}\left(\frac{1}{n-1}\right)\mathbf{Z}_X'\mathbf{z}_y$$

where $\mathbf{W} \equiv (\mathbf{R}_{XX} + h\mathbf{I}_k)^{-1}$ and $[1/(n-1)]\mathbf{Z}_X'\mathbf{z}_y = \mathbf{r}_{Xy}$. Consequently,

$$\begin{aligned} V(\mathbf{b}_h^*) &= \frac{1}{(n-1)^2}(\mathbf{W}\mathbf{Z}_X')\sigma_\varepsilon^{*2}\mathbf{I}_n(\mathbf{Z}_X\mathbf{W}') \\ &= \frac{\sigma_\varepsilon^{*2}}{n-1}\mathbf{W}\mathbf{R}_{XX}\mathbf{W} \\ &= \frac{\sigma_\varepsilon^{*2}}{n-1}(\mathbf{R}_{XX} + h\mathbf{I}_k)^{-1}\mathbf{R}_{XX}(\mathbf{R}_{XX} + h\mathbf{I}_k)^{-1} \end{aligned} \tag{3.17}$$

As h increases, the inverted term $(\mathbf{R}_{XX} + h\mathbf{I}_k)$ becomes increasingly dominated by $h\mathbf{I}_k$. The sampling variance of the ridge estimator, therefore, is a decreasing function of h; this result is reasonable, since the estimator itself is driven towards zero.

The mean-squared error of the ridge estimator is the sum of its squared bias and sampling variance. The relationship of h to these quantities is shown schematically in Figure 3.3. Hoerl and Kennard prove that it is always possible to choose a positive value of h so that $MSE(\mathbf{b}_h^*) < MSE(\mathbf{b}^*)$. The optimal value of h, however, depends upon the (unknown) parameter vector $\boldsymbol{\beta}^*$ as well as upon the structure of correlations relating the regressors.

The central problem in applying ridge regression is to find a value of h for which the trade-off of bias against variance is favorable. In deriving the properties of the ridge estimator, we treated h as fixed. If h is determined from the data, however, it becomes a random variable, casting serious doubt upon the conceptual basis for the ridge estimator. A number of methods have been proposed for selecting h. Some of these methods are rough and qualitative, while others incorporate specific formulas or procedures for estimating the optimal value of h. Problem 3.4 describes a qualitative method presented by Hoerl and Kennard in their 1970 papers.

There have been a number of random-sampling simulation experiments exploring the properties of ridge estimation as well as those of other methods meant to cope with collinear data. While these studies are by no means unanimous in their conclusions, the ridge estimator often performs well by comparison with least-squares estimation and by comparison with other biased estimation methods. On the basis of evidence from simulation experiments, it would, however, be premature to recommend a particular procedure for selecting h, and, indeed, the dependence of the optimal value of h on the unknown regression parameters makes it unlikely that there is generally a best way of finding h. Several authors critical of ridge regression (e.g., Draper and

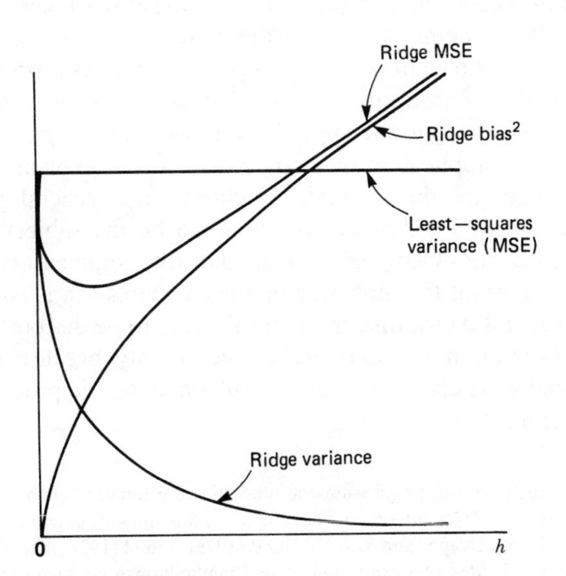

FIGURE 3.3. Bias, sampling variance, and mean-square error of the ridge-regression estimator. (Adapted with permission from Hoerl and Kennard, 1970a: 61).

Smith, 1981: 324) have noted that simulation studies supporting the method generally incorporate restrictions on parameter values particularly suited to ridge regression (see the concluding remarks in this section).[10]

Because the ridge estimator is biased, the standard errors from equation (3.17) cannot be used in the normal manner for statistical inference concerning the population regression coefficients. Indeed, as Obenchain (1977b) has pointed out, under the assumptions of the linear model, confidence intervals centered at the least-squares estimators retain their optimal properties regardless of the degree of collinearity: In particular, they are the shortest possible intervals at the stated level of confidence (see Scheffé, 1959: Chapter 2). An interval centered at the ridge estimator of a regression coefficient is therefore *wider* than the corresponding least-squares interval, even if the ridge estimator has a smaller mean-squared error than the least-squares estimator. Thus, we may think of biased estimation as a method for improving point estimation when the data are collinear, but not as a means for obtaining more acute confidence intervals or hypothesis tests.

Concluding Comments Although we presented them separately, there are similarities among the three general strategies that we described for dealing with collinear data. Variable selection, for example, sets particular regression coefficients to zero, and therefore constrains the estimation process. If the coefficients for deleted regressors are not zero, then the estimators employed following variable selection are biased.

The ridge-regression method may be used for variable selection, deleting regressors whose coefficients are driven rapidly to zero as h increases (Marquardt and Snee, 1975; Hocking, 1976). Furthermore, as Draper and Smith (1981: 320) demonstrate, choice of h in ridge regression implies a prior constraint on the length of $\mathbf{b}_h^*$, a constraint that typically is not made explicit: In effect, coefficients of large magnitude are judged unreasonable *a priori*.

When they are applicable, methods based on the explicit introduction of prior information are the methods of choice. The general utility of ridge regression and related techniques continues to be the subject of controversy (see, e.g., Smith and Campbell, 1980, and the commentary following it). Indeed, it is important to emphasize that ridge regression is *not* a panacea for collinearity ills, and its routine, mechanical use is to be discouraged. Nevertheless, biased estimation methods are useful, if only because they provide a means for judging whether anomalous results may be the product of collinearity (see Problem 3.4).

[10] Simulation studies of ridge regression and other biased estimation methods are too numerous to cite individually here. References to and comments on this simulation literature may be found in many sources, including Draper and Van Nostrand (1979), Vinod (1978), and Hocking (1976). An extensive treatment of ridge regression and related methods may be found in Vinod and Ullah (1981).

PROBLEMS

3.1.# The data shown below were constructed by Mandel to illustrate the problem of collinearity (reprinted with permission from Mandel, 1982: 16):

X_1	X_2	Y
16.85	1.46	41.38
24.81	−4.61	31.01
18.85	−0.21	37.41
12.63	4.93	50.05
21.38	−1.36	39.17
18.78	−0.08	38.86
15.58	2.98	46.14
16.30	1.73	44.47

(a) Compute the mean and standard deviation of each variable. Find the correlations among X_1, X_2, and Y, and use these correlations to calculate the standardized coefficients for the regression of Y on X_1 and X_2. Compute the unstandardized regression coefficients B_0, B_1, and B_2.

(b) Perform a principal-components analysis for X_1 and X_2. Draw the geometric vector representation of the principal-components analysis.

(c) Find the variance-inflation factor for the regression in part (a), and calculate the proportional contribution of each principal component to the VIF. Compute the condition number K.

(d) Use the second principal component to approximate the near-collinear relation between the standardized regressors Z_1 and Z_2. Express this relation as a linear relationship between the unstandardized regressors X_1 and X_2.

(e) Now regress X_1 on X_2. How does the fitted regression equation compare with the linear relationship found in part (d)?

3.2. The time-series data on French imports shown in Table 3.5 are taken from Malinvaud (1970), who used the data to illustrate the problem of collinearity. These data are also discussed in Chatterjee and Price (1977: Chapter 7). Malinvaud regresses the value of imports (Y) on domestic production (X_1), stock formation (X_2), consumption (X_3), and a trend variable (X_4) meant to capture the increased growth of trade after the

TABLE 3.5. French Import Data

Year	Imports[a]	Gross[a] Domestic Production	Stock[a] Formation	Consumption[a]	Common Market
1949	15.9	149.3	4.2	108.1	0
1950	16.4	161.2	4.1	114.8	0
1951	19.0	171.5	3.1	123.2	0
1952	19.1	175.5	3.1	126.9	0
1953	18.8	180.8	1.1	132.1	0
1954	20.4	190.7	2.2	137.7	0
1955	22.7	202.1	2.1	146.0	0
1956	26.5	212.4	5.6	154.1	0
1957	28.1	226.1	5.0	162.3	0
1958	27.6	231.9	5.1	164.3	0
1959	26.3	239.0	0.7	167.6	0
1960	31.1	258.0	5.6	176.8	0
1961	33.3	269.8	3.9	186.6	1
1962	37.0	288.4	3.1	199.7	2
1963	43.3	304.5	4.6	213.9	3
1964	49.0	323.4	7.0	223.8	4
1965	50.3	336.8	1.2	232.0	5
1966	56.6	353.9	4.5	242.9	6

Source: Malinvaud, *Statistical Methods of Econometrics*, 2nd Rev. Ed, © North-Holland Publishing Company, 1970. Reprinted with permission from Malinvaud (1970: Table 1).
[a]In milliards of new francs at 1959 prices.

formation of the European Common Market in 1960.

(a) Compute the least-squares regression of Y on X_1, X_2, X_3, and X_4.

(b) Use the methods of this section to diagnose the presence of collinearity in Malinvaud's data.

3.3. Collinearity in Cross-Sectional Data:

(a) Use the methods described in this section to diagnose the presence of collinearity in Blau and Duncan's data (Problem 1.25). Is collinearity a problem here?

(b) Construct a table showing various levels of multiple correlation R_j ranging between zero and .999 (say, 0, .2, .4, .6, .8, .9, .95, .99, and .999); the $\text{VIF}_j = 1/(1 - R_j^2)$ associated with each R_j; and the square root of the VIF_j. (Note that $\sqrt{\text{VIF}_j}$ is the relative increase due to collinearity in the standard error of the jth regression coefficient.) Do you think that collinearity is likely to be a serious, common problem for linear models fit to cross-sectional social-survey data?

3.4.† Finding h in Ridge Regression: Hoerl and Kennard suggest plotting the entries of $\mathbf{b}_h^*$ against values of h ranging between zero and one. The resulting graph, called a *ridge trace*, both furnishes a visual representation of instability due to collinearity and provides a basis for selecting a value of h. When data are collinear, we generally observe dramatic changes in regression coefficients as h is gradually increased from zero. As h is increased further, the coefficients eventually stabilize, and then are driven slowly toward zero. During this process, the variance-inflation factors of the coefficients [the diagonal entries of $\mathbf{WR}_{XX}\mathbf{W}$, see equation (3.17)] decrease, at first quickly and then more gradually. The estimated variance of the errors, S_E^{*2}, which is minimized at the least-squares solution ($h = 0$), rises slowly with increased h.

We may use the ridge trace, together with information on the VIFs and error variance, to select h. Hoerl and Kennard recommend choosing h so that the regression coefficients are stabilized and the error variance is not unreasonably inflated from its minimum value. Marquardt and Snee (1975) suggest that h be chosen so that the maximum VIF is less than 10 and, preferably, not much larger than one (as it would be for nearly orthogonal data).

Construct a ridge trace, including the standard error S_E^*, for B. Fox's regression data (Table 3.1); also graph $\log_{10}\text{VIF}_j$ against h. Use this information to select a value of h [B. Fox chose $h = 0.05$], and compare the resulting ridge estimates of the regression parameters to the least-squares estimates (equation (3.14)). Make this comparison both for standardized and unstandardized coefficients.

3.2. ANALYSIS OF RESIDUALS AND INFLUENTIAL DATA

The linear models developed in Chapter 1 and extended in Chapter 2 are based on the assumption that the errors are normally and independently distributed with zero expectations and common variance; for random regressors, we assumed further that the regressors and errors are independent. We did not question these assumptions, except in passing. Yet, violation of the assumptions can, in certain circumstances, seriously compromise our analysis. We shall show in the present section how an examination of residuals can serve partially to confirm the adequacy of a linear model or, alternatively, to cast doubt upon its specification. Later in the chapter we shall consider how difficulties revealed in a residual analysis may be remedied.

Despite the diagnostic importance of residuals, it would be wrong to suppose that all errors of model specification are potentially reflected in the residuals. We are aware, for example, that the method of least squares insures that the mean residual is zero (so long as there is a constant in the model) and, hence, this mean cannot be used to test the assumption that the expectation of the error is zero. Likewise, the fact that least-squares residuals are orthogonal to the regressors makes it inappropriate to use the residuals to test for a linear relationship between the error and the regressors.

Nevertheless, because they are estimates of the errors, the residuals contain a great deal of valuable information. We begin by examining the leverage of different observations in determining a least-squares fit, a topic related to the analysis of residuals. Then we describe several transformations of the least-squares residuals that are useful in diagnosing model inadequacies. We proceed to discuss methods for the detection of outliers and influential observations, nonnormality, nonlinearity, and heteroscedasticity.

3.2.1. The Leverage of Observations

In Section 1.2.5, we wrote the fitted values

$$\hat{\mathbf{y}} = \mathbf{Xb} = \mathbf{X}(\mathbf{X}'\mathbf{X})^{-1}\mathbf{X}'\mathbf{y} = \mathbf{Hy}$$

calling $\mathbf{H} \equiv \mathbf{X}(\mathbf{X}'\mathbf{X})^{-1}\mathbf{X}'$ the *hat matrix*. The i, jth element of $\mathbf{H}$, h_{ij}, indicates the contribution of the jth observation to the fit for the ith observation. Let $\mathbf{h}_i$ represent the ith column of $\mathbf{H}$; because $\mathbf{H}$ is symmetric and idempotent, the diagonal element

$$h_{ii} = \mathbf{h}_i'\mathbf{h}_i = \sum_{j=1}^{n} h_{ij}^2 \tag{3.18}$$

not only gives the potential influence of Y_i on $\hat{Y}_i$, but also summarizes the potential influence of Y_i on all fitted values. We may therefore think of the *hat value* h_{ii} as the *leverage* of the ith observation. $\mathbf{H}$ is a function of $\mathbf{X}$, and thus the leverage of each observation depends solely upon the configuration of the independent-variable scores.

As a sum of squares, h_{ii} is nonnegative, and since from equation (3.18) $h_{ii} = h_{ii}^2 +$ squared terms, it cannot exceed one; thus, $0 \leq h_{ii} \leq 1$. Hoaglin and Welsch (1978) show that the average value of h_{ii} is

$$\bar{h} = \frac{k+1}{n} \tag{3.19}$$

where k, we recall, is the number of regressors excluding the constant. From equation (3.19) we see that the leverage of individual observations declines on average when the sample size increases, as is intuitively reasonable. Based on approximate distributional results, Belsley, Kuh, and Welsch (1980: 17) adopt the rule of thumb that an h_{ii} is large when it exceeds twice the average value, $2(k+1)/n$.

It is instructive to examine leverage in the context of the simple-regression model (cf. Hoaglin and Welsch, 1978; Behnken and Draper, 1972). Straightforward substitution into $h_{ii} = \mathbf{x}_i'(\mathbf{X}'\mathbf{X})^{-1}\mathbf{x}_i$, where $\mathbf{x}_i'$ is the ith row of $\mathbf{X}$, produces

$$h_{ii} = \frac{1}{n} + \frac{(x_i - \bar{x})^2}{\Sigma_j (x_j - \bar{x})^2}$$

Design points far from the mean $\bar{x}$, therefore, are relatively influential, as if the regression line were a rod balanced at $\bar{x}$ and the observations were pulling the rod towards them.

An idealized illustration of this result appears in Figures 3.4(a) and (b). The white dot in each figure represents a discrepant or outlying observation—an observation that departs from the linear pattern suggested by the remainder of the data. The solid line shows the least-squares fit to the full dataset, while the broken line shows the fit when the discrepant data point is eliminated. In Figure 3.4(a), where the outlying observation has an X-value equal to $\bar{x}$, its deletion has little effect on the fit; in contrast, in Figure 3.4(b), the outlying observation is far from $\bar{x}$ and its presence affects the fit substantially.

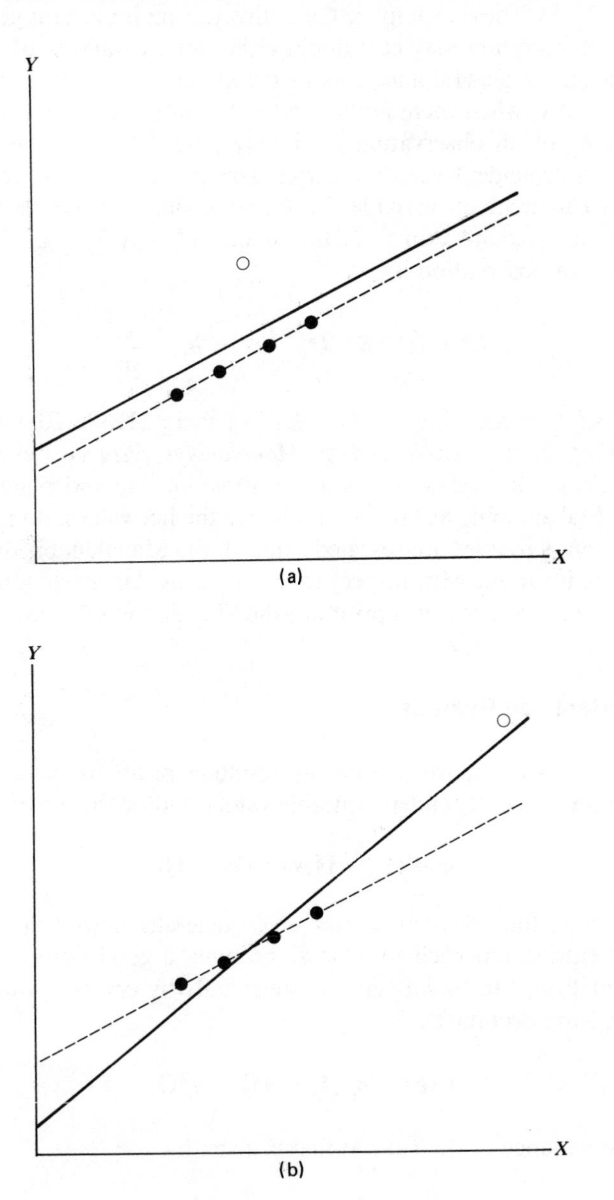

FIGURE 3.4. Leverage in simple regression. (a) Low leverage. (b) High leverage.

The interpretation of Figures 3.4(a) and (b) differs importantly in at least one respect: The discrepant observation in the first illustration clearly does not belong with the others—perhaps it resulted from an error in data collection or was generated differently from the remainder of the data. The second illustration is more ambiguous—here, the fact that the data point on the right is out of line could simply mean that the relationship between Y and X is nonlinear (see Section 3.2.5). These examples, then, illustrate an important general point: An outlying observation may cast doubt either on the validity of the observation itself or on the general adequacy of the model.

More generally, when there are several independent variables in the model, the leverage h_{ii} of an observation is directly related to its distance from the center of the independent-variable scatter. To demonstrate this property of the hat values, let us write all variables in the regression in mean-deviation form, eliminating the constant term from the model: $\mathbf{y}^* = \mathbf{X}^*\boldsymbol{\beta}_1 + \boldsymbol{\varepsilon}$. Then, the hat value for the ith observation is

$$h_{ii}^* = \mathbf{x}_i^{*\prime}(\mathbf{X}^{*\prime}\mathbf{X}^*)^{-1}\mathbf{x}_i^* = h_{ii} - \frac{1}{n}$$

where $\mathbf{x}_i^{*\prime} = (x_{1i} - \bar{x}_1, \ldots, x_{ki} - \bar{x}_k)$. As Weisberg (1980: 105) has pointed out, $(n-1)h_{ii}^*$ is the *generalized* or *Mahalanobis distance* between $\mathbf{x}_i$ and $\bar{\mathbf{x}} = (\bar{x}_1, \ldots, \bar{x}_k)'$, the mean vector or *centroid* of the independent-variable values. The Mahalanobis distances, and hence the hat values, do not change if the independent variables are rescaled; indeed, the Mahalanobis distances and hat values are invariant with respect to any nonsingular linear transformation of $\mathbf{X}$. Some computer programs print out the Mahalanobis distances in place of h_{ii} as an index of leverage.

3.2.2. Studentized Residuals

In Section 1.2.5, we discovered that the residuals result from the same linear transformation of the dependent-variable values and of the errors:

$$\mathbf{e} = (\mathbf{I}_n - \mathbf{H})\mathbf{y} = \mathbf{Q}\mathbf{y} = \mathbf{Q}\boldsymbol{\varepsilon}$$

When n is large, the diagonal entries of $\mathbf{Q}$ generally approach one and the off-diagonal entries approach zero, so E_i becomes a good estimator of ε_i; but when n is small, E_i can be influenced nontrivially by errors in other observations. Again from Section 1.2.5,

$$V(\mathbf{e}) = \sigma_\varepsilon^2(\mathbf{I}_n - \mathbf{H}) = \sigma_\varepsilon^2\mathbf{Q}$$

and thus the variance of the ith residual is given by

$$\sigma_{E_i}^2 = \sigma_\varepsilon^2(1 - h_{ii}) = \sigma_\varepsilon^2 q_{ii}$$

Because we do not have access to σ_ε^2, we must employ the estimator $S_E^2 = \Sigma E_i^2/(n-k-1)$. *Standardized residuals* are obtained by dividing each least-squares residual by its estimated standard deviation; denoting the standardized residual by $\tilde{E}_i$, we have

$$\tilde{E}_i \equiv \frac{E_i}{S_{E_i}} = \frac{E_i}{S_E\sqrt{1-h_{ii}}} = \frac{E_i}{S_E\sqrt{q_{ii}}} \tag{3.20}$$

Although it may appear as if $\tilde{E}_i$ should follow a t-distribution, this is not the case: E_i is a component of both the numerator and denominator of equation (3.20)—E_i enters the denominator through S_E—and therefore the numerator and denominator are not independent.

Suppose that we fit the model

$$Y = \beta_0 + \beta_1 X_1 + \cdots + \beta_k X_k + \gamma D_i + \varepsilon$$

where D_i is a dummy regressor coded one for the ith observation and zero for all others. (To motivate this specification, imagine that we suspect *a priori* that observation i is special.) Let $E_i^* \equiv C/S_C$ represent the t-statistic for testing H_0: $\gamma = 0$ (where C is the estimate of γ and S_C is its estimated standard error). Note that E_i^* has $n-k-2$ degrees of freedom. E_i^*, called the *studentized residual*, may be calculated in turn for each observation.

Hoaglin and Welsch (1978) arrive at the studentized residuals by successively omitting each observation, calculating its residual based on the regression coefficients obtained for the remaining sample, and dividing the resulting residual by its standard error. Belsley, Kuh, and Welsch (1980: 19–20) equivalently find $E_i^* = E_i/S_{E(-i)}\sqrt{1-h_{ii}}$, where $S_{E(-i)}$ is the standard error of the estimate calculated for a regression from which the ith observation has been deleted. Finally, Beckman and Trussell (1974) show that

$$E_i^* = \frac{\tilde{E}_i\sqrt{n-k-2}}{\sqrt{n-k-1-\tilde{E}_i^2}} \tag{3.21}$$

If n is large, then $E_i^* \simeq \tilde{E}_i \simeq E_i/S_E$. In small samples, therefore, we should prefer to examine E_i^*, but in large samples $E_i^{**} \equiv E_i/S_E$ will suffice and is easier to compute. We shall call the E_i^{**} *normed residuals*.[11]

[11]Unfortunately, usage varies. For example, the term "standardized residual" has been used by different authors to refer both to $\tilde{E}_i$ and to E_i^{**}, and the term "studentized residual" has been used both for E_i^* and $\tilde{E}_i$.

3.2.3. Detecting Outliers and Influential Observations

Thus far, we have proceeded formally under the assumptions of the linear model. There is, of course, a variety of ways in which these assumptions may be violated, perhaps causing the model to be inadequate. In the remainder of Section 3.2, we shall describe practical methods for detecting certain sorts of model inadequacy. Since the errors are unobservable, our primary clues are contained in the residuals.

A first step in analyzing the residuals from a linear model is to examine their distribution. In this examination we may detect unusual or discrepant observations, often termed *outliers*. Outliers are important because they can (1) have "undue" influence on the overall fit of the model, (2) reveal general model inadequacy, and (3) point the researcher toward noteworthy, perhaps anomalous, features of the data.

To illustrate the detection of outliers, we shall examine the data shown in Table 3.6, taken from Sahlins's (1972) treatment of agricultural production in a primitive community. This example was introduced in Section 1.1, where we fit a simple-regression model to the data. Rather than displaying diagnostic statistics such as hat values and studentized residuals in a table (as we have done here), it is often easier to examine these statistics by plotting them against the observation indices. (See Problem 3.5 for an application of *index plots*.) Other sorts of graphic display are also possible.

The studentized residuals given in Table 3.6 are graphed in Figure 3.5, using Tukey's (1977: Chapter 1) *stem-and-leaf* plot. The stem-and-leaf plot is a modified frequency plot or histogram, well adapted to small samples, in which the values of the observations (here to two digits) may be read directly from the plot. The most significant digit (or digits) is provided by the "stem"—the

TABLE 3.6. Diagnostic Statistics for Sahlins's Data

							$\mathbf{d}_i$	
i	X_i	Y_i	E_i	h_{ii}	E_i^*	D_i	D_{Ai}	D_{Bi}
1	1.00	1.71	-0.1820	0.1695	-0.4294	0.0197	-0.0872	0.0500
2	1.08	1.52	-0.4133	0.1357	-0.9772	0.0752	-0.1648	0.0925
3	1.15	1.29	-0.6794	0.1108	-1.6616	0.1566	-0.2278	0.1245
4	1.15	3.09	1.1206	0.1108	3.2285	0.4261	0.3756	-0.2053
5	1.20	2.21	0.2148	0.0956	0.4864	0.0131	0.0629	-0.0335
6	1.30	2.26	0.2131	0.0717	0.4764	0.0092	0.0455	-0.0224
7	1.37	2.40	0.3170	0.0602	0.7098	0.0166	0.0512	-0.0225
8	1.37	2.10	0.0170	0.0602	0.0375	0.0000	0.0027	-0.0012
9	1.43	1.96	-0.1540	0.0538	-0.3398	0.0034	-0.0182	0.0066
10	1.46	2.09	-0.0395	0.0517	-0.0867	0.0002	-0.0038	0.0011
11	1.52	2.02	-0.1405	0.0500	-0.3091	0.0026	-0.0077	0.0002
12	1.57	1.31	-0.8763	0.0510	-2.1755	0.1053	-0.0173	-0.0190
13	1.65	2.17	-0.0576	0.0570	-0.1269	0.0005	0.0021	-0.0034
14	1.65	2.28	0.0524	0.0570	0.1155	0.0004	-0.0019	0.0031
15	1.65	2.41	0.1824	0.0570	0.4038	0.0052	-0.0067	0.0107
16	1.66	2.23	-0.0027	0.0582	-0.0060	0.0000	0.0001	-0.0002
17	1.87	3.04	0.6988	0.1026	1.7080	0.1507	-0.1408	0.1180
18	2.03	2.06	-0.3638	0.1623	-0.8688	0.0741	0.1247	-0.0961
19	2.05	2.73	0.2959	0.1713	0.7054	0.0529	-0.1073	0.0822
20	2.30	2.36	-0.2032	0.3137	-0.5289	0.0666	0.1382	-0.1005

leftmost figure in each row of the plot—while the final digits for the observations in each row are provided by the "leaves"—the figures given to the right of the stem. For example, the largest studentized residual, 3.229, is displayed with a stem of 3 and a leaf of 2. For large samples, we should prefer a traditional histogram, where residuals are grouped into class intervals, and are then graphed as bars whose areas are proportional to the frequencies in the intervals.

It is clear from the stem-and-leaf plot that the residual 3.229 (for the fourth observation) is relatively large. Is it, however, abnormally large given the assumptions of the linear model? This is a potentially important question, because the value of $h_{44} = 0.1108$ is not small (though it does not exceed $2(k + 1)/n = 0.2$): The fourth observation is a moderate-leverage point in the design. Moreover, as is apparent in Figure 1.5, the fourth observation "pulls" the regression line away from the origin and decreases its slope, thus lending substantive support to Sahlins's argument (described in Section 1.1) concerning the nature of primitive economics.

Influence The actual importance of an outlying observation, then, depends both upon the size of its residual and upon its leverage. Several authors have developed methods for identifying influential data, the most comprehensive treatments to date appearing in Belsley, Kuh, and Welsch (1980: Ch. 2) and Cook and Weisberg (1982). Beckman and Cook's (1983) review article on outliers, and the commentary following it, also include much relevant material.

One straightforward approach to assessing influence is to examine the impact on the fitted regression coefficients of deleting each observation in turn. Let $\mathbf{b}_{(-i)}$ denote the vector of estimated coefficients with the ith observation eliminated; then $\mathbf{d}_i \equiv \mathbf{b} - \mathbf{b}_{(-i)}$ shows the influence of the ith observation on the least-squares coefficients. The computational burden of finding the $\mathbf{d}_i$ (and, indeed, of calculating many of the quantities discussed in this section—see Belsley, Kuh, and Welsch, 1980: 69–83; and Velleman and Welsch, 1981) is not as great as it appears, for it may be shown that

$$\mathbf{d}_i = (\mathbf{X}'\mathbf{X})^{-1}\mathbf{x}_i \frac{E_i}{1 - h_{ii}} \tag{3.22}$$

where $\mathbf{x}_i$ is the ith row of $\mathbf{X}$ written as a column vector.

Stem	Leaf
−2	2
−1	07
−0	01133459
0	0145577
1	7
2	
3	2

FIGURE 3.5. Stem-and-leaf plot of studentized residuals from Sahlins's regression.

A disadvantage of this approach is that it generates a relatively large number ($n(k+1)$) of diagnostic statistics. Furthermore, since the elements of $\mathbf{d}_i$ are generally incomensurable, it is difficult to assess the overall impact of deleting an observation, even though each of the entries of $\mathbf{d}_i$ is directly interpretable. Cook (1977a, 1979) has suggested a method that summarizes $\mathbf{d}_i$ in a single index. The heuristic motivation for Cook's approach comes from hypothesis testing: To test the hypothesis H_0: $\boldsymbol{\beta} = \boldsymbol{\beta}_0$ that the vector of regression coefficients is equal to a particular set of values, we may calculate the F-statistic (see Section 1.2.6)

$$F_0 = \frac{(\mathbf{b} - \boldsymbol{\beta}_0)'\mathbf{X}'\mathbf{X}(\mathbf{b} - \boldsymbol{\beta}_0)}{(k+1)S_E^2}$$

Cook finds the F-value that would result if $\mathbf{b}_{(-i)}$ were tested as a hypothesis, and he uses this value as a scale-invariant measure of the squared distance between $\mathbf{b}$ and $\mathbf{b}_{(-i)}$:

$$D_i \equiv \frac{(\mathbf{b} - \mathbf{b}_{(-i)})'\mathbf{X}'\mathbf{X}(\mathbf{b} - \mathbf{b}_{(-i)})}{(k+1)S_E^2} = \frac{\mathbf{d}_i'\mathbf{X}'\mathbf{X}\mathbf{d}_i}{(k+1)S_E^2}$$

Using equation (3.22)

$$D_i = \frac{E_i^2}{S_E^2(k+1)} \times \frac{h_{ii}}{(1-h_{ii})^2} = \frac{\tilde{E}_i^2}{k+1} \times \frac{h_{ii}}{1-h_{ii}} \tag{3.23}$$

and Cook's distance is thus a simple increasing function of the residual and the leverage of the ith observation. The reader should be cautioned (as Cook, 1977b, and Obenchain, 1977a, have pointed out) that the D_i cannot be interpreted literally as F-statistics.

Table 3.6 indicates that, for Sahlins's data, by far the largest D_i belongs to the fourth observation, an impression that is reinforced by examining the $\mathbf{d}_i = (D_{Ai}, D_{Bi})'$, which are also shown in this table. A reasonable general approach, especially helpful when n and k are larger, is to examine $\mathbf{d}_i$ only for those observations with the several biggest D_i-values.

Belsley, Kuh, and Welsch (1980: Chapter 2) elaborate a strategy for detecting influential observations that examines the extent to which regression statistics are altered by deleting observations or by changing the weight accorded to them in estimation. Besides determining the influence of observations on regression coefficients, these authors consider regression outputs such as fitted values, residual variance, and sampling variances of coefficients. The effect on each of these statistics of deleting an observation depends, in a more or less complex manner, upon the observation's residual and leverage, as is the case for Cook's D-statistic in equation (3.23), and for $\mathbf{d}_i$ in equation (3.22).

Testing for Outliers We have shown that Sahlins's fourth observation is relatively influential, but is it so discrepant that it should be discarded or otherwise accorded special treatment? There has been much recent discussion about the detection of outlying observations in data (see Barnett and Lewis, 1978, and Hawkins, 1980, for comprehensive reviews). Some of this work is specifically applicable to linear models. One approach to outlier detection—indeed, the approach most frequently employed—is to test the statistical significance of the most discrepant observation (here, the largest absolute studentized residual), rejecting the observation as an outlier if its occurrence is sufficiently improbable given the model that has been specified to generate the data.

It was mentioned in Section 3.2.2 that each studentized residual follows a t-distribution with $n - k - 2$ degrees of freedom. If we suspect *a priori* that a *particular* observation is an outlier, a t-test for the corresponding E_i^* is appropriate. In practice, however, it is unusual to predict that a specific observation is discrepant, and we rather wish to test the statistical significance of the largest absolute E_i^*, regardless of the observation to which it belongs.

An upper bound for the critical value of $E_m^* \equiv \max|E_i^*|$ for a test at the $100a\%$ level may be obtained from the Bonferroni inequality: Testing the largest $|E_i^*|$ is equivalent to performing simultaneous, separate two-tailed tests on the n studentized residuals. Suppose that we do each of these tests at the $100b\%$ level. Then, under the null hypothesis, the probability of rejecting at least one of the n residuals (i.e., the probability of finding an outlier) is no greater than nb. Recall that the overall probability for the test is to be a; b must be picked so that $nb \leq a$, and thus we take $b = a/n$. Since b is divided equally between the two tails of the t-distribution, the critical value of E_m^* is given by $t_{b/2}(n - k - 2) = t_{a/2n}(n - k - 2)$—the t-value corresponding to a right-tail probability of $a/2n$.

Evidence presented by Prescott (1975) suggests that this bound closely approximates the actual critical value of E_m^*; indeed, it has been shown that in most instances the Bonferroni bound is exact (Beckman and Cook, 1983; also see Tietjen, Moore, and Beckman, 1973; Ellenberg, 1973, 1976; Lund, 1975; and Cook and Prescott, 1981). A practical difficulty in applying the result is that it employs nonstandard percentage points of the t-distribution, and therefore requires special tables. Approximate critical values for E_m^*, computed by a method outlined in Bailey (1977), are given in Appendix D. Alternatively, we may multiply the two-tail p-value for E_m^* by n.

For the Sahlins data, $E_m^* = 3.229$, which just exceeds the critical value 3.222 for $a = .10$. Roughly, then, a data point as discrepant as the fourth observation would occur in 10 percent of samples, if the model is correct.

Anscombe (1960) has suggested a conceptually different approach to the rejection of outliers, developing an analogy with the purchase of insurance. Rather than testing the statistical significance of a suspected outlier, Anscombe evaluates the increase in estimator mean-squared error that results from incorrectly discarding an outlying observation of particular size (i.e., an observation that, though unusual, is in fact produced by the mechanism assumed to generate the data). Under the assumptions of the linear model, including the assumption of normality, the least-squares estimator is the best unbiased estimator. An outlier-rejection rule that is indifferent to the sign of residuals will not introduce bias and, hence, cannot improve the behavior of the least-squares estimator *if* the assumptions hold. Consequently, an increase in mean-squared error is the "premium" that the researcher pays for a specific policy of outlier rejection. In return for this premium, the researcher obtains protection against invalid observations, that is "insurance" against the harmful influence of discrepant data values not produced by the putative data-generating process. Higher levels of protection are bought at the cost of higher premiums.

Anscombe and Tukey (1963) describe a method for calculating approximate rejection values for standardized residuals from linear models.[12] Let p denote the desired premium, expressed as a proportion; $p = .05$, for example, implies an outlier rejection policy resulting in a five-percent increase in estimator mean-squared error for a correct model. Find the unit-normal deviate z corresponding to a tail probability of $p(n - k - 1)/n$, and define $m = 1.40 + 0.85z$. Let $\tilde{E}_p$ represent the value beyond which the largest standardized residual will be identified as an outlier at the specified premium; then

$$\tilde{E}_p = m\left[1 - \frac{m^2 - 2}{4(n - k - 1)}\right]\sqrt{\frac{n - k - 1}{n}}$$

Using equation (3.21), we may express this rejection rule in terms of the largest studentized residual:

$$E_p^* = \frac{\tilde{E}_p\sqrt{n - k - 2}}{\sqrt{n - k - 1 - \tilde{E}_p^2}} \tag{3.24}$$

For Sahlins's simple-regression data, $n = 20$ and $k = 1$; using a premium of five percent, $z = 1.695$, $m = 2.841$, and $\tilde{E}_{.05} = 2.468$. Thus, from equation (3.24), $E_{.05}^* = 2.949$. The fourth observation, therefore, with a studentized residual of 3.229, is identified as an outlier. When the simple-regression model is estimated with this observation deleted, the fit becomes

$$\hat{Y} = \underset{(0.396)}{1.00} + \underset{(0.251)}{0.722}\,X \qquad r^2 = .326 \qquad S_E = 0.368$$

Compared with the original fit

$$\hat{Y} = \underset{(0.468)}{1.38} + \underset{(0.300)}{0.517}\,X \qquad r^2 = .141 \qquad S_E = 0.454$$

[12]Anscombe and Tukey employ normed residuals (in our terminology), but discuss balanced analysis-of-variance designs in which the q_{ii}, and hence the variances of the residuals, are identical. They suggest, however, using standardized residuals (again, our terminology) when the q_{ii} differ.

the intercept is now smaller, the slope is larger (and statistically significant), and the correlation is stronger. Sahlins's expectation of a positive intercept holds up, however.

In light of Sahlins's characterization of primitive economics, it would incidentally be interesting to ascertain the social position of the fourth household in the village. Thus, the detection of outliers may result in positive findings, not just in discarding discrepant observations.[13] Indeed, we should be loath simply to throw out an observation without understanding why it is discrepant—whether the outlier is due to an error in the data or reflects the substantive unusualness of an observation. Although this sort of knowledge of individual observations is hard to come by except in small-sample research, it is precisely in small samples that outliers have the greatest potential for adversely affecting results.

Multiple Outliers and Influential Subsets Although it is possible to develop tests for a specific number of outlying observations (see Barnett and Lewis, 1978), it would be unusual to specify beforehand the expected number of outliers. (See Beckman and Cook, 1983, for a review of current work on multiple-outlier methods.) The most common approach (suggested, for example, by Anscombe, 1960) is to test for a single outlier; if an outlier is found, it is removed from the sample, the model is refit, and once more the largest studentized residual is tested as an outlier. This procedure is repeated until no more outliers are detected. To test instead for multiple outliers employing the original fit together with a single-outlier rejection rule would be potentially misleading, since the original fit could be markedly affected by the largest outlier. A disadvantage of the sequential approach is that multiple outliers may "mask" each other's presence (see the example of joint influence given in Figure 3.6). For Sahlins's regression, the sequential approach detects just one outlier—observation four.

A generally more satisfactory but more complex alternative is to examine how regression statistics are affected by the deletion of subsets of observations, thus extending the concept of influence to multiple observations. This approach, practical in relatively small samples, is adopted by Belsley, Kuh, and Welsch (1980: Ch. 2), Cook and Weisberg (1980), and Draper and John (1981). For instance, Cook and Weisberg (1980) extend the D-statistic to a subset of observations $\mathbf{i} = (i_1, i_2, \ldots, i_p)'$, defining

$$D_{\mathbf{i}} \equiv \frac{\mathbf{d}'_{\mathbf{i}}(\mathbf{X}'\mathbf{X})\mathbf{d}_{\mathbf{i}}}{(k+1)S_E^2}$$

[13]An alternative to discarding outliers is to use an estimation method that accords them relatively little weight—unlike least squares, which accentuates the effect of outliers on the fit by squaring deviations from the regression surface. See, for example, Mosteller and Tukey (1977: Chapter 14).

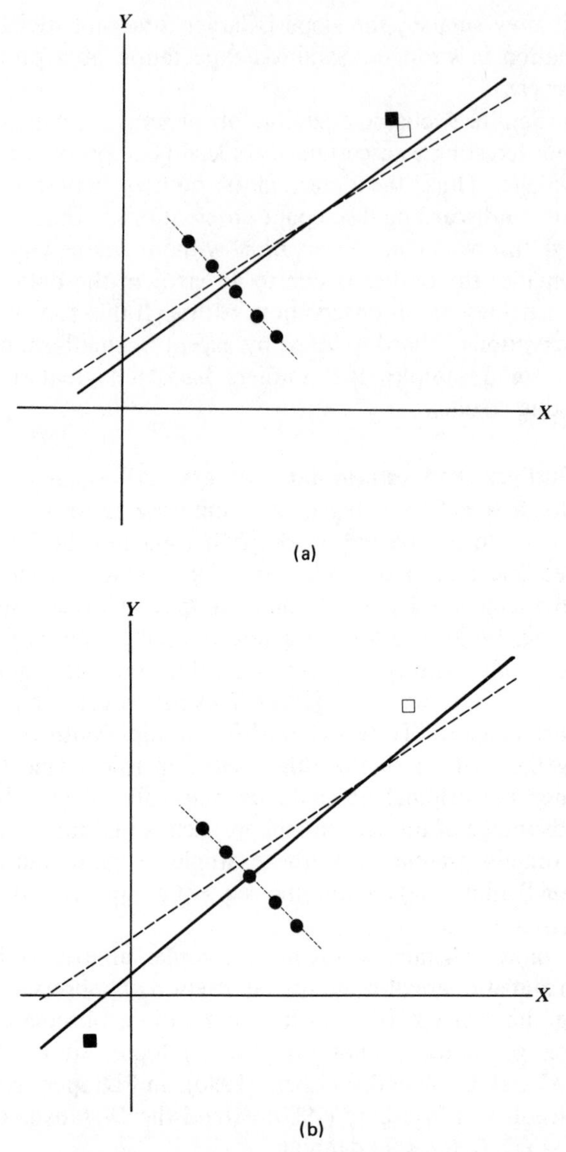

FIGURE 3.6. Joint influence in simple regression. (a) A jointly influential pair of observations. (b) A widely separated jointly influential pair. (c) An individually but not jointly influential pair.

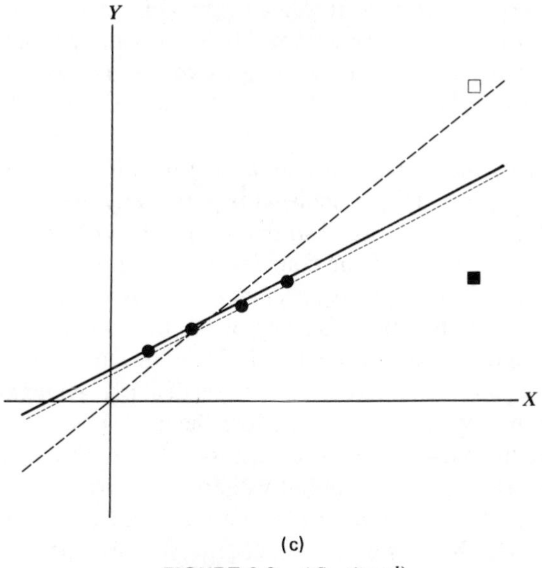

(c)

FIGURE 3.6. (*Continued*).

where $\mathbf{d_i} \equiv \mathbf{b} - \mathbf{b}_{(-\mathbf{i})}$ gives the impact on the regression coefficients of deleting the subset **i**.

Observations can be *jointly influential* without having large individual D_i-values. For example, two high-leverage observations with similar independent-variable values and similar dependent-variable scores may substantially alter the fitted coefficients if the observations are removed simultaneously but not when they are deleted individually. High-leverage observations on opposite sides of the data can behave similarly. Two extreme cases are illustrated in Figure 3.6(a) and (b), where the jointly influential observations are plotted as squares. Conversely, Figure 3.6(c) shows that a pair of observations may be influential individually but not jointly. In these figures, the solid line indicates the regression of Y on X for the full dataset; the heavier broken line shows the regression when the solid square is deleted; and the lighter broken line shows the regression when both squares are deleted. Plots of residuals against independent variables (discussed in Section 3.2.5) are often useful for revealing jointly influential subsets of observations.

3.2.4. Detecting Nonnormality of Errors

Many of the properties of least-squares estimation developed in Section 1.2.5 do not require the assumption of normally distributed errors: The unbias of the least-squares estimator, the derivation of its sampling variance, and even the Gauss–Markov theorem do not depend upon normality. Although the usual t- and F-tests employed for least-squares estimation are derived under the assumption of normal errors, if residual degrees of freedom are large (i.e., if the sample size is large), then these tests will be approximately valid even if the errors are not normally distributed, so long as very broad and nonrestrictive conditions are met (see Scheffé, 1959: 334–335; and Arnold, 1981: Chapter 10).

It is often stated, then, that least-squares estimation is "robust" under departures from normality. This conclusion is potentially misleading, however, because although the least-squares estimator may be *robust of validity* (the p-values for hypothesis tests and the confidence levels for confidence intervals are not grossly distorted), it is not *robust of efficiency*: When the assumptions of the linear model are violated, the efficiency of the least-squares estimator may decline drastically and become markedly inferior to the efficiency of alternative estimators. Least-squares estimation seems particularly vulnerable to error distributions with heavy tails (see, e.g., Mosteller and Tukey, 1977: Chapter 1).[14] It is therefore of some interest to determine whether the data are roughly consistent with the assumption of normally distributed errors.

As we might expect, the sample residuals are our key to the distribution of the errors. Although we have established that, under linear-model assumptions, **e** is *not* a sample from a (common) normally distributed population, differences in the variances of the E_i and correlations between them will be negligible even for relatively small samples. Moreover, we may examine the studentized residuals, perhaps comparing them to the t-distribution rather than to the normal distribution. Another, though more complex, approach is to transform the least-squares residuals to a normally and identically distributed set (**z**), using an orthonormal basis for the error space, as explained in Section 1.2.5.

Probably the most generally useful methods for detecting gross departures from normality are graphical. As mentioned in Section 3.2.3, we can derive a general impression of the residual distribution from a histogram that dissects the range of the residuals into class intervals. Nevertheless, an examination of the cumulative distribution of the residuals is a more promising approach here, for it does not require the establishment of arbitrary class boundaries, and it retains more resolution in the tails of the distribution, where data tend to be sparse, and where serious departures from normality may occur.

Quantile–quantile (Q–Q) cumulative normal-probability plots (Wilk and Gnanadesikan, 1968; Gnanandesikan, 1977: Chapter 6) provide a convenient method for comparing the distribution of the residuals to the normal distribution. Suppose that we are working with the studentized residuals and that the sample size is large enough to ignore differences between the normal and t-distributions. As a first step, arrange the residuals in ascending order: $E^*_{(1)} \leq E^*_{(2)} \leq \cdots \leq E^*_{(n)}$. Then $E^*_{(i)}$ has $(i - 1/2)/n$ proportion of the data

[14]Error distributions with heavy tails are especially likely to give rise to outliers. Screening data for outliers, therefore, should improve the performance of least-squares estimation when the error distribution has heavy tails.

below it.[15] Next, find the value z_i from the unit-normal distribution $Z \sim N(0,1)$ so that $Pr(Z < z_i) = (i - \frac{1}{2})/n$. Finally, plot the z_i as horizontal coordinates against the $E^*_{(i)}$ as vertical coordinates.[16] If the E_i^* are independently drawn from a normal distribution (which, of course, is only approximately true under the assumptions of the model), then the Q–Q plot should be roughly linear. Departures from normality are manifest as various sorts of nonlinearity. In fitting a line visually to a normal-probability plot, we pay attention to the middle of the distribution—say, that portion between the first and third quartiles. Analytic methods for fitting a line to a Q–Q plot are discussed in Mage (1982).

Some artificially constructed illustrative normal-probability plots are shown in Figure 3.7. All are for samples of size 100. Figure 3.7(a) is for a sample drawn from the unit-normal distribution; Figure 3.7(b) is for a sample drawn from the positively skewed $\chi^2(2)$ distribution; and Figure 3.7(c) is for a sample from $t(2)$, a distribution that has much heavier tails than the normal distribution. (Graphs of chi-square and t-distributions appear in Appendix C, Figures C.4 and C.5.)

A normal-probability plot for the studentized residuals from Sahlins's simple regression is displayed in Figure 3.8(a). Note that the outlying observation appears clearly in this plot as a point skewed to the right. Normally, we deal with outliers before proceeding to further analysis. Figure 3.8(b) shows a residual normal plot for Sahlins's regression after the outlying observation has been deleted; here, the plot appears quite straight.

A word of caution is in order concerning the interpretation of normal-probability plots: In small samples, where there are few residual degrees of freedom, even radical departures from normality of errors may give rise to apparently normally distributed residuals; Andrews (1979) presents an example of this phenomenon. On the other hand, there can be a great deal of noise in small-sample plots, giving the appearance of nonnormality even when the errors are in fact normal (see the illustrations in Daniel and Wood, 1980: 34–43).

A variety of methods are available for formally testing normality of errors in linear models. Since formal tests are likely to be less useful than graphical examination of residuals, however, these methods will be described only briefly here.

One approach to testing normality of errors is to treat the least-squares residuals as an independent sample drawn from a common distribution. This approach is reasonable for large samples, where covariances between residuals

[15] The fraction $(i - \frac{1}{2})/n$ is chosen by convention, to avoid cumulative proportions of zero and one. There are other reasonable choices (see Mage, 1982, for a comprehensive discussion). Anscombe and Tukey (1963), for example, suggest using $(3i - 1)/(3n - 1)$. The differences, of course, are slight unless n is very small.

[16] Some computer programs place normal quantiles on the vertical axis and ordered residuals on the horizontal axis. The user must therefore pay attention to the convention employed before attempting to interpret the plot.

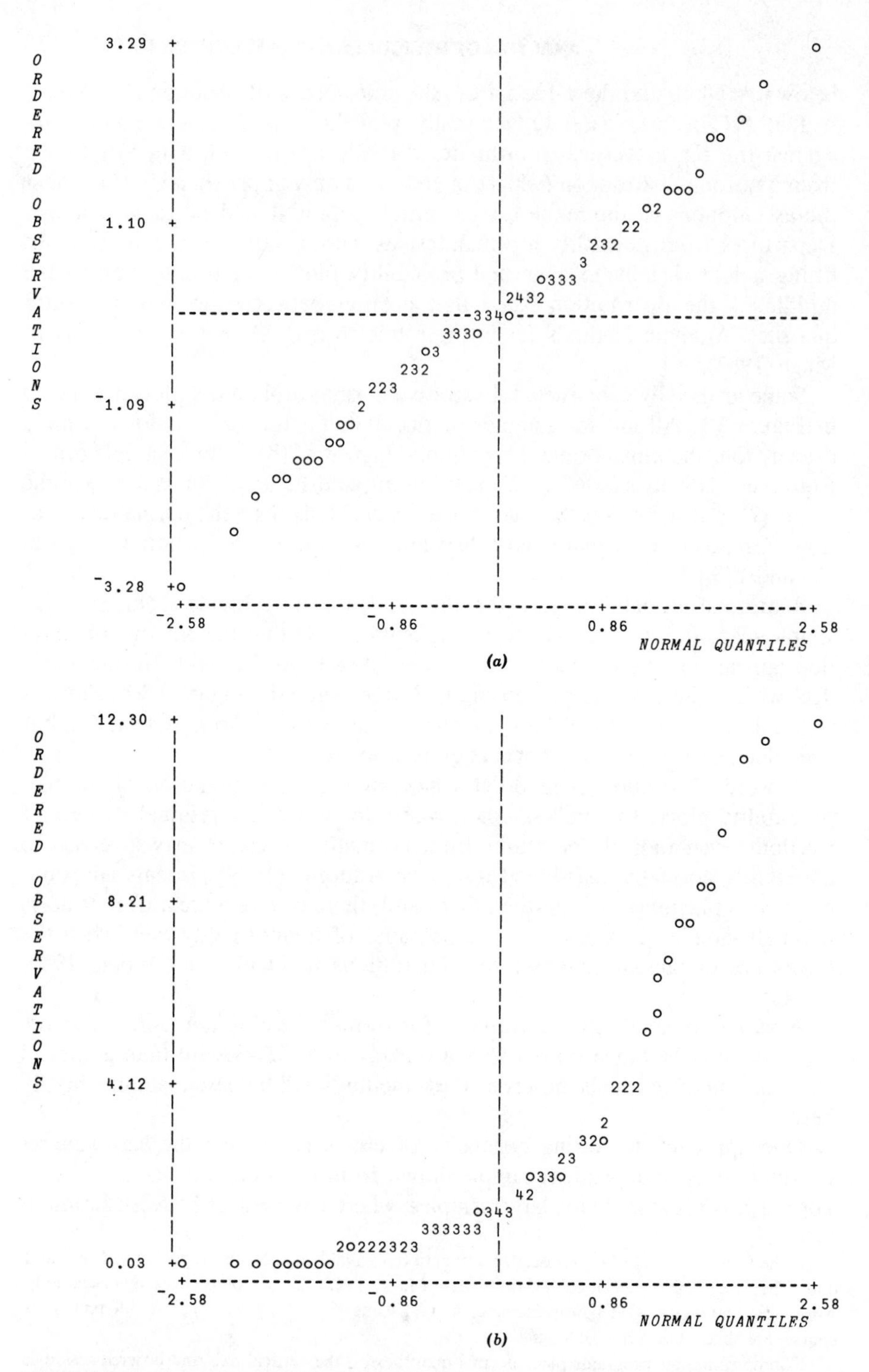

FIGURE 3.7. Illustrative normal Q–Q plots for samples of size 100. (a) Sample from $N(0, 1)$. (b) Sample from $\chi^2(2)$. (c) Sample from $t(2)$.

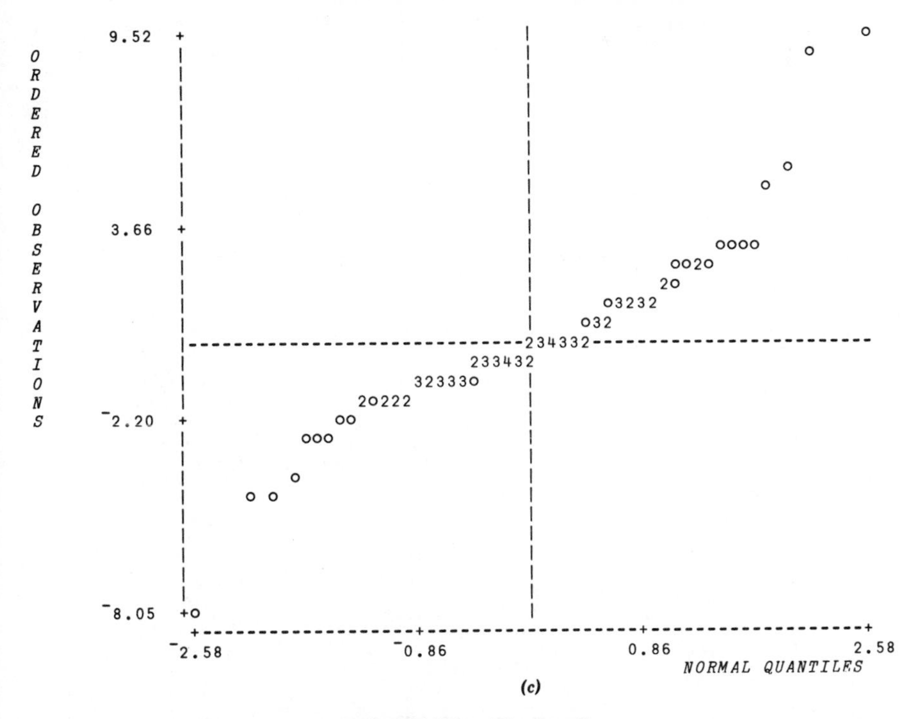

(c)

FIGURE 3.7. (*Continued*).

and differences between their variances are vanishingly small. A more strictly correct procedure would be first to transform the least-squares residuals employing an orthonormal error-space basis and then to proceed to a test. In either event, there are many tests for normality available, as described, for example, in Kendall and Stuart (1979: Chapter 30).

Another approach to testing normality is to employ test statistics specifically constructed for least-squares residuals. Anscombe (1961; 1981: Appendix 2) describes two such tests, based upon the skewness (i.e., asymmetry) and kurtosis (relative flatness) of the residual distribution. Although fairly complex to compute, these tests correct the usual normal-distribution skewness and kurtosis tests for the dependencies among the least-squares residuals.

3.2.5. Detecting Nonlinearity

Although least-squares estimation constrains residuals to be linearly uncorrelated with regressor variables, nonlinear relationships between residuals and regressors are possible. It is therefore useful and important to plot residuals against independent variables to detect departures from linearity. As we shall explain in Section 3.3.1, nonlinearity revealed by a residual plot may often be accommodated by modifying the model: For example, nonlinear terms in an independent variable may be included in the model, or the independent

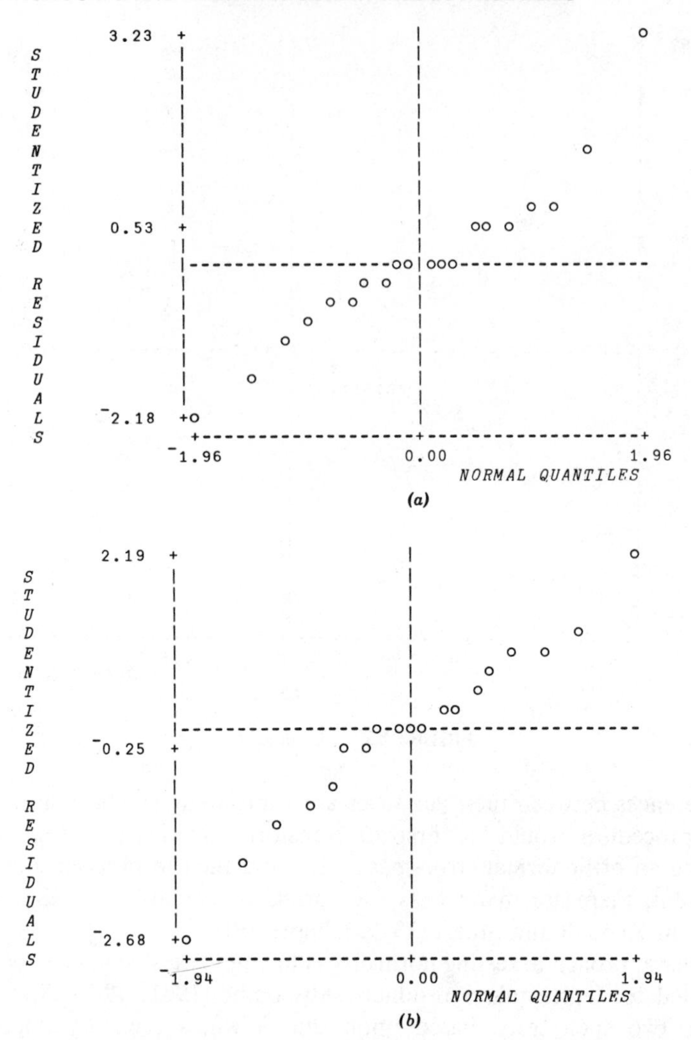

FIGURE 3.8. Normal Q–Q plots for studentized residuals from Sahlins's regression. (a) Full sample. (b) Deleting the fourth household.

variable in question may be transformed. The common practice of plotting the dependent variable (as opposed to residuals) separately against each of several independent variables can prove misleading when the independent variables are correlated, and, therefore, is a poor substitute for residual plots (see Daniel and Wood, 1980: 50–53, for an example).

Residual plots are sensitive instruments for detecting nonlinearity, but in determining how nonlinearity is to be modeled it is helpful to examine the linear component of a relationship along with the nonlinear one. Larsen and McCleary (1972), Wood (1973), and others have suggested plotting each independent variable against the residuals plus the fit for that independent variable. Larsen and McCleary term these graphs *partial-residual plots*.

The partial residual for the jth regressor is defined by

$$\mathbf{e}_j \equiv \mathbf{e} + B_j \mathbf{x}_j$$

where $\mathbf{e}$ is the vector of least-squares residuals (as before), $\mathbf{x}_j$ is the vector of scores for the jth regressor, and B_j is the estimated coefficient for this regressor. In examining the plot, it may be helpful to "adjust" the partial residuals to the typical level of the dependent variable. To this end, Larsen and McCleary suggest the following modification:

$$\tilde{\mathbf{e}}_j \equiv \mathbf{e} + B_j\left(\mathbf{x}_j - \overline{X}_j \mathbf{1}\right) + \overline{Y}\mathbf{1}$$

Here, $\overline{X}_j$ is the mean of the jth regressor, $\overline{Y}$ is the dependent-variable mean, and $\mathbf{1}$ is an $(n \times 1)$ vector of ones. Note that, by construction, the linear regression of $\mathbf{e}_j$ (or of $\tilde{\mathbf{e}}_j$) on $\mathbf{x}_j$ has slope B_j.

We shall use Angell's data on the moral integration of American cities (first discussed in Section 1.2.1) to illustrate the detection of nonlinearity. Although we found some evidence of regional effects in Angell's data (Section 2.1.2), we shall work with his original formulation, in which moral integration is specified to be a linear function of ethnic heterogeneity and geographic mobility (and, of course, error). The fitted model (recalled from Section 1.2) is

$$\underset{(2.19)}{\hat{Y} = 21.8} - \underset{(0.0552)}{0.167}\, X_1 - \underset{(0.0514)}{0.214}\, X_2 \qquad R^2 = .415 \tag{3.25}$$

where Y is integration, X_1 is heterogeneity, and X_2 is mobility.

Plots of studentized residuals from Angell's model against the two independent variables are shown in Figures 3.9(a) and (b). Note that there is a strong suggestion of nonlinearity in the relationship between integration and mobility. A partial-residual plot for mobility is shown in Figure 3.9(c). Here we can see that although integration appears to decline with mobility (controlling for heterogeneity), the decline becomes less pronounced as mobility increases, and may even be reversed at the highest levels of mobility. We shall return to Angell's data in Section 3.3.1.

More broadly, we are interested in detecting systematic departures from the specified functional form, even when the model under consideration is not a linear-regression model. Plotting residuals against independent variables is useful in this general context, although certain model deficiencies, such as failure to include important interactions, will not be revealed in such plots. "Nonlinearity," broadly interpreted, is the most serious of the problems considered in this chapter.

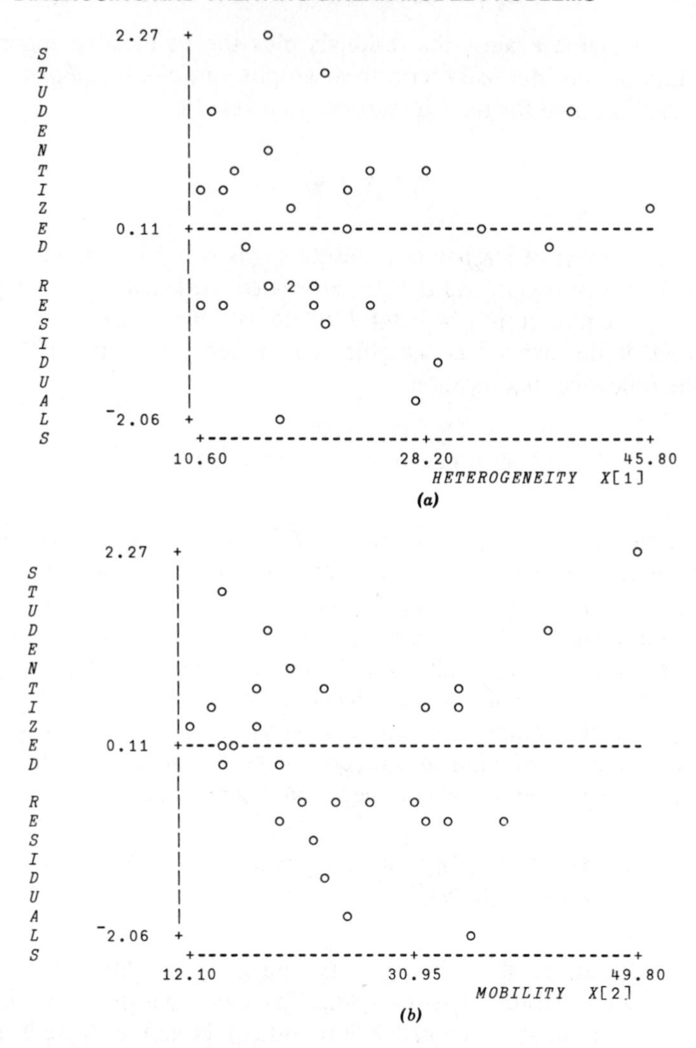

FIGURE 3.9. Residual plots for Angell's multiple regression. (a) Studentized residuals by heterogeneity. (b) Studentized residuals by mobility. (c) Partial residuals by mobility.

3.2.6. Detecting Heteroscedasticity

Like nonlinearity, *heteroscedasticity* (inequality of error variance) can occur in a variety of ways. Frequently, however, heteroscedasticity takes the form of a systematic increasing (or decreasing) relationship between error magnitude and the expected value of the dependent variable, or between error magnitude and an independent variable.

In the previous section, we introduced plots of residuals against independent variables as a means of detecting nonlinearity. The same plots may be

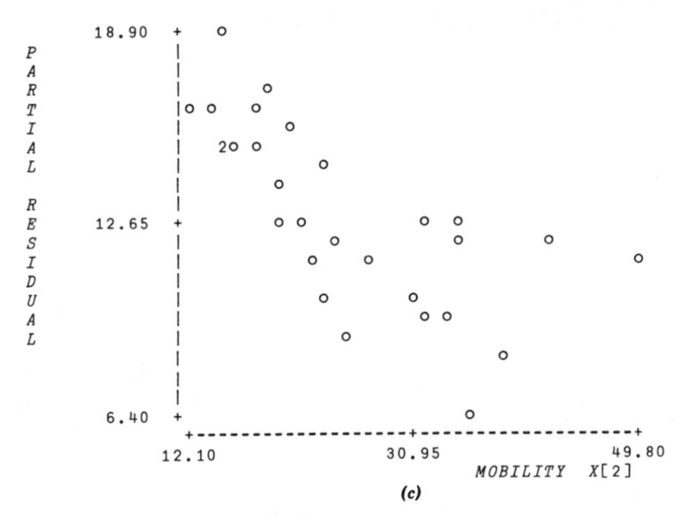

(c)

FIGURE 3.9. (*Continued*).

examined for evidence of heteroscedasticity—in particular, we may see a tendency of the residual scatter to become wider or narrower as the independent variable increases. To detect heteroscedasticity, however, it is more generally useful to plot residuals against fitted dependent-variable values.[17] Sometimes, a pattern of unequal residual magnitudes can be more clearly discerned in a plot of squared residuals (or absolute values of residuals) against $\hat{Y}$.

If heteroscedasticity is detected and is sufficiently extreme to cause concern, it is often possible to improve the behavior of the residuals by appropriately transforming the dependent variable. We shall develop this point in Section 3.3.2. Alternatively, *weighted-least-squares* estimation, which places relatively greater emphasis on low-variance observations, may be substituted for ordinary least squares (see Problem 3.13).

A variety of tests for heteroscedasticity have been suggested. Some of these are specifically applicable to discrete independent variables and will be discussed in Section 3.2.7. Several authors (Putter, 1967; Ramsey, 1969; Theil, 1971; cf. Hedayat and Robson, 1970) have proposed tests that use orthonormal error-space bases.

Anscombe (1961; 1981: Appendix 2) describes a test of heterogeneity of error variance that is useful in selecting a variance-stabilizing transformation. Anscombe assumes an exponential relation between the variance of the error and the expectation of the dependent variable: $V(\varepsilon_i) = \alpha e^{\gamma E(Y_i)}$. An estimate

[17]We do not plot residuals against Y because the residuals are a component of $Y = \hat{Y} + E$; Y and E therefore are correlated, making it difficult to examine a plot for evidence of heteroscedasticity.

of γ that is valid if this parameter is small is given by

$$H = \frac{\sum_{i=1}^{n} E_i^{**2}(\hat{Y}_i - \bar{Y})}{\sum_{i=1}^{n}\sum_{j=1}^{n} q_{ij}^2(\hat{Y}_i - \bar{Y})(\hat{Y}_j - \bar{Y})}$$

$$= \frac{\Sigma E_i^{**2}(\hat{Y}_i - \bar{Y})}{D}$$

where q_{ij} is the i, jth entry of the $\mathbf{Q}$ matrix (recall that $\mathbf{Q} = \mathbf{I}_n - \mathbf{H}$), E_i^{**} is the normed residual, and $D \equiv \Sigma\Sigma q_{ij}^2(\hat{Y}_i - \bar{Y})(\hat{Y}_j - \bar{Y})$. The sampling variance of H (conditional on the fitted values, $\hat{Y}_i$) is

$$V(H) = \frac{2(n-k-1)}{(n-k+1)D} \tag{3.26}$$

Anscombe and Tukey (1963) remark that H is a type of slope coefficient for the regression of the squared residuals on the fitted values. Indeed, if the sample size is large, then in most instances $q_{ii} \simeq 1$ and $q_{ij} \simeq 0$ (for $i \neq j$). Thus

$$H \simeq \frac{\Sigma E_i^{**2}(\hat{Y}_i - \bar{Y})}{\Sigma(\hat{Y}_i - \bar{Y})^2} \tag{3.27}$$

which is literally the least-squares slope for the regression of E^{**2} on $\hat{Y}$. An approximate normal-distribution test for heteroscedasticity may be obtained by dividing H by its standard error (i.e., $\sqrt{V(H)}$).

To illustrate some of the methods described in this section, we shall examine the data from Problem 2.2 (Table 2.2) on interlocking directorates among major Canadian corporations. These data were assembled by Ornstein (1976), who fit a linear model in which the number of executive and director interlocks maintained by a firm was the dependent variable, and the firm's assets, industry, and nation of control were independent variables. Assets was represented in the model as millions of dollars, and as the logarithm of this quantity.[18] Both industry and nation of control appeared as sets of dummy-variable regressors. The results of Ornstein's linear-model analysis will be presented and discussed in Section 3.3.2 (see Table 3.10; cf. Problem 2.2). For the present, we concern ourselves with the detection of heteroscedasticity in Ornstein's data.

A plot of normed residuals against fitted dependent-variable values is displayed in Figure 3.10(a). The magnitude of the residuals appears to increase

[18] To use both assets and log assets in the same equation is unusual.

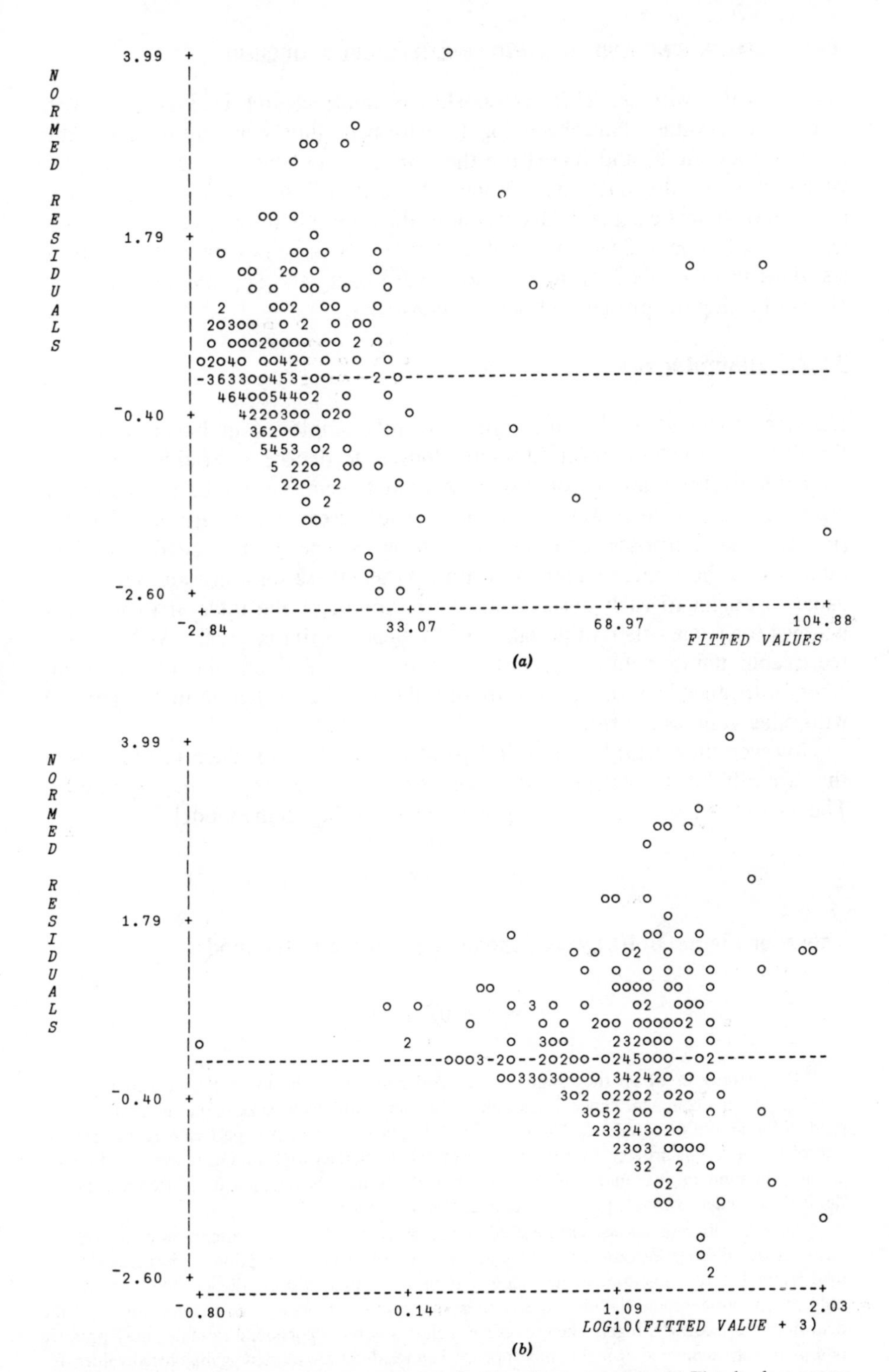

FIGURE 3.10. Residual plots for Ornstein's interlocking-directorate data. (a) Fitted values versus normed residuals. (b) Log fitted values versus normed residuals.

systematically with $\hat{Y}$. This relationship is made clearer in Figure 3.10(b), where the $\hat{Y}$-values have been log transformed, thus compensating for the positive skew in $\hat{Y}$, and increasing the resolution on the left side of the plot where most of the data lie.[19] Since the "sample" size is fairly large here ($n = 248$), H was calculated according to the approximation given in equation (3.27), yielding $H \simeq 0.047$. This value of H is very large in comparison with its standard error (0.0068, from equation (3.26) using the approximate value of D), confirming the presence of heteroscedasticity.

3.2.7. Discreteness

Discrete independent variables are frequently employed in linear models.[20] Recall that we use the term "discrete" loosely to denote a variable that takes on a relatively small set of scores. Although some variables are naturally discrete, it is not uncommon in social-science research for variables that are (for practical purposes) continuous, such as income, to be coded into class intervals at the point of data collection. When these data are analyzed, it is usual to assign to each observation the midpoint of the interval to which it belongs (or some other value taken as "typical" of its interval). As long as a reasonable number and range of categories are retained, the measurement errors introduced by this procedure are likely to be negligible in comparison with other sources of error.

However they arise, discrete independent variables produce residual plots that are difficult to interpret. Two contrived examples appear in Figure 3.11. The data for Figure 3.11(a) were generated according to the model

$$Y_i = 2 + 3X_i + \varepsilon_i$$

Those for Figure 3.11(b) were generated according to the model

$$Y_i = 2 + 0.5X_i^2 + \varepsilon_i$$

[19]The manner in which the residuals line up diagonally in the lower left corner of Figure 3.10(a) is worth a comment. Since the minimum number of interlocks is zero, the smallest possible residual for a given fitted value is $E = 0 - \hat{Y} = -\hat{Y}$. The heteroscedastic pattern here, however, is not solely due to this artifact. In extreme cases (as when a large proportion of scores are observed at the minimum or maximum value), special methods may be employed to accommodate a 'limited' dependent variable (see, for example, Maddala, 1983).

[20]Although discrete *independent* variables are accommodated by the assumptions of the general linear model, discrete *dependent* variables necessarily violate the assumption of normally distributed errors. Discrete quantitative dependent variables are often used in practice, however, and are unlikely to cause practical difficulties, unless the number of categories is very small or the distribution of scores is highly skewed. Nevertheless, discrete dependent variables may produce odd plots; for example, diagonal bands appear when residuals are plotted against fitted values, for a reason similar to that explained in footnote 19. When the dependent variable has very few categories or is qualitative, methods such as logistic regression (Section 5.1) may be employed.

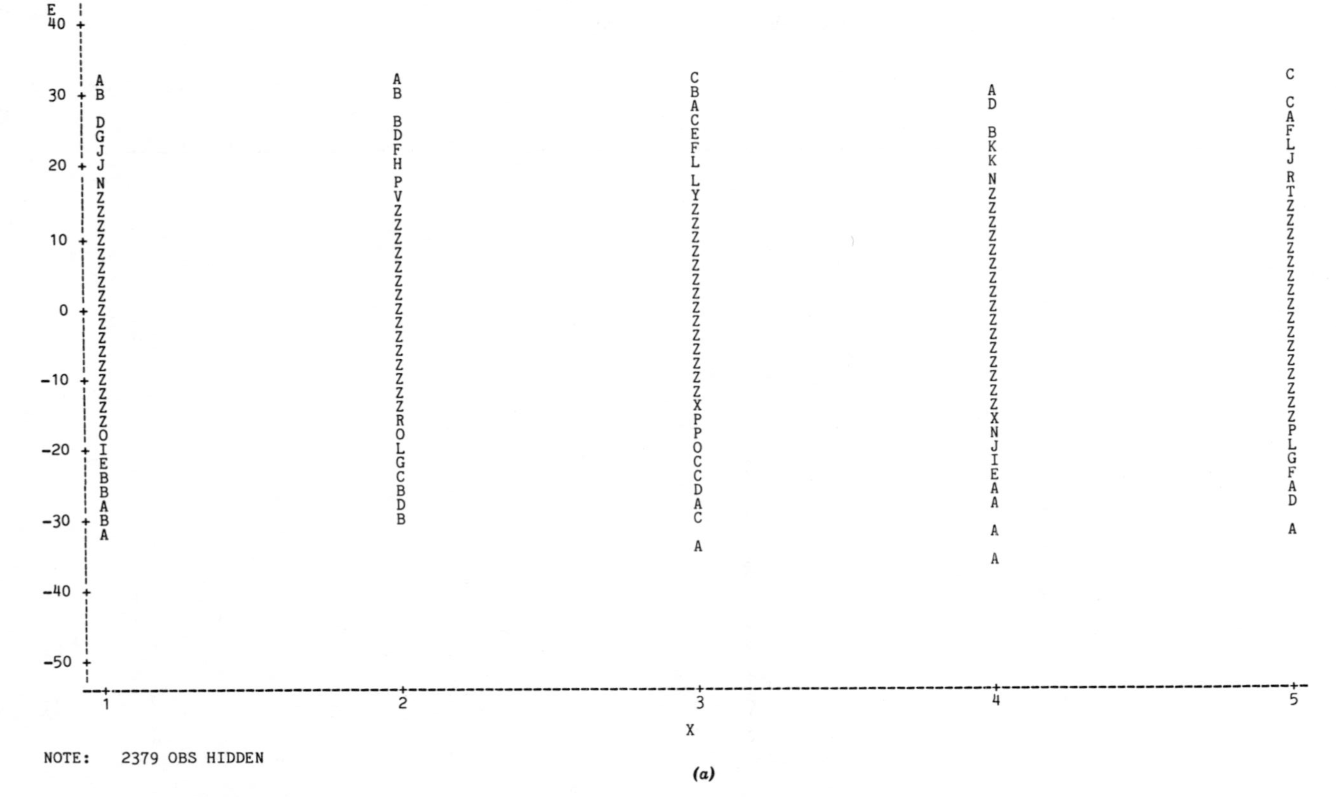

(a)

FIGURE 3.11. Uniformative residual plots for a discrete independent variable. (a) Data generated according to the linear model $Y = 2 + 3X + \varepsilon$. (b) Data generated according to the nonlinear model $Y = 2 + 0.5X^2 + \varepsilon$.

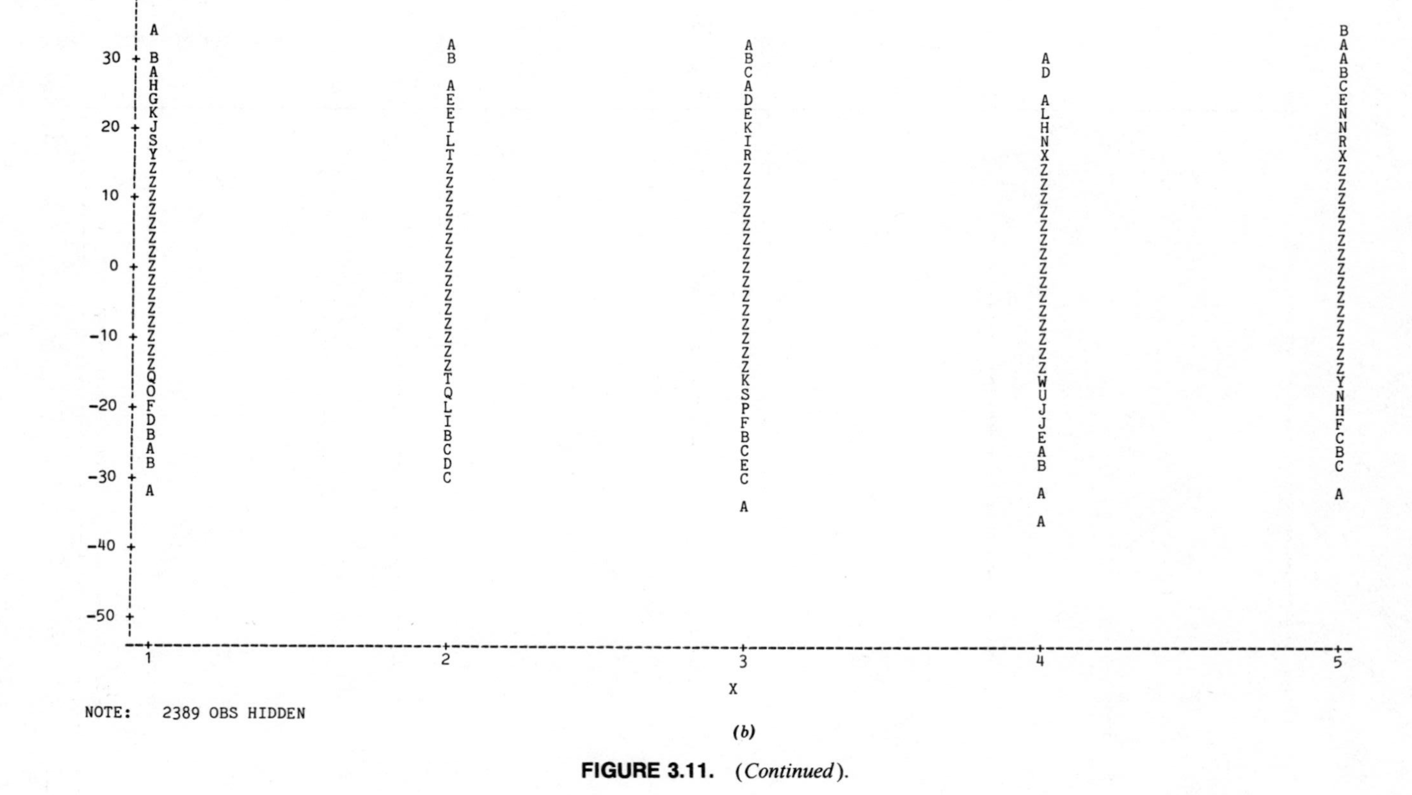

(b)

FIGURE 3.11. (*Continued*).

In each instance, we sampled errors from the normal distribution $N(0, 100)$; the same errors were used to construct the two datasets. Both datasets consist of 5000 observations, 1000 for each integral value of X between one and five. In both cases, we fit the simple-regression model $Y_i = \alpha + \beta X_i + \varepsilon_i$ to the data, producing the residuals plotted against X in Figure 3.11. This model is, of course, appropriate for the first dataset and inappropriate for the second, facts that are impossible to discern in the residual plots: So many data points fall on top of one another that it is very hard to form an impression of the distribution of the residuals for each value of X. The problem is exacerbated when a discrete device such as a line printer or printing terminal is used to produce scatterplots. Despite their general deficiencies, however, plots of this sort will reveal extreme outliers.

These difficulties are not insurmountable: Indeed, replicated observations at each of the several levels of a discrete independent variable present us with opportunities for checking the adequacy of the model fit to a set of data. One attractive approach is to examine side-by-side plots of the residual distributions in the different independent-variable categories. Summary graphs, such as Tukey's (1977: Chapter 2) "box plots" (see Section 3.2.8), which display a variety of distributional information, would serve nicely here. If computational facilities are limited, it is simpler and often adequate to calculate the mean and standard deviation of the residuals within independent-variable categories; these statistics may be transferred to the scatterplot, if desired. Results for the two illustrative datasets are shown in Table 3.7. Although the model appears adequate for the first dataset, the residual means for the second dataset display an unmistakable nonlinear pattern: The means systematically decrease and then increase.

Discrete independent variables present other diagnostic possibilities besides summary statistics and plots. By modeling a discrete numerical independent variable as a set of categories, it is possible to test simply for nonlinearity. First we fit the model as originally formulated, say

$$Y = \cdots + \beta_j X_j + \cdots + \varepsilon \tag{3.28}$$

TABLE 3.7. Means and Standard Deviations for the Residuals Plotted in Figures 3.11(a) and (b)

	Dataset(a)		Dataset(b)	
X	Mean	Standard Deviation	Mean	Standard Deviation
1	0.1821	9.9512	1.1821	9.9512
2	−0.2452	9.9440	−0.7452	9.9440
3	−0.0899	10.2050	−1.0899	10.2050
4	0.1872	9.9177	−0.3128	9.9177
5	−0.0341	10.3770	0.9659	10.3770

where X_j is the independent variable in question. Then we represent this independent variable by a set of zero/one dummy-variable regressors:

$$Y = \cdots + \gamma_1 D_1 + \gamma_2 D_2 + \cdots + \gamma_{m-1} D_{m-1} + \cdots + \varepsilon \tag{3.29}$$

where the dummy variable D_l is coded one if X_j takes on its lth of m values and is zero otherwise. The explained sum of squares for model (3.29) less that for model (3.28) provides us with an $m - 2$ degrees-of-freedom sum of squares for nonlinearity. This application of the incremental sum of squares approach takes advantage of the fact that a linear relationship between Y and X_j is a special case of a general (i.e., possibly nonlinear) relationship between these two variables. The analyses of variance given in Table 3.8 apply this procedure to our illustrative datasets.

If all independent variable are discrete, then the full sample divides into independent subsamples within combinations of independent-variable categories (cells). This is the case, for example, for the ANOVA designs treated in Chapter 2. The mean within-cell residuals should not differ greatly from zero; substantial nonzero values here imply inadequacies in the functional form of the model (e.g., nonlinearities or omitted important interactions).

There are, as well, a variety of tests for equality of variance that may be used to compare residual variances across cells of the design. The most commonly used test is one developed by Bartlett (1937). We shall describe the test for a one-way ANOVA—it is readily generalized to higher-order designs. The hypothesis to be tested is

$$H_0\colon \sigma_1^2 = \sigma_2^2 = \cdots = \sigma_m^2$$

where σ_i^2 is the population variance in the ith of m groups. Bartlett's test statistic

$$X^2 = \frac{n \log S_E^2 - \sum_{i=1}^{m} n_i \log S_i^2}{1 + \frac{1}{3(m+1)}\left(\sum_{i=1}^{m} \frac{1}{n_i} - \frac{1}{n}\right)} \tag{3.30}$$

TABLE 3.8. Nonlinearity Tests for the Data Plotted in Figures 3.11(a) and (b)

		Dataset(a)			Dataset(b)		
Source	*df*	Sum of Squares	*F*	*p*	Sum of Squares	*F*	*p*
X	4	94,857	233	< .001	98,890	243	< .001
Linear	1	94,719	932	< .001	94,719	932	< .001
Nonlinear	3	138	0.45	.72	4,171	13.7	< .001
Residuals	4995	507,584			507,584		
Total	4999	602,441			606,474		

is approximately distributed as $\chi^2(m-1)$ under the null hypothesis. In equation (3.30), following the notation of Chapter 2, n is the total number of observations; n_i is the number of observations in the ith group; $S_E^2 = \text{RSS}/(n-m)$ is the estimated error variance; and $S_i^2 = \sum_{j=1}^{n_i}(Y_{ij} - \bar{Y}_i)^2/(n_i - 1)$ is the sample variance in the ith group. Box (1953) shows that Bartlett's test is sensitive to nonnormality, decreasing its usefulness as a preliminary to analysis of variance.

Draper and Hunter (1969) have suggested a simple test that—ironically in the present context—finds its justification in the robustness of analysis of variance. The test can be applied whenever observations are partitioned into groups. Let $E_{i1}, E_{i2}, \ldots, E_{in_i}$ denote the n_i residuals in the ith of m groups; and let $\bar{E}_i = \sum E_{ij}/n_i$ be the mean residual in group i. (If the E_{ij} originate from an analysis of variance, where $E_{ij} = Y_{ij} - \bar{Y}_i$, then $\bar{E}_i = 0$.) Define $U_{ij} \equiv |E_{ij} - \bar{E}_i|$. A one-way analysis of variance over the groups, treating U as the dependent variable, tests the null hypothesis of equal error variances in the several groups. In small samples for which the hat values differ (which is *not* the case for balanced ANOVA designs), we may use $\tilde{E}_{ij}$ or E_{ij}^* in place of E_{ij}.

Further discussion of equality-of-variance tests may be found in many sources, including Scheffé (1959: 83–87), and Seber (1977: 146–149).

3.2.8. Summary Plots and Smoothed Plots

Scatterplots are sometimes difficult to interpret even when variables are not discrete. A common problem is uneven distribution on the variable defining the horizontal axis of the plot. In Figure 3.10(a), we recall, the skewed distribution of the fitted values caused low resolution in the portion of the plot containing most of the data. Moreover, where data are sparse, extreme residuals are less likely to be observed, even if the variability of the residuals is constant, or higher in the sparse region. Thus, unequal variances may be masked or equal variances made to appear different (see Cleveland and Kleiner, 1975, for an example). In addition, the systematic component in a plot may be overwhelmed visually by "noise."

One way of dealing with such problems is to divide the horizontal axis into intervals—say, by grouping the observations into equal-frequency classes based upon their horizontal coordinates. Average within-class vertical coordinates and measures of spread may then be plotted against average within-class horizontal coordinates. This approach is crude, but quick and simple. One difficulty, besides the loss of information produced by grouping and averaging, is the spurious increase in variation that occurs when values are averaged over a region in which they are rising or falling.

Parallel summary plots for each class retain more information about the residual distributions. An example showing the relationship between $\hat{Y}$ and E^{**} for Ornstein's data (cf. Figure 3.10) is given in Figure 3.12. Here, the data are divided into fifths based upon their $\hat{Y}$ values, the horizontal coordinate is given by the median $\hat{Y}$ in each fifth, and the within-class residual distribution is shown as a *box plot*. The following conventions (from McNeil, 1977: 6–9) are used to plot each distribution: (1) the central box is drawn between the quartiles; (2) the median is shown as a horizontal line within this box; (3) the dashed lines terminated by ×'s extend the distance of the box (i.e., the interquartile range, IR) beyond each end of the box or to the most extreme data value, whichever is closer; (4) data points outside of the ×'s are shown individually as O's, or if they are more extreme (beyond 1.5 × IR from the box), as ●'s; and (5) the frequency of multiple outlying points falling at the same print position is noted. The increasing spread of the residuals is clear in this plot, as is their tendency towards positive skew.

There are several methods of "smoothing" scatterplots that are more satisfactory, but that impose a greater computational burden (Cleveland and Kleiner, 1975; Tukey, 1977: Ch. 8, 9; Cleveland, 1979). These methods, incidentally, have application beyond residual plots.

Cleveland and Kleiner's technique is as follows: Let the horizontal coordinates be given by $X_1, X_2, \ldots, X_n$, and the corresponding vertical coordinates by $Y_1, Y_2, \ldots, Y_n$. (In the present context, these "Y's" are residuals, but we use a more general notation.) It is convenient, but not essential, to order the data by X-values. Let $\mathbf{x}_i = (X_{1(i)}, X_{2(i)}, \ldots, X_{m(i)})'$ denote the vector of m X-scores

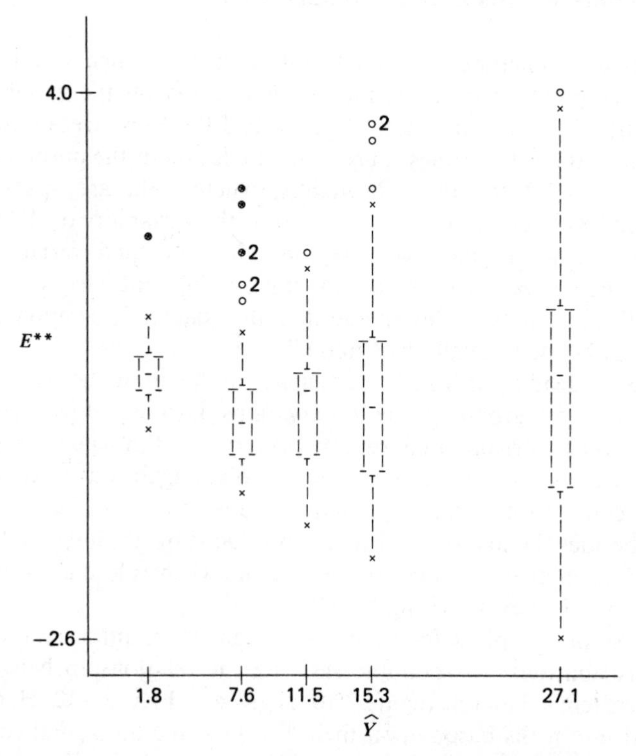

FIGURE 3.12. Schematic residual plot for Ornstein's interlocking-directorate data.

closest in value to X_i; and let $\mathbf{y}_i$ denote the vector of Y-scores for the same observations. Define $S_i \equiv \text{midmean}(\mathbf{x}_i)$; that is, arranging the entries of $\mathbf{x}_i$ in ascending order, delete the lowest fourth and the highest fourth, and take the mean of the remaining central half of these scores. Similarly, let $M_i \equiv \text{midmean}(\mathbf{y}_i)$. S_i and M_i are, respectively, the smoothed horizontal and vertical coordinates for the ith observation.

Next, subtract from each Y-score its smoothed value, obtaining $Z_i = Y_i - M_i$. Let $\mathbf{z}_i$ be the vector of Z's for the same observations as in $\mathbf{x}_i$; that is $\mathbf{z}_i = (Z_{1(i)}, Z_{2(i)}, \ldots, Z_{m(i)})'$. Define

$$L_i \equiv M_i + \text{lower semi-midmean}(\mathbf{z}_i)$$

and

$$H_i \equiv M_i + \text{upper semi-midmean}(\mathbf{z}_i)$$

The lower semi-midmean is the midmean of the lower half of the ordered $\mathbf{z}_i$, and the upper semi-midmean is similarly defined. Finally, S is plotted against M, L, and H, which represent a running median and quartiles. It is sometimes useful to superimpose the smoothed plot on the original scatterplot. The relatively complex definition of L and H prevents spurious increases in spread when Y is rising or falling.

An example "enhanced" scatterplot of this sort for the residual plot of Figure 3.10(a) is shown in Figure 3.13. To produce this plot, we took $m = 40$.

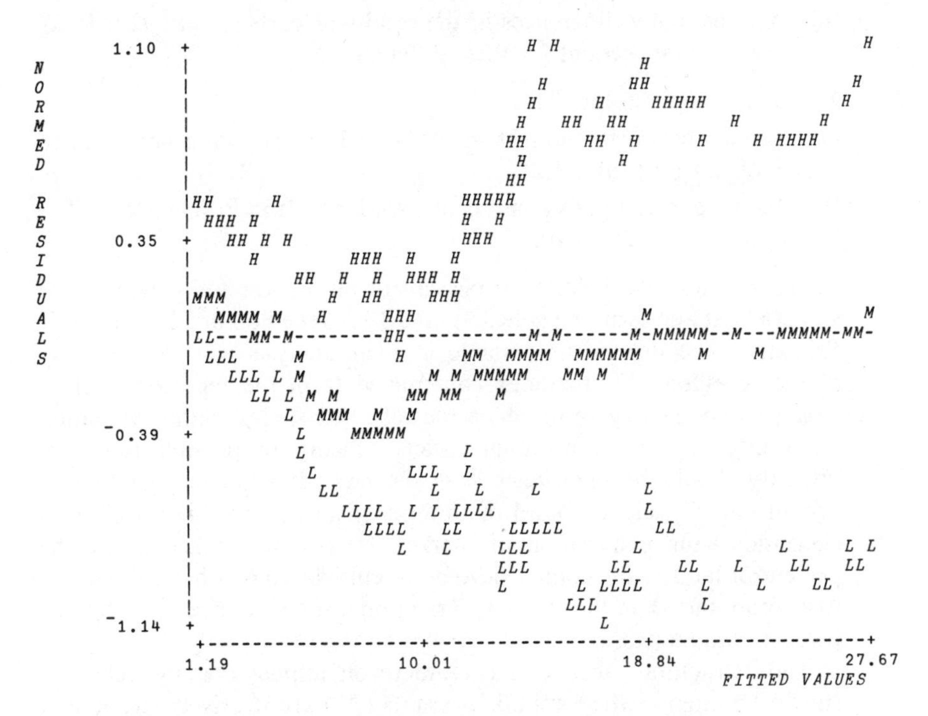

FIGURE 3.13. Enhanced residual plot for Ornstein's interlocking-directorate data.

Recall that, in this instance, we were able to achieve satisfactory visual results simply by log-transforming $\hat{Y}$ (in Figure 3.10(b)). Gentleman (1978) lists a FORTRAN computer program for producing enhanced scatterplots by a slightly more sophisticated version of the Cleveland and Kleiner method described here.

PROBLEMS

3.5. Perform a complete analysis of residuals for Angell's urban moral-integration data, checking for outliers, high leverage and influential observations, nonlinearity, and so on. If you find influential observations, explore the impact of removing them. Construct index plots for hat values, studentized residuals, D_i values, and the components of $\mathbf{d}_i$, graphing each of these statistics against the observation indices $(1, 2, \ldots, n)$.

(a) Employ residuals for the model fit in Problem 1.10, in which integration was regressed on ethnic heterogeneity and geographic mobility.

(b) Employ residuals for the model fit in Problem 2.1, in which integration was regressed on heterogeneity, mobility, and region.

(c) Are there any differences in the results of parts (a) and (b)? If so, how do you account for these differences?

3.6. Repeat Problem 3.5 for:

(a) Anscombe's regression fit in Problem 1.11. The data are given in Problem 1.3 (Table 1.2).

(b) Moore and Krupat's analysis of covariance fit in Problem 2.13. The data are in Table 2.10.

3.7. In order to test two theories of peasant revolt, Chirot and Ragin (1975) gathered data (shown in Table 3.9) on a 1907 rebellion in 32 counties of Romania. The dependent variable in their analysis was the intensity of the rebellion (I), an index constructed from the reported level of violence and the degree to which the rebellion spread within a county. According to the "transitional society" theory of peasant rebellion, intensity should be high when *both* the level of commercialization of agriculture (C) and the level of traditionalism (T) are high. Commercialization—the penetration of market forces—was measured by the percent of land in the county devoted to cultivation of wheat, the major cash crop raised in the region. Traditionalism was measured by the percent of illiterates.

The "structural" theory of peasant revolt implies that the rebellion should be intense where middle peasants (M) are relatively strong and

TABLE 3.9. Data on the 1907 Romanian Peasant Rebellion

County	I^a	C	T	M	G
1	-1.39	13.8	86.2	6.2	0.60
2	0.65	20.4	86.7	2.9	0.72
3	1.89	27.6	79.3	16.9	0.66
4	-0.15	18.6	90.1	3.4	0.74
5	-0.86	17.2	84.5	9.0	0.70
6	0.11	21.5	81.5	5.2	0.60
7	-0.51	11.6	82.6	5.1	0.52
8	-0.86	20.4	82.4	6.3	0.64
9	-0.24	19.5	87.5	4.8	0.68
10	-0.77	8.9	85.6	9.5	0.58
11	-0.24	25.8	82.2	10.9	0.68
12	-1.57	24.1	83.5	8.4	0.74
13	-0.51	22.0	88.3	6.2	0.70
14	-1.57	24.2	84.9	6.1	0.62
15	-0.51	30.6	76.1	1.3	0.76
16	-1.13	33.9	85.5	5.8	0.70
17	-1.22	28.6	84.2	2.9	0.58
18	-1.22	36.5	78.1	4.3	0.72
19	-0.86	40.9	84.4	2.3	0.64
20	-1.39	6.8	76.3	3.6	0.58
21	2.81	41.9	89.7	6.6	0.66
22	-1.04	25.4	83.2	2.5	0.68
23	1.57	30.5	80.2	4.1	0.76
24	4.32	48.2	91.0	4.2	0.70
25	3.79	46.0	90.5	3.7	0.68
26	3.79	45.1	85.5	5.1	0.64
27	-1.75	12.5	83.8	7.2	0.50
28	0.82	39.3	85.6	4.9	0.60
29	2.59	47.7	87.6	5.2	0.58
30	-0.86	15.2	87.3	10.8	0.42
31	-1.84	11.7	82.3	81.7	0.42
32	-1.84	25.6	80.1	68.4	0.26
Mean	0.0	26.31	84.27	10.17	0.627
Standard Deviation	1.77	11.92	3.86	17.38	0.110

Source: Reprinted with permission from Chirot and Ragin (1975: Table 1).
[a]Corrected from the original.

where the inequality of land tenure (G) is high. The strength of the middle peasantry was assessed by the percent of rural households owning between seven and 50 hectares of land. Inequality of land tenure was measured by a Gini coefficient.

Chirot and Ragin tested the two theories by regressing I on C, T, the product of C and T ($C \times T$), M and G. The first theory predicts a positive coefficient for $C \times T$, while the second predicts positive coefficients for M and G.

(a) Redo Chirot and Ragin's linear-model analysis, commenting on the results.

(b) Perform an analysis of residuals for the model fit in part (a), proceeding as in Problems 3.5 and 3.6.

3.8. Apply Bartlett's test and Draper and Hunter's test for equality of variance to the ANOVA datasets used as examples and exercises in Chapter 2.

3.9.† Partial Regression Plots: Consider the fitted regression equation

$$Y = B_0 + B_1X_1 + B_2X_2 + \cdots + B_kX_k + E$$

Problem 1.15 demonstrated that the least-squares coefficient B_1 could be calculated by: (i) regressing Y on X_2 through X_k, obtaining residuals $E_{Y:2\ldots k}$; (ii) regressing X_1 on X_2 through X_k, obtaining residuals $E_{1:2\ldots k}$; and (3) regressing $E_{Y:2\ldots k}$ on $E_{1:2\ldots k}$, producing B_1. Several statisticians have suggested a diagnostic plot of $E_{Y:2\ldots k}$ against $E_{1:2\ldots k}$, repeating this procedure for each of the k independent variables. Belsley, Kuh, and Welsch (1980: 30) call these graphs *partial-regression plots*; some authors (e.g., Atkinson, 1982) use the term "partial-residual plot" to refer both to these graphs and to those described in Section 3.2.5. Welsch (commenting on Atkinson, 1982) has suggested that partial-residual plots (of the sort discussed in Section 3.2.5) are superior for detecting nonlinearity, but partial-regression plots are generally better for discovering high-leverage and influential observations. Construct and interpret partial-regression plots for the data analyzed in Problems 3.5, 3.6, and 3.7, comparing these graphs to the corresponding partial-residual plots.

3.3. TRANSFORMATIONS OF VARIABLES

Deficiencies revealed by an analysis of residuals motivate model respecification. A linear model can often be improved by transforming one or more of the variables appearing in the model. In this section, we shall show how variable transformations can be employed to correct problems of nonlinearity and heteroscedasticity.

3.3.1. Transforming Nonlinearity

When we specify that a dependent variable Y is a linear function of an independent variable X, we usually choose the linear form strictly for simplicity and convenience. We may expect Y to increase (or decrease) with X, but we generally lack specific knowledge of the functional form of this relationship. Nevertheless, the linear specification often proves adequate, at least within the limited range in which the variables are observed and to an acceptable degree of approximation. At other times, however, a residual analysis reveals that although the *direction* of the relationship between Y and X is as expected, this relationship does not appear to be even roughly linear in *form*. In these instances, we may wish to specify Y as some nonlinear but monotone function of X. A particularly simple way of accomplishing this purpose is to replace X in the linear model by a monotone transformation of X.

Suppose, for instance, that the simple-regression model $Y = \alpha + \beta X + \varepsilon$ appears inadequate because increases in X are attended by more-than-linear increases in $E(Y)$, as depicted in Figure 3.14(a). If all X-values are positive, then $X' = X^2$ is a monotone transformation of X, and the model

$$Y = \alpha + \beta X' + \varepsilon = \alpha + \beta X^2 + \varepsilon \tag{3.31}$$

more adequately represents the relationship between Y and X. Figure 3.14(b) shows why the transformation $X' = X^2$ straightens the relationship: The transformation differentially "stretches" the X-axis, affecting larger values of X more than smaller values. Of course, the data in Figure 3.14 may be better represented by some other transformation of X than $X' = X^2$, or by a function of form different from $Y = \alpha + \beta f(X) + \varepsilon$; for the moment, however, let us suppose that equation (3.31) does a good job of capturing the relation between Y and X. Given the shape of this relation, it is certainly true that use of $X' = X^2$ in the simple-regression model is preferable to use of X.

Although the model in equation (3.31) is nonlinear in the independent variable X, it is linear in the parameters α and β, and is therefore representable as a linear equation in the transformed variable X'. Model (3.31), and indeed any model linear in the parameters, may be fit to data by our usual linear-model procedures, as long as the linear-model assumptions concerning the errors are reasonable.

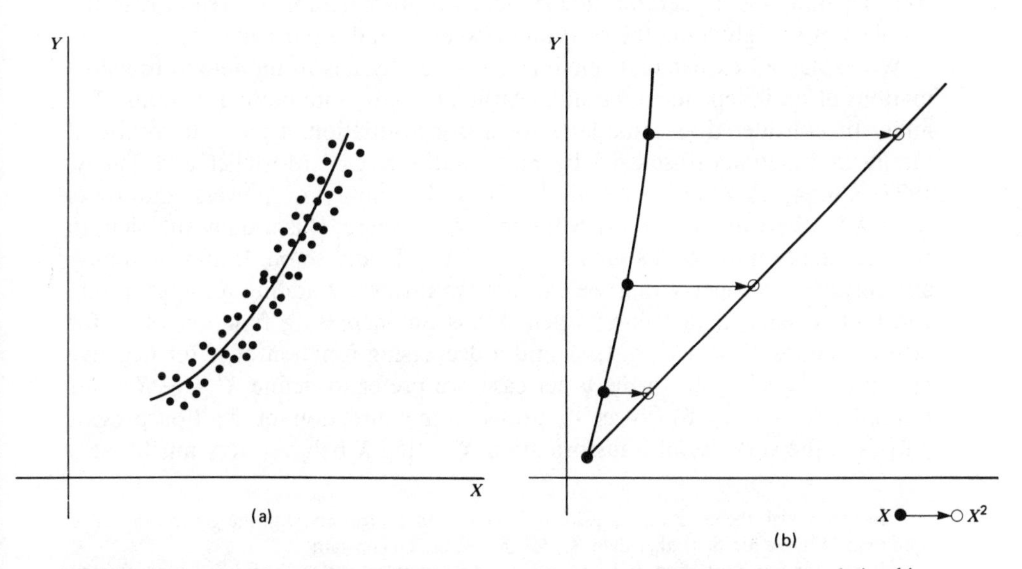

FIGURE 3.14. Straightening a nonlinear relationship. (a) Nonlinear monotone relationship between Y and X. (b) Transforming X to X^2.

In its least restrictive form, then, we may write the general linear model as

$$\begin{aligned} f(Y) &= \beta_1 f_1(\mathbf{x}) + \beta_2 f_2(\mathbf{x}) + \cdots + \beta_p f_p(\mathbf{x}) + \varepsilon \\ Y' &= \beta_1 X_1' + \beta X_2' + \cdots + \beta_p X_p' + \varepsilon \end{aligned} \tag{3.32}$$

where Y is the dependent variable, $\mathbf{x}_{(k \times 1)}$ is a vector of k independent variables, and the functions $f, f_1, \ldots, f_p$ do not involve unknown parameters. We require, of course, that $X_1', \ldots, X_p'$ be linearly independent, and that the errors satisfy the usual assumptions. Each $X_j' = f_j(\mathbf{x})$ may be a function of more than one independent variable in $\mathbf{x}$, thus encompassing models such as[21]

$$Y = \beta_1 + \beta_2 X_1 + \beta_3 X_2 + \beta_4 (X_1^2) + \beta_5 (X_2^2) + \beta_6 (X_1 X_2) + \varepsilon$$

Indeed, we made implicit use of this sort of specification in Chapter 2, where we defined dummy-variable regressors, interaction regressors, and so on, each of which may be regarded as a function of a set of (not necessarily numerical) independent variables.

Note that in equation (3.32), we may consider transformations $Y' = f(Y)$ of the dependent variable as well as functions $X_j' = f_j(\mathbf{x})$ of the independent variables. In seeking to linearize the relationship between the dependent variable and a particular independent variable, however, we should generally restrict ourselves to transformations $X_j' = f(X_j)$ of the independent variable in question. To transform the dependent variable in this context is to risk disturbing its relationship to other independent variables in the model; and to consider functions of several independent variables here is not relevant to the problem of straightening the relation between Y and a particular X_j.

We explained earlier that our interest generally lies in monotone transformations of an independent variable. Although many monotone functions of X might be considered as candidates for a transformation, a particularly useful family of functions (discussed by many authors, e.g., Mosteller and Tukey, 1977: Ch. 4; Box and Tidwell, 1962) is the family of powers and roots $X' = X^m$, where m may be any real number, save zero; it is usually sufficient to restrict attention to powers such as $\pm \frac{1}{2}$, ± 1, ± 2, and so on. If all values of X are positive, a property that we can always insure by adding an appropriate constant to each data value,[22] then X^m is an increasing function of X for positive values of m (e.g., $\frac{1}{2}, 1, 2$), and a decreasing function of X for negative m (e.g., $-\frac{1}{2}, -1, -2$). In the latter case, we prefer to define $X' = -X^m$ (for example, $X' = -1/X$) so as to preserve the direction of X. For present purposes, the very useful transformation $X' = \log X$ behaves very much like a

[21] In this model, the constant regressor is $f_1(\mathbf{x}) = 1$. In conformity with the general model of equation (3.32), we use β_1 (rather than β_0) for the regression constant.

[22] For power transformations to be effective, it is also necessary that the ratio of the largest to the smallest X-value not be too small. If we are dealing with the variable time coded in years, for instance, the values $1950, 1951, \ldots, 1980$ are not good candidates for transformation, since $1980/1950 = 1.015$. Here, we may subtract 1949 from each value to produce $1, 2, \ldots, 31$ prior to applying a transformation.

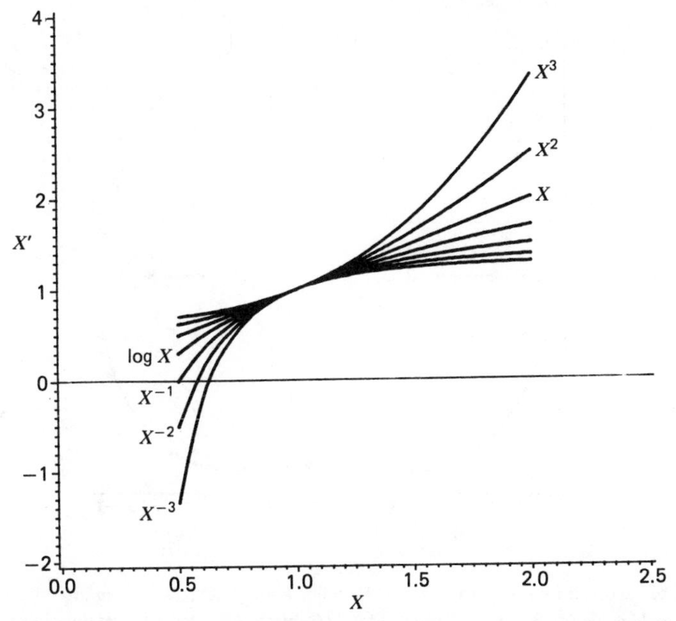

FIGURE 3.15. Ladder of powers, including log X. The transformation labelled "X^m" is $(X^m/m) - (1/m) + 1$. (Tukey, *Exploratory Data Analysis*, page 90, © 1977. Addison-Wesley, Reading, MA. Exhibit 18. Adapted with permission.)

"zeroth power" (indeed, $\lim_{m\to 0}[(X^m - 1)/m] = \log X$), and will be included as such in our family of transformations. Some of the commonly used power transformations, including log X, are shown in Figure 3.15.[23]

Simple nonlinear monotone relationships may be increasing or decreasing, and may be concave upward or concave downward. The four possibilities are shown schematically in Figure 3.16. In cases (a) and (c), we may straighten the relationship between Y and X by moving down the "ladder" of powers (Tukey's, 1977: 172, term) from X to $X^{1/2}$, log X, $X^{-1/2}$, X^{-1}, and so on; how far down the ladder we proceed depends upon the curvature of the relation between Y and X, and may be determined by trial and error. In cases (b) and (d), we proceed up the ladder to X^2, X^3, and so on. This procedure makes intuitive sense, for in (a) and (c) the change in Y is less than linear in X, while in (b) and (d) the change in Y is more than linear in X. Our usual tool for determining the form of the relationship between Y and a particular X_j is the partial-residual plot described in Section 3.2.5. An alternative analytic procedure for selecting a power transformation, based on the method of maximum likelihood, is explained in Problem 3.18 (following Section 3.4 on nonlinear models).

[23]In Figure 3.15, the curve labeled X^m is actually $(X^m/m) - (1/m) + 1$; for log X we plot log $X + 1$. As Mosteller and Tukey (1977: 102–103) explain, this device permits us to match the various powers at the value $X = 1$.

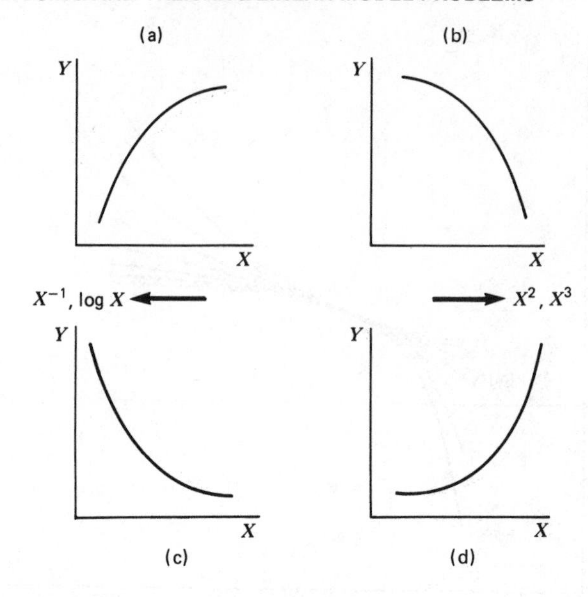

FIGURE 3.16. Choice of transformation. (Mosteller and Tukey, *Data Analysis and Regression*, page 84, © 1977. Addison-Wesley, Reading, MA. Exhibit 4. Adapted with permission.)

In our residual analysis for Angell's data, we noted a nonlinear relationship between the moral integration of cities and the geographic mobility of their residents (Figure 3.9). We suggested that the generally negative relationship between integration and mobility appears to level off as mobility increases, and may in fact be reversed at the highest levels of mobility occurring in the data. Assuming a monotone relationship between integration and mobility, we must move down the ladder of powers to straighten the relation between these two variables. A little experimentation reveals that the transformation $X_2' = -X_2^{-1}$ works well, providing the following fit:

$$\hat{Y} = \underset{(1.16)}{10.4} - \underset{(0.0497)}{0.194}\, X_1 - \underset{(26.9)}{146}\left(-\frac{1}{X_2}\right) \qquad R^2 = .542$$

Including both linear and quadratic terms in X_2 permits Y first to decrease and then to increase with X_2; this specification produces the fit[24]

$$\hat{Y} = \underset{(1.06)}{15.9} - \underset{(0.0472)}{0.194}\, X_1 - \underset{(0.0514)}{0.308}\left(X_2 - \bar{X}_2\right) + \underset{(0.00360)}{0.0123}\left(X_2 - \bar{X}_2\right)^2$$

$$R^2 = .600 \tag{3.33}$$

[24] X_2 is transformed to mean deviations $(X_2 - \bar{X}_2)$ to reduce the correlation between the linear and quadratic regressors in the model. This transformation, called *centering*, does not change the meaning or fit of the model, but by decreasing collinearity it tends to improve the accuracy of calculations.

Note that both of these models fit appreciably better than the original specification [equation (3.25)], for which $R^2 = .415$. The quadratic term in equation (3.33) is more than three times its standard error, indicating that this term leads to a statistically significant increment in the explained sum of squares.

Despite the improved fit of the curvilinear models, there is an alternative interpretation of these data. Figures 3.9(b) and (c) reveal that for most cities in Angell's sample there appears to be a strong, negative, and roughly linear relationship between moral integration and mobility. Five cities in the upper right of the plots (Des Moines, Denver, Peoria, Wichita, and San Diego) are somewhat detached from the rest and are jointly influential in determining the fit of the models that we have entertained: These observations decrease the magnitude of the mobility slope in the original model (equation (3.25)) and are responsible for the improved fit of the nonlinear specifications. Because the five cities are not obviously related, however, there appears to be no substantive justification for according them separate treatment.

3.3.2. Transforming Heteroscedasticity

Heteroscedasticity may often be eliminated by replacing the dependent variable with some monotone transformation $Y' = f(Y)$. As in the previous

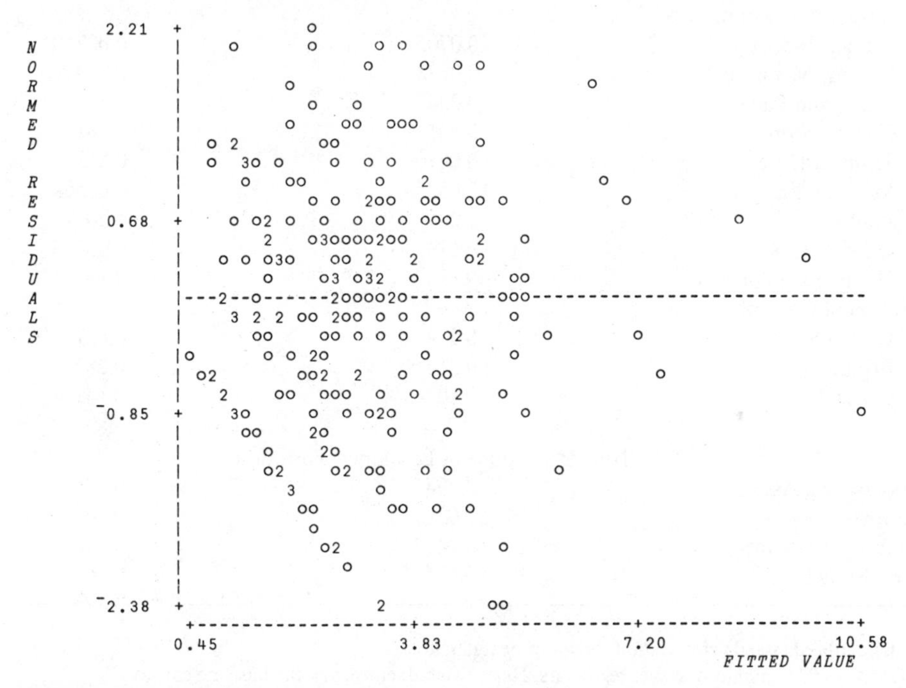

FIGURE 3.17. Residual plot for Ornstein's interlocking-directorate data after square-root transformation of Y.

section, the family of power transformations, including $\log Y$, proves generally useful for transforming positive Y. When large values of $E(Y)$ are associated with relatively large errors, we tend to stabilize the error variance by moving down the ladder of powers: Taking $Y' = \log Y$, for example, decreases differences between large values of Y relative to those between small values of Y. For the same reason, incidentally, moving down the ladder of powers tends to correct positive skew in the residuals. It is possible, of course, that transforming Y will introduce nonlinearity where none existed before, but we should generally be able to compensate by suitably transforming the independent variables in the model.

TABLE 3.10. Linear-Model Analysis for Ornstein's Data

	Dependent Variable[a]	
Regressor	Interlocks/16.083	$\sqrt{\text{Interlocks}}/2.023$
	(a) Estimated Coefficients	
Assets	0.000415	0.000238
Log Assets	0.423	0.678
Industry[b]		
Agriculture, Food, Light Industry	−0.0705	−0.00210
Mining, Metals, etc.	0.0878	0.213
Wood and Paper	0.334	0.397
Construction	−0.305	−0.281
Transport, etc.	0.0870	0.228
Merchandizing	−0.0525	0.0855
Banks	−1.28	−1.20
Other Financials	−0.123	0.00813
Holding Companies	−0.132	−0.180
Nation of Control[c]		
United States	−0.514	−0.621
Britain	−0.314	−0.282
Other	−0.103	−0.141
	(b) Net Proportions of Explained Variation	
Assets, Log Assets	0.294	0.225
Industry	0.035	0.040
Nation of Control	0.048	0.070
Full Model	0.675	0.569

[a]Standardized (standard deviation shown in denominator).
[b]Coded as 0/1 dummy-variable regressors, Heavy Manufacturing as the baseline category.
[c]Coded as 0/1 dummy-variable regressors, Canada as the baseline category.

Anscombe (1961) suggests employing his statistic H (introduced in Section 3.2.6 as a test for heteroscedasticity) to obtain a variance-stabilizing transformation of Y. First, insure that all Y-values are positive, adding a constant to the data, if necessary. Second, calculate $P = 1 - \frac{1}{2}H\overline{Y}$. Finally, use P for a power transformation of the dependent variable: $Y' = Y^P$; if $P = 0$, then take $Y' = \log Y$. An alternative, maximum-likelihood method for choosing P is developed in Problem 3.14.

For Ornstein's interlocking-directorate data, we found (in Section 3.2.6) that $H = 0.047$. Using Anscombe's rule, we obtain the power $P = 0.68$. Both square-root ($P = 0.5$) and log ($P =$ "0") transformations were tried, and the square root transformation proved more satisfactory. A plot of E^{**} against $\hat{Y}$ for the transformed data is shown in Figure 3.17. The value of H calculated for the transformed regression ($H \simeq 0.026$) is smaller than its approximate standard error (0.059); the suggested power transformation corresponding to this value of H is $P = 0.96$, confirming that the magnitude of the residuals no longer increases systematically with $\hat{Y}$.

Note that H does not have absolute meaning, but depends upon the level ($\overline{Y}$) of the dependent variable. This is clear in the formula for P. Since the square-root transformation makes the Y-scores smaller, the H of 0.026 for the transformed data is effectively much smaller than the H of 0.047 for the untransformed data.

The results of fitting Ornstein's model to the transformed data are shown in the right-hand panel of Table 3.10; these results do not differ greatly from those of the original analysis, which are given in the left-hand panel of the table. In this table, the dependent variables (Y and $Y^{1/2}$) have been standardized to facilitate comparisons between the two analyses.

PROBLEMS

3.10. Employ variable transformations in an attempt to correct any problems encountered in the residuals analysis for Angell's moral-integration data (Problem 3.5).

3.11. Repeat Problem 3.10 for Anscombe's regression analysis of state education expenditures [Problem 3.6(a)]. [These data are also analyzed by Chatterjee and Price (1977: 108) employing weighted least squares, described in Problem 3.13.]

3.12.† Ordinal Data in Regression: Ordinal variables—that is, quantitative variables without units of measurement (such as an attitude item coded "strongly disagree," "disagree," "neutral," "agree," and "strongly agree")—frequently appear in social research. These variables are often given arbitrary numerical scores prior to employing them in regression analysis. (For the attitude-item example, the numbers one through five would likely be assigned to the ranked categories.)

(a) Evaluate the validity of the following argument: The potential danger of arbitrarily scoring an ordinal independent variable X is that relationships involving this variable will be distorted. In particular, a relationship that is linear (for a "correct" assignment of numbers to the categories of X) may appear nonlinear, though still increasing or decreasing. If we plot residuals, however, the induced nonlinearity will be apparent, and can be corrected by transformation of X. Indeed, since we are rarely in a position to predict that a relationship is linear as opposed to simply monotone (i.e., increasing or decreasing), we need to entertain the possibility of nonlinearity in any event. These considerations suggest that, *for purposes of data analysis*, the distinction between numerical and ordinal quantitative independent variables is moot.

(b) Can this argument be extended to ordinal dependent variables?

3.13.*† Weighted Least Squares (An Alternative Approach to Heteroscedasticity): Suppose that the errors in the general linear model $\mathbf{y} = \mathbf{X}\boldsymbol{\beta} + \boldsymbol{\varepsilon}$ are heteroscedastic. Let σ_i^2 denote the variance of the error ε_i for the ith observation.

(a) Show that if the ε_i are normally and independently distributed with zero expectations, then the likelihood is maximized when the *weighted* sum of squared errors

$$\sum_{i=1}^{n} \frac{1}{\sigma_i^2}\varepsilon_i^2 = \sum \frac{1}{\sigma_i^2}\left(Y_i - \mathbf{x}_i'\boldsymbol{\beta}\right)^2$$

is minimized. This method of estimation is called *weighted-least-squares (WLS) regression*.

(b) Now suppose more specifically that σ_i is proportional to the value of one of the independent variables, say X_1; that is, $\sigma_i = \alpha X_{1i}$. We proceed to divide both sides of the model by X_{1i}, obtaining

$$\frac{Y_i}{X_{1i}} = \beta_0 \frac{1}{X_{1i}} + \beta_1 + \beta_2 \frac{X_{2i}}{X_{1i}} + \cdots + \beta_k \frac{X_{ki}}{X_{1i}} + \frac{\varepsilon_i}{X_{1i}}$$

$$Y_i' = \beta_0 X_{0i}' + \beta_1 + \beta_2 X_{2i}' + \cdots + \beta_k X_{ki}' + \varepsilon_i'$$

where $Y_i' \equiv Y_i/X_{1i}$, $X_{0i}' \equiv 1/X_{1i}$, and so on. Show that (i) the variance of ε_i' is constant; and (ii) fitting the transformed model by ordinary least-squares regression is equivalent to fitting the original model by weighted least squares. [*Hints*: (i) Note that

$X_{1i} = \sigma_i/\alpha$. (ii) Prove that minimizing $\Sigma E_i'^2$ is equivalent to minimizing $\Sigma E_i^2/\sigma_i^2$.]

(c) Use the result in part (b) to re-analyze Ornstein's interlocking-directorate data (Problem 2.2) under the assumption that σ_i is proportional to assets. More information on weighted least squares may be found, for example, in Wonnacott and Wonnacott (1979: 194–198, 427–435).

3.14.*† Box-Cox Transformations of Y: Box and Cox (1964) have applied the method of maximum likelihood to the problem of finding a transformation of Y that (simultaneously) linearizes the model, corrects heteroscedasticity, and normalizes the distribution of the errors. One model entertained by Box and Cox is

$$\underset{(n\times 1)}{\mathbf{y}^{(\lambda)}} = \underset{(n\times k+1)}{\mathbf{X}} \; \underset{(k+1\times 1)}{\boldsymbol{\beta}} + \underset{(n\times 1)}{\boldsymbol{\varepsilon}}$$

where

$$Y_i^{(\lambda)} = \begin{cases} \dfrac{Y_i^{\lambda} - 1}{\lambda} & \text{for } \lambda \neq 0 \\ \log Y_i & \text{for } \lambda = 0 \end{cases}$$

and $\boldsymbol{\varepsilon} \sim N_n(\mathbf{0}, \sigma_\varepsilon^2 \mathbf{I}_n)$. Except for an inessential change in scale, used here so that the transformation is a continuous function at $\lambda = 0$, this is precisely the family of powers and roots considered in the preceding section.

(a) Show that the probability density for the observations is given by

$$p(\mathbf{y}) = \frac{1}{(2\pi\sigma_\varepsilon^2)^{n/2}} \exp\left[- \frac{\sum_{i=1}^{n} \left(Y_i^{(\lambda)} - \mathbf{x}_i'\boldsymbol{\beta}\right)^2}{2\sigma_\varepsilon^2} \right] \prod_{i=1}^{n} Y_i^{\lambda-1}$$

where $\mathbf{x}_i'$ is the ith row of $\mathbf{X}$. (*Hint*: $Y_i^{\lambda-1}$ is the Jacobian of the transformation from Y_i to ε_i.)

(b) For a given value of λ, the *conditional* maximum-likelihood estimator of $\boldsymbol{\beta}$ is the least-squares estimator $\mathbf{b}_\lambda = (\mathbf{X}'\mathbf{X})^{-1}\mathbf{X}'\mathbf{y}^{(\lambda)}$. (Why?) Show that the maximized log likelihood may be written

$$\log L(\lambda) = -\frac{n}{2}(1 + \log 2\pi) - \frac{n}{2}\log \hat{S}_E^2(\lambda) + (\lambda - 1)\sum_{i=1}^{n} \log Y_i$$

where $\hat{S}_E^2(\lambda) = \mathbf{e}_\lambda'\mathbf{e}_\lambda/n$, and $\mathbf{e}_\lambda$ is the vector of residuals from the regression of $\mathbf{y}^{(\lambda)}$ on $\mathbf{X}$. (Because the first term in the equation for $\log L(\lambda)$ is a constant, it may be disregarded in what follows.)

(c) Box and Cox suggest computing $\log L(\lambda)$ for a range of values of λ, say between -2 and 2. (If this range proves unsatisfactory, we can always expand it in one direction or the other). By plotting $\log L(\lambda)$ against λ, we can locate the approximate value of λ that maximizes the likelihood (i.e., the maximum-likelihood estimate of λ). Apply this method to Ornstein's interlocking-directorate data, discussed in Section 3.3.2. Compare the maximum-likelihood transformation with that obtained by Anscombe's method.

(d) Let $\hat{\lambda}$ represent the maximum-likelihood estimator of λ, and suppose that we wish to test the hypothesis $H_0 : \lambda = \lambda_0$. (A hypothesis of general interest is the null hypothesis that no transformation is required, $H_0 : \lambda = 1$.) The likelihood-ratio test statistic for the hypothesis, $G_0^2 = -2[\log L(\lambda_0) - \log L(\hat{\lambda})]$, is asymptotically distributed as $\chi^2(1)$. (Why?) Using this result, explain why a $100(1 - a)\%$ confidence interval for λ includes all values for which $\log L(\lambda) > \log L(\hat{\lambda}) - \frac{1}{2}z_{a/2}^2$, where $z_{a/2}$ is the unit-normal deviate with probability $a/2$ to the right. Locate $\log L(\hat{\lambda}) - \frac{1}{2}z_{.05}^2$ on the graph constructed in part (c). What do you conclude about the need to transform Y?

3.4. NONLINEAR MODELS AND NONLINEAR LEAST SQUARES

As explained in the previous section, linear statistical models effectively encompass nonlinear models that are linear in the parameters. The forms of nonlinear relationship that may be expressed as linear models are therefore very diverse. In certain circumstances, however, theory dictates that we fit models nonlinear in their parameters. This is a relatively rare necessity in the social sciences, primarily because our theories are seldom mathematically concrete, although nonlinear models are employed in some areas of demography, economics, and psychology. Moreover, the methods of this section may be adapted to the problem of finding good transformations for the independent variables in a linear model (see Problem 3.18).

3.4.1. Transformable Nonlinearity

Some models that are nonlinear in the parameters may be transformed into linear models, and consequently may be fit to data by linear least squares. A model of this sort is the so-called "gravity model" of migration, employed in human geography (see Abler, Adams, and Gould, 1971: 221–233). Let Y_{ij} represent the number of migrants moving from city j to city i; let D_{ij} represent the distance between these cities; and let P_i and P_j represent the respective populations of the two cities. The gravity model of migration is built in rough analogy to the Newtonian formula for gravitational attraction between two bodies, where population plays the role of mass and migration plays the role of gravitational attraction. This analogy is far from perfect, in part because there are two migration streams of generally different sizes between a pair of cities: one from city i to city j, and the other from j to i.

The gravity model is given by the equation

$$Y_{ij} = \alpha \frac{P_i^{\beta} P_j^{\gamma}}{D_{ij}^{\delta}} \varepsilon_{ij} = \tilde{Y}_{ij} \varepsilon_{ij} \tag{3.34}$$

where α, β, γ, and δ are unknown parameters to be estimated from the data, and ε_{ij} is a necessarily positive multiplicative error term that reflects the imperfect determination of migration by distance and population size. When ε_{ij} is one, Y_{ij} is equal to its predicted value $\tilde{Y}_{ij}$ given by the systematic part of the model; when ε_{ij} is less than one, Y_{ij} is smaller than $\tilde{Y}_{ij}$; and when ε_{ij} is greater than one, Y_{ij} exceeds $\tilde{Y}_{ij}$. We shall say more about the error presently. Note that because of the multiplicative form of the model, $\tilde{Y}_{ij}$ is not $E(Y_{ij})$.

Although equation (3.34) is nonlinear in its parameters, it may be transformed into a linear equation by taking logs.[25] That is

$$\begin{aligned} \log Y_{ij} &= \log \alpha + \beta \log P_i + \gamma \log P_j - \delta \log D_{ij} + \log \varepsilon_{ij} \\ Y'_{ij} &= \alpha' + \beta P'_i + \gamma P'_j + \delta D'_{ij} + \varepsilon'_{ij} \end{aligned} \tag{3.35}$$

defining

$$\alpha' \equiv \log \alpha$$

$$P'_i \equiv \log P_i$$

$$P'_j \equiv \log P_j$$

$$D'_{ij} \equiv -\log D_{ij}$$

$$\varepsilon'_{ij} \equiv \log \varepsilon_{ij}$$

If we may make the usual linear-model assumptions about the transformed errors ε'_{ij}, then we may fit the transformed model (3.35) by linear least squares. In the present instance, it would probably be unrealistic to assume that the transformed errors are independent, since individual cities are involved in many different migration streams; a particularly attractive city, for example, might have positive errors for each of its in-migration streams and negative errors for each of its out-migration streams. Illustrative data for the gravity model are given in Problem 3.15.

[25] The log transformation requires that Y_{ij}, α, P_i, P_j, D_{ij}, and ε_{ij} be positive, as is the case for the gravity model of migration.

Our ability to linearize model (3.34) by a log transformation is dependent upon the presence of multiplicative errors in this model. The multiplicative form of the error specifies that the general magnitude of the difference between Y_{ij} and $\tilde{Y}_{ij}$ is proportional to the size of the latter: The model tends to make larger absolute mistakes in predicting large migration streams than in predicting small ones. This assumption appears reasonable here. In most cases, we should prefer to specify a form of error—additive or multiplicative—that leads to a simple statistical analysis, supposing, of course, that the specification is reasonable; a subsequent analysis of residuals permits us to subject these assumptions to scrutiny.

Another form of multiplicative model is

$$\begin{aligned} Y_i &= \beta_0 e^{\beta_1 X_{1i}} e^{\beta_2 X_{2i}} \ldots e^{\beta_k X_{ki}} \varepsilon_i \\ &= \beta_0 \exp(\beta_1 X_{1i} + \beta_2 X_{2i} + \cdots + \beta_k X_{ki}) \varepsilon_i \end{aligned} \tag{3.36}$$

Taking logs produces the linear equation

$$Y_i' = \beta_0' + \beta_1 X_{1i} + \beta_2 X_{2i} + \cdots + \beta_k X_{ki} + \varepsilon_i'$$

where $Y_i' \equiv \log Y_i$, $\beta_0' \equiv \log \beta_0$, and $\varepsilon_i' \equiv \log \varepsilon_i$. In model (3.36), the impact on Y of increasing X_j by one unit is proportional to the level of Y. The effect of rescaling Y by taking logs is to eliminate interaction among the X's. A similar result is at times achievable empirically through other power transformations of Y.

Multiplicative models provide the most common instance of transformable nonlinearity, but there are also other models to which this approach is applicable. Consider, for example, the model

$$Y_i = \frac{1}{\alpha + \beta X_i + \varepsilon_i} \tag{3.37}$$

where Y is the dependent variable, X is the independent variable, α and β are parameters, and ε is an error random variable satisfying the standard assumptions. Then, if we take $Y_i' = 1/Y_i$, we may rewrite equation (3.37) as the linear model $Y_i' = \alpha + \beta X_i + \varepsilon_i$.

3.4.2. Essential Nonlinearity and Nonlinear Least Squares

Models that are nonlinear in the parameters and that cannot be rendered linear by a transformation are called *essentially nonlinear*. The general nonlinear model is given by the equation

$$Y_i = f\Big(\underset{(p \times 1)}{\boldsymbol{\beta}}, \underset{(1 \times k)}{\mathbf{x}_i'} \Big) + \varepsilon_i \tag{3.38}$$

In this equation, Y_i is the dependent-variable value for the ith of n observations, $\boldsymbol{\beta}$ is a vector of p unknown parameters, $\mathbf{x}'_i$ is a row vector of scores for k independent variables, and ε_i is the error for the ith observation. For the full sample of n observations, we shall find it convenient to write the model in matrix form as

$$\underset{(n\times1)}{\mathbf{y}} = \mathbf{f}\Big(\underset{(p\times1)}{\boldsymbol{\beta}},\ \underset{(n\times k)}{\mathbf{X}}\Big) + \underset{(n\times1)}{\boldsymbol{\varepsilon}}$$

We assume, as in the general linear model, that $\boldsymbol{\varepsilon} \sim N_n(\mathbf{0}, \sigma_\varepsilon^2 \mathbf{I}_n)$.[26]

An illustrative essentially nonlinear model is the logistic population-growth model (Shryock and Siegel, 1973: 382–385)

$$Y_i = \frac{\beta_1}{1 + e^{\beta_2 + \beta_3 X_i}} + \varepsilon_i \tag{3.39}$$

In equation (3.39), Y_i is population size and X_i is time; for equally-spaced observations, it is usual to take $X_i = i - 1$. Since the logistic growth model is fit to time-series data, the assumption of independent errors is problematic (cf. our discussion in Section 3.1 of Fox's female labor force time-series data). The additive form of the error is also questionable here, since errors may well grow larger in magnitude as population size increases.

Under the assumption of normally and identically distributed independent errors, the general nonlinear model in equation (3.38) has likelihood

$$\begin{aligned} L(\boldsymbol{\beta}, \sigma_\varepsilon^2) &= \frac{1}{(2\pi\sigma_\varepsilon^2)^{n/2}} \exp\left\{-\frac{\sum[Y_i - f(\boldsymbol{\beta}, \mathbf{x}'_i)]^2}{2\sigma_\varepsilon^2}\right\} \\ &= \frac{1}{(2\pi\sigma_\varepsilon^2)^{n/2}} \exp\left[-\frac{1}{2\sigma_\varepsilon^2} S(\boldsymbol{\beta})\right] \end{aligned}$$

where $S(\boldsymbol{\beta})$ is the sum-of-squares function

$$S(\boldsymbol{\beta}) \equiv \sum [Y_i - f(\boldsymbol{\beta}, \mathbf{x}'_i)]^2$$

As in the case of the general linear model (Section 1.2.5), we maximize the likelihood when the sum of squared errors $S(\boldsymbol{\beta})$ is minimized.

To derive estimating equations for the nonlinear model, we differentiate $S(\boldsymbol{\beta})$, obtaining

$$\frac{\partial S(\boldsymbol{\beta})}{\partial \boldsymbol{\beta}} = -2\sum [Y_i - f(\boldsymbol{\beta}, \mathbf{x}'_i)] \frac{\partial f(\boldsymbol{\beta}, \mathbf{x}'_i)}{\partial \boldsymbol{\beta}}$$

[26] For multiplicative errors, we may put the model in the form of equation (3.38) by taking logs.

Setting these partial derivatives to zero, and replacing the unknown parameters $\boldsymbol{\beta}$ with the estimator $\mathbf{b}$, produces the nonlinear-least-squares normal equations. It is convenient to write these equations in matrix form as

$$[\mathbf{F}(\mathbf{b}, \mathbf{X})]'[\mathbf{y} - \mathbf{f}(\mathbf{b}, \mathbf{X})] = \mathbf{0} \tag{3.40}$$

where $\underset{(n \times p)}{\mathbf{F}}(\mathbf{b}, \mathbf{X})$ is the matrix of derivatives with i, jth entry

$$F_{ij} \equiv \frac{\partial f(\mathbf{b}, \mathbf{x}_i')}{\partial B_j}$$

The roots $\mathbf{b}$ of equation (3.40) are the maximum-likelihood estimates of $\boldsymbol{\beta}$. If there is more than one solution to the estimating equations, then we choose the solution that minimizes the sum of squares $S(\mathbf{b})$.

Because the estimating equations (3.40) arising from a nonlinear model are in general themselves nonlinear, their solution is often difficult. It is, in fact, unusual to obtain nonlinear-least-squares estimates by explicitly solving the normal equations. It is instead more common to work directly with the sum-of-squares function. There are several methods for obtaining nonlinear-least-squares estimates; we shall pursue in some detail a technique called *steepest descent*. Although steepest descent usually performs poorly relative to alternative procedures, the rationale of the method is simple. Furthermore, many general aspects of nonlinear-least-squares calculations may be explained in terms of the steepest-descent procedure. Because of the practical limitations of steepest descent, we shall also briefly describe two other procedures, the *Gauss–Newton* method and the *Marquardt* method, without developing their rationales. Further discussion of nonlinear least squares may be found in many sources, including Gallant (1975), Draper and Smith (1981: Chapter 10), and Bard (1974), in order of increasing detail.

The method of steepest descent, like other methods for calculating nonlinear-least-squares estimates, begins with a vector $\mathbf{b}_0$ of initial estimates. These initial estimates may be obtained in a variety of ways. We may, for example, choose p "typical" observations, substitute their values into the model equation (3.38), and solve the resulting system of p nonlinear equations for the parameters. Alternatively, we may select a set of reasonable trial values for each parameter, find the residual sum of squares for every combination of trial values, and select as initial estimates the combination associated with the smallest residual sum of squares. It is often possible to choose initial estimates on the basis of prior research, hypothesis, or substantive knowledge of the process being modeled. It is unfortunately the case that the choice of starting values may prove significant: Iterative methods such as steepest descent generally converge to a solution more quickly for initial values that are close to the final values; and, even more importantly, the sum-of-squares function $S(\mathbf{b})$ may have local minima different from the global minimum (see, for example, Figure 3.18).

We denote the gradient vector for the sum-of-squares function by

$$\mathbf{d}(\mathbf{b}) \equiv \frac{\partial S(\mathbf{b})}{\partial \mathbf{b}}$$

The vector $\mathbf{d}(\mathbf{b}_0)$ gives the direction of maximum increase of the sum-of-squares function from the initial point $[\mathbf{b}_0, S(\mathbf{b}_0)]$; the negative of this vector, $-\mathbf{d}(\mathbf{b}_0)$, therefore gives the direction of steepest descent. Figure 3.18 illustrates these relations for the particularly simple case of one parameter, where we can move either left or right from the initial point. If we move in the direction of steepest descent, we can find a new estimated parameter vector $\mathbf{b}_1 = \mathbf{b}_0 - M_0\mathbf{d}(\mathbf{b}_0)$ for which $S(\mathbf{b}_1) < S(\mathbf{b}_0)$. Since $S(\mathbf{b})$ is by definition decreasing in the direction of steepest descent, we can always choose a scalar M_0 small enough to improve the residual sum of squares. We may, for instance, first try $M_0 = 1$; if this choice does not lead to a decrease in $S(\mathbf{b})$, we may then take $M_0 = \frac{1}{2}$; and so on. Our new estimate $\mathbf{b}_1$ may then be improved in the same manner, by finding $\mathbf{b}_2 = \mathbf{b}_1 - M_1\mathbf{d}(\mathbf{b}_1)$ so that $S(\mathbf{b}_2) < S(\mathbf{b}_1)$. This procedure continues iteratively until it converges upon a solution $\mathbf{b} = \mathbf{b}_m$; that is, until the changes in $S(\mathbf{b}_l)$ and $\mathbf{b}_l$ from one iteration to the next are insignificantly small. The method of steepest descent often converges much more slowly than alternative procedures, and at times falls prey to other computational difficulties.

At each iteration, we need to compute the gradient vector $\mathbf{d}(\mathbf{b})$ for the current value of $\mathbf{b} = \mathbf{b}_l$. From our previous work in this section, we have

$$\begin{aligned} -\mathbf{d}(\mathbf{b}) &= 2[\mathbf{F}(\mathbf{b}, \mathbf{X})]'[\mathbf{y} - \mathbf{f}(\mathbf{b}, \mathbf{X})] \\ &= 2\sum \left[\frac{\partial f(\mathbf{b}, \mathbf{x}_i')}{\partial \mathbf{b}}\right][Y_i - f(\mathbf{b}, \mathbf{x}_i')] \end{aligned} \tag{3.41}$$

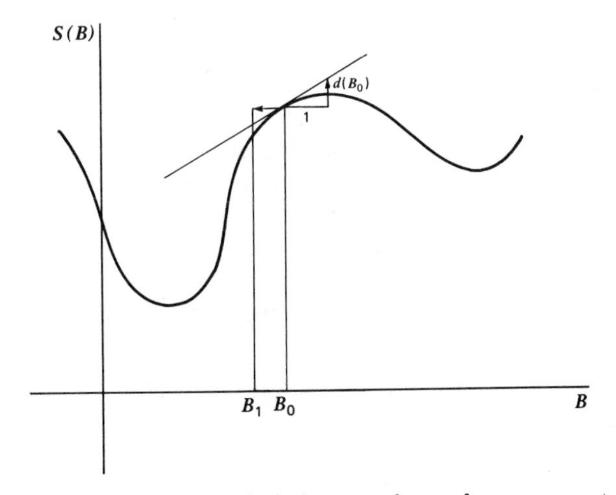

FIGURE 3.18. The method of steepest descent for one parameter.

The partial derivatives $\partial f(\mathbf{b}, \mathbf{x}_i')/\partial B_j$ either may be given analytically (which is preferable), or may be evaluated numerically (i.e., approximated by finding the slope of $f(\mathbf{b}, \mathbf{x}_i')$ in a small interval around the current value of B_j). For example, for the logistic growth model (3.39) discussed earlier in this section, the analytic derivatives are

$$\frac{\partial f(\mathbf{b}, X_i)}{\partial B_1} = \left(1 + e^{B_2 + B_3 X_i}\right)^{-1}$$

$$\frac{\partial f(\mathbf{b}, X_i)}{\partial B_2} = -B_1\left(1 + e^{B_2 + B_3 X_i}\right)^{-2} e^{B_2 + B_3 X_i}$$

$$\frac{\partial f(\mathbf{b}, X_i)}{\partial B_3} = -B_1\left(1 + e^{B_2 + B_3 X_i}\right)^{-2} e^{B_2 + B_3 X_i} X_i$$

In the method of steepest descent, we take

$$\mathbf{b}_{l+1} = \mathbf{b}_l + M_l \mathbf{F}_l' \mathbf{e}_l$$

where $\mathbf{F}_l \equiv \mathbf{F}(\mathbf{b}_l, \mathbf{X})$ and $\mathbf{e}_l = \mathbf{y} - \mathbf{f}(\mathbf{b}_l, \mathbf{X})$ (and the constant two in equation (3.41) is absorbed into M_l). The Gauss–Newton method calculates

$$\mathbf{b}_{l+1} = \mathbf{b}_l + M_l\left(\mathbf{F}_l'\mathbf{F}_l\right)^{-1}\mathbf{F}_l'\mathbf{e}_l$$

As for steepest descent, M_l is chosen so that $S(\mathbf{b}_{l+1}) < S(\mathbf{b}_l)$. We first try $M_l = 1$, then $M_l = \frac{1}{2}$, and so on. The direction chosen in the Gauss–Newton procedure is based upon a first-order Taylor series expansion of $S(\mathbf{b})$ around $S(\mathbf{b}_l)$.

In the Marquardt procedure,

$$\mathbf{b}_{l+1} = \mathbf{b}_l + \left(\mathbf{F}_l'\mathbf{F}_l + M_l \mathbf{I}_p\right)^{-1}\mathbf{F}_l'\mathbf{e}_l$$

Initially, M_0 is set to some small number, such as 10^{-8}. If $S(\mathbf{b}_{l+1}) < S(\mathbf{b}_l)$, then we accept the new value of $\mathbf{b}_{l+1}$ and proceed to the next iteration, with $M_{l+1} = M_l/10$; if $S(\mathbf{b}_{l+1}) > S(\mathbf{b}_l)$, then we increase M_l by a factor of 10 and try again. When M is small, the Marquardt procedure is similar to Gauss–Newton; as M grows larger, Marquardt approaches steepest descent. Marquardt's method is thus an adaptive compromise between the other two approaches.

Estimated asymptotic sampling covariances for the parameter estimates may be obtained by the maximum-likelihood approach (see Bard, 1974: 176–179) and are given by

$$\widehat{\mathscr{V}(\mathbf{b})} = S_E^2\left\{[\mathbf{F}(\mathbf{b}, \mathbf{X})]'\mathbf{F}(\mathbf{b}, \mathbf{X})\right\}^{-1}$$

$$= S_E^2(\mathbf{F}'\mathbf{F})^{-1} \qquad (3.42)$$

We may estimate the error variance from the residuals $\mathbf{e} = \mathbf{y} - \mathbf{f}(\mathbf{b}, \mathbf{X})$, accord-

TABLE 3.11. United States Population Growth Data

Year	X_i	Population (millions) Y_i	Logistic Growth Model $\hat{Y}_i$	E_i
1790	0	3.895	4.8940	-0.9990
1800	1	5.267	6.4265	-1.1595
1810	2	7.182	8.4221	-1.2401
1820	3	9.566	11.0086	-1.4426
1830	4	12.834	14.3412	-1.5072
1840	5	16.985	18.6020	-1.6170
1850	6	23.069	23.9963	-0.9273
1860	7	31.278	30.7417	0.5363
1870	8	38.416	39.0466	-0.6306
1880	9	49.924	49.0788	0.8452
1890	10	62.692	60.9223	1.7697
1900	11	75.734	74.5309	1.2031
1910	12	91.812	89.6900	2.1220
1920	13	109.806	106.0037	3.8023
1930	14	122.775	122.9208	-0.1458
1940	15	131.669	139.8022	-8.1332
1950	16	150.697	156.0145	-5.3175
1960	17	178.464	171.0206	7.4434

Source: Shryock and Siegel (1973: 383).

ing to the formula $S_E^2 = \mathbf{e}'\mathbf{e}/(n - p)$. Note the similarity of equation (3.42) to the familiar linear-least-squares result, $\widehat{V(\mathbf{b})} = S_E^2(\mathbf{X}'\mathbf{X})^{-1}$. Indeed, $\mathbf{F}(\mathbf{b}, \mathbf{X}) = \mathbf{X}$ for the linear model $\mathbf{y} = \mathbf{X}\boldsymbol{\beta} + \boldsymbol{\varepsilon}$.

Decennial population data for the United States are given by Shryock and Siegel (1973: 383) for the period from 1790 to 1960. These data are reproduced in Table 3.11 and are plotted in Figure 3.19. We wish to fit the logistic growth model (3.39) to these data using nonlinear least squares.

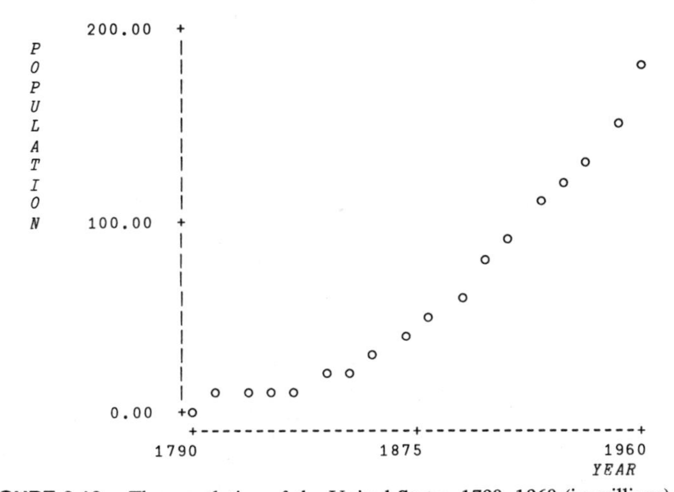

FIGURE 3.19. The population of the United States, 1790–1960 (in millions).

TABLE 3.12. Gauss – Newton Estimates for Logistic Growth Model Fit to United States Population Data

Iteration	Residual Sum of Squares	Parameter Estimates B_1	B_2	B_3
0	14,318.4599	200.00	4.0000	−0.40000
1	1,085.9707	192.26	3.5980	−0.28646
2	333.3442	235.52	3.8915	−0.27908
3	186.4973	244.01	3.8886	−0.27883
4	186.4972	243.99	3.8889	−0.27886
Final	186.4972	243.99	3.8888	−0.27886
Standard Error		17.97	0.0937	0.01559

The parameter β_1 of the logistic growth model gives the asymptote that expected population approaches as time goes to infinity. In 1960, when $Y = 178$ (million), population did not appear to be near an asymptote; so as not to extrapolate too far beyond the data, we arbitrarily set $B_{1,0} = 200$. At time $X_1 = 0$, we have

$$Y_1 = \frac{\beta_1}{1 + e^{\beta_2 + \beta_3 0}} + \varepsilon_1 \tag{3.43}$$

Ignoring the error, using $B_{1,0} = 200$, and substituting the observed value of $Y_1 = 3.895$ into equation (3.43), we get $e^{B_{2,0}} = (200/3.895) - 1$, or $B_{2,0} = \log 50.35 = 3.919 \simeq 4$. At time $X_2 = 1$,

$$Y_2 = \frac{\beta_1}{1 + e^{\beta_2 + \beta_3 1}} + \varepsilon_2$$

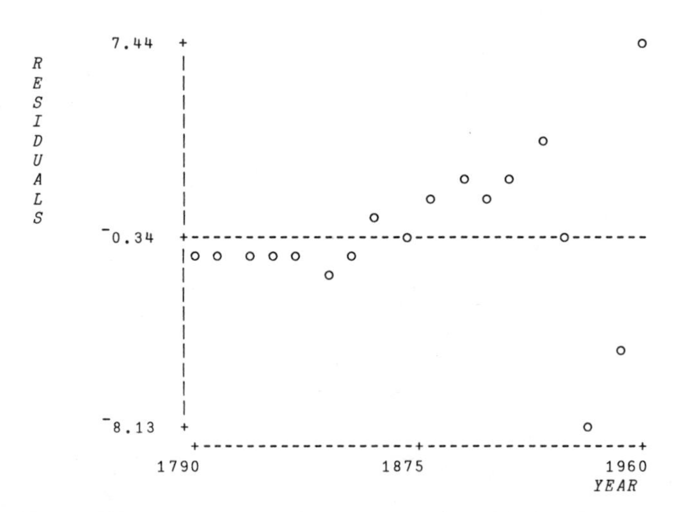

FIGURE 3.20. Residuals from the logistic growth model plotted against time.

Again, ignoring the error and making appropriate substitutions, $e^{4+B_{3,0}} = (200/5.267) - 1$, or $B_{3,0} = \log 36.97 - 4 = -0.3899 \simeq -0.4$. The Gauss–Newton iterations based upon these start values are shown in Table 3.12. Estimated asymptotic standard errors also appear in this table. Residuals for the least-squares fit, given in Table 3.11 and plotted against time in Figure 3.20, suggest that the error variance is not constant.

PROBLEMS

3.15. Table 3.13 reports inter-provincial migration in Canada for the period 1966 to 1971. Also shown in this table are 1966 and 1971 provincial population, and road distances among the major cities in the provinces. Averaging the 1966 and 1971 population figures, fit the gravity model of migration (3.34) to the inter-provincial migration data. Display the residuals from the fitted model in a 10 × 10 table. Can you account for the pattern of the residuals? How might the model be modified to provide a more satisfactory fit?

3.16. Using a nonlinear least-squares program, refit the logistic growth model to the U.S. population data given in Table 3.11, assuming multiplicative rather than additive errors:

$$Y_i = \frac{\beta_1}{1 + e^{\beta_2 + \beta_3 X_i}} \varepsilon_i$$

Which form of the model appears more adequate for these data?

3.17. The following population data for Canada (excluding Newfoundland) are taken from Urquhart and Buckley (1965) and from Canada (1972: 1369) (reprinted with permission from Urquhart and Buckley, 1965: 14):

Year	Population (millions)	Year	Population (millions)
1851	2.436	1921	8.788
1861	3.230	1931	10.377
1871	3.689	1941	11.507
1881	4.325	1951	13.648
1891	4.833	1961	17.780
1901	5.371	1971	21.046
1911	7.207		

Fit a logistic growth model to these data, assuming (a) additive errors

TABLE 3.13. Canadian Inter-Provincial Migration Data

1971 Residence	(a) Migration Streams 1966 Residence									
	1	2	3	4	5	6	7	8	9	10
1. Newfoundland		255	2380	1140	2145	6295	215	185	425	425
2. Prince Edward Island	340		1975	1310	755	3060	400	95	185	330
3. Nova Scotia	3340	2185		8310	6090	18805	1825	840	2000	2490
4. New Brunswick	1740	1335	7635		9315	12455	1405	480	1130	1195
5. Quebec	2235	635	4350	7905		48370	4630	1515	3305	4740
6. Ontario	17860	3570	25730	18550	99430		23785	11805	17655	21205
7. Manitoba	680	265	1655	1355	4330	18245		16365	7190	6310
8. Saskatchewan	280	125	620	495	1570	6845	9425		10580	6090
9. Alberta	805	505	3300	2150	7750	23550	17410	41910		27765
10. British Columbia	1455	600	6075	3115	16740	47395	26910	29920	58915	
1966 Population	493396	108535	756039	616788	5780845	6960870	963066	955344	1463203	1873674
1971 Population	522104	111641	788960	534557	6027764	7703106	988247	926242	1627874	2184621

(b) Road Distances in Miles Among Major Canadian Cities

City	1	2	3	4	5	6	7	8	9	10
1. St. John, Nfld.	0	924	952	1119	1641	1996	3159	3542	4059	4838
2. Charlottetown, P.E.I.	924	0	164	252	774	1129	2293	2675	3192	3972
3. Halifax, N.S.	952	164	0	310	832	1187	2351	2733	3250	4029
4. Fredericton, N.B.	1119	252	310	0	522	877	2041	2423	2940	3719
5. Montreal, Que.	1641	774	832	522	0	355	1519	1901	2418	3197
6. Toronto, Ont.	1996	1129	1187	877	355	0	1380	1763	2281	3059
7. Winnipeg, Man.	3159	2293	2351	2041	1519	1380	0	382	899	1679
8. Regina, Sask.	3542	2675	2733	2423	1901	1763	382	0	517	1297
9. Edmonton, Alta.	4059	3192	3250	2940	2418	2281	899	517	0	987
10. Vancouver, B.C.	4838	3972	4029	3719	3197	3059	1679	1297	987	0

Sources: Canada (1962; 1971: Vol. 1, Part 2, Table 32; 1972: 1369).

[equation (3.39)], and (b) multiplicative errors (as in the previous problem). Compare the results of the two analyses.

3.18.† Box–Tidwell Transformations of Independent Variables: Box and Tidwell (1962) employ the method of maximum likelihood to find power transformations of the independent variables in the regression model

$$Y_i = \beta_0 + \beta_1 X_{1i}^{\gamma_1} + \cdots + \beta_k X_{ki}^{\gamma_k} + \varepsilon_i$$

in which the errors ε_i follow the usual assumptions, and we take logs when $\gamma_j = 0$. This model may be fit by nonlinear least squares. Reasonable start values for the parameter estimates may be obtained by setting the B_j's equal to the linear-least-squares estimates and the C_j's (i.e., the estimates of the γ_j's) equal to one. To speed convergence, we may first examine partial-residual plots to select guessed C-values, using these in a least-squares regression to obtain initial estimates of the B's.

(a) Find the derivative of $E(Y_i)$ with respect to each parameter of the model: $\beta_0, \beta_1, \ldots, \beta_k, \gamma_1, \ldots, \gamma_k$.

(b) Use a nonlinear-least-squares program to fit the Box–Tidwell model to Angell's regression of moral integration on ethnic heterogeneity and geographic mobility. (Transformations for Angell's data are discussed in Section 3.3.1; use the subset of non-Southern cities from Table 1.3.) Construct 90 percent confidence intervals for the transformation parameters γ_1 and γ_2. What do you conclude about the need to transform X_1 and X_2? How do these results compare to those of Section 3.3.1?

(c) Box and Tidwell suggest an iterative estimation method for their model that usually converges more quickly than a general nonlinear-least-squares procedure. The method is based upon a first-order Taylor-series approximation for $E(Y)$. (i) Regress Y on $X_1, \ldots, X_k$, obtaining coefficients $B_0, B_1, \ldots, B_k$. (ii) Regress Y on $X_1, \ldots, X_k$, and $X_1 \log X_1, \ldots, X_k \log X_k$, obtaining coefficients $B_0', B_1', \ldots, B_k'$, and $D_1, \ldots, D_k$. For each independent variable, calculate the estimate $C_j = 1 + D_j/B_j$ of the power-transformation parameter. Note that, in general, $B_j \neq B_j'$ and that the B_j' are ignored. (iii) Apply the transformations found in step (ii), forming $X_j' = X_j^{C_j}$. Repeat steps (i) and (ii), substituting X_j' for X_j, and continue this procedure until the transformations converge. Apply Box and Tidwell's method to Angell's regression data. Do you obtain the same estimates as found in part (b)?

3.19.† Interpreting Effects in Nonlinear Models: For simplicity, disregard the error, and let Y represent the systematic part of the dependent variable.

Let $Y = f(X_1, X_2)$. The *metric effect* of X_1 on Y is defined as the partial derivative $\partial Y/\partial X_1$; the *effect of proportional change* in X_1 on Y is defined as $X_1(\partial Y/\partial X_1)$; the *instantaneous rate of return* of Y with respect to X_1 is $(\partial Y/\partial X_1)/Y$; and the *point elasticity* of Y with respect to X_1 is $(\partial Y/\partial X_1)(X_1/Y)$. Find each of these four measures of effect for X_1 in the following models. Which measure yields the simplest result in each case? How may the several measures be interpreted?

(a) $Y = \beta_0 + \beta_1 X_1 + \beta_2 X_2$

(b) $Y = \beta_0 + \beta_1 X_1 + \beta_2 X_1^2 + \beta_3 X_2$

(c) $Y = \beta_0 + \beta_1 X_1 + \beta_2 X_2 + \beta_3 X_1 X_2$

(d) $Y = e^{\beta_0 + \beta_1 X_1 + \beta_2 X_2}$

(e) $Y = \beta_0 X_1^{\beta_1} X_2^{\beta_2}$

This question is based upon Stolzenberg (1979).

3.5. SAMPLING CONSIDERATIONS AND MISSING DATA IN THE ANALYSIS OF SOCIAL SURVEYS

Many of our data in the social sciences are collected in sample surveys, which are generally conducted according to complex sampling designs, and which almost invariably generate missing data. Missing data, of course, can also occur in other sorts of research. Although we cannot deal at length with these topics, it is important at least briefly to mention the problems raised by them for linear-model analysis. Blalock (1979: Ch. 21) includes a good, brief introduction to survey sampling. Kish (1965) presents a detailed treatment of sampling oriented toward the social sciences. Reviews of the extensive statistical literature on the treatment of missing data may be found in Afifi and Elashoff (1966) and in Kim and Curry (1977).

3.5.1. Sampling

The procedures for statistical inference developed in this book assume independently sampled observations. This assumption is met approximately when a *simple random sample* is selected from a large population. In simple random sampling, each possible sample of n elements has the same probability of selection. The trivial departure from strict independence of observations is the consequence of sampling without replacement from a finite population (i.e., each element in the population cannot be selected more than once). For reasons of cost and for other practical considerations, however, simple random sampling is rarely used in large-scale survey research. Social surveys typically are conducted according to *multistage* sampling designs that employ *stratification* and *clustering*. Frequently, different elements of the population are selected for the sample with unequal probabilities.

In a multistage sampling design, elements—generally individual respondents—are not chosen until the final stage of the sampling process. In a national survey, for instance, the first stage of sampling might be the random selection of a sample of communities; a second stage might involve the selection of census tracts within the chosen communities; then blocks within tracts; households within blocks; and, finally, an individual in each household. What is crucial to scientific sampling is that the probability of selection be known at each stage of the sampling process.

It is often possible to improve the efficiency of a survey sample (i.e., to increase the probability of selecting a representative sample) by randomly sampling within predefined homogeneous strata of the population. A national sample may, for example, be stratified by region so that each region is represented in the final sample in proportion to its population size. In cluster sampling, several sampling units—for example, a cluster of adjacent households—are selected simultaneously. Clustering generally decreases the efficiency of the sample relative to a simple random sample of comparable size, but it also decreases survey-administration costs.

Although correct procedures exist for carrying out statistical inference on data collected according to complex sampling designs (see, e.g., Kish, 1965; Finifter, 1972; Mosteller and Tukey, 1977: Chapters 8, 9; and Efron, 1982), the usual practice in the social sciences is to treat the data as if they were collected by independent sampling. In certain circumstances, a sampling statistician may be able to provide an approximate correction factor, called a *design effect*, for adjusting the sampling variances computed in the usual manner.[27] When the combined effects of stratification and clustering are taken into account, real survey samples generally prove less efficient than simple random samples of the same size (though more efficient than simple random samples of the same *cost*). The practice of ignoring the design effect, therefore, leads us to understate sampling variability, a shortcoming that should at least be acknowledged in reporting research findings.

When different observations are selected for the sample with unequal probabilities, this fact must be taken into account if we wish to compute unbiased estimators of population parameters. In general, we need to weight each observation in inverse proportion to its probability of selection. Thus, for example, in calculating the (weighted) sample mean of a variable Y, we take

$$\overline{Y} = \frac{\sum_{i=1}^{n} w_i Y_i}{\sum_{i=1}^{n} w_i}$$

[27]Since clustering and stratification affect different variables differently, no single correction factor can prove perfectly accurate for all analyses.

and to calculate the (weighted) correlation of X and Y we take

$$r_{XY} = \frac{\Sigma w_i (X_i - \bar{X})(Y_i - \bar{Y})}{\sqrt{\Sigma w_i (X_i - \bar{X})^2 \Sigma w_i (Y_i - \bar{Y})^2}}$$

The weights w_i are generally scaled so that $\Sigma w_i = n$. Computation of appropriate weights is the responsibility of the sampling statistician who designs a survey sample, and most computer-program packages oriented towards survey research (e.g., SPSS, OSIRIS) make general provision for case weights in their calculations. It is important to realize, however, that unless the weights are calculated to include a design effect (in which case generally $\Sigma w_i < n$), the use of weights does not compensate for the nonnegligible dependencies introduced into a survey sample by stratification and clustering.

3.5.2. Missing Data

Missing data are typically generated in social-survey research when a respondent refuses to answer a question, or by an error in data collection or data management. Often, too, certain variables are undefined or inappropriate for specific respondents—a respondent who has no children, for example, cannot report their ages—but undefined data are not missing in the same sense as data that exist but fail to be collected or properly recorded. Undefined data do not threaten the representativeness of the sample as missing data do. We also differentiate missing data from the related problem of global nonresponse, where for one reason or another a respondent selected for the sample is not interviewed. Since we cannot generally rely on the randomness of nonresponse, its presence may destroy the representativeness of the sample. If certain characteristics of nonrespondents are known, however, a sampling statistician can often adjust case weights to compensate for the pattern of nonresponse, increasing the relative weights for categories of respondents whose rate of response is low.

Three general (and overlapping) strategies that have been adopted for dealing with missing data are: (1) to employ formal statistical methods (see, e.g., Beale and Little, 1975, for an application of maximum likelihood) that take explicit account of missing data; (2) to replace missing scores with reasonable guessed or predicted values; and (3) to delete observations with missing data from calculations. The most commonly employed strategy in social research is the third.

The risk entailed by deleting observations with missing data is that their deletion will render the sample unrepresentative. In fact, some risk is run regardless of the strategy adopted for coping with missing data: If we knew the missing values, they would not be missing; to proceed in their absence, we need to make some (if only implicit) assumption about them. As King Lear says, "Nothing will come of nothing."

In linear-model analysis, two common procedures for eliminating missing data are: (1) to delete an observation if its value is missing for any of the variables in the linear model; or (2) to deal separately with each pair of variables in the model, computing the covariance or correlation between that pair on the basis of all cases that have valid data for the pair; then the linear model is fit from the correlations or covariances. The first procedure is termed "casewise" (or "listwise") deletion of missing data; the second procedure is called "pairwise" deletion of missing data. The casewise procedure has the advantage of calculating all statistics for the same subset of observations; the pairwise procedure generally retains more data, but can lead to inconsistent calculations (such as negative eigenvalues for $\mathbf{R}_{XX}$).

Although there are several methods for coping with missing data, we can have little confidence in any set of results if different reasonable procedures lead to substantially different findings. Monte-Carlo studies comparing casewise and pairwise missing-data deletion have been conducted by Haitovsky (1968) and by Kim and Curry (1977), but with divergent conclusions.

PROBLEM

3.20. Secure access to a large-scale social-survey dataset that includes case weights and for which there are missing data. (If a suitable dataset is not available locally, one may be obtained from a data archive, such as those maintained by the Inter-University Consortium for Political and Social Research at the University of Michigan; the National Opinion Research Center in Chicago; and the Institute for Behavioural Research at York University, Toronto.) After examining the codebook for the dataset, specify a reasonable linear model and explore fitting the model to the data with and without case weights. Determine whether the fit is affected by using different methods for dealing with missing data.

4

Linear Structural-Equation Models

Structural-equation models enable the researcher to examine simultaneous relationships among a number of variables, some of which may exert mutual influence on each other. These models are therefore differentiated from the single-equation linear models discussed in the first three chapters of this text, which treat the relationship of one dependent variable to one or more of its causes. Structural-equation models likewise differ from direct multivariate extensions of the general linear model which, although they take into account the correlation of several dependent variables, treat these variables in parallel rather than distinguishing causal relations among them (see, for example, Morrison, 1976: Chapter 5).

In the social sciences, structural-equation models are frequently termed "causal models," and although this terminology accurately reflects the purpose of the models, it is potentially misleading for two quite different reasons. On the one hand, most data analysis in the social sciences seeks to discover causal relations: Few studies are simply correlational or predictive in their purpose. On the other hand, structural-equation models in no way avoid the pitfalls of drawing causal inferences from observational data. The term "causal model," then, at once promises too much and is nonspecific.

Yet, one of the great virtues of structural-equation models is the light they shed on the process of causal interpretation of correlational data, making explicit the assumptions underlying causal inference. These insights apply not only to the formal application of structural-equation models, but also to other methods of data analysis.

This chapter begins with a consideration of the form and specification of structural-equation models. Section 4.2 develops the method of instrumental-variables estimation, which provides us with a tool for analyzing and estimating structural-equation models. Having specified a structural-equation model, it is necessary to determine whether the model is estimable, an issue termed the identification problem. In Section 4.3 we show how the instrumental-variables method and other approaches lead to a solution of the identification problem, after which, in Section 4.4, we describe several estimation methods applicable to structural-equation models. In Section 4.5 we explain how an estimated structural-equation model may be used for causal interpretation of statistical relationships, and we take an opportunity to draw general methodological lessons for causal inference from this discussion.

Variables in the social sciences are frequently measured with error; likewise, there is often a less-than-perfect relation between theoretical constructs and their measured indicators. There has been a consequent interest to incorporate measurement errors and multiple indicators in structural-equation models. In Section 4.6, after considering some simple examples of models with measurement errors, we present a very general model for variables measured with error. The final section of the chapter takes up the evaluation of structural-equation models that have been fit to data.

Outside of economics, the majority of applications of structural-equation models have appeared in the literature on social stratification. These applications are reflected in the illustrations and exercises of this chapter.

4.1. SPECIFICATION OF STRUCTURAL-EQUATION MODELS

This section begins by distinguishing the different categories of variables that enter into a structural-equation model, and develops graphic and equation representations of the model, introducing notational conventions along the way. We discuss the assumptions underlying the model, and define two important varieties of structural-equation models, termed recursive and block-recursive models. The section concludes by defining what is called the reduced form of a structural-equation model.

Structural-equation models include three broad classes of variables: endogenous variables, exogenous variables, and disturbances. *Endogenous variables*, as their name implies, are determined within the model, and may be influenced by other endogenous variables, by exogenous variables, and by disturbances. *Exogenous variables*, in contrast, are treated as "givens": They may appear as causes in the model, but not as effects. *Disturbance variables*, sometimes termed *errors* or *errors in equations*, represent most importantly the aggregated omitted causes of the endogenous variables, and, thus, play a role similar to that of the error variable in the general linear model. Disturbance variables are taken to be independent of the exogenous variables in the model.

4.1.1. Path Diagrams

One useful way of representing the structural relations of a model is in the form of a causal graph or *path diagram*. Consider, for example, the model

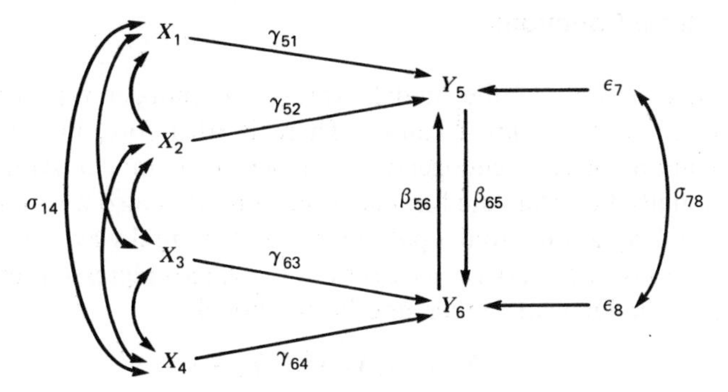

FIGURE 4.1. Nonrecursive structural-equation model for the Duncan, Haller, and Portes peer-influences data. X_1, respondent's intelligence; X_2, respondent's family SES; X_3, best friend's family SES; X_4, best friend's intelligence; Y_5, respondent's occupational aspiration; Y_6, best friend's occupational aspiration. (Source: see Table 4.2).

shown in Figure 4.1, from work done by Duncan, Haller, and Portes (1968) on the occupational aspirations of high-school boys. The exogenous variables in the model are represented by X's, the endogenous variables by Y's, and the disturbances by ε's. The directed (i.e., one-way) arrows in the model indicate the direct effect of one variable on another: For example, each boy's intelligence is specified to affect directly his own aspirations, but not those of the other boy. The double-headed arrows indicate statistical relationships that are not given causal interpretation. Thus, the model, and structural-equation models in general, permit the exogenous variables to be correlated with one another. The disturbances, similarly, are not assumed to be uncorrelated: In this model, then, the aggregated omitted causes of the respondent's aspirations may be correlated with the omitted causes of his best friend's aspirations—as appears substantively sensible. Note, furthermore, that the lack of correlation[1] between exogenous variables and disturbances is reflected in the omission of double arrows linking variables in these two classes.

Each directed arrow in the path diagram is labeled with a symbol representing a *structural coefficient* of the model. As we shall see in Section 4.1.2, structural coefficients are simply regression coefficients interpreted as direct effects. γ's are used to represent the effects of exogenous variables on endogenous variables, while β's give the effects of endogenous variables on each other. The two subscripts of each structural parameter specify respectively the index of the effect and of its cause. γ_{51}, therefore, is the direct effect of X_1 on Y_5, and β_{56} is the effect of Y_6 on Y_5. The double-headed arrows are labeled with σ's, standing for the covariances of the variables attached by the arrows. For notational convenience, each variable in the model has been assigned a unique index.

[1]Throughout this chapter, we shall employ the assumptions that the disturbances and exogenous variables are independent, uncorrelated, or asymptotically uncorrelated interchangeably, according to convenience.

4.1.2. Structural Equations

The structural equations of the model express the endogenous variables as linear functions[2] of their direct causes. There is, therefore, in general one structural equation for each endogenous variable in the model. Although the model may initially be formulated in equation form, it is also a simple matter to read structural equations from a path diagram. A natural way of writing the structural equations of a model is as a series of (related) regression equations; for example, for the Duncan, Haller, and Portes model:

$$\begin{aligned} Y_5 &= \gamma_{51}X_1 + \gamma_{52}X_2 + \beta_{56}Y_6 + \varepsilon_7 \\ Y_6 &= \gamma_{63}X_3 + \gamma_{64}X_4 + \beta_{65}Y_5 + \varepsilon_8 \end{aligned} \tag{4.1}$$

To eliminate constant terms from the structural equations, we simply stipulate that each endogenous and exogenous variable be measured as deviations from its expectation.[3] Frequently, in sociological applications, structural-equation models are specified for standardized variables, so that the coefficients of the model are standardized structural parameters. This is the case, for example, in Duncan, Haller, and Portes's research. So as not to proliferate notation, we shall not distinguish explicitly between the standardized and unstandardized cases.

Another representation of the model places every endogenous and exogenous variable in each structural equation. Variables that do not appear in a particular structural equation are given zero coefficients, and the dependent variable is given a coefficient of one. This form of the model, though more cumbersome than the regression format, has the virtue of showing explicitly which variables have been excluded from each structural equation, information that will be useful to us later on. So that variables may be aligned vertically, we shift all but the disturbances to the left side of the structural equations. For the illustrative model,

$$\begin{aligned} 1Y_5 - \beta_{56}Y_6 - \gamma_{51}X_1 - \gamma_{52}X_2 + 0X_3 + 0X_4 &= \varepsilon_7 \\ -\beta_{65}Y_5 + 1Y_6 + 0X_1 + 0X_2 - \gamma_{63}X_3 - \gamma_{64}X_4 &= \varepsilon_8 \end{aligned}$$

Finally, for compactness and generality, we write the structural-equation model as a matrix equation:

$$\underset{(q\times q)}{\mathbf{B}}\ \underset{(q\times 1)}{\mathbf{y}_i} + \underset{(q\times m)}{\mathbf{\Gamma}}\ \underset{(m\times 1)}{\mathbf{x}_i} = \underset{(q\times 1)}{\boldsymbol{\varepsilon}_i} \tag{4.2}$$

[2]As in the single-equation linear model, we require only that the structural equations be linear in the parameters. Essentially nonlinear structural-equation models are beyond the scope of this chapter: see Amemiya (1974, 1977) and Gallant (1977).

[3]So as to avoid complicating the notation in this chapter, unless otherwise noted, we do not use asterisks to indicate that variables are in mean-deviation form. To specify a constant term for a structural equation, it is merely necessary to leave variables in raw-score form and to include as a regressor a dummy exogenous variable coded one for each observation. Constants are rarely of substantive interest, however.

The vectors $\mathbf{y}_i$, $\mathbf{x}_i$, and $\boldsymbol{\varepsilon}_i$ contain endogenous variables, exogenous variables, and disturbances, each for the ith observation of a sample. $\mathbf{B}$ contains the structural coefficients relating the endogenous variables to each other, while $\boldsymbol{\Gamma}$ contains the coefficients relating the endogenous to the exogenous variables. Each row of the parameter matrices includes the coefficients for one structural equation of the model, and we order the equations so that ones appear on the diagonal of $\mathbf{B}$. The matrix representation of the Duncan, Haller, and Portes model is shown in equation (4.3). As a matter of convenience, we have omitted the subscript i for observation.

$$\begin{pmatrix} 1 & -\beta_{56} \\ -\beta_{65} & 1 \end{pmatrix}\begin{pmatrix} Y_5 \\ Y_6 \end{pmatrix} + \begin{pmatrix} -\gamma_{51} & -\gamma_{52} & 0 & 0 \\ 0 & 0 & -\gamma_{63} & -\gamma_{64} \end{pmatrix}\begin{pmatrix} X_1 \\ X_2 \\ X_3 \\ X_4 \end{pmatrix} = \begin{pmatrix} \varepsilon_7 \\ \varepsilon_8 \end{pmatrix} \tag{4.3}$$

Sometimes we shall require the structural equations for a sample of n observations:

$$\underset{(n\times q)}{\mathbf{Y}}\ \underset{(q\times q)}{\mathbf{B}'} + \underset{(n\times m)}{\mathbf{X}}\ \underset{(m\times q)}{\boldsymbol{\Gamma}'} = \underset{(n\times q)}{\mathbf{E}}$$

Here, we have transposed the matrices of structural parameters, writing equations as columns, so that each observation comprises a row of $\mathbf{Y}$, $\mathbf{X}$, and $\mathbf{E}$.

4.1.3. Assumptions Underlying the Model

The assumptions underlying a structural-equation model are of two general types: first, assumptions of causal structure captured in the structural equations of the model; and second, distributional assumptions regarding the errors. Assumptions of causal structure are implicit in the specification of $\mathbf{B}$ and $\boldsymbol{\Gamma}$, certain of whose entries are prespecified to be zero, and in the choice of endogenous and exogenous variables.

We have already remarked that the exogenous variables and disturbances are defined (i.e., in an application, assumed) to be uncorrelated. It is often convenient to express this assumption in terms of a probability limit:

$$\operatorname{plim}\frac{1}{n}\mathbf{X}'\mathbf{E} = \underset{(m\times q)}{\mathbf{0}}$$

The remaining assumptions about the distribution of the disturbances are analogous to the assumptions concerning the error in the general linear model: that the observations on each disturbance are independently and normally distributed with expectation zero and common variance. Note that although we

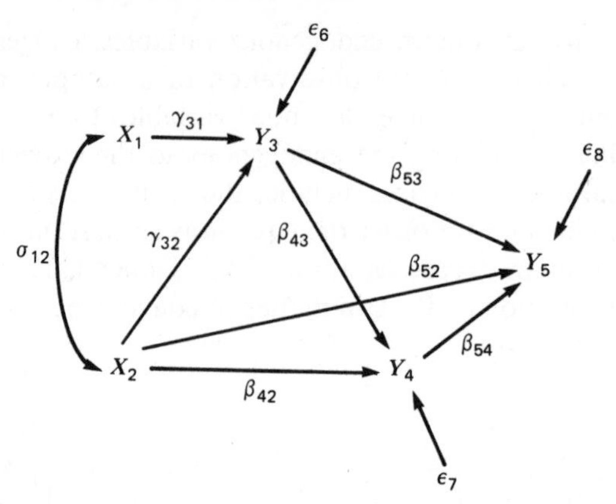

FIGURE 4.2. Recursive structural-equation model for the Blau and Duncan stratification data. X_1, father's education; X_2, father's occupational status; Y_3, education; Y_4, status of first job; Y_5, 1962 occupational status.

assume independent observations, we in general expect *different* disturbance variables to be correlated. The joint distribution of $\boldsymbol{\varepsilon}_i$ is assumed to be multivariate normal with covariance matrix $\boldsymbol{\Sigma}_{\varepsilon\varepsilon}$.

4.1.4. Recursive and Block-Recursive Models

Although structural-equation models do not in general require that different disturbance variables be independent, such assumptions may be made. In conjunction with special patterns of restrictions on the structural coefficients of the model, restrictions on disturbance covariances serve to define two important varieties of structural-equation models: *recursive* and *block-recursive* models. Models that do not satisfy the special requirements of recursive and block-recursive structures are termed *nonrecursive*. As we shall discover in Sections 4.3 and 4.4, the classification of a model has implications for its identification and estimation.

An example of a recursive structural-equation model, taken from work on stratification by Blau and Duncan (1967), is shown in Figure 4.2. This model is recursive because it meets two special conditions: (1) Different disturbance variables are specified to be uncorrelated—a characteristic reflected in the absence of bidirectional arrows linking the disturbances; and (2) the causal

FIGURE 4.3. Relations ruled out by recursive structure. (a) Reciprocal effects. (b) Causal loop.

structure of the model is unidirectional—there are no reciprocal paths or causal loops of the sort illustrated in Figure 4.3.

In the matrix representation of the model, the uncorrelated disturbances of a recursive model imply a diagonal covariance matrix of disturbances. The unidirectional causal structure implies a lower-triangular **B** matrix, or a **B** matrix that can be made triangular by a reordering of the endogenous variables. For the Blau and Duncan stratification model

$$\mathbf{B} = \begin{pmatrix} 1 & 0 & 0 \\ -\beta_{43} & 1 & 0 \\ -\beta_{53} & -\beta_{54} & 1 \end{pmatrix}$$

$$\boldsymbol{\Sigma}_{\varepsilon\varepsilon} = \begin{pmatrix} \sigma_6^2 & 0 & 0 \\ 0 & \sigma_7^2 & 0 \\ 0 & 0 & \sigma_8^2 \end{pmatrix}$$

It should be stressed that the special requirements of a recursive model, including the stipulation of uncorrelated disturbances, must be justifiable on substantive grounds, as is the case generally for the application of statistical models. In the Blau and Duncan model, for example, we may question the independence of ε_7 and ε_8, for Y_4 and Y_5 are likely to have common omitted causes.

It may be the case that a structural-equation model is not recursive, but that the requirements for a recursive model are met for subsets (termed *blocks*) of the endogenous variables and associated disturbances, rather than for these variables treated individually. That is, if we partition the endogenous variables and disturbances into blocks: (1) causation is unidirectional between blocks; and (2) errors are uncorrelated between blocks. Within blocks, mutual causation and correlated disturbances are permitted.

A block-recursive model is shown in Figure 4.4; this model is a modification of one specified by Duncan, Haller, and Portes (1968). Here Y_5 and Y_6 together with the associated errors ε_9 and ε_{10} comprise the first block, and Y_7, Y_8, ε_{11}, and ε_{12} comprise the second block. For this model,

$$\mathbf{B} = \left(\begin{array}{cc|cc} 1 & -\beta_{56} & 0 & 0 \\ -\beta_{65} & 1 & 0 & 0 \\ \hline -\beta_{75} & 0 & 1 & -\beta_{78} \\ 0 & -\beta_{86} & -\beta_{87} & 1 \end{array}\right)$$

$$\boldsymbol{\Sigma}_{\varepsilon\varepsilon} = \left(\begin{array}{cc|cc} \sigma_{99} & \sigma_{9,10} & 0 & 0 \\ \sigma_{9,10} & \sigma_{10,10} & 0 & 0 \\ \hline 0 & 0 & \sigma_{11,11} & \sigma_{11,12} \\ 0 & 0 & \sigma_{11,12} & \sigma_{12,12} \end{array}\right)$$

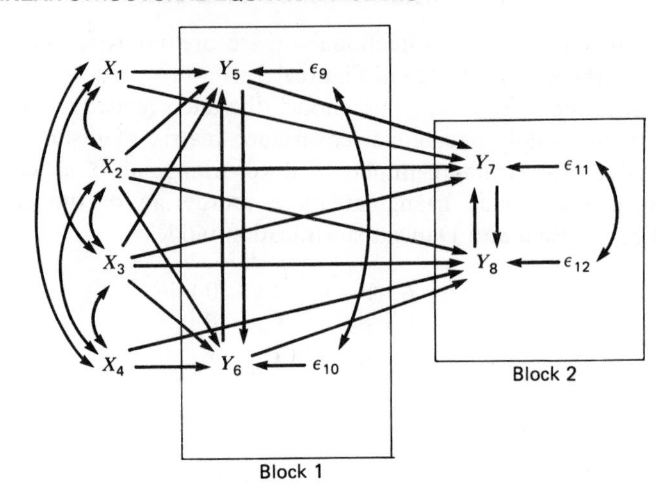

FIGURE 4.4. Block-recursive model for the Duncan, Haller, and Portes peer-influences data. X_1, respondent's intelligence; X_2, respondent's family SES; X_3, best friend's family SES; X_4, best friend's intelligence; Y_5, respondent's occupational aspiration; Y_6, best friend's occupational aspiration; Y_7, respondent's educational aspiration; Y_8, best friend's educational aspiration. (Source: see Table 4.2.)

Note that $\mathbf{B}$, though not triangular, is *block triangular* when partitioned according to blocks of endogenous variables, and that $\mathbf{\Sigma}_{\varepsilon\varepsilon}$ is *block diagonal* when partitioned by blocks of disturbances:

$$\mathbf{B} = \begin{pmatrix} \mathbf{B}_{11} & \mathbf{0} \\ \mathbf{B}_{12} & \mathbf{B}_{22} \end{pmatrix}$$

$$\mathbf{\Sigma}_{\varepsilon\varepsilon} = \begin{pmatrix} \mathbf{\Sigma}_{11} & \mathbf{0} \\ \mathbf{0} & \mathbf{\Sigma}_{22} \end{pmatrix}$$

These two characteristics, of course, may be extended to models with more than two blocks and with different numbers of endogenous variables in each block. For the peer-influences data, the specification of uncorrelated errors between blocks is questionable, for the residual causes of a boy's educational aspirations are likely also to affect his occupational aspirations.

4.1.5. The Reduced Form of the Model

Thus far, we have dealt with the structural equations of a simultaneous-equation model, equations that specify the direct causal relations among the variables in the model. The *reduced form* of a structural-equation model expresses the endogenous variables in terms of the exogenous variables and disturbances, which comprise the ultimate inputs to the system under study.

Solving the structural equations (4.2) for $\mathbf{y}$ is straightforward:

$$\begin{aligned}\mathbf{y} &= -\mathbf{B}^{-1}\mathbf{\Gamma}\mathbf{x} + \mathbf{B}^{-1}\boldsymbol{\varepsilon} \\ &= \underset{(q\times m)}{\mathbf{\Pi}}\ \underset{(m\times 1)}{\mathbf{x}} + \underset{(q\times 1)}{\boldsymbol{\delta}}\end{aligned} \tag{4.4}$$

where $\mathbf{\Pi} \equiv -\mathbf{B}^{-1}\mathbf{\Gamma}$ is the matrix of *reduced-form coefficients*, and $\boldsymbol{\delta} \equiv \mathbf{B}^{-1}\boldsymbol{\varepsilon}$ is the vector of *reduced-form errors*. Note that $\mathbf{\Pi}$ is a function of the structural parameters ($\mathbf{B}$ and $\mathbf{\Gamma}$), and that $\boldsymbol{\delta}$ results from a linear transformation of the structural disturbances $\boldsymbol{\varepsilon}$.

In solving for $\mathbf{y}$ we have implicitly assumed that $\mathbf{B}$ is nonsingular, and we now make this condition a requirement for a well formed structural-equation model. This requirement is not problematic, however, since our method of constructing structural equations, which places ones on the main diagonal of $\mathbf{B}$, virtually assures that $\mathbf{B}$ will be nonsingular.[4]

In the reduced form, $\mathbf{x}$ and $\boldsymbol{\delta}$ are uncorrelated, because $\mathbf{x}$ contains the exogenous variables, while $\boldsymbol{\delta}$ is a linear transformation of the structural disturbances $\boldsymbol{\varepsilon}$. The reduced form, therefore, meets the assumptions of ordinary least-squares (*OLS*) estimation, since the independent variables in the reduced form ($\mathbf{x}$) are uncorrelated with the errors ($\boldsymbol{\delta}$). We shall see later (Section 4.4) that OLS estimation does not in general provide consistent estimators of the structural-form parameters.

The reduced form has several uses: (1) It traces the indirect, as well as the direct impact of the exogenous variables on the endogenous variables (see Section 4.5); (2) it is useful in certain forecasting applications;[5] and (3) it is useful in deriving a procedure for determining the estimability of a structural-equation model, a topic taken up in Section 4.3.

We gave the structural form of the Duncan, Haller, and Portes peer-influences model in equation (4.3). The reduced form of this model is

$$\begin{pmatrix} Y_5 \\ Y_6 \end{pmatrix} = -\begin{pmatrix} 1 & -\beta_{56} \\ -\beta_{65} & 1 \end{pmatrix}^{-1} \begin{pmatrix} -\gamma_{51} & -\gamma_{52} & 0 & 0 \\ 0 & 0 & -\gamma_{63} & -\gamma_{64} \end{pmatrix} \begin{pmatrix} X_1 \\ X_2 \\ X_3 \\ X_4 \end{pmatrix} + \begin{pmatrix} 1 & -\beta_{56} \\ -\beta_{65} & 1 \end{pmatrix}^{-1} \begin{pmatrix} \varepsilon_7 \\ \varepsilon_8 \end{pmatrix}$$

[4] The assignment of a coefficient of one to an endogenous variable is each structural equation is sometimes called a *normalization rule*. If we view a structural equation as simply specifying a relation among the endogenous and exogenous variables of the model, the normalization employed may be regarded as arbitrary. For models representable as causal diagrams, it is natural to normalize for the dependent variable in each structural equation. Some estimation methods (such as 2SLS, see Section 4.4.1) are sensitive to the normalization applied, while others (e.g., FIML, also in Section 4.4.1) are not.

[5] In economics, structural-equation models are typically applied to time series, rather than to cross-sectional data.

which, upon manipulation, yields

$$\begin{aligned}
Y_5 &= \frac{\gamma_{51}}{1-\beta_{56}\beta_{65}}X_1 + \frac{\gamma_{52}}{1-\beta_{56}\beta_{65}}X_2 + \frac{\beta_{56}\gamma_{63}}{1-\beta_{56}\beta_{65}}X_3 \\
&\quad + \frac{\beta_{56}\gamma_{64}}{1-\beta_{56}\beta_{65}}X_4 + \left(\frac{1}{1-\beta_{56}\beta_{65}}\varepsilon_7 + \frac{\beta_{56}}{1-\beta_{56}\beta_{65}}\varepsilon_8\right) \\
&= \pi_{51}X_1 + \pi_{52}X_2 + \pi_{53}X_3 + \pi_{54}X_4 + \delta_5 \\
Y_6 &= \frac{\beta_{65}\gamma_{51}}{1-\beta_{56}\beta_{65}}X_1 + \frac{\beta_{65}\gamma_{52}}{1-\beta_{56}\beta_{65}}X_2 + \frac{\gamma_{63}}{1-\beta_{56}\beta_{65}}X_3 \\
&\quad + \frac{\gamma_{64}}{1-\beta_{56}\beta_{65}}X_4 + \left(\frac{\beta_{65}}{1-\beta_{56}\beta_{65}}\varepsilon_7 + \frac{1}{1-\beta_{56}\beta_{65}}\varepsilon_8\right) \\
&= \pi_{61}X_1 + \pi_{62}X_2 + \pi_{63}X_3 + \pi_{64}X_4 + \delta_6
\end{aligned}$$

PROBLEMS

4.1. For each of the path diagrams shown in Figure 4.5: (i) write out the structural equations of the model; (ii) determine whether the model is nonrecursive, recursive, or block recursive; and (iii) find the reduced form of the model.

4.2. The model in Figure 4.5(g) was employed by Rindfuss, Bumpass, and St. John (1980) to study the possibly reciprocal relationship between women's education and fertility. The variables in the model are:

- X_1 respondent's father's occupational status
- X_2 respondent's race, coded one for blacks and zero otherwise
- X_3 number of respondent's siblings
- X_4 farm background, coded one if the respondent grew up on a farm
- X_5 regional background, coded one if the respondent grew up in the South
- X_6 household composition when the respondent was 14 years old, coded zero if both parents were present in the household and one otherwise
- X_7 religion, coded one if the respondent is Catholic
- X_8 smoking, coded one if the respondent smoked prior to age 16
- X_9 fecundity, coded one if the respondent had a miscarriage prior to the birth of her first child
- Y_{10} respondent's education at first marriage, in years
- Y_{11} age at first birth

This model was fit to data from the 1970 National Fertility Survey, for a sample of 1766 black and white American women, age 35 to 40, who had one or more children. Comment on the specification of the model, paying particular attention to its causal structure and to the distributional assumptions concerning the errors.

4.3. The model diagrammed in Figure 4.5(h) was specified by Berk and Berk (1978) to account for the division of household labor among the members of a family. The model was estimated for a sample of 184 households in Evanston, Illinois, an affluent suburb of Chicago. The variables

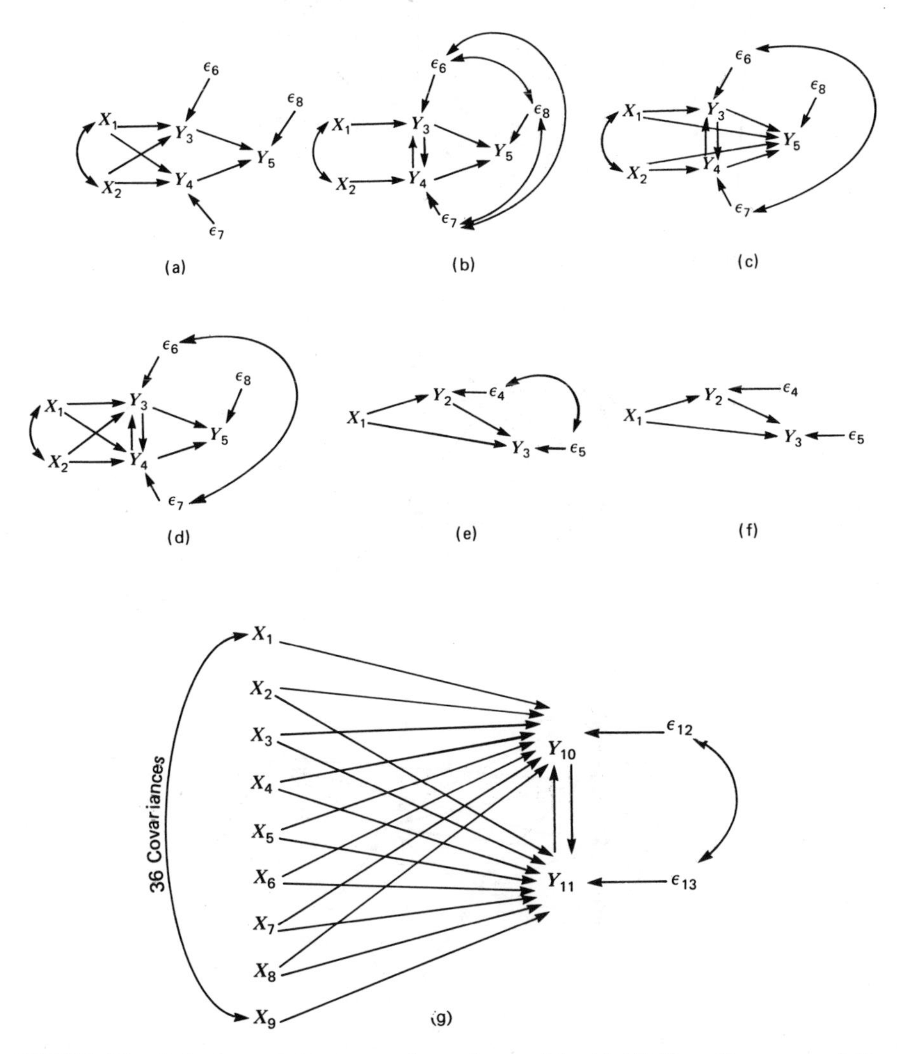

FIGURE 4.5. Path diagrams for structural-equation models. (g) Rindfuss, Bumpass, and St. John's model. (h) Berk and Berk's model. (i) Duncan, Featherman, and Duncan's model (adapted with permission from Duncan, Featherman, and Duncan, 1972). (j) Lincoln's model.

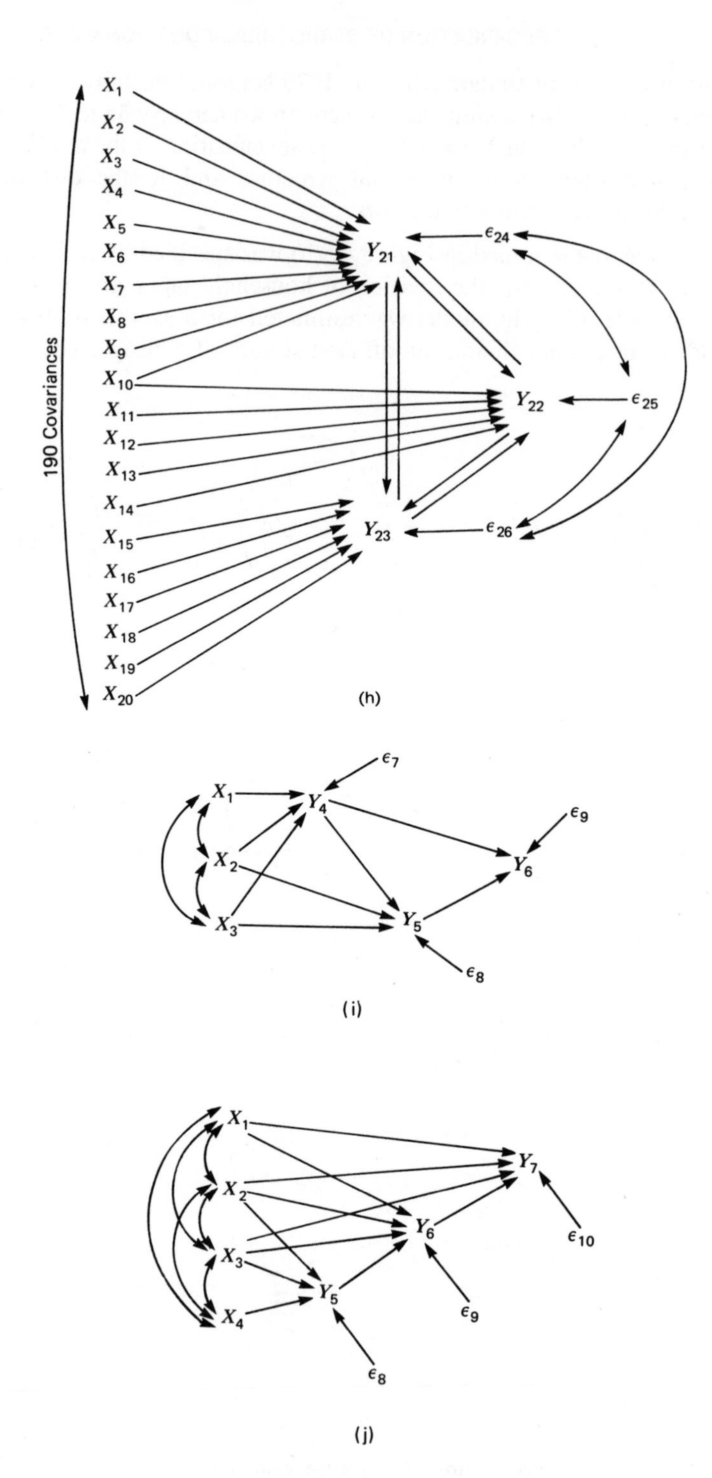

FIGURE 4.5. (*Continued*).

appearing in the model include:

X_1	wife's wages, in $1000s/month
X_2	coded one if the wife is employed as a professional or technical worker, zero otherwise
X_3	coded one if the wife is employed as a manager or proprietor
X_4	coded one if there is a boy under one-year old in the home
X_5	coded one if there is a girl under one-year old
X_6	coded one if there is a one-year-old boy
X_7	coded one if there is a one-year-old girl
X_8	coded one if there is a two-year-old boy
X_9	coded one if there is a two-year-old girl
X_{10}	year married
X_{11}	husband's monthly income
X_{12}	coded one if the husband is employed as a professional or technical worker
X_{13}	coded one if the husband is employed as a manager or proprietor
X_{14}	coded one if, in the recent past, the husband decided to do more housework
X_{15}	coded one if there is a boy 11–15
X_{16}	coded one if there is a girl 11–15
X_{17}	coded one if there is a boy 16–20
X_{18}	coded one if there is a girl 16–20
X_{19}	husband's education, measured in seven levels
X_{20}	coded one if, in the recent past, a child decided to do more housework
Y_{21}	proportion of household tasks generally done by the wife
Y_{22}	proportion of tasks generally done by the husband
Y_{23}	proportion of tasks generally done by children

Note: Y_{21}, Y_{22}, and Y_{23} are *not* constrained to sum to one, since different members of the household can regularly contribute to the same tasks. As in Problem 4.2, comment on the specification of the model.

4.4. The model in Figure 4.5(i) appears in Duncan, Featherman, and Duncan's (1972) monograph on socioeconomic background and achievement. This model, which was fit separately to each of several age groups of men, was estimated for the same 1962 sample as employed in Blau and Duncan's (1967) study. The variables are defined as follows:

X_1	father's education
X_2	father's occupational status
X_3	number of respondent's siblings
Y_4	respondent's education
Y_5	respondent's occupational status
Y_6	respondent's income

Discuss the specification of Duncan, Featherman, and Duncan's model.

4.5. In a study of strike activity in metropolitan areas of the United States, Lincoln (1978) specified the structural-equation model shown in Figure 4.5(j). The model was estimated using data for 78 metropolitan areas. The variables in the model are:

X_1	the degree of concentration of union staff in the metropolitan area; this index is high when most union staff members work for a relatively small number of large unions
X_2	the degree of concentration in employment
X_3	the log of the number of employed workers
X_4	the proportion of workers who are in unionized establishments
Y_5	the number of strikes in the period 1963–1969
Y_6	the number of strikers
Y_7	the number of person-days idle due to strikes

Comment on the specification of Lincoln's model.

4.2. INSTRUMENTAL-VARIABLES ESTIMATION

After specifying a structural-equation model, it is necessary to determine whether the parameters of the model may be estimated. This issue is called the identification problem, and it will be taken up in the next section. The present section is, therefore, a necessary detour, for instrumental-variables estimation will provide us with an approach to the identification problem as well as with a means for approaching the topic of estimation (Section 4.4). Although our ultimate interest is in applying the method of instrumental variables to structural-equation models, we shall develop the method in a more general context.

Suppose, to begin, that we wish to estimate the simple-regression model

$$Y = \alpha + \beta X + \varepsilon \tag{4.5}$$

and that we make the usual assumptions that $E(\varepsilon) = 0$, $V(\varepsilon) = \sigma_\varepsilon^2$, and X and ε are independent. (Here, X and Y are in raw-score form.) In Chapter 1, this model and these assumptions were employed in conjunction with least-squares estimation. We shall now derive the ordinary-least-squares estimator of β in an alternative manner, termed the *expectation method*. Let us express both X and Y as deviations from their expectations; that is, $Y^* \equiv Y - E(Y)$, and $X^* \equiv X - E(X)$. Then, because of the assumption that $E(\varepsilon) = 0$, the model (4.5) becomes $Y^* = \beta X^* + \varepsilon$. Multiplying this equation through by X^* and taking expectations of both sides, we get[6]

$$\begin{aligned} E(X^*Y^*) &= \beta E(X^{*2}) + E(X^*\varepsilon) \\ \sigma_{XY} &= \beta\sigma_X^2 + \sigma_{X\varepsilon} \end{aligned} \tag{4.6}$$

Here, σ_{XY} is the population covariance of X and Y, σ_X^2 is the population variance of X, and $\sigma_{X\varepsilon}$ is the population covariance of X and ε; this last quantity is zero due to the stipulation that X and ε are independent. Solving equation (4.6) for β gives us $\beta = \sigma_{XY}/\sigma_X^2$.

We cannot of course apply this result without knowledge of the population quantities σ_{XY} and σ_X^2, knowledge that is generally unavailable. We can, however, estimate these parameters from sample data, employing the sample variance $S_X^2 = \Sigma(X_i - \bar{X})^2/(n-1)$ and the sample covariance $S_{XY} = \Sigma[(X_i - \bar{X})(Y_i - \bar{Y})]/(n-1)$. We know that S_X^2 and S_{XY} are consistent estimators; that is, plim $S_X^2 = \sigma_X^2$, and plim $S_{XY} = \sigma_{XY}$. The estimator $B = S_{XY}/S_X^2$ is, therefore, also consistent, for

$$\text{plim } B = \frac{\text{plim } S_{XY}}{\text{plim } S_X^2} = \frac{\sigma_{XY}}{\sigma_X^2} = \beta$$

Thus far, we have shown nothing new, because we recognize B as the usual OLS estimator of β.

Suppose, however, that we cannot assume the uncorrelation of X and ε in model (4.5), which justified the crucial elimination of $\sigma_{X\varepsilon}$ from equation (4.6), but that there is some variable Z for which it may reasonably be assumed that plim $S_{Z\varepsilon} = \sigma_{Z\varepsilon} = 0$, and that plim $S_{ZX} = \sigma_{ZX} \neq 0$. In words, Z and ε are uncorrelated in the population, but Z and X are correlated. Then, following the previous development, but multiplying through by $Z^* \equiv Z - E(Z)$ rather

[6] We assume throughout that expectations, variances, and covariances exist.

than by X^*, we obtain

$$E(Z^*Y^*) = \beta E(Z^*X^*) + E(Z^*\varepsilon)$$

$$\sigma_{ZY} = \beta\sigma_{ZX} + \sigma_{Z\varepsilon} \tag{4.7}$$

$$\beta = \frac{\sigma_{ZY}}{\sigma_{ZX}}$$

Replacing the population covariances in equation (4.7) with their consistent estimators produces a consistent estimator of β: $B = S_{ZY}/S_{ZX}$. Here, B is called an *instrumental-variable* (*IV*) *estimator*, which is generally distinct from the OLS estimator, and Z is an *instrumental variable*. Recall that the two critical requirements for an instrumental variable are uncorrelation with the error ($\sigma_{Z\varepsilon} = 0$) and nonzero correlation with the independent variable ($\sigma_{ZX} \neq 0$). OLS, then, may be thought of as a type of IV estimation, for which the instrumental variable and the independent variable are one and the same.

The method of instrumental variables may be generalized to models with several regressors, for which purpose we cast the model in matrix form:

$$Y^* = \left(X_1^*, X_2^*, \ldots, X_k^*\right)\begin{pmatrix}\beta_1\\ \beta_2\\ \vdots\\ \beta_k\end{pmatrix} + \varepsilon$$

$$= \mathbf{x}^{*\prime}\boldsymbol{\beta} + \varepsilon \tag{4.8}$$

where Y^* and the X_j^* are written as deviations from their expectations, eliminating the constant term from $\boldsymbol{\beta}$. Suppose that we have available k instrumental variables in a vector $\underset{(k\times 1)}{\mathbf{z}^*}$, the entries of which are also in deviation form. We require that

$$\text{plim} \underset{(k\times 1)}{\mathbf{s}_{Z\varepsilon}} = \boldsymbol{\sigma}_{Z\varepsilon} = \mathbf{0}$$

$$\text{plim} \underset{(k\times k)}{\mathbf{S}_{ZX}} = \boldsymbol{\Sigma}_{ZX} \text{ nonsingular}$$

where $\mathbf{s}_{Z\varepsilon}$ and $\mathbf{S}_{ZX}$ contain sample covariances, and $\boldsymbol{\sigma}_{Z\varepsilon}$ and $\boldsymbol{\Sigma}_{ZX}$ contain the corresponding population covariances. The first criterion specifies that the instrumental variables are uncorrelated with the error in the population; the second criterion requires that the instrumental variables are correlated with the independent variables and that there is not perfect collinearity. Later on, to obtain an asymptotic covariance matrix for the IV estimators, we shall also require the existence of the population covariance matrix for the instrumental variables: $\text{plim}\, \mathbf{S}_{ZZ} = \mathbf{\Sigma}_{ZZ}$.

Proceeding as before, we premultiply both sides of equation (4.8) by $\mathbf{z}^*$ and take expectations:

$$E(\mathbf{z}^* Y^*) = E(\mathbf{z}^* \mathbf{x}^{*\prime})\boldsymbol{\beta} + E(\mathbf{z}^* \varepsilon)$$

$$\boldsymbol{\sigma}_{ZY} = \mathbf{\Sigma}_{ZX}\boldsymbol{\beta} + \boldsymbol{\sigma}_{Z\varepsilon}$$

$$\boldsymbol{\beta} = \mathbf{\Sigma}_{ZX}^{-1}\boldsymbol{\sigma}_{ZY}$$

Substituting sample covariances for population covariances, we then obtain

$$\mathbf{b} = \mathbf{S}_{ZX}^{-1}\mathbf{s}_{ZY} = (\mathbf{Z}^{*\prime}\mathbf{X}^*)^{-1}\mathbf{Z}^{*\prime}\mathbf{y}^* \tag{4.9}$$

as the IV estimator of $\boldsymbol{\beta}$. Because $\mathbf{S}_{ZX}$ and $\mathbf{s}_{ZY}$ are consistent estimators, so is $\mathbf{b}$. In equation (4.9), $\underset{(n\times k)}{\mathbf{Z}^*}$, $\underset{(n\times k)}{\mathbf{X}^*}$, and $\underset{(n\times 1)}{\mathbf{y}^*}$ are data matrices of variables in mean-deviation form.

The asymptotic covariance matrix of the IV estimator is given by

$$\mathscr{V}(\mathbf{b}) = \frac{\sigma_\varepsilon^2}{n}\mathbf{\Sigma}_{ZX}^{-1}\mathbf{\Sigma}_{ZZ}\mathbf{\Sigma}_{XZ}^{-1} \tag{4.10}$$

(A relatively simple but flawed proof of equation (4.10) is given in Johnston (1972: Chapter 9); see McCallum (1973) for the correction.) Because in applications we are not in a position to know the population quantities in equation (4.10), the covariance matrix for $\mathbf{b}$ must be estimated. We may proceed as follows:[7]

$$\begin{aligned}
\mathbf{e} &= \mathbf{y} - \mathbf{X}\mathbf{b} \\
S_E^2 &= \frac{\mathbf{e}'\mathbf{e}}{n-k-1} \\
\widehat{\mathscr{V}(\mathbf{b})} &= \frac{S_E^2}{n-1}\mathbf{S}_{ZX}^{-1}\mathbf{S}_{ZZ}\mathbf{S}_{XZ}^{-1} \\
&= S_E^2(\mathbf{Z}^{*\prime}\mathbf{X}^*)^{-1}\mathbf{Z}^{*\prime}\mathbf{Z}^*(\mathbf{X}^{*\prime}\mathbf{Z}^*)^{-1}
\end{aligned} \tag{4.11}$$

For the sampling variances of $\mathbf{b}$ to be small, there must be large covariances between the instrumental variables and the regressors. For example, in the

[7]Since these results are asymptotic, it is also reasonable to calculate $S_E^2 = \mathbf{e}'\mathbf{e}/n$. Using "degrees of freedom" $n - k - 1$ rather than n in the denominator of S_E^2 produces a larger estimate of error variance and, hence, is conservative.

simple-regression model (4.5), for which $B = S_{ZY}/S_{ZX}$, the asymptotic variance of B is

$$\mathcal{V}(B) = \frac{\sigma_\varepsilon^2}{n}\left(\frac{\sigma_Z^2}{\sigma_{ZX}^2}\right) = \frac{\sigma_\varepsilon^2}{n}\left(\frac{1}{\rho_{ZX}^2\sigma_X^2}\right)$$

where ρ_{ZX} is the population correlation of Z and X.

PROBLEMS

4.6. Show how the expectation method may be used to derive the OLS estimator in multiple regression. Show that OLS estimation is a type of IV estimation.

4.7.† Consider the multiple-regression model $\mathbf{y}^* = \mathbf{X}^*\boldsymbol{\beta} + \boldsymbol{\varepsilon}$, for n observations and k independent variables. Let $\mathbf{Z}^*$ be an $(n \times k)$ matrix of instrumental variables, producing the IV estimator $\mathbf{b} = (\mathbf{Z}^{*\prime}\mathbf{X}^*)^{-1}\mathbf{Z}^{*\prime}\mathbf{y}^*$. Let $\mathbf{T}$ be any $(k \times k)$ nonsingular matrix. Show that

(a) $\tilde{\mathbf{Z}}^* \equiv \mathbf{Z}^*\mathbf{T}$ is also a matrix of instrumental variables; and that

(b) using $\tilde{\mathbf{Z}}^*$ in place of $\mathbf{Z}^*$ produces the same IV estimates.

(c) Why is it therefore valid to conclude that what is important about a set of instrumental variables is the subspace that it spans and not the basis selected for this subspace?

(d)* Figure 4.6 shows the vector geometry of the one-independent variable case, that is for the model $\mathbf{y}^* = \beta\mathbf{x}^* + \boldsymbol{\varepsilon}$. In this figure, $\mathbf{x}^*$ is the independent-variable vector for a *particular* sample; $\bar{\boldsymbol{\varepsilon}} \equiv E(\boldsymbol{\varepsilon}|\mathbf{x}^*)$ (why is it nonzero?); $\bar{\mathbf{z}}^* \equiv E(\mathbf{z}^*|\mathbf{x}^*)$ (why, from the figure, is Z qualified to be an IV?); and $\bar{\mathbf{y}}^* \equiv E(\mathbf{y}^*|\mathbf{x}^*) = \beta\mathbf{x}^* + \bar{\boldsymbol{\varepsilon}}$ (note that $E(\mathbf{y}^*) \neq \bar{\mathbf{y}}^*$; why?). In drawing Figure 4.6, we take a line of sight perpendicular to the $\mathbf{x}^*, \bar{\mathbf{z}}^*$ plane; in other words, we may think of $\bar{\boldsymbol{\varepsilon}}$ and $\bar{\mathbf{y}}^*$ as orthogonal projections onto this plane. On the basis of this figure, and working with the population analogs of the estimators, explain why (i) the OLS estimator is biased, and (ii) the IV estimator is not. [Hint: Show that the triangles $(\mathbf{0}, \tilde{\mathbf{y}}^*, \beta\mathbf{x}^*)$ and $(\mathbf{0}, \tilde{\mathbf{x}}^*, \mathbf{x}^*)$ are similar and thus $\beta = \|\beta\mathbf{x}^*\|/\|\mathbf{x}^*\| = \|\tilde{\mathbf{y}}^*\|/\|\tilde{\mathbf{x}}^*\|$; find expressions for $\tilde{\mathbf{y}}^*$ and $\tilde{\mathbf{x}}^*$ using the fact that they are orthogonal projections onto $\bar{\mathbf{z}}^*$, and substitute these into the formula for β.] Note that in this problem we are dealing with the population analogs of the estimators. Since $\mathbf{x}^*$ is not fixed over repeated sampling, these results expressed in terms of expectations hold only roughly in finite samples. Further information on the geometry of IV estimation may be found in Wonnacott and Wonnacott (1979: 453–455).

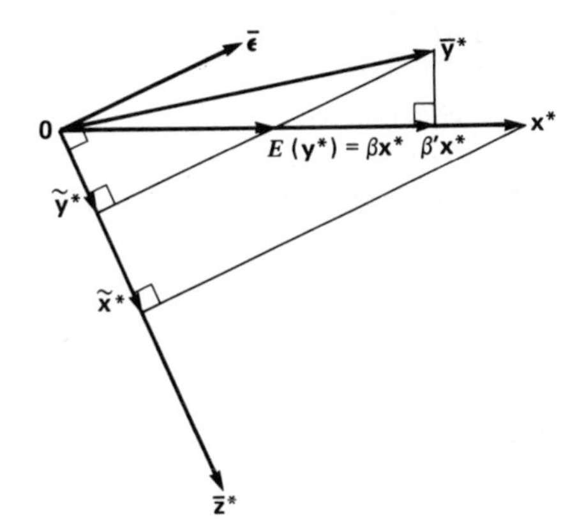

FIGURE 4.6. Vector geometry of instrumental-variable estimation. (Adapted with permission from Wonnacott and Wonnacott, 1979: 454.)

4.3. THE IDENTIFICATION PROBLEM AND ITS SOLUTION

As we have mentioned, once a structural-equation model is specified, it is necessary to determine whether the parameters of the model can be estimated. This issue of estimability is called the *identification problem*. In single-equation linear models with full-rank design matrices, the general assumptions of the model assure that its parameters may be estimated. In structural-equation models, the distributional assumptions of the model are generally insufficient to guarantee identification of its parameters; to assure identification, additional assumptions, taking the form of *a priori* restrictions on the parameters of the model, are necessary.

In general, two sorts of prior restrictions are placed on the model: (1) restrictions on structural parameters, typically specifying that certain parameters are zero; and (2) restrictions on covariances between disturbances, typically specifying that certain of these covariances are zero. We shall first consider the identification of nonrecursive models, where no restrictions are placed on disturbance covariances, deriving general rules for determining whether a model is identified. Then we shall take up the identification of recursive and block-recursive models which, as we know, specify that certain disturbance covariances are zero. Finally, we shall examine the identification status of nonrecursive models that place restrictions on covariances between disturbances.

A parameter in a structural-equation model is *identified* if it can be estimated, and *underidentified* (or *unidentified*) otherwise. If more than one estimator of the parameter can be obtained, then the parameter is *overidentified*; if just one estimator can be obtained, then the parameter is *exactly* (or *just*) *identified*. These distinctions are illustrated in Figure 4.7. The same terminology is applicable to structural equations and to the structural-equation model as a whole. Thus, a structural equation is just identified if there is one and only one way of estimating its parameters. Likewise, a model is overidentified if all its structural equations are identified, and if at least one structural

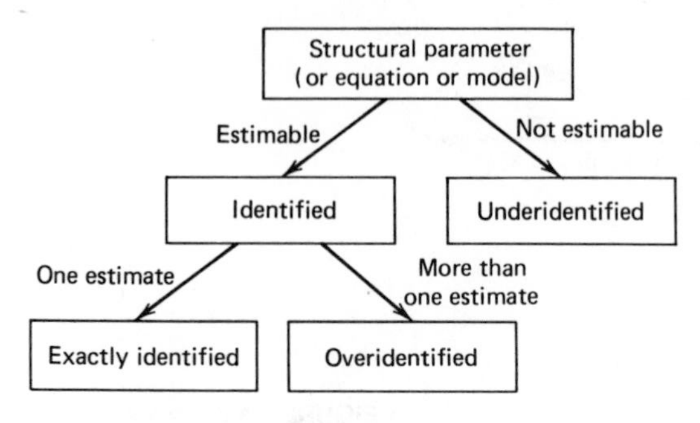

FIGURE 4.7. Identification terminology.

equation is overidentified. In practice, the identification status of a model is generally determined one structural equation at a time.

4.3.1. Identification of Nonrecursive Models

There are several approaches that can be taken to the identification problem. The simplest approach employs the method of instrumental variables, and it is with this approach that we begin. The IV approach yields the so-called order condition, which is a necessary-but-not-sufficient condition for identification. After pursuing the IV approach, we develop a method based on transformations of the structural equations and on the relationship of the structural equations to the reduced form. This method produces a necessary-and-sufficient condition for identification called the rank condition.

The Instrumental-Variables Approach Because they are independent of the disturbances, the exogenous variables of a structural-equation model provide a pool of instrumental variables for estimating the structural parameters of the model. Consider the first structural equation of the Duncan, Haller, and Portes model, given in equation (4.1) and repeated here:

$$Y_5 = \gamma_{51}X_1 + \gamma_{52}X_2 + \beta_{56}Y_6 + \varepsilon_7 \tag{4.12}$$

Multiplying equation (4.12) through by the exogenous variables, taking expectations, and substituting sample covariances for population covariances produces four IV estimating equations:

IVs	Estimating Equations
X_1	$S_{15} = C_{51}S_{11} + C_{52}S_{12} + B_{56}S_{16}$
X_2	$S_{25} = C_{51}S_{12} + C_{52}S_{22} + B_{56}S_{26}$
X_3	$S_{35} = C_{51}S_{13} + C_{52}S_{23} + B_{56}S_{36}$
X_4	$S_{45} = C_{51}S_{14} + C_{52}S_{24} + B_{56}S_{46}$

(4.13)

Here there are four estimating equations in only three unknowns (C_{51}, C_{52}, and B_{56}), since the sample covariances represent known coefficients calculable from sample data. In general, therefore, the estimating equations (4.13) will be overdetermined: For an actual sample, there will in general be no set of values c_{51}, c_{52}, and b_{56} that simultaneously satisfies all four equations.

Our problem, however, is not that we have too little information, but too much: By arbitrarily discarding one IV estimating equation from (4.13), we can obtain consistent estimators of the structural parameters. The surplus of instrumental variables indicates that the structural equation is overidentified.

The essential nature of overidentification is clarified by considering the population analogs of the estimating equations (4.13):

$$\begin{aligned}
\sigma_{15} &= \gamma_{51}\sigma_{11} + \gamma_{52}\sigma_{12} + \beta_{56}\sigma_{16} \\
\sigma_{25} &= \gamma_{51}\sigma_{12} + \gamma_{52}\sigma_{22} + \beta_{56}\sigma_{26} \\
\sigma_{35} &= \gamma_{51}\sigma_{13} + \gamma_{52}\sigma_{23} + \beta_{56}\sigma_{36} \\
\sigma_{45} &= \gamma_{51}\sigma_{14} + \gamma_{52}\sigma_{24} + \beta_{56}\sigma_{46}
\end{aligned} \tag{4.14}$$

If the model is correctly specified, and indeed the X's and ε's are independent, then equations (4.14) hold precisely and simultaneously.

It is helpful to think geometrically about the issue of overidentification. To simplify the geometry, imagine that we wish to estimate the structural equation

$$Y_5 = \gamma_{51}X_1 + \beta_{54}Y_4 + \varepsilon_7 \tag{4.15}$$

and have available three exogenous variables, X_1, X_2, and X_3 to serve as instruments. Applying these instrumental variables produces three estimating equations, each with a population analog. As we have pointed out with respect to the Duncan, Haller, and Portes model, if the model (4.15) is correctly specified, all three population equations hold simultaneously, despite the fact that there are but two unknown structural parameters. As depicted in Figure 4.8(a), each equation represents a line in the $\gamma_{51} \times \beta_{54}$ space. Since the equations hold simultaneously, the three lines intersect at a point, determining the true values of the parameters.

In the sample, however, the estimating equations are perturbed by sampling error; that is, while $\boldsymbol{\sigma}_{X\varepsilon_7} = \mathbf{0}$, and while the average value (ignoring small-sample bias) of $\mathbf{s}_{X\varepsilon_7}$ over many samples is zero, in a particular sample it is unlikely that $\mathbf{s}_{X\varepsilon_7}$ is precisely zero. Geometrically, the lines corresponding to the three estimating equations do not in general intersect at a point, as shown in Figure 4.8(b), even if the model if correctly specified. Of course, if the estimating equations are highly inconsistent with one another (i.e., if the lines in Figure 4.8(b) enclose a large triangle), then we should suspect the specification of the model.

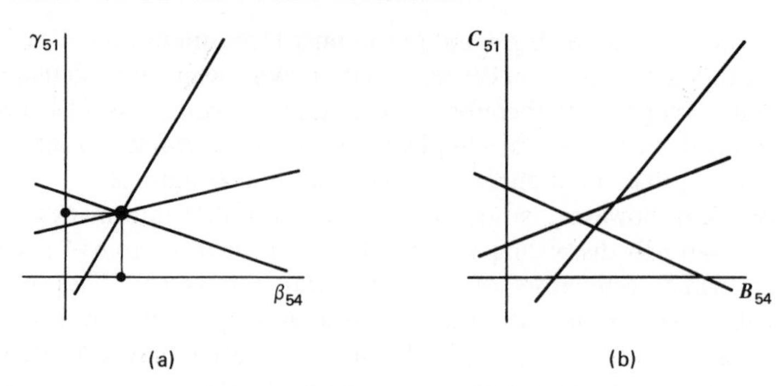

FIGURE 4.8. Overidentification in the population and the sample. (a) Population equations. (b) Sample estimating equations.

Returning to the peer-influences model, suppose that a path is added from X_3 to Y_5, altering the first structural equation:

$$Y_5 = \gamma_{51}X_1 + \gamma_{52}X_2 + \gamma_{53}X_3 + \beta_{56}Y_6 + \varepsilon_7 \tag{4.16}$$

There are now four structural parameters to estimate. Because there are four instrumental variables available (the exogenous variables X_1, X_2, X_3, and X_4), we may derive as many estimating equations as unknowns. We shall generally be able to solve uniquely for C_{51}, C_{52}, C_{53}, and B_{56}, and, therefore, the structural equation (4.16) is just identified. Note that we need not actually derive the IV estimating equations in order to draw this conclusion: We may simply compare the number of IVs to the number of parameters to be estimated, as in a balance sheet:

IVs ("credits")	Parameters ("debits")
X_1	γ_{51}
X_2	γ_{52}
X_3	γ_{53}
X_4	β_{56}
4	4

Now imagine that a path is added from X_4 to Y_5 (though this hardly makes substantive sense for the peer-influences data):

$$Y_5 = \gamma_{51}X_1 + \gamma_{52}X_2 + \gamma_{53}X_3 + \gamma_{54}X_4 + \beta_{56}Y_6 + \varepsilon_7 \tag{4.17}$$

Comparing the number of instrumental variables to the number of structural parameters to be estimated indicates that there is a deficit of IVs:

IVs	Parameters
X_1	γ_{51}
X_2	γ_{52}
X_3	γ_{53}
X_4	γ_{54}
	β_{56}
4	5

There are five parameters to estimate, yet the pool of available instrumental variables yields but four estimating equations. The structural equation (4.17), therefore, is underidentified.

The *order condition* for identification is easily abstracted from these examples: For a structural equation to be identified, there must be at least as many exogenous variables (IVs) in the model as there are parameters to estimate in the structural equation.

The Admissible-Transformation Approach[8] To understand why the order condition is insufficient to insure the identification of a structural equation, it is necessary to consider the equation not in isolation, but in relation to the other structural equations of the model. One way to accomplish this goal is to develop the relationship between the structural parameters and the parameters of the reduced form of the model.

In deriving the reduced form, we determined the following relation between reduced-form and structural parameters: $\mathbf{\Pi} = -\mathbf{B}^{-1}\mathbf{\Gamma}$ (see equation (4.4)). Knowing the structural parameters, then, we can find the reduced-form parameters. It is simple to show that, in general, it is not possible to reverse this relation: that is, to determine the structural parameters uniquely from the reduced form. Since the reduced form represents the directly observable empirical relationships of the endogenous to the exogenous variables, if two structures can give rise to the same reduced form, then it will be impossible to choose between these alternative structures on empirical grounds alone.

Let us multiply both sides of the structural equations (4.2) by a nonsingular $(q \times q)$ transformation matrix $\mathbf{T}$, producing a new set of equations:

$$\begin{aligned} \mathbf{TBy} + \mathbf{T\Gamma x} &= \mathbf{T\varepsilon} \\ \mathbf{B^*y} + \mathbf{\Gamma^*x} &= \mathbf{\varepsilon^*} \end{aligned} \tag{4.18}$$

where $\mathbf{B}^* \equiv \mathbf{TB}$, $\mathbf{\Gamma}^* \equiv \mathbf{T\Gamma}$, and $\boldsymbol{\varepsilon}^* \equiv \mathbf{T}\boldsymbol{\varepsilon}$. Equation (4.18) only *resembles* a

[8] The general approach in this section is from Fisher (1966).

structural-equation model, because in general each row of $\mathbf{B}^*$ and $\mathbf{\Gamma}^*$ combines parameters from *different* structural equations (different rows of $\mathbf{B}$ and $\mathbf{\Gamma}$). Yet, the reduced form corresponding to the "pseudo structure" (4.18) is the same as that corresponding to the "true structure" (4.2): Solving equation (4.18) for $\mathbf{y}$ produces

$$\begin{aligned}\mathbf{y} &= -(\mathbf{TB})^{-1}\mathbf{T\Gamma x} + (\mathbf{TB})^{-1}\mathbf{T}\boldsymbol{\varepsilon}\\ &= -\mathbf{B}^{-1}\mathbf{T}^{-1}\mathbf{T\Gamma x} + \mathbf{B}^{-1}\mathbf{T}^{-1}\mathbf{T}\boldsymbol{\varepsilon}\\ &= -\mathbf{B}^{-1}\mathbf{\Gamma x} + \mathbf{B}^{-1}\boldsymbol{\varepsilon}\\ &= \mathbf{\Pi x} + \boldsymbol{\delta}\end{aligned}$$

The true and transformed structures, therefore, are *observationally indistingushable* (i.e., they imply the same pattern of empirical relationships among the variables of the model).

We may, however, be able to distinguish true from transformed structures on the basis of the prior restrictions placed on the structural equations of the model. Suppose, for example, that some entries of $\mathbf{\Gamma}$ are prespecified to be zero. Then we may rule out any parameter matrix $\mathbf{\Gamma}^* = \mathbf{T\Gamma}$ that contains nonzero entries where zeroes should appear. Following Fisher (1966), a transformation $\mathbf{T}$ that produces a structure satisfying *all* prior restrictions placed on the model is termed an *admissible transformation*.

If we are able to show that the only admissible transformation is the identity transformation $\mathbf{T} = \mathbf{I}_q$, then the structural-equation model is identified. In fact, we need not be quite so stringent, for problems of underidentification only occur when we mix coefficients from different structural equations. It is therefore sufficient to require that the only admissible transformations are diagonal, multiplying each structural equation by a nonzero constant. Because the dependent variable in an equation has a coefficient of one, we can always recover the original structural equation by "renormalizing," dividing through by the same constant that we multiplied by (see footnote 4).

Before deriving a simple rule for determining whether the only admissible transformations of a structural-equation model are diagonal, let us consider two examples in some detail. First, we take another look at the Duncan, Haller, and Portes peer-influences model. There are no prior restrictions on $\mathbf{B}$ in this model (other than the diagonal entries of one, which we have already taken account of), so any nonsingular transformation $\mathbf{B}^* = \mathbf{TB}$ is admissible from the point of view of $\mathbf{B}$. $\mathbf{\Gamma}$, however, has four zero entries, and, therefore, if $\mathbf{T}$ is an admissible transformation, then

$$\begin{aligned}\mathbf{\Gamma}^* = \mathbf{T\Gamma} &= \begin{pmatrix} t_{11} & t_{12} \\ t_{21} & t_{22} \end{pmatrix} \begin{pmatrix} -\gamma_{51} & -\gamma_{52} & 0 & 0 \\ 0 & 0 & -\gamma_{63} & -\gamma_{64} \end{pmatrix}\\ &= \begin{pmatrix} -\gamma_{51}^* & -\gamma_{52}^* & 0 & 0 \\ 0 & 0 & -\gamma_{63}^* & -\gamma_{64}^* \end{pmatrix}\end{aligned} \tag{4.19}$$

Equation (4.19) requires that $t_{12} = t_{21} = 0$, and, thus, only diagonal transformations **T** are admissible.

For a contrasting example, examine the model diagrammed in Figure 4.9. For this model,

$$\mathbf{B} = \begin{pmatrix} 1 & -\beta_{34} & 0 \\ -\beta_{43} & 1 & 0 \\ 0 & -\beta_{54} & 1 \end{pmatrix}$$

$$\boldsymbol{\Gamma} = \begin{pmatrix} -\gamma_{31} & 0 \\ 0 & 0 \\ 0 & -\gamma_{52} \end{pmatrix}$$

It is simple to show that there are nondiagonal admissible transformations that confuse the first structural equation with the second, despite the fact that all three structural equations of the model meet the order condition. That is,

$$\mathbf{T} = \begin{pmatrix} t_{11} & t_{12} & 0 \\ 0 & t_{22} & 0 \\ 0 & 0 & t_{33} \end{pmatrix}$$

is admissible because

$$\mathbf{B}^* = \mathbf{TB} = \begin{pmatrix} t_{11} - t_{12}\beta_{43} & -t_{11}\beta_{34} + t_{12} & 0 \\ -t_{22}\beta_{43} & t_{22} & 0 \\ 0 & -t_{33}\beta_{54} & t_{33} \end{pmatrix}$$

$$\boldsymbol{\Gamma}^* = \mathbf{T\Gamma} = \begin{pmatrix} -t_{11}\gamma_{31} & 0 \\ 0 & 0 \\ 0 & -t_{33}\gamma_{52} \end{pmatrix}$$

meet the prior restrictions placed on **B** and **Γ**. Upon renormalizing the first

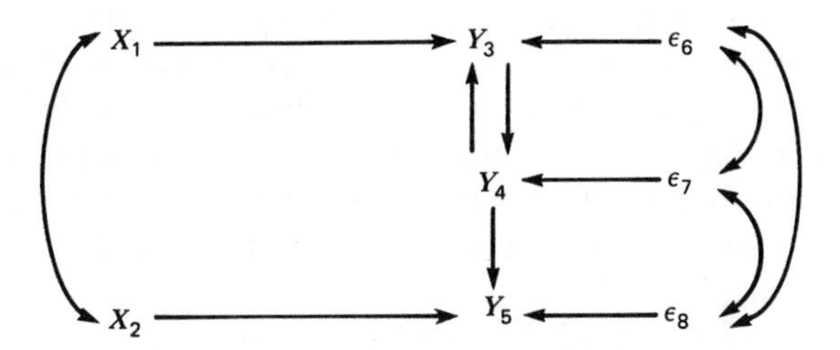

FIGURE 4.9. An underidentified nonrecursive model that meets the order condition.

equation, we obtain

$$1Y_3 - \frac{t_{11}\beta_{34} - t_{12}}{t_{11} - t_{12}\beta_{43}} Y_4 + 0Y_5 - \frac{t_{11}\gamma_{31}}{t_{11} - t_{12}\beta_{43}} X_1 + 0X_2 = \frac{t_{11}\varepsilon_6 + t_{12}\varepsilon_7}{t_{11} - t_{12}\beta_{43}}$$

$$1Y_3 - \beta_{34}^* Y_4 + 0Y_5 - \gamma_{31}^* X_1 + 0X_2 = \varepsilon_6^*$$

Thus, if we obtain estimates for the first structural equation, we cannot be sure whether we have in fact estimated the structural parameters of interest or some confused combination of parameters from the first two structural equations. These difficulties occur because the second structural equation has zeroes in the same places as the first (the additional zero in equation two is irrelevant when we consider the identification of the first equation); if we take a linear combination of the first two equations, therefore, the zeroes still appear as specified in the prior restrictions on the first equation.

These observations suggest a relatively simple procedure for determining whether the only admissible transformations are diagonal. We examine each structural equation in turn, insuring that the corresponding row of **T** has zeroes except in the diagonal position. Without loss of generality, let us consider the first structural equation of a model. Let $\mathbf{t}_1'$ denote the first row of **T**. The first structural equation is identified if and only if all entries except t_{11} in every admissible $\mathbf{t}_1'$ are zero. Other entries in $\mathbf{t}_1'$ may be nonzero only if linear combinations of the other equations meet the restrictions placed on the first structural equation.

We collect all structural coefficients of the model in a single matrix $\underset{(q \times q+m)}{\mathbf{A}} = [\mathbf{B}, \mathbf{\Gamma}]$. From this matrix, extract those columns that have zeroes in the first row, and then delete the first row, calling the resulting matrix $\mathbf{A}_1$. For example, for the Duncan, Haller, and Portes model, we have

$$\mathbf{A} = \left(\begin{array}{cc|cccc} 1 & -\beta_{56} & -\gamma_{51} & -\gamma_{52} & 0 & 0 \\ -\beta_{65} & 1 & 0 & 0 & -\gamma_{63} & -\gamma_{64} \end{array} \right)$$

$$\mathbf{A}_1 = (-\gamma_{63} \quad -\gamma_{64})$$

More generally, $\mathbf{A}_1$ is of order $(q - 1 \times r_1)$, where q is, as before, the number of equations in the model, and r_1 is the number of restrictions on (i.e., the number of variables excluded from) the first structural equation. If $\mathbf{A}_1$ is of full-row rank (that is, if $\text{rank}(\mathbf{A}_1) = q - 1$), then, by the definition of matrix rank, there is no linear combination of rows equalling the zero vector: The restrictions on the first structural equation cannot be duplicated from the other equations of the model. In the example, this is obviously the case since unless $\gamma_{63} = \gamma_{64} = 0$, $\text{rank}(\mathbf{A}_1) = 1 = q - 1$.

Note that for the rank of $\mathbf{A}_1$ to be $q - 1$, $\mathbf{A}_1$ must have at least $q - 1$ columns. This is the order condition for identification in a new guise (and, indeed, the name "order condition" refers to the column order of the $\mathbf{A}_1$ matrix): There must be at least as many restrictions on a structural equation as one less than the number of endogenous variables in the model. That this condition is equivalent to the earlier statement of the order condition is easily verified by comparing the number of instrumental variables (i.e., exogenous variables) to the *potential* number of unknown parameters in a structural equation:

Number of IVs	Potential Number of Parameters
m	$m + q - 1$

There are $m + q - 1$ potential parameters because the coefficient of the dependent variable is fixed at one. Thus at least $q - 1$ potential independent variables must be excluded from a structural equation to reduce the number of parameters to or below the number of IVs:

$$r_1 \geq (m + q - 1) - m = q - 1$$

The *rank condition* must be met, of course, not just by the first structural equation, but by each structural equation of the model. That is,

$$\text{rank}(\mathbf{A}_j) = q - 1$$

for $j = 1,\ldots,q$. Only then are all admissible transformations diagonal. In practice, for models of the form considered here, the rank condition will be met if the order condition is met and if no structural equation duplicates the restrictions placed on any other.

4.3.2. The Identification of Recursive and Block-Recursive Models

As we shall show presently, the pool of instrumental variables for estimating an equation in a recursive or block-recursive model includes not only the exogenous variables but also prior endogenous variables. Put alternatively, the restrictions on disturbance covariances in recursive and block-recursive models may help to identify the models. We shall see, in fact, that all recursive models are identified.

Recall the recursive Blau and Duncan stratification model shown in Figure 4.2. The first structural equation of this model is

$$Y_3 = \gamma_{31} X_1 + \gamma_{32} X_2 + \varepsilon_6 \tag{4.20}$$

This equation has two parameters to be estimated, and two instrumental variables are available for estimation, the exogenous variables X_1 and X_2. The structural equation (4.20), therefore, is just identified. More generally, the first structural equation in a recursive model can contain only exogenous independent variables, for there are no endogenous variables causally prior to the first. Because the exogenous variables are also available as IVs, if all exogenous variables are included in the first equation,then the equation is just identified; if some exogenous variables are excluded, then the first structural equation is overidentified.

The second structural equation of the Blau and Duncan model is

$$Y_4 = \gamma_{42} X_2 + \beta_{43} Y_3 + \varepsilon_7 \tag{4.21}$$

Here, as in the first structural equation, there are two parameters to be estimated. The exogenous variables, as always, are available as IVs. In addition, the prior endogenous variable Y_3 may be used as an IV, since Y_3 is a linear combination of X_1, X_2, and ε_6 (as given in the first structural equation (4.20)), each of which is uncorrelated with the disturbance of the second structural equation, ε_7: Because they are exogenous, X_1 and X_2 are uncorrelated with ε_7; ε_6 and ε_7 are uncorrelated because, in recursive models, different disturbance variables are specified to be independent. Y_3, therefore, is uncorrelated with ε_7. There are, then, three IVs for the structural equation (4.21), rendering this equation overidentified. Note that, more generally, the first endogenous variable in a recursive model is a linear combination of exogenous variables and a disturbance, and thus may be used as an IV in estimating the second equation of the model. This second equation, consequently, is overidentified if any prior variables are excluded and just identified otherwise.

Returning to our example, there are three parameters to be estimated in the third structural equation of the Blau and Duncan model:

$$Y_5 = \gamma_{52} X_2 + \beta_{53} Y_3 + \beta_{54} Y_4 + \varepsilon_8 \tag{4.22}$$

X_1 and X_2 may be employed as instrumental variables because they are exogenous. Y_3 is an eligible IV because it is composed of X_1, X_2, and ε_6, each of which is uncorrelated with ε_8. Y_4, similarly, has components (X_2, Y_3, and ε_7) that are uncorrelated with ε_8, and is a fourth IV. The structural equation (4.22) is consequently overidentified.

To generalize: Exogenous and all prior endogenous variables are eligible IVs for estimating a structural equation in a recursive model; if all of these variables are independent variables as well, the equation is just identified; if one or more prior variables are excluded, the equation is overidentified.[9]

Strictly speaking, we should set out to show that the only admissible transformations of a recursive model are diagonal. This may, in fact, be

[9]In certain cases, the causal ordering of endogenous variables in a recursive model is partial rather than complete. When this happens, each of two variables may count as "prior" relative to each other (i.e., uncorrelated with the disturbance in the other's equation). See Problem 4.8(a) for an example.

demonstrated by taking into account the restrictions placed on disturbance covariances as well as those placed on structural parameters (see Problem 4.9). Notice that in the context of recursive models, the rank condition derived in Section 4.3.1 is sufficient but not necessary to insure the identification of a model, for this condition fails to take account of restrictions on disturbance covariances.

An illustrative block-recursive model was given in Figure 4.4. There are four instrumental variables available for estimating the structural equations in the first block of this model: X_1, X_2, X_3, and X_4. Because each equation in the first block of the model has four parameters to be estimated, each equation is just identified. To estimate the structural equations in the second block, the pool of instrumental variables expands to include the endogenous variables Y_5 and Y_6 in the first block. This expansion occurs because Y_5 and Y_6 may be written (in reduced form) as linear functions of the exogenous variables and first-block errors, all of which are uncorrelated with the errors of the second block. Each structural equation in the second block has five parameters to be estimated, and therefore is overidentified. In the absence of the block-recursive restrictions on disturbance covariances, the second-block equations in this model would be *under*identified.

In general, to identify a structural equation in a block-recursive model, we may employ endogenous variables in prior blocks (along with the exogenous variables) as IVs. A necessary and sufficient condition for identification may be obtained by suitably modifying the rank condition:

$$\text{rank}(\mathbf{A}_j^*) = q_j - 1$$

where $\mathbf{A}_j^*$ is formed from $\mathbf{A}_j$ by deleting equations (rows) for other blocks and by deleting endogenous variables (columns) in subsequent blocks. q_j is the number of endogenous variables in the block containing Y_j.

4.3.3. Restrictions on Disturbance Covariances: The General Case

The preceding section demonstrated how prior restrictions on disturbance covariances help to identify recursive and block-recursive structural-equation models. Restrictions of this type can also assist in identifying nonrecursive models, although there is not, unfortunately, a general rule (such as the order or rank condition) to apply in these cases. Instead, each model must be examined individually, employing, for example, the admissible-transformation approach.

We return to the general structural-equation model transformed by a nonsingular matrix $\mathbf{T}$ (repeating equation (4.18)):

$$\mathbf{TBy} + \mathbf{T\Gamma x} = \mathbf{T\varepsilon}$$

$$\mathbf{B^*y} + \mathbf{\Gamma^*x} = \boldsymbol{\varepsilon}^*$$

As we pointed out earlier, for $\mathbf{T}$ to be an admissible transformation, $\mathbf{B}^*$ and $\mathbf{\Gamma}^*$

must meet the prior restrictions placed on $\mathbf{B}$ and $\boldsymbol{\Gamma}$. Because $\boldsymbol{\varepsilon}^*$ results from a linear transformation of $\boldsymbol{\varepsilon}$, the covariance matrix of the transformed disturbances may also be derived employing $\mathbf{T}$:

$$\boldsymbol{\Sigma}^*_{\varepsilon\varepsilon} = \mathbf{T}\boldsymbol{\Sigma}_{\varepsilon\varepsilon}\mathbf{T}'$$

For $\mathbf{T}$ to be admissible, $\boldsymbol{\Sigma}^*_{\varepsilon\varepsilon}$ must satisfy the prior restrictions placed on $\boldsymbol{\Sigma}_{\varepsilon\varepsilon}$ (i.e., must have zeroes in the right places). We were able to ignore $\boldsymbol{\Sigma}^*_{\varepsilon\varepsilon}$ previously because we stipulated that there were *no* restrictions placed on the disturbance covariance matrix.

To provide an illustration, we shall examine the model shown in Figure 4.10, the structural equations of which are

$$\begin{pmatrix} 1 & -\beta_{23} \\ -\beta_{32} & 1 \end{pmatrix}\begin{pmatrix} Y_2 \\ Y_3 \end{pmatrix} + \begin{pmatrix} 0 \\ -\gamma_{31} \end{pmatrix} X_1 = \begin{pmatrix} \varepsilon_4 \\ \varepsilon_5 \end{pmatrix}$$

(This model is adapted from Johnston, 1972: 365.) The two disturbance variables are specified to be uncorrelated:

$$\boldsymbol{\Sigma}_{\varepsilon\varepsilon} = \begin{pmatrix} \sigma_{44} & 0 \\ 0 & \sigma_{55} \end{pmatrix}$$

Note that the second structural equation of the model would be underidentified if not for the restriction $\sigma_{45} = 0$.

To show that the model is in fact identified, it is necessary to demonstrate that only diagonal transformations are admissible. Let

$$\mathbf{T} = \begin{pmatrix} t_{11} & t_{12} \\ t_{21} & t_{22} \end{pmatrix}$$

There are no restrictions on $\mathbf{B}$, so any $\mathbf{B}^* = \mathbf{TB}$ will do. Because $\gamma_{21} = 0$, however, $t_{12} = 0$, for otherwise $\gamma^*_{21} = t_{11}0 - t_{12}\gamma_{31} \neq 0$, making $\mathbf{T}$ inadmissible. Because $\sigma_{45} = 0$, the off-diagonal entries of $\boldsymbol{\Sigma}^*_{\varepsilon\varepsilon}$ must be zero. Forming this product, we have

$$\boldsymbol{\Sigma}^*_{\varepsilon\varepsilon} = \mathbf{T}\boldsymbol{\Sigma}_{\varepsilon\varepsilon}\mathbf{T}' = \begin{pmatrix} t_{11} & 0 \\ t_{21} & t_{22} \end{pmatrix}\begin{pmatrix} \sigma_{44} & 0 \\ 0 & \sigma_{55} \end{pmatrix}\begin{pmatrix} t_{11} & t_{21} \\ 0 & t_{22} \end{pmatrix}$$

$$= \begin{pmatrix} t_{11}^2\sigma_{44} & t_{11}t_{21}\sigma_{44} \\ t_{11}t_{21}\sigma_{44} & t_{21}^2\sigma_{44} + t_{22}^2\sigma_{55} \end{pmatrix}$$

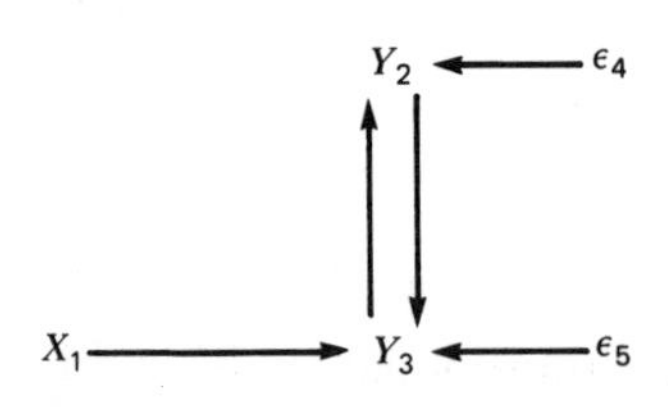

FIGURE 4.10. A nonrecursive model with restricted disturbance covariance.

It is clear that the off-diagonal entries are zero only if $t_{21} = 0$ (since $t_{11} \neq 0$ because $\mathbf{T}$ is nonsingular). Thus, all admissible transformations are of the form

$$\mathbf{T} = \begin{pmatrix} t_{11} & 0 \\ 0 & t_{22} \end{pmatrix}$$

and the model is identified.

PROBLEMS

4.8. Determine the identification status of each of the models given in Problem 4.1.

4.9.* Using the admissible-transformation approach, demonstrate that a recursive model (i.e., a model for which $\mathbf{B}$ is triangular and $\mathbf{\Sigma}_{\varepsilon\varepsilon}$ is diagonal) is necessarily identified. (*Hints*: Employ the general approach used in Section 4.3.3. Work with a relatively simple system, such as Blau and Duncan's model modified so that all prior variables appear in each structural equation; the generalization to any recursive model is obvious.)

4.4. ESTIMATION OF STRUCTURAL-EQUATION MODELS

Having specified a structural-equation model, and having determined that it is identified, we wish to estimate the parameters of the model—that is, to fit the model to the data. If each of the structural equations of the model is just identified, estimation is not problematic, since the instrumental-variables estimating equations may then be solved for unique estimates of the structural parameters.

An overidentified model, however, presents difficulties: As we noted in Section 4.3.1, the sample estimating equations for an overidentified structural equation are overdetermined, even if the model is correctly specified. Recall the situation illustrated in Figure 4.8(b), where there are two parameters to be estimated and three estimating equation. To obtain consistent estimators of the structural parameters we could delete one of the estimating equations.

Though discarding surplus estimating equations is not an unreasonable response to overidentification, there are at least two factors that recommend against it. First, in the absence of a justifiable rule for determining which estimating equation is to be deleted, we must proceed arbitrarily, causing different investigators to obtain different estimates from the same data. Second, the surplus of information available when a model is overidentified might be used to improve the efficiency of estimation; discarding this information is statistically wasteful. In this section, we shall deal first with the estimation of nonrecursive models, and then turn to a consideration of recursive and block-recursive models.

4.4.1. Estimating Overidentified Nonrecursive Models

There are two general approaches to estimating an overidentified structural-equation model. One approach, called *single-equation* or *limited-information estimation*, is to estimate each structural equation separately. Ordinary least squares is an example of a single-equation method. OLS, however, is generally inconsistent when applied to nonrecursive models, because endogenous independent variables are correlated with the disturbance of the structural equation in which they appear. There are other single-equation methods that produce consistent estimators. One such method, *two-stage least squares* (*2SLS*), is developed in this section.

A second general approach, termed *systems* or *full-information estimation*, estimates all of the parameters of the model (including the covariance matrix of the disturbances) at once. Later in this section, we shall take up the *full-information maximum-likelihood* (*FIML*) method.

Although a lengthy comparative discussion of different estimation methods is beyond the scope of this chapter, a few remarks are in order. Further summary information may be found in various sources, including Christ (1966: 464–481), Malinvaud (1970: 718–722), Kmenta (1971: 581–586), Johnston (1972: 408–420), and Wonnacott and Wonnacott (1979: 518–521).

The asymptotic (i.e., large-sample) properties of the various estimation methods have been determined analytically, and the full-information methods are asymptotically more efficient than the limited-information techniques. Determining the small-sample properties of structural-equation estimators analytically generally proves infeasible, and therefore these properties have primarily been explored empirically through "Monte-Carlo" (random-sampling) simulation experiments. The disadvantage of this approach, aside from its relative inelegance, is that conclusions may depend in an undetermined way on the specific conditions of a Monte-Carlo study (i.e., on the models, true parameter values, variable distributions, and sample sizes employed in the experiment). The random element in Monte-Carlo studies also introduces statistical uncertainty. Over the course of a number of studies, however, patterns tend to emerge.

To summarize the results of such studies briefly is possibly misleading. It is, nevertheless, fair to say that overall 2SLS appears to be the best of the limited-information methods, and FIML the best of the full-information methods. Moreover, FIML generally is superior to 2SLS, except when the estimated system has a misspecified equation, in which case single-equation methods like 2SLS tend to perform relatively well. Intuitively, full-information estimation proliferates a specification error throughout an equation system, while the limited-information approach isolates the error in a single equation.

A final point may be made with respect to OLS estimation. Though OLS is generally inconsistent in nonrecursive models, the OLS estimator nevertheless has smaller sampling variance than the consistent estimators. Even apart from the possible small-sample bias of the consistent estimators, their larger sampling variance may depress their small-sample efficiency below that of OLS. OLS estimation, therefore, cannot simply be dismissed in very small samples on the grounds of inconsistency.[10]

Two-Stage Least Squares (2SLS) Aside from its desirable properties as an estimation method, 2SLS is worth studying because its rationale is relatively simple, it is computationally straightforward and inexpensive, and it is the method most frequently employed in practice. The 2SLS method was originally formulated in the 1950s by Theil (cited in Theil, 1971: 452) and Basmann (1957). We shall approach 2SLS by developing an example before proceeding to the general case.

We return to the first structural equation of the Duncan, Haller, and Portes peer-influences model (originally given in equations (4.1)):

$$Y_5 = \gamma_{51} X_1 + \gamma_{52} X_2 + \beta_{56} Y_6 + \varepsilon_7 \tag{4.23}$$

Recall that this structural equation is overidentified because of the exclusion of the exogenous variables X_3 and X_4. We therefore have four instrumental variables but only three structural parameters to estimate. 2SLS may be thought of as a method for reducing the number of instrumental variables to the number of parameters.

Aside from being uncorrelated with the error, a good instrumental variable should be as highly correlated as possible with the independent variables in the equation to be estimated. We may apply this criterion individually to the independent variables in equation (4.23). X_1 and X_2, being perfectly correlated with themselves, are therefore their own best instruments. We might choose as an instrument for Y_6 the remaining exogenous variable (X_3 or X_4) that has the higher correlation with Y_6. We can do better, however, by regressing Y_6 on *both* X_3 and X_4, using the fitted values $\hat{Y}_6$ that result as an optimal instrument for Y_6.

An equivalent, and ultimately more convenient, result is obtained by regressing Y_6 on *all* of the exogenous variables (X_1, X_2, X_3, and X_4) in the model.[11] That is, in the first stage of 2SLS, we fit the reduced-form relation

$$Y_6 = \pi_{61} X_1 + \pi_{62} X_2 + \pi_{63} X_3 + \pi_{64} X_4 + \delta_6$$

[10]"Small samples" employed in econometric Monte-Carlo studies are small indeed by sociological standards—often in the neighborhood of 20. This is because economists frequently work with relatively short time series rather than with cross-sectional sample surveys. A social scientist employing a sample of several hundred or more observations almost certainly can rely on asymptotic results, if the assumptions underlying the results (e.g., the independence of exogenous variables and disturbances) are realistic.

[11]Recall from Problem 4.7 that what is significant about a set of IVs is the subspace that they span, not the basis selected for this subspace. In the present context, since X_1 and X_2 are included among the set of IVs employed, it does not matter whether $\hat{Y}_6$ is defined in terms of all four X's or only in terms of X_3 and X_4; the same subspace is spanned by X_1, X_2, and $\hat{Y}_6$ in both cases.

obtaining (from OLS estimates of the π's)

$$\hat{Y}_6 = P_{61}X_1 + P_{62}X_2 + P_{63}X_3 + P_{64}X_4$$

$\hat{Y}_6$, as a linear combination of the exogenous variables, is uncorrelated with the structural error ε_7 of equation (4.23), and, therefore, may legitimately be used as an instrumental variable in estimating this equation.[12] Moreover, $\hat{Y}_6$ is as highly correlated as possible with Y_6 while still remaining uncorrelated with ε_7.

In the second stage of 2SLS, we apply the IVs obtained in the first stage (the exogenous independent variables X_1 and X_2, and the fitted endogenous independent variable $\hat{Y}_6$) to estimate the structural equation (4.23). Using an obvious notation, we derive IV estimating equations

$$\begin{aligned} S_{15} &= C_{51}S_{11} + C_{52}S_{12} + B_{56}S_{16} \\ S_{25} &= C_{51}S_{12} + C_{52}S_{22} + B_{56}S_{26} \\ S_{5\hat{6}} &= C_{51}S_{1\hat{6}} + C_{52}S_{2\hat{6}} + B_{56}S_{6\hat{6}} \end{aligned} \tag{4.24}$$

which, given data, may be solved for 2SLS estimates of the structural parameters.

The name "two-stage least squares" derives from an alternative but equivalent approach employing an OLS regression in the second stage. The structural equation (4.23) could be fit directly by OLS but for the correlation of Y_6 with ε_7. If we substitute for Y_6 from the first-stage reduced-form regression, we get

$$\begin{aligned} Y_5 &= \gamma_{51}X_1 + \gamma_{52}X_2 + \beta_{56}(\hat{Y}_6 + D_6) + \varepsilon_7 \\ &= \gamma_{51}X_1 + \gamma_{52}X_2 + \beta_{56}\hat{Y}_6 + \varepsilon_7^* \end{aligned} \tag{4.25}$$

Here, $\varepsilon_7^* \equiv \beta_{56}D_6 + \varepsilon_7$ is a linear combination of errors, and hence is uncorrelated with X_1 and X_2, which are exogenous, and with $\hat{Y}_6$, which is a linear combination of exogenous variables (see footnote 12). OLS estimation, therefore, may justifiably be applied to equation (4.25), producing estimating equations

$$\begin{aligned} S_{15} &= C_{51}S_{11} + C_{52}S_{12} + B_{56}S_{1\hat{6}} \\ S_{25} &= C_{51}S_{12} + C_{52}S_{22} + B_{56}S_{2\hat{6}} \\ S_{5\hat{6}} &= C_{51}S_{1\hat{6}} + C_{52}S_{2\hat{6}} + B_{56}S_{\hat{6}\hat{6}} \end{aligned} \tag{4.26}$$

Comparing the OLS estimating equations (4.26) with the IV estimating equa-

[12] $\hat{Y}_6$ depends upon the P's, which, in turn, depend upon the structural error ε_7 (a component of δ_6). $\hat{Y}_6$ and ε_7, therefore, are not, strictly speaking, independent. Since the P's are consistent estimators of the π's, however, $\hat{Y}_6$ and ε_7 are independent in the limit.

tions (4.24) for the second stage, we need to show that $S_{16} = S_{1\hat{6}}$, $S_{26} = S_{2\hat{6}}$, and $S_{6\hat{6}} = S_{\hat{6}\hat{6}}$, in order to demonstrate the equivalence of the two approaches. This equivalence will be proven shortly, but for the general case, to which we now turn.

Let us consider the jth structural equation in a model, writing this equation in the following format:

$$\underset{(n\times1)}{\mathbf{y}_j} = \underset{(n\times q_j-1)}{\mathbf{Y}_j}\;\underset{(q_j-1\times1)}{\boldsymbol{\beta}_j} + \underset{(n\times m_j)}{\mathbf{X}_j}\;\underset{(m_j\times1)}{\boldsymbol{\gamma}_j} + \underset{(n\times1)}{\boldsymbol{\varepsilon}_j} \tag{4.27}$$

The symbols used in this equation, some of which are familiar, are explained in Table 4.1, which employs the first structural equation of the peer-influences model as an illustration. Equation (4.27) may be written more compactly as

$$\mathbf{y}_j = [\mathbf{Y}_j, \mathbf{X}_j]\begin{pmatrix}\boldsymbol{\beta}_j\\ \boldsymbol{\gamma}_j\end{pmatrix} + \boldsymbol{\varepsilon}_j \tag{4.28}$$

In the first stage of 2SLS, we regress the endogenous independent variables in $\mathbf{Y}_j$ on *all* of the exogenous variables in the model:

$$\underset{(n\times q_j-1)}{\mathbf{Y}_j} = \underset{(n\times m)}{\mathbf{X}}\;\underset{(m\times q_j-1)}{\boldsymbol{\Pi}_j'} + \underset{(n\times q_j-1)}{\boldsymbol{\Delta}_j}$$

TABLE 4.1. Notation for a Structural Equation to be Estimated by 2SLS

Symbol	Meaning	Example[a]
j	jth structural equation	1
n	Number of observations	329
q_j	Number of included endogenous variables	2
m_j	Number of exogenous independent variables	2
$\underset{(n\times1)}{\mathbf{y}_j}$	Dependent variable vector	$\mathbf{y}_5$
$\underset{(n\times q_j-1)}{\mathbf{Y}_j}$	Endogenous independent variable matrix	$\mathbf{y}_6$
$\underset{(n\times m_j)}{\mathbf{X}_j}$	Exogenous independent variable matrix	$[\mathbf{x}_1, \mathbf{x}_2]$
$\underset{(n\times1)}{\boldsymbol{\varepsilon}_j}$	Disturbance vector	$\boldsymbol{\varepsilon}_7$
$\underset{(q_j-1\times1)}{\boldsymbol{\beta}_j}$	Structural parameters for endogenous independent variables	(β_{56})
$\underset{(m_j\times1)}{\boldsymbol{\gamma}_j}$	Structural parameters for exogenous independent variables	$(\gamma_{51}, \gamma_{52})'$

[a] First equation of Duncan, Haller, and Portes nonrecursive model.

obtaining the OLS reduced-formed estimator

$$\mathbf{P}_j' = (\mathbf{X}'\mathbf{X})^{-1}\mathbf{X}'\mathbf{Y}_j$$

and fitted values

$$\hat{\mathbf{Y}}_j = \mathbf{X}\mathbf{P}_j' = \mathbf{X}(\mathbf{X}'\mathbf{X})^{-1}\mathbf{X}'\mathbf{Y}_j \tag{4.29}$$

In the second stage, we apply $\hat{\mathbf{Y}}_j$ and $\mathbf{X}_j$ as instrumental variables to equation (4.28):

$$\begin{pmatrix} \mathbf{b}_j \\ \mathbf{c}_j \end{pmatrix} = \left([\hat{\mathbf{Y}}_j, \mathbf{X}_j]'[\mathbf{Y}_j, \mathbf{X}_j]\right)^{-1}[\hat{\mathbf{Y}}_j, \mathbf{X}_j]'\mathbf{y}_j$$

$$= \begin{pmatrix} \hat{\mathbf{Y}}_j'\mathbf{Y}_j & \hat{\mathbf{Y}}_j'\mathbf{X}_j \\ \mathbf{X}_j'\mathbf{Y}_j & \mathbf{X}_j'\mathbf{X}_j \end{pmatrix}^{-1} \begin{pmatrix} \hat{\mathbf{Y}}_j'\mathbf{y}_j \\ \mathbf{X}_j'\mathbf{y}_j \end{pmatrix} \tag{4.30}$$

Alternatively, we may proceed by the regression approach, substituting $\mathbf{Y}_j = \hat{\mathbf{Y}}_j + \mathbf{D}_j$ into equation (4.27):

$$\mathbf{y}_j = (\hat{\mathbf{Y}}_j + \mathbf{D}_j)\boldsymbol{\beta}_j + \mathbf{X}_j\boldsymbol{\gamma}_j + \boldsymbol{\varepsilon}_j$$

$$= \hat{\mathbf{Y}}_j\boldsymbol{\beta}_j + \mathbf{X}_j\boldsymbol{\gamma}_j + (\mathbf{D}_j\boldsymbol{\beta}_j + \boldsymbol{\varepsilon}_j)$$

$$= [\hat{\mathbf{Y}}_j, \mathbf{X}_j]\begin{pmatrix} \boldsymbol{\beta}_j \\ \boldsymbol{\gamma}_j \end{pmatrix} + \boldsymbol{\varepsilon}_j^* \tag{4.31}$$

where $\boldsymbol{\varepsilon}_j^* \equiv \mathbf{D}_j\boldsymbol{\beta}_j + \boldsymbol{\varepsilon}_j$. We may apply OLS to equation (4.31) because, by the reasoning outlined earlier, $\hat{\mathbf{Y}}_j$ and $\mathbf{X}_j$ are both uncorrelated (in the limit) with $\boldsymbol{\varepsilon}_j^*$:

$$\begin{pmatrix} \mathbf{b}_j \\ \mathbf{c}_j \end{pmatrix} = \left([\hat{\mathbf{Y}}_j, \mathbf{X}_j]'[\hat{\mathbf{Y}}_j, \mathbf{X}_j]\right)^{-1}[\hat{\mathbf{Y}}_j, \mathbf{X}_j]'\mathbf{y}_j$$

$$= \begin{pmatrix} \hat{\mathbf{Y}}_j'\hat{\mathbf{Y}}_j & \hat{\mathbf{Y}}_j'\mathbf{X}_j \\ \mathbf{X}_j'\hat{\mathbf{Y}}_j & \mathbf{X}_j'\mathbf{X}_j \end{pmatrix}^{-1} \begin{pmatrix} \hat{\mathbf{Y}}_j'\mathbf{y}_j \\ \mathbf{X}_j'\mathbf{y}_j \end{pmatrix} \tag{4.32}$$

It is clear that the two approaches, equations (4.30) and (4.32), produce identical results if

$$\begin{aligned} \hat{\mathbf{Y}}_j'\mathbf{Y}_j &= \hat{\mathbf{Y}}_j'\hat{\mathbf{Y}}_j \\ \mathbf{X}_j'\mathbf{Y}_j &= \mathbf{X}_j'\hat{\mathbf{Y}}_j \end{aligned} \tag{4.33}$$

To see that these equations hold, substitute $\mathbf{Y}_j = \hat{\mathbf{Y}}_j + \mathbf{D}_j$ into the left side of each, obtaining

$$\hat{\mathbf{Y}}_j'(\hat{\mathbf{Y}}_j + \mathbf{D}_j) = \hat{\mathbf{Y}}_j'\hat{\mathbf{Y}}_j + \hat{\mathbf{Y}}_j'\mathbf{D}_j$$

$$\mathbf{X}_j'(\hat{\mathbf{Y}}_j + \mathbf{D}_j) = \mathbf{X}_j'\hat{\mathbf{Y}}_j + \mathbf{X}_j'\mathbf{D}_j$$

$\hat{\mathbf{Y}}_j$ and $\mathbf{D}_j$ are, respectively, the fitted dependent-variable matrix and the residual matrix from an OLS regression; they are, therefore, orthogonal, since each column of $\hat{\mathbf{Y}}_j$ lies in the $\mathbf{X}$ subspace, and each column of $\mathbf{D}_j$ is orthogonal to this subspace. For a similar reason, $\mathbf{X}_j$ (a column subset of $\mathbf{X}$) and $\mathbf{D}_j$ are orthogonal. Thus, $\hat{\mathbf{Y}}_j'\mathbf{D}_j$ and $\mathbf{X}_j'\mathbf{D}_j$ both vanish, and the equalities in (4.33) are demonstrated.

The estimated asymptotic covariance matrix of the 2SLS estimator follows from the observation that 2SLS is a type of instrumental-variables estimation. We may, consequently, apply the result given in Section 4.2, equations (4.11). In the present context, $\mathbf{y}_j$ plays the role of the dependent variable ($\mathbf{y}$, in the general case given in (4.11)), $[\mathbf{Y}_j, \mathbf{X}_j]$ is the independent-variable matrix ($\mathbf{X}$ in equations (4.11)), and $[\hat{\mathbf{Y}}_j, \mathbf{X}_j]$ is the matrix of instrumental variables ($\mathbf{Z}$ in equations (4.11)). Straightforward substitution produces the desired result:

$$\mathbf{e}_j = \mathbf{y}_j - \mathbf{X}_j\mathbf{c}_j - \mathbf{Y}_j\mathbf{b}_j$$

$$S_{E_j}^2 = \frac{\mathbf{e}_j'\mathbf{e}_j}{n - q_j - m_j} \tag{4.34}$$

$$\widehat{\mathscr{V}\begin{pmatrix}\mathbf{b}_j\\ \mathbf{c}_j\end{pmatrix}} = S_{E_j}^2\left([\hat{\mathbf{Y}}_j, \mathbf{X}_j]'[\mathbf{Y}_j, \mathbf{X}_j]\right)^{-1}\left([\hat{\mathbf{Y}}_j, \mathbf{X}_j]'[\hat{\mathbf{Y}}_j, \mathbf{X}_j]\right)\left([\mathbf{Y}_j, \mathbf{X}_j]'[\hat{\mathbf{Y}}_j, \mathbf{X}_j]\right)^{-1}$$

This expression may be simplified by multiplying out the partitioned matrices and taking advantage of the identities given in equations (4.33):

$$\widehat{\mathscr{V}\begin{pmatrix}\mathbf{b}_j\\ \mathbf{c}_j\end{pmatrix}} = S_{E_j}^2\begin{pmatrix}\hat{\mathbf{Y}}_j'\mathbf{Y}_j & \mathbf{Y}_j'\mathbf{X}_j\\ \mathbf{X}_j'\mathbf{Y}_j & \mathbf{X}_j'\mathbf{X}_j\end{pmatrix}^{-1}$$

The square roots of the diagonal entries of this matrix are the standard errors of the estimated structural coefficients, which may be used, therefore, to test hypotheses and construct confidence intervals for individual β's and γ's.

Although we have developed 2SLS as a two-step procedure, the first stage (4.29) may be substituted into the second (4.30), bypassing separate calculation of the first-stage regression:

$$\begin{pmatrix} \mathbf{b}_j \\ \mathbf{c}_j \end{pmatrix} = \begin{pmatrix} \mathbf{Y}_j'\mathbf{X}(\mathbf{X}'\mathbf{X})^{-1}\mathbf{X}'\mathbf{Y}_j & \mathbf{Y}_j'\mathbf{X}_j \\ \mathbf{X}_j'\mathbf{Y}_j & \mathbf{X}_j'\mathbf{X}_j \end{pmatrix}^{-1} \begin{pmatrix} \mathbf{Y}_j'\mathbf{X}(\mathbf{X}'\mathbf{X})^{-1}\mathbf{X}'\mathbf{y}_j \\ \mathbf{X}_j'\mathbf{y}_j \end{pmatrix} \tag{4.35}$$

In fact, because of the location of matrix inverses in equation (4.35), every sum-of-squares-and-products matrix in this equation may be replaced by the corresponding sample covariance matrix:

$$\begin{pmatrix} \mathbf{b}_j \\ \mathbf{c}_j \end{pmatrix} = \begin{pmatrix} \mathbf{S}_{Y_jX}\mathbf{S}_{XX}^{-1}\mathbf{S}_{XY_j} & \mathbf{S}_{Y_jX_j} \\ \mathbf{S}_{X_jY_j} & \mathbf{S}_{X_jX_j} \end{pmatrix}^{-1} \begin{pmatrix} \mathbf{S}_{Y_jX}\mathbf{S}_{XX}^{-1}\mathbf{s}_{Xy_j} \\ \mathbf{s}_{X_jy_j} \end{pmatrix} \tag{4.36}$$

where, for example, $\mathbf{S}_{Y_jX} = [1/(n-1)]\mathbf{Y}_j'\mathbf{X}$. For models with standardized variables, the covariances in equation (4.36) are, of course, correlations.

A correlation matrix for the peer-influences data is given in Table 4.2. (This correlation matrix includes variables not employed in the current example, but which will be used later.) 2SLS estimates of the standardized structural

TABLE 4.2. Correlation Matrix for Peer-Influences Data, $n = 329$

	1	2	3	4	5	6	7	8	9
2	.6247								
3	.3269	.3669							
4	.4216	.3275	.6404						
5	.2137	.2742	.1124	.0839					
6	.4105	.4043	.2903	.2598	.1839				
7	.3240	.4047	.3054	.2786	.0489	.2220			
8	.2930	.2407	.4105	.3607	.0186	.1861	.2707		
9	.2995	.2863	.5191	.5007	.0782	.3355	.2302	.2950	
10	.0760	.0702	.2784	.1988	.1147	.1021	.0931	−.0438	.2087

1 Respondent's occupational aspiration score
2 Respondent's educational aspiration score
3 Best friend's occupational aspiration score
4 Best friend's educational aspiration score
5 Respondent's parental aspiration
6 Respondent's intelligence
7 Respondent's family socio-economic status (SES)
8 Best friend's family SES
9 Best friend's intelligence
10 Best friend's parental aspiration

Source: Duncan, Haller, and Portes (1968: Table 1). Reprinted from the *American Journal of Sociology* by permission of the University of Chicago Press. Copyright 1968, the University of Chicago Press.

parameters of the model are shown with parenthetical standard errors in equations (4.37):

$$\begin{aligned}
Y_5 &= \underset{(0.104)}{0.403}\ Y_6 + \underset{(0.053)}{0.272}\ X_1 + \underset{(0.054)}{0.151}\ X_2 + 0.841E_7' \\
Y_6 &= \underset{(0.125)}{0.341}\ Y_5 + \underset{(0.054)}{0.157}\ X_3 + \underset{(0.055)}{0.352}\ X_4 + 0.805E_8'
\end{aligned} \qquad (4.37)$$

The coefficients associated with the estimated disturbances require comment. It is usual practice, in a standardized model, to set the standard deviation of the estimated disturbance variables to one. Here, for example, $E_7' \equiv E_7/S_7$. By way of compensation, S_7, often called a *residual path*, becomes the coefficient of E_7' in the estimated structural equation.

The estimates for the peer-influences model appear generally reasonable. All structural coefficients are positive, as expected, and corresponding coefficients in the two structural equations are similar in magnitude. It is, perhaps, surprising that the estimated structural coefficients for peer influences, B_{56} and B_{65}, are as large as they are relative to the coefficients for intelligence (C_{51} and C_{64}) and for SES (C_{52} and C_{63}).

Having obtained 2SLS estimates of the structural parameters of a model, we are generally interested in estimating as well the covariances among disturbances from different structural equations, if only to verify that these values are reasonable. Equation (4.34) gives us the variance of each estimated disturbance. Their covariances can be obtained similarly, say $S_{E_iE_j} = \mathbf{e}_i'\mathbf{e}_j/n$ (disregarding degrees of freedom). It is generally simpler, however, to compute the covariance matrix for the disturbances directly from structural coefficients and from covariances of endogenous and exogenous variables. According to the general structural-equation model (4.2), $\boldsymbol{\varepsilon} = \mathbf{By} + \boldsymbol{\Gamma}\mathbf{x}$. Thus, since $E(\boldsymbol{\varepsilon}) = \mathbf{0}$,

$$\begin{aligned}
\boldsymbol{\Sigma}_{\varepsilon\varepsilon} &= E(\boldsymbol{\varepsilon}\boldsymbol{\varepsilon}') = E[(\mathbf{By} + \boldsymbol{\Gamma}\mathbf{x})(\mathbf{By} + \boldsymbol{\Gamma}\mathbf{x})'] \\
&= \mathbf{B}E(\mathbf{yy}')\mathbf{B}' + \mathbf{B}E(\mathbf{yx}')\boldsymbol{\Gamma}' + \boldsymbol{\Gamma}E(\mathbf{xy}')\mathbf{B}' + \boldsymbol{\Gamma}E(\mathbf{xx}')\boldsymbol{\Gamma}' \\
&= \mathbf{B}\boldsymbol{\Sigma}_{YY}\mathbf{B}' + \mathbf{B}\boldsymbol{\Sigma}_{YX}\boldsymbol{\Gamma}' + \boldsymbol{\Gamma}\boldsymbol{\Sigma}_{XY}\mathbf{B}' + \boldsymbol{\Gamma}\boldsymbol{\Sigma}_{XX}\boldsymbol{\Gamma}'
\end{aligned}$$

The result we are seeking follows upon substituting sample covariances and estimated structural coefficients for their population counterparts:

$$\mathbf{S}_{EE} = \mathbf{BS}_{YY}\mathbf{B}' + \mathbf{BS}_{YX}\mathbf{C}' + \mathbf{CS}_{XY}\mathbf{B}' + \mathbf{CS}_{XX}\mathbf{C}' \qquad (4.38)$$

Estimated disturbance correlations may be calculated from the covariances in equation (4.38) in the usual manner; that is,

$$r_{E_iE_j} = \frac{S_{E_iE_j}}{S_{E_i}S_{E_j}}$$

For the Duncan, Haller, and Portes model estimated in equations (4.37), the disturbance correlation is $r_{78} = -.476$. A negative correlation between disturbances makes little substantive sense in this case, for we expect similar omitted causes of respondent's and best friend's aspirations. The negative correlation between disturbances, therefore, casts doubt upon the specification of the model. This point is developed in Gillespie and Fox (1980), and is pursued briefly in Section 4.6.2.

Full-Information Maximum Likelihood (FIML) It is surprising that the full-information maximum-likelihood method of estimation (Koopmans, Rubin, and Leipnik, 1950) antedates simpler estimation methods such as 2SLS. Application of FIML was not generally practical, however, until electronic computers became available to take on the formidable computational burden imposed by the method.

The derivation of the FIML estimator follows the usual maximum-likelihood approach.[13] We begin with the general structural-equation model [from equation (4.2)]

$$\mathbf{B}\mathbf{y}_i + \boldsymbol{\Gamma}\mathbf{x}_i = \boldsymbol{\varepsilon}_i \tag{4.39}$$

and with the following distributional assumptions regarding the errors:

$$\boldsymbol{\varepsilon}_i \sim N_q(\mathbf{0}, \boldsymbol{\Sigma}_{\varepsilon\varepsilon})$$

$$\boldsymbol{\varepsilon}_i, \boldsymbol{\varepsilon}_j \text{ independent for } i \neq j$$

$$\mathbf{x}_i, \boldsymbol{\varepsilon}_i \text{ independent}$$

From the formula for the multivariate-normal distribution, we have

$$p(\boldsymbol{\varepsilon}_i) = \frac{1}{(2\pi)^{q/2}|\boldsymbol{\Sigma}_{\varepsilon\varepsilon}|^{1/2}} \exp\left(-\tfrac{1}{2}\boldsymbol{\varepsilon}_i'\boldsymbol{\Sigma}_{\varepsilon\varepsilon}^{-1}\boldsymbol{\varepsilon}_i\right) \tag{4.40}$$

We cannot apply equation (4.40) directly, because the disturbance vector $\boldsymbol{\varepsilon}_i$ is unobservable. We may, however, use the model (4.39) to transform $\boldsymbol{\varepsilon}_i$ to $\mathbf{y}_i$,

[13]For an alternative approach to the derivation of the FIML estimator, see Christ (1966: 395–405). It is possible, moreover, to arrive at the same estimator by applying a heuristic variance-minimizing criterion (Wonnacott and Wonnacott, 1979: 521–526), providing a justification for the FIML estimator without the necessity for making the strong distributional assumptions required by the maximum-likelihood method.

treating $\mathbf{x}_i$ as conditionally fixed and employing the Jacobian of the transformation:

$$\begin{aligned} p(\mathbf{y}_i|\mathbf{x}_i) &= p(\boldsymbol{\varepsilon}_i)\left|\frac{\partial \boldsymbol{\varepsilon}_i}{\partial \mathbf{y}_i}\right|_+ \\ &= p(\boldsymbol{\varepsilon}_i)\left|\frac{\partial(\mathbf{B}\mathbf{y}_i + \boldsymbol{\Gamma}\mathbf{x}_i)}{\partial \mathbf{y}_i}\right|_+ \\ &= p(\boldsymbol{\varepsilon}_i)|\mathbf{B}|_+ \\ &= \frac{|\mathbf{B}|_+}{(2\pi)^{q/2}|\boldsymbol{\Sigma}_{\varepsilon\varepsilon}|^{1/2}}\exp\left[-\tfrac{1}{2}(\mathbf{B}\mathbf{y}_i + \boldsymbol{\Gamma}\mathbf{x}_i)'\boldsymbol{\Sigma}_{\varepsilon\varepsilon}^{-1}(\mathbf{B}\mathbf{y}_i + \boldsymbol{\Gamma}\mathbf{x}_i)\right] \end{aligned}$$

Because the n observations are independent, their joint probability density conditional on the exogenous variables is given by the product of their marginal probability densities:

$$p(\mathbf{Y}|\mathbf{X}) = \frac{|\mathbf{B}|_+^n}{(2\pi)^{nq/2}|\boldsymbol{\Sigma}_{\varepsilon\varepsilon}|^{n/2}}\exp\left[-\frac{1}{2}\sum_{i=1}^{n}(\mathbf{B}\mathbf{y}_i + \boldsymbol{\Gamma}\mathbf{x}_i)'\boldsymbol{\Sigma}_{\varepsilon\varepsilon}^{-1}(\mathbf{B}\mathbf{y}_i + \boldsymbol{\Gamma}\mathbf{x}_i)\right]$$

and the logarithm of the likelihood function is

$$\begin{aligned} \log L(\mathbf{B}, \boldsymbol{\Gamma}, \boldsymbol{\Sigma}_{\varepsilon\varepsilon}) = n\log|\mathbf{B}|_+ - \frac{nq}{2}\log(2\pi) \\ - \frac{n}{2}\log|\boldsymbol{\Sigma}_{\varepsilon\varepsilon}| - \frac{1}{2}\sum_{i=1}^{n}(\mathbf{B}\mathbf{y}_i + \boldsymbol{\Gamma}\mathbf{x}_i)'\boldsymbol{\Sigma}_{\varepsilon\varepsilon}^{-1}(\mathbf{B}\mathbf{y}_i + \boldsymbol{\Gamma}\mathbf{x}_i) \end{aligned} \tag{4.41}$$

Note that the joint density for $\mathbf{X}$ and $\mathbf{Y}$ is given by $p(\mathbf{X}, \mathbf{Y}) = p(\mathbf{X})p(\mathbf{Y}|\mathbf{X})$. If the distribution of $\mathbf{X}$ does not depend upon the parameters $\mathbf{B}$, $\boldsymbol{\Gamma}$, and $\boldsymbol{\Sigma}_{\varepsilon\varepsilon}$, then maximizing $L(\mathbf{B}, \boldsymbol{\Gamma}, \boldsymbol{\Sigma}_{\varepsilon\varepsilon})$ is equivalent to maximizing the joint likelihood for $\mathbf{X}$ *and* $\mathbf{Y}$.

Because of the prior restrictions on the model, some of the entries of $\mathbf{B}$ and $\boldsymbol{\Gamma}$ (and possibly of $\boldsymbol{\Sigma}_{\varepsilon\varepsilon}$) are constrained to be zero. Likewise, the diagonal entries of $\mathbf{B}$ are fixed to one. Maximum-likelihood estimators of $\mathbf{B}$, $\boldsymbol{\Gamma}$, and $\boldsymbol{\Sigma}_{\varepsilon\varepsilon}$ maximize equation (4.41) subject to these constraints. The partial derivatives of the log likelihood with respect to the parameters are nonlinear, and therefore equation (4.41) must be maximized numerically. This is why FIML estimation is computationally burdensome.

As for maximum-likelihood estimation generally, the estimated asymptotic covariance matrix for the FIML estimator may be obtained from the inverse of the information matrix evaluated at the estimated parameter values. Since maximum-likelihood estimators are asymptotically normally distributed, we may employ estimated standard errors for normal-distribution tests of the model parameters. Moreover, for an overidentified model, the general likelihood-ratio criterion yields a test of the overidentifying restrictions, as explained in Section 4.7.2.

FIML estimates for the Duncan, Haller, and Portes model are shown in equations (4.42):

$$\begin{aligned} Y_5 &= \underset{(0.053)}{0.237}\, X_1 + \underset{(0.047)}{0.176}\, X_2 + \underset{(0.104)}{0.398}\, Y_6 + 0.890E_7' \\ Y_6 &= \underset{(0.047)}{0.219}\, X_3 + \underset{(0.056)}{0.311}\, X_4 + \underset{(0.131)}{0.422}\, Y_5 + 0.847E_8' \end{aligned} \tag{4.42}$$

These estimates are in reasonable agreement with the 2SLS estimates given in equations (4.37). The FIML method produces standard errors for estimated disturbance covariances, showing here that the embarrassing negative covariance between E_7 and E_8 is statistically highly significant: $S_{78} = -0.495$, with a standard error of 0.137.

4.4.2. Estimation of Recursive and Block-Recursive Models

In examining the identification status of recursive models (Section 4.3.2), we determined that all independent variables in a structural equation of a recursive model are uncorrelated with the disturbance of the equation. We may, therefore, consistently estimate any structural equation in a recursive model by OLS regression. Even if prior variables have been excluded from the equation in question, and there are consequently extra instrumental variables available, the Gauss-Markov theorem (Section 1.2.5) assures the optimality of the OLS estimator; by reasoning similar to that underlying 2SLS, each independent variable is its own best instrumental variable. Moreover, it may be shown that, for recursive models, the OLS and FIML estimators coincide (Land, 1973).

Correlations for the Blau and Duncan stratification data are shown in Table 4.3. OLS estimates for the Blau and Duncan recursive model appear in equations (4.43); standard errors are not given here because of the very large sample employed in Blau and Duncan's research.

$$\begin{aligned} Y_3 &= 0.310X_1 + 0.279X_2 + 0.859E_6' \\ Y_4 &= 0.224X_2 + 0.440Y_3 + 0.818E_7' \\ Y_5 &= 0.115X_2 + 0.394Y_3 + 0.281Y_4 + 0.753E_8' \end{aligned} \tag{4.43}$$

As for the nonrecursive peer-influences model, we have standardized the estimated disturbances, introducing residual paths into the structural equations. For a standardized structural equation estimated by OLS, the squared multiple correlation $R^2 \simeq 1 - S_E^2$ (disregarding degrees of freedom for error). The model, therefore, accounts for 26.2, 33.1, and 43.3 percent of the variation in Y_3, Y_4, and Y_5, consecutively. The estimated structural parameters are all positive, as expected, and assume reasonable values. In discussing the specification of the Blau and Duncan model, we were skeptical of the assumption that ε_7 and ε_8 are uncorrelated. A positive correlation between these disturbances would induce a positive correlation between Y_4 and ε_8, tending to inflate B_{54}. This coefficient, however, is not strikingly large.

Block-recursive models may be estimated straightforwardly by IV, FIML, 2SLS, or some other applicable technique. Direct IV estimation may reasonably be undertaken for a just-identified structural equation; here, exogenous and prior endogenous variables comprise the pool of available instrumental variables. In applying FIML, we merely need to indicate the prior restrictions on $\mathbf{\Sigma}_{\varepsilon\varepsilon}$ along with those on $\mathbf{B}$ and $\mathbf{\Gamma}$. For 2SLS estimation of an equation in a block-recursive model, endogenous variables in prior blocks should be treated as exogenous.

A block-recursive model for the peer-influences data was presented in Figure 4.4, and the correlations for the variables in this model were given in Table 4.2. 2SLS and FIML estimates for the model, along with their standard errors, appear in Table 4.4. Because the first two structural equations are just identified, the 2SLS estimates for these equations are the same as those obtained by direct application of the exogenous variables as instrumental variables. Furthermore, the 2SLS and FIML estimates for these equations are necessarily identical. Note that the two sets of estimates are similar, and that

TABLE 4.3. Correlations for Stratification Data, $n \simeq 20{,}700$

	X_1	X_2	Y_3	Y_4	Y_5
X_1	1.000				
X_2	.516	1.000			
Y_3	.453	.438	1.000		
Y_4	.332	.417	.538	1.000	
Y_5	.322	.405	.596	.541	1.000

X_1 Father's education
X_2 Father's occupational status
Y_3 Education
Y_4 First-job status
Y_5 1962 occupational status

Source: Blau and Duncan (1967: 169) (see Table 1.7).

TABLE 4.4. Block-Recursive Model for the Duncan, Haller, and Portes Peer-Influences Data (Figure 4.4)

Coefficient for	Structural Equation for			
	Y_5	Y_6	Y_7	Y_8
		(a) FIML Estimates		
X_1	0.2793	—	0.0939	—
	(0.0559)[a]		(0.0397)	
X_2	0.1535	0.0772	0.1865	−0.0470
	(0.0559)	(0.0599)	(0.0462)	(0.0535)
X_3	0.0843	0.2015	−0.0398	0.0697
	(0.0672)	(0.0553)	(0.0491)	(0.0480)
X_4	—	0.3574	—	0.1589
		(0.0567)		(0.0436)
Y_5	—	0.2819	0.4502	—
		(0.1590)	(0.0518)	
Y_6	0.2804	—	—	0.4202
	(0.1362)			(0.0522)
Y_7	—	—	—	0.3506
				(0.0900)
Y_8	—	—	0.2235	—
			(0.0875)	
		(b) 2SLS Estimates		
X_1	0.2793	—	0.1391	—
	(0.0563)		(0.0475)	
X_2	0.1535	0.0772	0.1864	−0.0428
	(0.0562)	(0.0603)	(0.0470)	(0.0544)
X_3	0.0843	0.2105	−0.0367	0.0707
	(0.0676)	(0.0556)	(0.0499)	(0.0489)
X_4	—	0.3574	—	0.1825
		(0.0571)		(0.0522)
Y_5	—	0.2819	0.4034	—
		(0.1600)	(0.0554)	
Y_6	0.2804	—	—	0.4063
	(0.1371)			(0.0576)
Y_7	—	—	—	0.3367
				(0.0910)
Y_8	—	—	0.2078	—
			(0.0876)	

[a]Standard errors in parentheses.

the unreasonable negative correlation between estimated disturbances appears in this model as well.

PROBLEMS

4.10. Indicate how you would estimate each structural-equation model specified in Problem 4.1.

4.11. Data for Rindfuss, Bumpass, and St. John's fertility model (Problem 4.2) are given in Table 4.5. Fit the model to these data using an appropriate estimation method. Comment on the results.

4.12. Table 4.6 shows a covariance matrix among the variables in Berk and Berk's model for the division of household labor (Problem 4.3). Use these covariances to estimate the model.

4.13. The correlations in Table 4.7 are for non-black men in the experienced civilian labor force, who were of non-farm background and 35–44 years old in 1962. Use these correlations to estimate Duncan, Featherman, and Duncan's stratification model (Problem 4.4).

4.14. The covariances in Table 4.8 were calculated from data presented by Lincoln. Using these covariances, estimate Lincoln's model for strike activity in metropolitan areas (Problem 4.5).

TABLE 4.5. Covariances for Rindfuss, Bumpass, and St. John's Fertility Data

	X_1	X_2	X_3	X_4	X_5	X_6	X_7	X_8
X_1	456.6769							
X_2	-0.9201	0.0894						
X_3	-15.8253	0.1416	9.2112					
X_4	-3.2442	0.0124	0.3908	0.2209				
X_5	-1.3205	0.0451	0.2181	0.0491	0.2294			
X_6	-0.4631	0.0174	-0.0458	-0.0055	0.0132	0.1498		
X_7	0.4768	-0.0191	0.0179	-0.0295	-0.0589	-0.0085	0.1772	
X_8	-0.3143	0.0031	0.0291	-0.0096	-0.0018	0.0089	-0.0014	0.1170
X_9	0.2356	0.0031	0.0018	-0.0045	-0.0039	0.0021	-0.0003	0.0009
Y_{10}	18.6603	-0.1567	-2.3493	-0.2052	-0.2385	-0.1434	-0.0119	-0.1380
Y_{11}	16.2133	-0.2305	-1.4237	-0.2262	-0.3458	0.1752	0.1683	-0.1702
Mean	30.209	0.099	3.889	0.330	0.357	0.183	0.231	0.136

	X_9	Y_{10}	Y_{11}
X_9	0.0888		
Y_{10}	0.0267	5.5696	
Y_{11}	0.2626	3.6580	16.6382
Mean	0.099	11.595	22.012

Source: Adapted with permission from Rindfuss, Bumpass, and St. John (1980: 436, 445).

TABLE 4.6. Covariances for Berk and Berk's Data on the Division of Household Labor

	X_1	X_2	X_3	X_4	X_5	X_6	X_7	X_8
X_1	0.01440							
X_2	-0.00288	0.02250						
X_3	-0.00898	0.00382	0.02890					
X_4	-0.00664	0.01226	0.00538	0.06946				
X_5	-0.01673	0.02214	0.00348	0.05079	0.16810			
X_6	-0.00433	0.00085	0.00807	0.00300	-0.00779	0.03610		
X_7	0.00281	-0.00135	-0.00520	0.00019	-0.00148	-0.00137	0.03240	
X_8	0.00228	0.00028	-0.00484	-0.00501	-0.00779	-0.00144	-0.00137	0.03610
X_9	-0.00114	0.00456	-0.00420	0.00015	0.00234	-0.00144	-0.00137	0.01444
X_{10}	0.00158	0.00033	-0.00636	-0.00638	0.00054	-0.00167	-0.00158	0.00334
X_{11}	0.00540	0.00225	-0.01122	-0.00949	-0.00984	-0.00342	-0.00324	0.01254
X_{12}	0.00437	-0.00195	-0.00707	-0.00274	-0.00426	-0.00247	-0.00234	0.00296
X_{13}	0.06754	0.19698	-0.52622	-0.09889	0.23075	0.03564	-0.33768	0.32080
X_{14}	-0.04208	-0.00928	0.01403	0.00815	0.04229	0.00784	-0.00743	-0.01960
X_{15}	-0.00360	0.01200	-0.00680	0.02504	0.06355	-0.00570	0.00180	-0.00066
X_{16}	-0.00027	-0.00270	0.00688	-0.01898	-0.03321	0.01026	0.00162	-0.00513
X_{17}	-0.00730	0.01104	0.00653	0.01434	0.01443	0.00669	-0.00346	-0.00426
X_{18}	-0.01214	-0.00828	0.03284	0.00242	-0.00189	0.01049	-0.00994	-0.01136
X_{19}	-0.00836	-0.00062	0.02927	0.00756	0.00336	0.00234	-0.00664	-0.00234
X_{20}	-0.00936	-0.00526	0.01392	0.00411	0.00799	0.00370	-0.00632	-0.00741
Y_{21}	-0.01426	-0.00648	0.02570	0.00474	-0.00074	0.00479	-0.00518	-0.00616
Y_{22}	-0.02534	0.05472	-0.01958	0.06072	0.17318	0.02189	0.00207	0.00730
Y_{23}	-0.01732	0.00936	0.02254	0.00925	0.00320	0.00889	-0.00632	-0.00741
Mean	0.84	0.21	0.16	0.49	0.21	0.04	0.03	0.04

	X_9	X_{10}	X_{11}	X_{12}	X_{13}	X_{14}	X_{15}	X_{16}
X_9	0.03610							
X_{10}	-0.00167	0.04840						
X_{11}	0.00684	0.00066	0.09000					
X_{12}	-0.00247	0.00172	-0.00702	0.06760				
X_{13}	0.44555	0.39208	0.78792	0.48776	87.98440			
X_{14}	-0.02352	-0.02042	-0.00309	-0.01609	-0.29023	1.06372		
X_{15}	-0.00095	0.01100	0.00600	0.02340	1.31320	0.10829	0.25000	
X_{16}	-0.00513	-0.00891	0.00945	-0.00936	-0.21105	0.06962	-0.12825	0.20250
X_{17}	-0.00426	0.00563	-0.01152	0.00250	0.15008	-0.00990	0.00960	-0.00576
X_{18}	-0.00524	-0.00911	-0.02346	-0.01555	-0.86296	0.00474	-0.02760	0.01863
X_{19}	-0.00779	-0.01082	-0.02091	-0.01492	-0.07692	0.02960	-0.02460	0.02583
X_{20}	-0.00741	-0.00858	-0.01872	-0.01318	-1.06088	0.01609	-0.03510	-0.00035
Y_{21}	-0.00616	-0.00792	-0.01512	-0.00562	-0.97927	0.04827	-0.02700	0.02430
Y_{22}	0.01094	0.02112	0.08064	0.02995	3.60192	0.79209	0.53760	-0.01728
Y_{23}	-0.00741	-0.00944	0.01872	-0.00811	-0.43898	0.06436	-0.00390	-0.00035
Mean	0.03	0.05	0.10	0.07	59.28	1.75	0.44	0.29

	X_{17}	X_{18}	X_{19}	X_{20}	Y_{21}	Y_{22}	Y_{23}
X_{17}	0.10240						
X_{18}	0.01619	0.21160					
X_{19}	0.00262	0.01320	0.16810				
X_{20}	-0.00499	0.05203	-0.00160	0.15210			
Y_{21}	0.00346	0.01325	0.02066	0.00842	0.12960		
Y_{22}	0.00614	-0.04416	-0.11808	-0.06739	-0.04147	3.68640	
Y_{23}	0.05990	0.03050	0.01919	0.02282	0.00842	0.00749	0.15210
Mean	0.11	0.29	0.27	0.19	0.16	4.94	0.19

Source: Reprinted from Richard A. Berk and Sarah Fenstermaker Berk, "A Simultaneous Equation Model for the Division of Household Labor," *Sociological Methods and Research*, Vol. 6, No. 4 (May 1978), pp. 431–468, with permission of Sage Publications, Inc.

TABLE 4.7. Duncan, Featherman, and Duncan's Stratification Data

	X_1	X_2	X_3	Y_4	Y_5	Y_6
X_1	1.0000					
X_2	0.5300	1.0000				
X_3	−0.2871	−0.2476	1.0000			
Y_4	0.4048	0.4341	−0.3311	1.0000		
Y_5	0.3194	0.3899	−0.2751	0.6426	1.0000	
Y_6	0.2332	0.2587	−0.1752	0.3759	0.4418	1.0000

Source: Reprinted with permission from Duncan, Featherman, and Duncan (1972: Table 3.1).

TABLE 4.8. Covariances for Lincoln's Strike-Activity Data

	X_1	X_2	X_3	X_4	Y_5	Y_6	Y_7
X_1	0.007744						
X_2	0.000635	0.000400					
X_3	0.052401	0.005077	1.065024				
X_4	0.006624	0.001471	0.066069	0.037636			
Y_5	0.054564	0.012024	0.823108	0.137249	1.809025		
Y_6	0.084675	0.015990	1.131609	0.171958	2.025220	2.496400	
Y_7	0.103616	0.019572	1.325756	0.184820	1.969703	2.567911	2.989441
Mean	0.509	0.752	12.332	0.649	5.009	6.182	8.832

Source: Adapted with permission from Lincoln (1978: 208).

4.5. PATH ANALYSIS OF RECURSIVE MODELS: CAUSAL INFERENCE AND DATA ANALYSIS

Path analysis is often taken to be synonymous with structural-equation modeling. We shall use the term in a more literal and delimited sense to mean the decomposition of statistical relationships between pairs of variables into causal and noncausal components.

In data analysis generally, when we assess the causal impact of one variable on another, we are motivated to control for some third variable or set of variables (in the absence of interaction effects) in two substantively different contexts: (1) the third variable (say, Z) intervenes causally between the other two (say, X and Y); (2) the third variable is a common prior cause of the other two. These two contrasting situations are illustrated in Figure 4.11. The broken arrows in this figure indicate that, in either causal structure, X may or may not exert a direct impact on Y.

Let us suppose, for the moment, that the direct effect is absent or negligible. It is perhaps surprising, but nonetheless true, that our statistical expectations in the two situations are identical: In both cases, the partial relationship between X and Y vanishes when we control for Z. Although the observable consequences of the two causal schemes are, then, identical, their interpretation is importantly different: In the first case we have *explained* the mechanism according to which X affects Y (that is, through the intervening variable Z), while in the second case we have *explained away* the empirical association between X and Y as "spurious" (that is, due to the common prior cause Z, not to the effect of X on Y).

It is instructive to examine the causal schemes in Figure 4.11 in an explicit structural-equation setting. In each case, suppose that the structure is recursive; then the equation for Y is

$$Y = \beta_1 X + \beta_2 Z + \varepsilon \tag{4.44}$$

Here we use β's for structural coefficients despite the fact that X is exogenous in the first model and Z is exogenous in the second. Now imagine that instead of fitting equation (4.44), we fit

$$Y = \beta_1 X + \varepsilon' \tag{4.45}$$

where $\varepsilon' \equiv \varepsilon + \beta_2 Z$. Including Z in the error will cause our OLS estimator of β_1 to be biased, since X and Z are correlated. That is,

$$\text{plim } B_1 = \frac{\sigma_{XY}}{\sigma_{XX}} = \beta_1 + \beta_2 \frac{\sigma_{XZ}}{\sigma_{XX}} \tag{4.46}$$

If, however, Z intervenes causally between X and Y [Figure 4.11(a)], the "bias" in equation (4.46) is simply the indirect effect of X on Y through Z. (Note that σ_{XZ}/σ_{XX} is the coefficient for the simple regression of Z on X; since X is causally prior to Z, this coefficient represents the effect of X on Z.) In other words, fitting equation (4.45) in place of equation (4.44), failing to control for an intervening variable, produces correct, albeit simplified, conclusions. In contrast, if Z is causally prior to both X and Y, then the bias term in equation (4.46) represents a spurious source of association between X and Y. To fail to control for a common antecedent cause, therefore, is to commit an error of causal inference.

Two important related conclusions may be drawn from this discussion. First, we cannot expect our data to mediate issues of causal priority, since very different causal structures have the same observable implications. Second, the conclusions that we draw from a data analysis depend centrally on the causal relations that are assumed to hold among the "independent" variables in the analysis.

In introducing structural-equation models to sociologists, Duncan (1966) argued that one of the strengths of the method "is that it produces a calculus

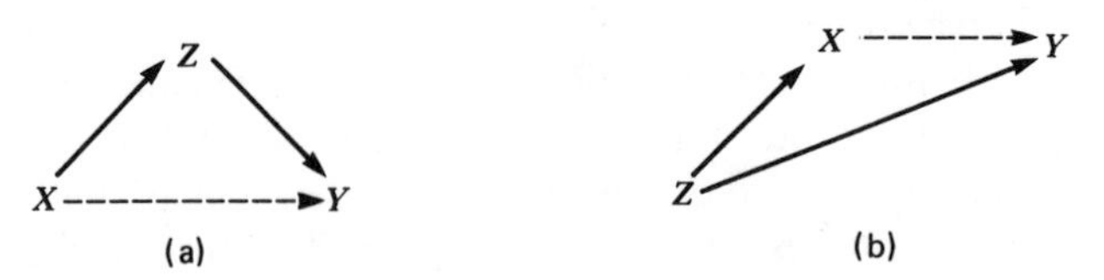

FIGURE 4.11. Two causal structures. (a) Intervening variable. (b) Common prior cause.

for indirect effects." This topic has been taken up by a number of researchers, including Finney (1972), Lewis-Beck (1974), Alwin and Hauser (1975), Lewis-Beck and Mohr (1976), Greene (1977), Fox (1980), and Sobel (1982). The approach adopted here is from Fox (1980). There has been some discussion of path-analytic decompositions for nonrecursive models (Lewis-Beck and Mohr, 1976; Heise, 1975; Fox, 1980), but we shall limit consideration to recursive models. Although decompositional methods have typically been applied to standardized models, we shall develop the more general unstandardized case.

We begin by noting that a structural-equation model implies a set of covariances among the endogenous variables of the model, and between endogenous and exogenous variables. These covariances may be derived by the now familiar expectation method. Multiplying the structural equations (4.2) through on the right by $\mathbf{x}'$ and taking expectations, we get

$$\begin{aligned} \mathbf{B}E(\mathbf{yx}') + \boldsymbol{\Gamma}E(\mathbf{xx}') &= E(\boldsymbol{\varepsilon}\mathbf{x}') \\ \mathbf{B}\boldsymbol{\Sigma}_{YX} + \boldsymbol{\Gamma}\boldsymbol{\Sigma}_{XX} &= \mathbf{0} \\ \boldsymbol{\Sigma}_{YX} &= -\mathbf{B}^{-1}\boldsymbol{\Gamma}\boldsymbol{\Sigma}_{XX} \end{aligned} \tag{4.47}$$

Similarly,

$$\begin{aligned} \boldsymbol{\Sigma}_{YY} &= E(\mathbf{yy}') \\ &= E\left[(-\mathbf{B}^{-1}\boldsymbol{\Gamma}\mathbf{x} + \mathbf{B}^{-1}\boldsymbol{\varepsilon})(-\mathbf{B}^{-1}\boldsymbol{\Gamma}\mathbf{x} + \mathbf{B}^{-1}\boldsymbol{\varepsilon})'\right] \\ &= \mathbf{B}^{-1}\boldsymbol{\Gamma}\boldsymbol{\Sigma}_{XX}\boldsymbol{\Gamma}'(\mathbf{B}^{-1})' + \mathbf{B}^{-1}\boldsymbol{\Sigma}_{\varepsilon\varepsilon}(\mathbf{B}^{-1})' \end{aligned} \tag{4.48}$$

The entries of $\boldsymbol{\Sigma}_{YX}$ and $\boldsymbol{\Sigma}_{YY}$ are shown in scalar form for the Blau and Duncan stratification model (Figure 4.2) in equations (4.49):

$$\begin{aligned} \sigma_{13} &= \gamma_{31}\sigma_{11} + \gamma_{32}\sigma_{12} \\ \sigma_{23} &= \gamma_{31}\sigma_{12} + \gamma_{32}\sigma_{22} \\ \sigma_{14} &= \gamma_{42}\sigma_{12} + \beta_{43}\gamma_{31}\sigma_{11} + \beta_{43}\gamma_{32}\sigma_{12} \\ \sigma_{24} &= \gamma_{42}\sigma_{22} + \beta_{43}\gamma_{31}\sigma_{12} + \beta_{43}\gamma_{32}\sigma_{22} \\ \sigma_{34} &= \gamma_{42}\gamma_{31}\sigma_{12} + \gamma_{42}\gamma_{32}\sigma_{22} + \beta_{43}\sigma_{33} \\ \sigma_{15} &= \gamma_{52}\sigma_{12} + \beta_{53}\gamma_{31}\sigma_{11} + \beta_{53}\gamma_{32}\sigma_{12} \\ &\quad + \beta_{54}\gamma_{42}\sigma_{12} + \beta_{54}\beta_{43}\gamma_{31}\sigma_{11} + \beta_{54}\beta_{43}\gamma_{32}\sigma_{12} \\ \sigma_{25} &= \gamma_{52}\sigma_{22} + \beta_{53}\gamma_{31}\sigma_{12} + \beta_{53}\gamma_{32}\sigma_{22} \\ &\quad + \beta_{54}\gamma_{42}\sigma_{22} + \beta_{54}\beta_{43}\gamma_{31}\sigma_{12} + \beta_{54}\beta_{43}\gamma_{32}\sigma_{22} \\ \sigma_{35} &= \gamma_{52}\gamma_{31}\sigma_{12} + \gamma_{52}\gamma_{32}\sigma_{22} + \beta_{53}\sigma_{33} \\ &\quad + \beta_{54}\gamma_{42}\gamma_{31}\sigma_{12} + \beta_{54}\gamma_{42}\gamma_{32}\sigma_{22} + \beta_{54}\beta_{43}\sigma_{33} \\ \sigma_{45} &= \gamma_{52}\gamma_{42}\sigma_{22} + \gamma_{52}\beta_{43}\gamma_{31}\sigma_{12} + \gamma_{52}\beta_{43}\gamma_{32}\sigma_{22} \\ &\quad + \beta_{53}\gamma_{42}\sigma_{12} + \beta_{53}\gamma_{42}\gamma_{32}\sigma_{22} + \beta_{53}\beta_{43}\sigma_{33} + \beta_{54}\sigma_{44} \end{aligned} \tag{4.49}$$

Note that we have not written down expressions for the variances (σ_{33}, σ_{44}, and σ_{55}) of the endogenous variables, nor have we substituted for these variances when they appear on the right-hand side of the equations. The equations in (4.49) are obtained by multiplying each structural equation of the model by all prior variables, taking expectations, and successively substituting for all covariances involving an endogenous variable, until only structural parameters, exogenous covariances, and variances appear on the right-hand side. To decompose σ_{15}, for instance, we proceed as follows:

$$E(X_1Y_5) = \gamma_{52}E(X_1X_2) + \beta_{53}E(X_1Y_3) + \beta_{54}E(X_1Y_4)$$

$$\begin{aligned}\sigma_{15} &= \gamma_{52}\sigma_{12} + \beta_{53}\sigma_{13} + \beta_{54}\sigma_{14}\\ &= \gamma_{52}\sigma_{12} + \beta_{53}(\gamma_{31}\sigma_{11} + \gamma_{32}\sigma_{12}) + \beta_{54}(\gamma_{42}\sigma_{12} + \beta_{43}\gamma_{31}\sigma_{11} + \beta_{43}\gamma_{32}\sigma_{12})\\ &= \gamma_{52}\sigma_{12} + \beta_{53}\gamma_{31}\sigma_{11} + \beta_{53}\gamma_{32}\sigma_{12} + \beta_{54}\gamma_{42}\sigma_{12} + \beta_{54}\beta_{43}\gamma_{31}\sigma_{11} + \beta_{54}\beta_{43}\gamma_{32}\sigma_{12}\end{aligned}$$

Dividing each covariance in equations (4.49) by the variance of the causally prior variable produces the population slope for the simple linear regression of each endogenous variable on each prior variable. For example, dividing σ_{25} by σ_{22}, and σ_{35} by σ_{33}, we obtain

$$\begin{aligned}\mu_{52} &= \frac{\sigma_{25}}{\sigma_{22}} = \gamma_{52} + \beta_{53}\gamma_{31}\frac{\sigma_{12}}{\sigma_{22}} + \beta_{53}\gamma_{32}\\ &\quad + \beta_{54}\gamma_{42} + \beta_{54}\beta_{43}\gamma_{31}\frac{\sigma_{12}}{\sigma_{22}} + \beta_{54}\beta_{43}\gamma_{32}\\ \mu_{53} &= \frac{\sigma_{35}}{\sigma_{33}} = \gamma_{52}\gamma_{31}\frac{\sigma_{12}}{\sigma_{33}} + \gamma_{52}\gamma_{32}\frac{\sigma_{22}}{\sigma_{33}} + \beta_{53}\\ &\quad + \beta_{54}\gamma_{42}\gamma_{31}\frac{\sigma_{12}}{\sigma_{33}} + \beta_{54}\gamma_{42}\gamma_{32}\frac{\sigma_{22}}{\sigma_{33}} + \beta_{54}\beta_{43}\end{aligned} \tag{4.50}$$

These simple-regression or "gross" slopes measure (in Goldberger's, 1973, terminology) the *empirical association* between each pair of variables.

Equations (4.50) illustrate how a gross slope may be decomposed into path components of three general types: (1) a *direct effect*, represented by a structural coefficient; (2) *indirect effects*, given by products of structural coefficients along a path linking a prior variable to an endogenous variable; and (3) *noncausal components*. For the association between an exogenous and endogenous variable, the noncausal components are termed *unanalyzed*, because they depend upon covariances among exogenous variables, for which a causal ordering is not distinguished. The noncausal components of the association between two endogenous variables are termed *spurious*, because they depend upon correlated causes of, or causes common to, both variables. Examples of the various sorts of components are given in Figure 4.12.

In the sample, we may estimate model-implied covariances by substituting for the population quantities appearing in equations (4.47) and (4.48):

$$\mathbf{S}_{YX}^{*} = -\mathbf{B}^{-1}\mathbf{C}\mathbf{S}_{XX}$$
$$\mathbf{S}_{YY}^{*} = \mathbf{B}^{-1}\mathbf{C}\mathbf{S}_{XX}\mathbf{C}'(\mathbf{B}^{-1})' + \mathbf{B}^{-1}\mathbf{S}_{EE}(\mathbf{B}^{-1})' \tag{4.51}$$

$\mathbf{S}_{YX}^{*}$ and $\mathbf{S}_{YY}^{*}$ are starred because, in an overidentified model, they may differ from covariances calculated directly from the data (i.e., $\mathbf{S}_{YX} = [1/(n-1)]\mathbf{Y}'\mathbf{X}$ and $\mathbf{S}_{YY} = [1/(n-1)]\mathbf{Y}'\mathbf{Y}$). In a recursive model, $\mathbf{B}$ and $\mathbf{C}$ are obtained by OLS regression, and $\mathbf{S}_{EE}$ is diagonal. Sample model-implied gross slopes are then given by

$$\mathbf{M}_{YX}^{*} = \mathbf{S}_{YX}^{*}\mathbf{V}_{X}^{-1}$$
$$\mathbf{M}_{YY}^{*} = \mathbf{S}_{YY}^{*}\mathbf{V}_{Y}^{*-1}$$

where $\mathbf{V}_X \equiv \text{diag}(\mathbf{S}_{XX})$ and $\mathbf{V}_Y^{*} \equiv \text{diag}(\mathbf{S}_{YY}^{*})$.

Let us denote by $\mathbf{E}_{YX}$ and $\mathbf{E}_{YY}$, respectively, the matrices of total effects (direct and indirect) of the exogenous on the endogenous variables, and of the

	Prior Variable: Exogenous	Prior Variable: Endogenous
Gross slope	$\mu_{52} = \sigma_{25}/\sigma_{22}$	$\mu_{53} = \sigma_{35}/\sigma_{33}$
Type of Component		
(1) Direct effect	γ_{52} $X_2 \to Y_5$	β_{53} $Y_3 \to Y_5$
(2) Indirect effect	$\beta_{54}\gamma_{42}$ $X_2 \to Y_4 \to Y_5$	$\beta_{54}\beta_{43}$ $Y_3 \to Y_4 \to Y_5$
(3) Noncausal		
(a) Unanalyzed	$\beta_{53}\gamma_{31}\sigma_{12}/\sigma_{22}$ $X_1 \to Y_3 \to Y_5$ (with $X_1 \leftrightarrow X_2$)	—
(b) Spurious	—	$\gamma_{52}\gamma_{32}\sigma_{22}/\sigma_{33}$
(i) Common prior cause		$X_2 \to Y_3$; $X_2 \to Y_5$
(ii) Correlated prior causes		$X_1 \to Y_3$; $X_2 \to Y_5$ (with $X_1 \leftrightarrow X_2$) $\gamma_{52}\gamma_{31}\sigma_{12}/\sigma_{33}$

FIGURE 4.12. Examples of path components from the Blau and Duncan model. (Adapted from Fox, 1980: Figure 2.)

endogenous variables on each other. Fox (1980) shows that[14]

$$\mathbf{E}_{YX} = -\mathbf{B}^{-1}\mathbf{C}$$

$$\mathbf{E}_{YY} = \mathbf{B}^{-1} - \mathbf{I}_q$$

Notice that $\mathbf{E}_{YX}$ is simply the reduced-form coefficient matrix obtained from the estimated structural parameters. Noncausal components may be calculated by subtraction:

$$\mathbf{N}_{YX} = \mathbf{M}^*_{YX} - \mathbf{E}_{YX}$$

$$\mathbf{N}_{YY} = \mathbf{M}^*_{YY} - \mathbf{E}_{YY}$$

where $\mathbf{N}_{YX}$ and $\mathbf{N}_{YY}$ are matrices of unanalyzed and spurious components, respectively. In a recursive model, $\mathbf{E}_{YY}$ is lower triangular (a later variable cannot affect a causally prior one); thus, the upper triangle of $\mathbf{M}^*_{YY}$ is necessarily wholly spurious and normally would not be shown.

Direct effects are given by the structural coefficients themselves:

$$\mathbf{D}_{YX} = -\mathbf{C}$$

$$\mathbf{D}_{YY} = -(\mathbf{B} - \mathbf{I}_q) = \mathbf{I}_q - \mathbf{B}$$

Indirect effects follow by subtraction:

$$\mathbf{I}_{YX} = \mathbf{E}_{YX} - \mathbf{D}_{YX}$$

$$\mathbf{I}_{YY} = \mathbf{E}_{YY} - \mathbf{D}_{YY}$$

An effect analysis for the standardized Blau and Duncan model appears in Table 4.9. (In applying the results of this section to a standardized model, it is merely necessary to substitute correlations for covariances.) Some of the information in Table 4.9 has been reorganized in Table 4.10 to show the sources of association between each prior variable in the Blau and Duncan model and the final endogenous variable, 1962 occupation, Y_5. The last column in the table, labeled B^*, gives the standardized partial regression coefficients for the regression of Y_5 on all prior variables; although this regression does not follow from the Blau and Duncan model, which sets $\gamma_{51} = 0$ *a priori*, we shall shortly have occasion to make reference to these coefficients. Note, for the present, that the regression coefficients in the last column of the table are

[14]Fox treats the structural-equation model as a directed network, with the value matrix of the network given by the structural coefficients of the model. Paths are traced by powering the value matrix, and effects are determined by summing the matrix powers, yielding the results given here after some algebraic manipulation.

nearly identical with the corresponding direct effects (obtained by regressing Y_5 on X_2, Y_3, and Y_4, but not X_1), suggesting that the restriction $\gamma_{51} = 0$ is reasonable.

A good deal might be said about Table 4.10, but we wish to use the results in this table to illustrate just two important points. First, the fact that a prior variable has negligible direct effects does not necessarily mean that it is causally unimportant. Father's education (X_1), for example, has no direct effect on Y_5, but it has nontrivial indirect effects. Likewise, the indirect effects of father's occupation (X_2) on Y_5 exceed the direct effects. Second, and in contrast to the first point, the fact that two variables have a strong empirical association does not imply that one exerts a strong causal impact on the other. For example, the implied slope relating Y_5 to first job (Y_4) is quite large, but nearly half of this association is spurious; indeed, in this case, the assumption that the disturbances ε_7 and ε_8 are uncorrelated is questionable (as we have noted), and even the direct-effect estimate B_{54} is probably inflated.

These points are not without consequence: Suppose that a naive investigator approaches the Blau and Duncan stratification data by regressing Y_5 on X_1, X_2, Y_3, and Y_4, as shown in the final column of Table 4.10. In light of the small (negative!) coefficient for X_1, he or she might conclude that father's education has no impact on son's eventual occupational status. Because Y_3 and Y_4 intervene between X_1 and Y_5, however, such a conclusion would be misleading:

TABLE 4.9. Effect Analysis for the Blau and Duncan Model

$\mathbf{S}^*_{YX}$	X_1	X_2	$\mathbf{S}^*_{YY}$	Y_3	Y_4	Y_5
Y_3	0.454[a]	0.439[a]	Y_3	1.001[a]		
Y_4	0.315	0.417[a]	Y_4	0.539[a]	1.000[a]	
Y_5	0.327	0.405[a]	Y_5	0.596[a]	0.541[a]	1.001[a]
$\mathbf{D}_{YX}$	X_1	X_2	$\mathbf{D}_{YY}$	Y_3	Y_4	Y_5
Y_3	0.310	0.279	Y_3			
Y_4	0.0	0.224	Y_4	0.440		
Y_5	0.0	0.115	Y_5	0.394	0.281	
$\mathbf{I}_{YX}$	X_1	X_2	$\mathbf{I}_{YY}$	Y_3	Y_4	Y_5
Y_3	0.0	0.0	Y_3			
Y_4	0.136	0.123	Y_4	0.0		
Y_5	0.161	0.207	Y_5	0.124	0.0	
$\mathbf{N}_{YX}$	X_1	X_2	$\mathbf{N}_{YY}$	Y_3	Y_4	Y_5
Y_3	0.144	0.160	Y_3			
Y_4	0.179	0.070	Y_4	0.098		
Y_5	0.166	0.083	Y_5	0.078	0.260	

Source: Fox (1980: Table 1).

[a] Necessarily equal to observed value (within rounding error).

TABLE 4.10. Effect Analysis for Relationships of Prior Variables With Occupation in 1962 (Y_5), Blau and Duncan Model

Prior Variable j	Implied Slope M_{5j}^*	Direct Effect D_{5j}	Indirect Effect I_{5j}	Total Effect E_{5j}	Noncausal N_{5j}	Multiple Regression Slope B_{5j}^*
X_1: Father's education	0.327	0.0	0.161	0.161	0.166	−0.014
X_2: Father's occupation	0.405	0.115	0.207	0.322	0.083	0.121
Y_3: Education	0.596	0.394	0.124	0.518	0.078	0.398
Y_4: First job	0.541	0.281	0.0	0.281	0.260	0.281

Source: Adapted from Fox (1980: Table 2).

Indirect effects transmitted through Y_3 and Y_4 are ignored. In contrast, the reduction in the relationship between Y_4 and Y_5 when X_1, X_2, and Y_3 are controlled is properly interpreted as reflecting spurious sources of association. Proper interpretation, therefore, depends crucially upon an explicit or implicit causal model.

PROBLEMS

4.15. Perform a path analysis for (a) the standardized Duncan, Featherman, and Duncan model estimated in Problem 4.13; and (b) the unstandardized Lincoln model estimated in Problem 4.14.

4.16. In light of the material presented in this section, what are the risks for causal interpretation of controlling for "all relevant factors" in examining the relationship between two variables?

4.6. LATENT VARIABLES IN STRUCTURAL-EQUATION MODELS

The recent literature on latent variables in structural-equation models represents a confluence of work in several disciplines (see Goldberger, 1971, 1972; Griliches, 1977): in economics, on measurement errors in structural-equation and regression models; in psychology, on factor analysis and test theory; in biology, on path analysis; and in sociology, on constructs and their indicators. Sociologists also deserve credit for many of the applications of latent-variable models, and for an interest in integrating the several streams of work.

Latent variables (also called *unobserved variables*, *true variables*, *factors*, and *constructs*) arise for several related reasons: (1) Our measurements, even of relatively straightforward quantities, such as income and education, are imperfect. That is, observed variables generally have a measurement-error component. If this component is relatively large, it may be important to take it explicitly into account when we estimate structural relations, as will be shown in this section. (2) A variable appearing in a structural-equation model may be an abstract construct, such as racial prejudice, which is not directly observable. The construct may, however, have an observable effect, or *indicator*, and, indeed, it is often the case that multiple indicators are available. In general, however, no indicator is perfect; that is, each contains a measurement-error component. (3) Just as a construct may have observable indicators without itself being observable, likewise a construct may have one or more observable causes. Racial prejudice, for example, may be affected by education.

Social scientists frequently employ multiple indicators to construct composite scales prior to undertaking model building. While this strategy is reasonable, there are advantages to combining the processes of scale and model construction. First, a structural model may contribute to the definition of better scales, employing multiple indicators in a more efficient fashion. Second, taking measurement error into account explicitly may improve our estimates of structural parameters. As we shall see, however, multiple indicators and measurement errors cannot be included in a structural-equation model in a haphazard fashion. To build identified models, we have to make careful specifications that incorporate strong assumptions about the behavior of measurement errors. Although this is often a difficult undertaking, the alternative of ignoring errors in measurement can distort our findings.

The construction of structural-equation models with latent variables is a complex subject that we shall only be able to take up briefly here. Among the growing literature in this area, Duncan (1975: Ch. 9–10) presents a clear introductory treatment; a number of important papers appear in volumes edited by Goldberger and Duncan (1973) and by Aigner and Goldberger (1977), each of which contains extensive bibliographies. The LISREL computer program manual (Jöreskog and Sörbom, 1978, 1981) is also a valuable source (see Section 4.6.2).

4.6.1. Consequences of Random Measurement Error

In this section, we trace the consequences of random measurement error, prior to introducing a general model that accommodates latent variables in the next section. We determine the implications of measurement error for our usual estimators of structural parameters, and show how measurement error may be explicitly taken into account in the process of estimation. We proceed by examining three simple structural-equation models. Additional examples of this type may be found in Duncan (1975: Chapters 9–10), to which the exposition in this section is indebted.

Consider first the model shown in Figure 4.13. The notational conventions employed in this diagram anticipate the usage that we shall adopt in Section 4.6.2, and require some comment. The X's and Y's, as before in this chapter, represent directly observed variables in mean-deviation form. In the present model, the X's are exogenous variables, assumed to be measured without error, and the Y's are fallible indicators of latent endogenous variables. The latent endogenous variables are symbolized by η's. ζ's are structural disturbances, while ε's represent measurement errors in the endogenous indicators. Covariances are represented by σ's. (Later on (in Section 4.6.2), we shall introduce additional notation for covariances among unobserved variables.)

The model in Figure 4.13 may conveniently be divided into two parts: First, there is the structural submodel:

$$\begin{aligned} \eta_5 &= \gamma_{51} X_1 + \beta_{56} \eta_6 + \zeta_7 \\ \eta_6 &= \gamma_{62} X_2 + \beta_{65} \eta_5 + \zeta_8 \end{aligned} \tag{4.52}$$

We make the usual distributional assumptions about the structural disturbances, and scale the latent variables so that they have zero expectations. Second, there is the measurement submodel:

$$Y_3 = \eta_5 + \varepsilon_9$$

$$Y_4 = \eta_6 + \varepsilon_{10}$$

We assume that the measurement errors, ε_9 and ε_{10}, are "well behaved," that is, each ε has an expectation of zero and is independent of all other variables in

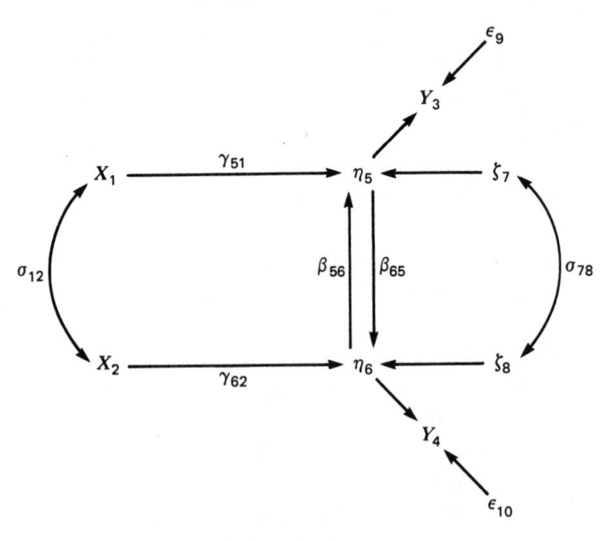

FIGURE 4.13. A structural-equation model in which the endogenous variables are measured with error.

the system, save the indicator with which it is associated. In other words, the measurement errors are "random," not systematic.[15]

One way of approaching latent-variable models is to eliminate the latent variables from the structural equations, substituting for these variables by employing the relations specified in the measurement submodel. This approach is particularly fruitful for our current purpose of determining the consequences of ignoring measurement error in the endogenous variables. We shall work with the first structural equation in (4.52), exploiting the symmetry of the model. Substituting for η_5 and η_6, we get

$$(Y_3 - \varepsilon_9) = \gamma_{51} X_1 + \beta_{56}(Y_4 - \varepsilon_{10}) + \zeta_7$$

which we may rewrite as

$$Y_3 = \gamma_{51} X_1 + \beta_{56} Y_4 + \zeta_7' \tag{4.53}$$

where $\zeta_7' \equiv \zeta_7 + \varepsilon_9 - \beta_{56}\varepsilon_{10}$. In effect, we merge the measurement errors of the endogenous indicators with the structural disturbance. Equation (4.53) may be estimated in the usual manner, because our instrumental variables, X_1 and X_2, are uncorrelated not only with the structural disturbance ζ_7 but also with the measurement errors ε_9 and ε_{10}, and consequently with the composite disturbance ζ_7'.

This result is general: In a nonrecursive model, with no disturbance-covariance restrictions, we may safely ignore random measurement error in the endogenous variables for purposes of estimating structural parameters of the model, so long as the exogenous variables are measured without error.

For a contrasting example, we shall examine the model in Figure 4.14. The structural equations for this model are

$$\begin{aligned} Y_4 &= \gamma_{46}\xi_6 + \gamma_{42} X_2 + \zeta_7 \\ Y_5 &= \gamma_{53} X_3 + \beta_{54} Y_4 + \zeta_8 \end{aligned} \tag{4.54}$$

and the measurement-submodel equation is

$$X_1 = \xi_6 + \delta_9 \tag{4.55}$$

Here, X_1 is a fallible indicator, with random measurement error δ_9, of the latent exogenous variable ξ_6. Proceeding as before, let us substitute for ξ_6 in the first structural equation:

$$\begin{aligned} Y_4 &= \gamma_{46}(X_1 - \delta_9) + \gamma_{42} X_2 + \zeta_7 \\ &= \gamma_{46} X_1 + \gamma_{42} X_2 + \zeta_7' \end{aligned} \tag{4.56}$$

[15] Under certain circumstances, it is possible to estimate models specifying measurement errors that are correlated with each other. The general model developed in Section 4.6.2 permits such a specification.

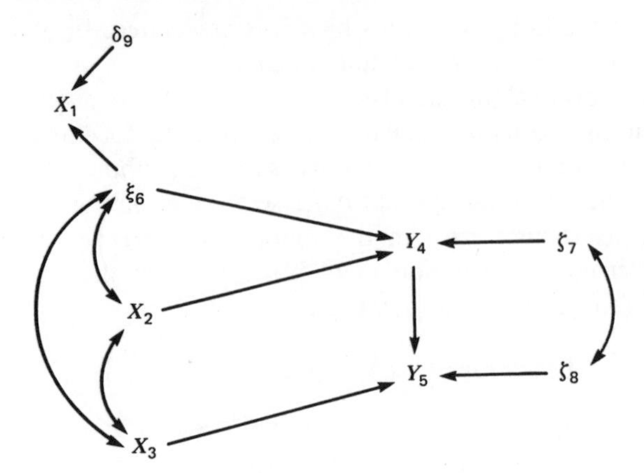

FIGURE 4.14. A structural-equation model in which an exogenous variable is measured with error.

where $\zeta_7' \equiv \zeta_7 - \gamma_{46}\delta_9$.

Multiplying equation (4.56) through by X_1 and X_2, and taking expectations, we get

$$\begin{aligned} \sigma_{14} &= \gamma_{46}\sigma_{11} + \gamma_{42}\sigma_{12} - \gamma_{46}\sigma_{99} \\ \sigma_{24} &= \gamma_{46}\sigma_{12} + \gamma_{42}\sigma_{22} \end{aligned} \tag{4.57}$$

Note that $E(X_1\zeta_7') = -\gamma_{46}\sigma_{99}$ because X_1 is the sum of ξ_6 and δ_9, both of which are uncorrelated with ζ_7, and because $\sigma_{19} = \sigma_{99}$ due to the uncorrelation of δ_9 and ξ_6. Similarly, $E(X_2\zeta_7') = 0$ because X_2 is uncorrelated with both ζ_7 and δ_9. Equations (4.57) may be solved for the structural parameters γ_{46} and γ_{42}; we find it convenient to write the solution in the following manner (after Duncan, 1975: 120):

$$\begin{aligned} \gamma_{46} &= \frac{\sigma_{14}\sigma_{22} - \sigma_{12}\sigma_{24}}{\sigma_{11}\sigma_{22} - \sigma_{12}^2 - \sigma_{99}\sigma_{22}} \\ \gamma_{42} &= \frac{\sigma_{11}\sigma_{24} - \sigma_{12}\sigma_{14}}{\sigma_{11}\sigma_{22} - \sigma_{12}^2} - \frac{\gamma_{46}\sigma_{12}\sigma_{99}}{\sigma_{11}\sigma_{22} - \sigma_{12}^2} \end{aligned} \tag{4.58}$$

Imagine, now, that we make the mistake of treating X_1 as if it were measured without error. In this instance, we would attempt to estimate the first structural equation by OLS, since both independent variables, X_1 and X_2, are "exogenous." In other words, we would apply X_1 and X_2 as instrumental variables, wrongly setting σ_{99} in equations (4.58) to zero. The population analog of the OLS estimator of γ_{46}, then, is given by

$$\gamma_{46}' = \frac{\sigma_{14}\sigma_{22} - \sigma_{12}\sigma_{24}}{\sigma_{11}\sigma_{22} - \sigma_{12}^2}$$

The denominator of γ_{46} is necessarily positive. Since the factor missing from the denominator of γ'_{46} (i.e., $-\sigma_{99}\sigma_{22}$) is necessarily negative, γ'_{46} is biased towards zero. In general, ignoring measurement error in an exogenous independent variable tends to attenuate its coefficient (although the combined effects of measurement errors in several exogenous independent variables are indeterminate for the individual coefficients—see the next paragraph).

The population analog of the OLS estimator of γ_{42} is

$$\gamma'_{42} = \frac{\sigma_{11}\sigma_{24} - \sigma_{12}\sigma_{14}}{\sigma_{11}\sigma_{22} - \sigma_{12}^2}$$

Put alternatively, $\gamma'_{42} = \gamma_{42} + \text{bias}$, where the sign of the bias term depends upon the signs of γ_{46} and σ_{12}. In general, if an exogenous independent variable is measured with error, ignoring that error will have an indeterminate effect on the coefficients of other independent variables in the structural equation.

In the present example, X_1 should not be used as an IV for estimating the first structural equation. Note, however, that X_1 is eligible as an IV for purposes of estimating the second structural equation. Because δ_9, the measurement-error component of X_1, is uncorrelated with all other variables in the system, the covariance of X_1 with any *other* variable is the same as the covariance of ξ_6 with that variable; for example, multiplying equation (4.55) by Y_4 and taking expectations produces

$$\sigma_{14} = \sigma_{46} + \sigma_{49} = \sigma_{46}$$

Although we cannot legitimately estimate the first structural equation in model (4.54) by OLS, this equation is identified because both X_2 and X_3 are available as IVs: Each of these variables is uncorrelated with ζ'_7 in equation (4.56). Note that the first structural equation would ordinarily be overidentified, but the overidentifying restriction has been "consumed" by the measurement error in X_1. It is often the case that strategically placed overidentifying structural restrictions may serve to identify a model with a measurement-error component.

In the current example, it is also possible to estimate the *measurement-error variance*, σ_{99}, along with the *true-score variance*, σ_{66}. Squaring the measurement-submodel equation (4.55), and taking expectations, we obtain

$$\sigma_{11} = \sigma_{66} + \sigma_{99} \tag{4.59}$$

due to the uncorrelation of ξ_6 and δ_9. From equation (4.57) we have

$$\sigma_{99} = \frac{\gamma_{46}\sigma_{11} + \gamma_{42}\sigma_{12} - \sigma_{14}}{\gamma_{46}} \tag{4.60}$$

The σ's on the right-hand side of equation (4.60) may be estimated directly from sample data, and we have already seen that we can obtain IV estimates of the structural parameters γ_{46} and γ_{42}. With an estimate of σ_{99} in hand, we may estimate σ_{66} by subtraction, using equation (4.59). For this model, then, the overidentifying restriction on the first structural equation not only permits us to estimate the structural parameters of the model, but also serves to identify the measurement submodel. Incidentally, notice that in the previous example (Figure 4.13), the variances of the measurement errors cannot be separated from the variances of the structural disturbances, rendering the measurement submodel (but not the structural parameters) underidentified.

The model shown in Figure 4.15 provides us with a third and final example, which incorporates multiple indicators X_1 and X_2 of a latent exogenous variable ξ_6. As before, we assume that the random measurement errors δ_9 and δ_{10} are well behaved—that is, have zero expectations, are uncorrelated with each other, and are uncorrelated with the other variables in the model except the indicators with which they are associated. The model consists of two structural equations,

$$Y_4 = \gamma_{46}\xi_6 + \beta_{45}Y_5 + \zeta_7$$

$$Y_5 = \gamma_{53}X_3 + \beta_{54}Y_4 + \zeta_8$$

and two measurement equations,

$$\begin{aligned} X_1 &= \xi_6 + \delta_9 \\ X_2 &= \lambda\xi_6 + \delta_{10} \end{aligned} \tag{4.61}$$

The coefficient for ξ_6 in the measurement equation for X_1 is implicitly one, while that in the equation for X_2 is an unknown parameter λ. By arbitrarily fixing the coefficient of the latent variable in one of the measurement-submodel equations, we in effect express the latent variable in the metric (i.e., units of measurement) of the corresponding indicator. This choice of scale is an essentially arbitrary normalization rule: We could equally well fix the coeffi-

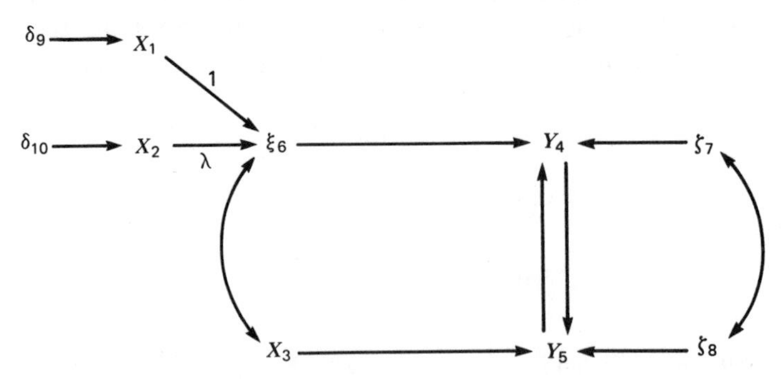

FIGURE 4.15. A structural-equation model with multiple indicators of an exogenous variable.

cient of ξ_6 to one in the measurement equation for X_2, in which case the coefficient for ξ_6 in the equation for X_1 becomes $1/\lambda$. Suppose, for example, that ξ_6 is socio-economic status, X_1 is education measured in years, and X_2 is income measured in dollars; according to equation (4.61), the latent variable ξ_6 is measured in years, and λ converts years to dollars.[16]

We may analyze the present model, as before, by substituting for the unobserved variable in the first structural equation. Now, however, because we have two indicators of ξ_6, we may substitute in two different ways:

$$\begin{aligned} Y_4 &= \gamma_{46}(X_1 - \delta_9) + \beta_{45}Y_5 + \zeta_7 \\ &= \gamma_{46}X_1 + \beta_{45}Y_5 + \zeta_7' \end{aligned} \tag{4.62}$$

where $\zeta_7' \equiv \zeta_7 - \gamma_{46}\delta_9$; and

$$\begin{aligned} Y_4 &= \gamma_{46}\left(\frac{X_2}{\lambda} - \frac{\delta_{10}}{\lambda}\right) + \beta_{45}Y_5 + \zeta_7 \\ &= \frac{\gamma_{46}}{\lambda}X_2 + \beta_{45}Y_5 + \zeta_7'' \end{aligned} \tag{4.63}$$

where $\zeta_7'' \equiv \zeta_7 - (\gamma_{46}/\lambda)\delta_{10}$.

Let us now multiply each of equations (4.62) and (4.63) by X_3 and take expectations, obtaining

$$\sigma_{34} = \gamma_{46}\sigma_{13} + \beta_{45}\sigma_{35}$$

$$\sigma_{34} = \frac{\gamma_{46}}{\lambda}\sigma_{23} + \beta_{45}\sigma_{35}$$

which we may solve for $\lambda = \sigma_{23}/\sigma_{13}$. This solution is not unique, however, for we may obtain alternative expressions for λ by taking expectations of equations (4.62) and (4.63) with Y_4 and Y_5 (rather than X_3); for example, for Y_4,

$$\sigma_{44} = \gamma_{46}\sigma_{14} + \beta_{45}\sigma_{45} + \sigma_{47}$$

$$\sigma_{44} = \frac{\gamma_{46}}{\lambda}\sigma_{24} + \beta_{45}\sigma_{45} + \sigma_{47}$$

which yields $\lambda = \sigma_{24}/\sigma_{14}$. Likewise, applying Y_5 produces $\lambda = \sigma_{25}/\sigma_{15}$. Y_4 and Y_5 serve to obtain expressions for λ because, although they are endogenous, they are uncorrelated with the measurement errors δ_9 and δ_{10}. If the model is correctly specified, then $\sigma_{23}/\sigma_{13} = \sigma_{24}/\sigma_{14} = \sigma_{25}/\sigma_{15}$; in the sample, however, we cannot expect these relations to hold precisely, and thus the parameter λ is overidentified.

[16]Although this example serves to illustrate clearly the arbitrary choice of scale for a latent variable, and though this type of specification for socio-economic status has been employed in applications, it is more sensible in this instance to conceive of the latent variable status as an *effect* of education and income rather than as a *cause* of these observable variables.

With knowledge of λ, we may proceed to determine the parameters in the first structural equation. Applying X_2 to equation (4.62), we get

$$\sigma_{24} = \gamma_{46}\sigma_{12} + \beta_{45}\sigma_{25} \tag{4.64}$$

Even though X_2 is measured with error, it is uncorrelated with δ_9. Similarly, applying X_1 to equation (4.63) produces

$$\sigma_{14} = \frac{\gamma_{46}}{\lambda}\sigma_{12} + \beta_{45}\sigma_{15} \tag{4.65}$$

Because we have already determined λ, equations (4.64) and (4.65) may be solved for the structural parameters γ_{46} and β_{45}. Since alternative estimates of λ are available, and since our estimates of γ_{46} and β_{45} depend upon which value we use, these structural parameters are also overidentified.

The following points concerning this last example are noteworthy: (1) If there were only one fallible indicator of ξ_6, the measurement submodel and the structural submodel would both be underidentified. (2) If ξ_6 were observed directly and measured without error, the structural submodel would be just identified. (3) The presence of two fallible indicators of ξ_6 in the example serves to overidentify both the measurement submodel *and* the structural submodel.

4.6.2. Specification and Estimation of LISREL Models

LISREL, an acronym for *li*near *s*tructural *rel*ations, refers both to a general structural-equation model with latent variables and multiple indicators (Jöreskog, 1973; Jöreskog and Sörbom, 1977), and to a computer program (Jöreskog and Sörbom, 1978, 1981) that provides full-information maximum-likelihood estimates for this model.[17] The highly general nature of the LISREL model permits a variety of specifications; for example, by specifying that all indicators are measured without error, and by establishing a one-to-one correspondence between indicators and latent variables, the LISREL model becomes an ordinary structural-equation model, for which the LISREL program computes the usual FIML estimates.

In this section, we consider the form of the general LISREL model and sketch its estimation; we also examine an illustrative application. Although we should generally establish the identification of a LISREL model prior to estimating its parameters, the identification of models with latent variables is a sufficiently involved topic to warrant separate treatment; we therefore take up this subject in the next section.

[17]The version of LISREL described in this section is LISREL IV (Jöreskog and Sörbom, 1978). The newer LISREL V (Jöreskog and Sörbom, 1981) incorporates a slightly different structural model. We employ the LISREL IV model because its format is more similar to that of the general structural-equation model presented in Section 4.1.

Because the LISREL model is complex, we adopt the notation established by Jöreskog and his colleagues, even though this notation conflicts to a degree with the conventions employed in this text. To do otherwise would be to invite confusion, for the reader will surely have occasion to consult other sources, such as the LISREL program manual. The symbols employed in the LISREL model are summarized in Table 4.11.

The LISREL model may conveniently be divided into two parts: (1) a *structural submodel*, specifying relations among latent variables; and (2) a *measurement submodel*, which links latent variables to their observed indicators. We shall examine each submodel in turn.

The structural submodel has the familiar form of a structural-equation model:

$$\underset{(m\times m)}{\mathbf{B}}\ \underset{(m\times 1)}{\boldsymbol{\eta}_i} = \underset{(m\times n)}{\boldsymbol{\Gamma}}\ \underset{(n\times 1)}{\boldsymbol{\xi}_i} + \underset{(m\times 1)}{\boldsymbol{\zeta}_i}$$

$\boldsymbol{\eta}_i$, $\boldsymbol{\xi}_i$, and $\boldsymbol{\zeta}_i$ are, consecutively, vectors of latent endogenous variables, latent exogenous variables, and structural disturbances, for the ith of N observations. Henceforth, we shall generally suppress the subscript i for observation. These, and indeed all, variables in the LISREL model are expressed as deviations from their expectations. $\mathbf{B}$ and $\boldsymbol{\Gamma}$ are matrices of structural parameters. Note that in the LISREL model, the exogenous variables appear on the right-hand side of the structural equations. The covariance matrix of the exogenous variables is given by $\underset{(n\times n)}{\boldsymbol{\Phi}}$, and the covariances of the structural disturbances by $\underset{(m\times m)}{\boldsymbol{\Psi}}$. We assume that $\mathbf{B}$ is nonsingular, that $\boldsymbol{\xi}_i \sim N_n(\mathbf{0}, \boldsymbol{\Phi})$, that $\boldsymbol{\zeta}_i \sim N_m(\mathbf{0}, \boldsymbol{\Psi})$, that $\boldsymbol{\zeta}_i$ is independent of $\boldsymbol{\xi}_i$, and that the observations are independent.[18]

The reduced form of the LISREL model is

$$\begin{aligned}\boldsymbol{\eta} &= \mathbf{B}^{-1}\boldsymbol{\Gamma}\boldsymbol{\xi} + \mathbf{B}^{-1}\boldsymbol{\zeta}\\ &= \mathbf{D}\boldsymbol{\xi} + \mathbf{B}^{-1}\boldsymbol{\zeta}\end{aligned}$$

where $\underset{(m\times n)}{\mathbf{D}} \equiv \mathbf{B}^{-1}\boldsymbol{\Gamma}$ is the matrix of reduced-form parameters. The covariance matrix $\underset{(m\times m)}{\mathbf{C}}$ of the latent endogenous variables is, therefore,

$$\begin{aligned}\mathbf{C} = E(\boldsymbol{\eta}\boldsymbol{\eta}') &= E\left[(\mathbf{D}\boldsymbol{\xi} + \mathbf{B}^{-1}\boldsymbol{\zeta})(\mathbf{D}\boldsymbol{\xi} + \mathbf{B}^{-1}\boldsymbol{\zeta})'\right]\\ &= \mathbf{D}E(\boldsymbol{\xi}\boldsymbol{\xi}')\mathbf{D}' + \mathbf{D}E(\boldsymbol{\xi}\boldsymbol{\zeta}')(\mathbf{B}^{-1})' + \mathbf{B}^{-1}E(\boldsymbol{\zeta}\boldsymbol{\xi}')\mathbf{D}' + \mathbf{B}^{-1}E(\boldsymbol{\zeta}\boldsymbol{\zeta}')(\mathbf{B}^{-1})'\\ &= \mathbf{D}\boldsymbol{\Phi}\mathbf{D}' + \mathbf{B}^{-1}\boldsymbol{\Psi}(\mathbf{B}^{-1})' \qquad (4.66)\end{aligned}$$

[18]For models in which the exogenous variables are measured without error (see below), we need not make assumptions about their distribution—as was the case for the observed-variable structural-equation models considered earlier in the chapter.

TABLE 4.11. Notation for the LISREL Model

Symbol	Meaning
N	Number of observations
m	Number of latent endogenous variables
n	Number of latent exogenous variables
p	Number of indicators of latent endogenous variables
q	Number of indicators of latent exogenous variables
$\underset{(m\times 1)}{\boldsymbol{\eta}}$	Vector of latent endogenous variables
$\underset{(n\times 1)}{\boldsymbol{\xi}}$	Vector of latent exogenous variables
$\underset{(m\times 1)}{\boldsymbol{\zeta}}$	Vector of structural disturbances (errors in equations)
$\underset{(m\times m)}{\mathbf{B}}$	Structural parameter matrix relating latent endogenous variables
$\underset{(m\times n)}{\boldsymbol{\Gamma}}$	Structural parameter matrix relating latent endogenous to latent exogenous variables
$\underset{(m\times n)}{\mathbf{D}}$	Reduced-form coefficient matrix
$\underset{(p\times 1)}{\mathbf{y}}$	Vector of indicators of latent endogenous variables
$\underset{(q\times 1)}{\mathbf{x}}$	Vector of indicators of latent exogenous variables
$\underset{(p\times 1)}{\boldsymbol{\varepsilon}}$	Vector of measurement errors in endogenous indicators
$\underset{(q\times 1)}{\boldsymbol{\delta}}$	Vector of measurement errors in exogenous indicators
$\underset{(p\times m)}{\boldsymbol{\Lambda}_y}$, $\underset{(q\times n)}{\boldsymbol{\Lambda}_x}$	Coefficient matrices relating indicators to latent variables
$\underset{(n\times n)}{\boldsymbol{\Phi}}$	Matrix of covariances among latent exogenous variables
$\underset{(m\times m)}{\boldsymbol{\Psi}}$	Matrix of covariances among structural disturbances
$\underset{(p\times p)}{\boldsymbol{\Theta}_\varepsilon}$, $\underset{(q\times q)}{\boldsymbol{\Theta}_\delta}$	Matrices of covariances among measurement errors
$\underset{(p+q\times p+q)}{\boldsymbol{\Sigma}}$	Matrix of covariances among observed (indicator) variables
$\underset{(m\times m)}{\mathbf{C}}$	Matrix of covariances among latent endogenous variables

Because the metric of the latent variables is frequently not substantively meaningful, the researcher may wish to standardize the latent variables to unit variance. Let us define diagonal matrices of standard deviations

$$\underset{(m\times m)}{\mathbf{A}_\eta} \equiv (\operatorname{diag} \mathbf{C})^{1/2}$$

$$\underset{(n\times n)}{\mathbf{A}_\xi} \equiv (\operatorname{diag} \mathbf{\Phi})^{1/2}$$

Then $\boldsymbol{\eta}^* \equiv \mathbf{A}_\eta^{-1}\boldsymbol{\eta}$ and $\boldsymbol{\xi}^* \equiv \mathbf{A}_\xi^{-1}\boldsymbol{\xi}$ are vectors of standardized latent variables. Each of the coefficient and covariance matrices previously defined may be simply adjusted to provide a standardized solution; for example

$$\mathbf{B}^* = \mathbf{A}_\eta^{-1}\mathbf{B}\mathbf{A}_\eta$$

$$\mathbf{\Gamma}^* = \mathbf{A}_\eta^{-1}\mathbf{\Gamma}\mathbf{A}_\xi$$

$$\mathbf{\Psi}^* = \mathbf{A}_\eta^{-1}\mathbf{\Psi}\mathbf{A}_\eta^{-1}$$

The LISREL measurement submodel consists of two matrix equations:

$$\underset{(p\times 1)}{\mathbf{y}_i} = \underset{(p\times m)}{\mathbf{\Lambda}_y}\ \underset{(m\times 1)}{\boldsymbol{\eta}_i} + \underset{(p\times 1)}{\boldsymbol{\varepsilon}_i}$$

$$\underset{(q\times 1)}{\mathbf{x}_i} = \underset{(q\times n)}{\mathbf{\Lambda}_x}\ \underset{(n\times 1)}{\boldsymbol{\xi}_i} + \underset{(q\times 1)}{\boldsymbol{\delta}_i}$$

Here, $\mathbf{y}_i$ and $\mathbf{x}_i$ are vectors of indicators of the latent endogenous and exogenous variables, respectively; $\boldsymbol{\varepsilon}_i$ and $\boldsymbol{\delta}_i$ are vectors of measurement-error variables, one for each indicator; and $\mathbf{\Lambda}_y$ and $\mathbf{\Lambda}_x$ are matrices of regression coefficients relating the indicators to the latent variables. In general, each column of $\mathbf{\Lambda}_y$ and $\mathbf{\Lambda}_x$ contains one unit entry, to fix the metric of the corresponding latent variable, as explained in the previous section [see equation (4.61)]. (Alternatively, the variance of a latent exogenous variable may be fixed, as may the variance of the disturbance associated with a latent endogenous variable.) It is, moreover, frequently the case that each *row* of $\mathbf{\Lambda}_y$ and $\mathbf{\Lambda}_x$ has but one nonzero entry: Each observed variable is an indicator of just one latent variable. In certain cases, however, it may make sense to treat an observed variable as the effect of more than one latent variable, and LISREL accommodates such a specification.

The covariances of the measurement errors appear in $\underset{(p\times p)}{\mathbf{\Theta}_\varepsilon}$ and $\underset{(q\times q)}{\mathbf{\Theta}_\delta}$. These matrices are not necessarily diagonal; that is, measurement errors may be correlated. Unless they are specified carefully and frugally, however, correlated measurement errors are likely to underidentify a model. The measurement errors $\boldsymbol{\varepsilon}_i$ and $\boldsymbol{\delta}_i$ are assumed to have zero expectations; to be independent of each other and of $\boldsymbol{\eta}_i$, $\boldsymbol{\xi}_i$, and $\boldsymbol{\zeta}_i$; and to be multivariately normally distributed: $\boldsymbol{\varepsilon}_i \sim N_p(\mathbf{0}, \boldsymbol{\Theta}_\varepsilon)$, $\boldsymbol{\delta}_i \sim N_q(\mathbf{0}, \boldsymbol{\Theta}_\delta)$.

An illustrative LISREL model (from Jöreskog and Sörbom, 1978) for the Duncan, Haller, and Portes peer-influences data is shown in Figure 4.16. In this model, there are multiple indicators for the latent endogenous variables, but the exogenous variables are assumed to be measured without error. We have, then, the following LISREL specifications for the exogenous indicators: $\underset{(6\times1)}{\mathbf{x}} = \underset{(6\times1)}{\boldsymbol{\xi}}$ (that is, $\boldsymbol{\Lambda}_x = \mathbf{I}_6$, $\boldsymbol{\delta} = \underset{(6\times1)}{\mathbf{0}}$); and consequently $\underset{(6\times6)}{\boldsymbol{\Phi}} = \boldsymbol{\Sigma}_{xx}$, $\boldsymbol{\Theta}_\delta = \underset{(6\times6)}{\mathbf{0}}$ (where $\boldsymbol{\Sigma}_{xx}$ is the covariance matrix of the exogenous variables). For the endogenous indicators,

$$\begin{pmatrix} Y_1 \\ Y_2 \\ Y_3 \\ Y_4 \end{pmatrix} = \begin{pmatrix} 1 & 0 \\ \lambda_{21}^y & 0 \\ 0 & 1 \\ 0 & \lambda_{42}^y \end{pmatrix} \begin{pmatrix} \eta_1 \\ \eta_2 \end{pmatrix} + \begin{pmatrix} \varepsilon_1 \\ \varepsilon_2 \\ \varepsilon_3 \\ \varepsilon_4 \end{pmatrix}$$

$$\underset{(4\times4)}{\boldsymbol{\Theta}_\varepsilon} = \text{diag}(\theta_{11}^\varepsilon, \theta_{22}^\varepsilon, \theta_{33}^\varepsilon, \theta_{44}^\varepsilon)$$

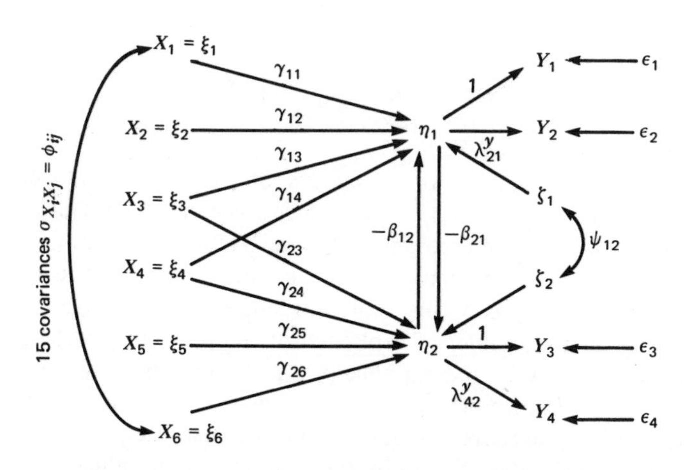

FIGURE 4.16. A LISREL model with latent endogenous variables for the Duncan, Haller, and Portes peer-influences data. X_1, respondent's parental aspiration; X_2, respondent's family SES; X_3, respondent's intelligence; X_4, best friend's intelligence; X_5, best friend's family SES; X_6, best friend's parental aspiration; Y_1, respondent's occupational aspiration; Y_2, respondent's educational aspiration; Y_3, best friend's occupational aspiration; Y_4, best friend's educational aspiration; η_1, respondent's general level of aspiration; η_2, best friend's general level of aspiration. (Source: Adapted with permission from Karl G. Jöreskog and Dag Sörbom, 1978, copyright National Educational Resources, Inc., 1978, 1981; and Duncan, Haller, and Portes, 1968, see Table 4.2.)

$\boldsymbol{\Theta}_\varepsilon$ is diagonal because the measurement errors in the endogenous indicators are specified to be uncorrelated. The structural submodel is given by

$$\begin{pmatrix} 1 & \beta_{12} \\ \beta_{21} & 1 \end{pmatrix}\begin{pmatrix} \eta_1 \\ \eta_2 \end{pmatrix} = \begin{pmatrix} \gamma_{11} & \gamma_{12} & \gamma_{13} & \gamma_{14} & 0 & 0 \\ 0 & 0 & \gamma_{23} & \gamma_{24} & \gamma_{25} & \gamma_{26} \end{pmatrix}\begin{pmatrix} \xi_1 \\ \xi_2 \\ \xi_3 \\ \xi_4 \\ \xi_5 \\ \xi_6 \end{pmatrix} + \begin{pmatrix} \zeta_1 \\ \zeta_2 \end{pmatrix}$$

From the LISREL structural and measurement submodels we may derive expressions for the covariances of the indicators. It is in this manner that a link is established between the parameters of the model and observable quantities. Let $\boldsymbol{\Sigma}$ represent the covariance matrix for the indicators,

$$\underset{(p+q\times p+q)}{\boldsymbol{\Sigma}} = \begin{pmatrix} \underset{(p\times p)}{\boldsymbol{\Sigma}_{yy}} & \underset{(p\times q)}{\boldsymbol{\Sigma}_{yx}} \\ \underset{(q\times p)}{\boldsymbol{\Sigma}_{xy}} & \underset{(q\times q)}{\boldsymbol{\Sigma}_{xx}} \end{pmatrix}$$

We have, for example,

$$\begin{aligned} \boldsymbol{\Sigma}_{yy} &= E(\mathbf{y}\mathbf{y}') = E\left[(\boldsymbol{\Lambda}_y\boldsymbol{\eta} + \boldsymbol{\varepsilon})(\boldsymbol{\Lambda}_y\boldsymbol{\eta} + \boldsymbol{\varepsilon})'\right] \\ &= \boldsymbol{\Lambda}_y E(\boldsymbol{\eta}\boldsymbol{\eta}')\boldsymbol{\Lambda}'_y + \boldsymbol{\Lambda}_y E(\boldsymbol{\eta}\boldsymbol{\varepsilon}') + E(\boldsymbol{\varepsilon}\boldsymbol{\eta}')\boldsymbol{\Lambda}'_y + E(\boldsymbol{\varepsilon}\boldsymbol{\varepsilon}') \\ &= \boldsymbol{\Lambda}_y \mathbf{C} \boldsymbol{\Lambda}'_y + \boldsymbol{\Theta}_\varepsilon \end{aligned}$$

Then, using equation (4.66):

$$\boldsymbol{\Sigma}_{yy} = \boldsymbol{\Lambda}_y\left[\mathbf{B}^{-1}\boldsymbol{\Gamma}\boldsymbol{\Phi}\boldsymbol{\Gamma}'(\mathbf{B}^{-1})' + \mathbf{B}^{-1}\boldsymbol{\Psi}(\mathbf{B}^{-1})'\right]\boldsymbol{\Lambda}'_y + \boldsymbol{\Theta}_\varepsilon \tag{4.67}$$

We may determine the other components of $\boldsymbol{\Sigma}$ in a similar manner, obtaining (see Problem 4.19)

$$\begin{aligned} \boldsymbol{\Sigma}_{yx} &= \boldsymbol{\Lambda}_y \mathbf{B}^{-1}\boldsymbol{\Gamma}\boldsymbol{\Phi}\boldsymbol{\Lambda}'_x \\ \boldsymbol{\Sigma}_{xy} &= \boldsymbol{\Sigma}'_{yx} = \boldsymbol{\Lambda}_x\boldsymbol{\Phi}\boldsymbol{\Gamma}'(\mathbf{B}^{-1})'\boldsymbol{\Lambda}'_y \\ \boldsymbol{\Sigma}_{xx} &= \boldsymbol{\Lambda}_x\boldsymbol{\Phi}\boldsymbol{\Lambda}'_x + \boldsymbol{\Theta}_\delta \end{aligned} \tag{4.68}$$

The covariances of the observed variables, therefore, are functions (albeit complex functions) of the parameters of the LISREL structural and measurement model. In addition, the assumptions of the model insure that the observed variables follow a multivariate normal distribution:

$$\begin{pmatrix} \mathbf{y} \\ \mathbf{x} \end{pmatrix} \sim N_{p+q}(\mathbf{0}, \mathbf{\Sigma})$$

Thus, the joint probability density for the observed variables in a sample of size N is given by

$$p(\mathbf{y}_1, \mathbf{x}_1; \ldots; \mathbf{y}_N, \mathbf{x}_N) = \frac{1}{(2\pi)^{N(p+q)/2} |\mathbf{\Sigma}|^{N/2}} \exp\left(-\frac{1}{2} \sum_{i=1}^{N} \begin{pmatrix} \mathbf{y}_i \\ \mathbf{x}_i \end{pmatrix}' \mathbf{\Sigma}^{-1} \begin{pmatrix} \mathbf{y}_i \\ \mathbf{x}_i \end{pmatrix} \right)$$

The logarithm of the likelihood function may be written (Problem 4.19)

$$\log L(\mathbf{B}, \mathbf{\Gamma}, \mathbf{\Phi}, \mathbf{\Psi}, \mathbf{\Theta}_\varepsilon, \mathbf{\Theta}_\delta, \mathbf{\Lambda}_y, \mathbf{\Lambda}_x)$$

$$= -\frac{N(p+q)}{2} \log(2\pi) - \frac{N}{2}\left[\log|\mathbf{\Sigma}| + \operatorname{trace}(\mathbf{S}\mathbf{\Sigma}^{-1})\right] \qquad (4.69)$$

where $\underset{(p+q \times p+q)}{\mathbf{S}} = (1/N)[\mathbf{Y}, \mathbf{X}]'[\mathbf{Y}, \mathbf{X}]$ is the sample covariance matrix for the indicators. Recall that $\mathbf{\Sigma}$ is a function of the model parameters (equations (4.67) and (4.68)). Equation (4.69) tells us, in essence, that the likelihood is large when the covariance matrix $\hat{\mathbf{\Sigma}}$ implied by the model is similar to the observed sample covariance matrix $\mathbf{S}$.

In estimating the model, it is necessary to take account of the prior constraints on the parameters. Some of these constraints are normalizations: Certain parameters are prespecified to be one. Other constraints are exclusions: Certain parameters are prespecified to be zero. (The LISREL computer program also permits equality constraints, where two or more parameters are prespecified to be equal to each other.) The maximum-likelihood estimators of $\mathbf{B}$, $\mathbf{\Gamma}$, and the other parameters maximize L subject to these prior constraints. As in the case of FIML estimation of observed-variable models, the log likelihood (4.69) for the LISREL model must be maximized numerically. Estimated asymptotic standard errors for estimators of all "free" (i.e., unconstrained) parameters may be obtained in the usual manner from the inverse of the information matrix, which the LISREL program computes.

The example model shown in Figure 4.16 was estimated for standardized indicators. Correlations among these indicators were given earlier in Table 4.2. The standardized maximum-likelihood solution appears in equations (4.70). (Note that although λ_{11}^y and λ_{32}^y originally were fixed to one, these values change along with the other entries of $\mathbf{\Lambda}_y$ when the model is standardized.)

Standard errors for free-parameter estimates are shown in parentheses beneath the estimates.

$$\hat{\mathbf{\Lambda}}_y^* = \begin{pmatrix} 0.7667 & 0 \\ \underset{(0.0691)}{0.8148} & 0 \\ 0 & 0.8299 \\ 0 & \underset{(0.0583)}{0.7716} \end{pmatrix}$$

$$\hat{\mathbf{\Theta}}_\varepsilon^* = \operatorname{diag}\begin{pmatrix} \underset{(0.0512)}{0.4121} & \underset{(0.0521)}{0.3361} & \underset{(0.0459)}{0.3112} & \underset{(0.0462)}{0.4046} \end{pmatrix}$$

$$\hat{\mathbf{B}}^* = \begin{pmatrix} 1 & \underset{(0.1027)}{-0.1994} \\ \underset{(0.1103)}{-0.2176} & 1 \end{pmatrix} \tag{4.70}$$

$$\hat{\mathbf{\Gamma}}^* = \begin{pmatrix} \underset{(0.0506)}{0.2103} & \underset{(0.0574)}{0.3256} & \underset{(0.0576)}{0.2848} & \underset{(0.0648)}{0.0937} & 0 & 0 \\ 0 & 0 & \underset{(0.0624)}{0.0746} & \underset{(0.0532)}{0.2758} & \underset{(0.0546)}{0.4205} & \underset{(0.0468)}{0.1922} \end{pmatrix}$$

$$\hat{\mathbf{\Psi}}^* = \begin{pmatrix} \underset{(0.0786)}{0.4780} & \\ \underset{(0.0804)}{-0.0355} & \underset{(0.0648)}{0.3830} \end{pmatrix}$$

These results seem to be substantively acceptable: All of the structural and measurement parameter estimates are of the anticipated sign, and all but $\hat{\gamma}_{14}$ and $\hat{\gamma}_{23}$ are statistically significant (at or beyond approximately the 2.5 percent level, one-tail). Since γ_{14} and γ_{23} represent the effects of the other boy's SES on each boy's aspirations, it is reasonable that these coefficients be small; indeed, in an earlier model for the peer-influences data (Figure 4.1), we set these parameters to zero *a priori*. When we estimated that earlier model, we found a large negative correlation between the structural disturbances. Notice that for the present LISREL model, the covariance between the disturbances, $\hat{\psi}_{12}$, is reassuringly close to zero, although a positive value would be even more reasonable. Apparently, the introduction of additional exogenous variables (parental aspirations, X_1 and X_6), and the recognition of measurement error in the indicators of the endogenous variables (made possible by multiple indicators) has had a salutary effect on our estimates (see Gillespie and Fox, 1980).

4.6.3. Identification of Models With Latent Variables

The identification of structural-equation models with latent variables is a complex problem, and one that does not yet have a straightforward, general solution. Some progress, however, has been made. Geraci (1977), for example, discusses general conditions for identification of single-indicator models in which (some) exogenous variables are measured with error. Further discussion of the identification problem for latent-variable models may be found in Wiley (1973).

It is always possible to demonstrate the identification of a model, if indeed it is identified, by showing that each of its parameters may be expressed in at least one way as a function of covariances of observed variables. To do this, we may use methods such as those employed in Section 4.6.1 and later in the present section. This process is a tedious one, and if we fail to demonstrate the identifiability of a model, we often cannot be certain that it is not our imagination that has failed rather than the model.

There are, however, necessary conditions for identification, which may show us that a model is underidentified; and there are sufficient conditions which insure that a model is identified. What is missing is a condition, analogous to the rank condition for nonrecursive observed-variable models, that is both necessary and sufficient. Through the application of known necessary conditions and sufficient conditions, we hope to avoid having to identify a model on an individual basis.

In practice, we may proceed to attempt to estimate a model without having established its identification. An underidentified model produces a singular information matrix, because the infinity of solutions implied by underidentification is reflected in a flat likelihood function at the maximum.[19] The LISREL computer program calculates the information matrix, and therefore is generally able to detect an attempt to estimate an underidentified model.

A global necessary condition for identification is that there be no more free parameters to estimate than there are unique covariances among the observed variables in the system. Put another way, the number of unknowns in the estimating equations cannot exceed the number of known quantities. There are $(p + q + 1)(p + q)/2$ unique observable variances and covariances, a number that is greatly exceeded by the number of entries in $\mathbf{B}$, $\mathbf{\Gamma}$, $\mathbf{\Phi}$, $\mathbf{\Psi}$, $\mathbf{\Theta}_\varepsilon$, $\mathbf{\Theta}_\delta$, $\mathbf{\Lambda}_x$, and $\mathbf{\Lambda}_y$, even after normalizations are taken into account. Many prior restrictions, therefore, are needed to identify a LISREL model. Some of these restrictions may derive trivially from the particular application, as when a variable is specified to be measured without error.

The model in Figure 4.16, for example, has 40 free parameters: λ^y_{21}, λ^y_{42}, β_{12}, β_{21}, eight γ's, ψ_{11}, ψ_{12}, ψ_{22}, four θ^ε's, and $6(7)/2 = 21$ ϕ's (which necessarily equal the corresponding covariances among observed exogenous variables). There are $(4 + 6 + 1)(4 + 6)/2 = 55$ unique entries in $\mathbf{\Sigma}$; if the model is identified, therefore (and it is), there are 15 overidentifying restrictions. This counting rule does not guarantee the identification of a model: Although restrictions are sufficiently numerous, they may be injudiciously located.

[19]In this respect, underidentification is analogous to perfect collinearity, which also leads to underdetermined estimating equations.

It is often possible to establish the identification of a model by treating the measurement and structural submodels separately. We then seek to show (1) that the restrictions on the measurement submodel are sufficient to identify all covariances among latent variables, and (2) that, given these covariances, the structural submodel is identified. For a nonrecursive structure, the conditional identification of the structural submodel may be assessed by the rank condition. One useful rule of thumb (i.e., sufficient condition) is that the measurement submodel is identified if (but not only if): (1) there are at least two unique indicators for each latent variable, or if there is just one indicator, it is measured without error; and (2) measurement errors are uncorrelated. It must be stressed that the separation of the measurement and structural submodels, and hence the scope of this rule of thumb, are restrictive: As we saw in the previous section, measurement and structural submodels may contribute to each other's identification; likewise, models with single fallible indicators or with correlated measurement errors may frequently be identified.

We demonstrate first that the variances and covariance of two latent variables are identified if each has two fallible indicators, and if measurement errors are independent. We refer to the model diagrammed in Figure 4.17. Here, there are 10 observable covariances among the four indicators. There are nine unknown parameters: λ^x_{21}, λ^x_{42}, ϕ_{11}, ϕ_{12}, ϕ_{22}, θ^δ_{11}, θ^δ_{22}, θ^δ_{33}, and θ^δ_{44}. To show that the parameters may be expressed in terms of observed covariances, we expand these covariances by the expectation method:

$$(a)\quad \sigma_{11} = \phi_{11} + \theta^\delta_{11}$$

$$(b)\quad \sigma_{22} = \phi_{11}\lambda^x_{21} + \theta^\delta_{22}$$

$$(c)\quad \sigma_{33} = \phi_{22} + \theta^\delta_{33}$$

$$(d)\quad \sigma_{44} = \phi_{22}\lambda^x_{42} + \theta^\delta_{44}$$

$$(e)\quad \sigma_{12} = \phi_{11}\lambda^x_{21}$$

$$(f)\quad \sigma_{34} = \phi_{22}\lambda^x_{42}$$

$$(g)\quad \sigma_{13} = \phi_{11}\phi_{12}\phi_{22}$$

$$(h)\quad \sigma_{14} = \phi_{11}\phi_{12}\phi_{22}\lambda^x_{42}$$

$$(i)\quad \sigma_{23} = \lambda^x_{21}\phi_{11}\phi_{12}\phi_{22}$$

$$(j)\quad \sigma_{24} = \lambda^x_{21}\phi_{11}\phi_{12}\phi_{22}\lambda^x_{42}$$

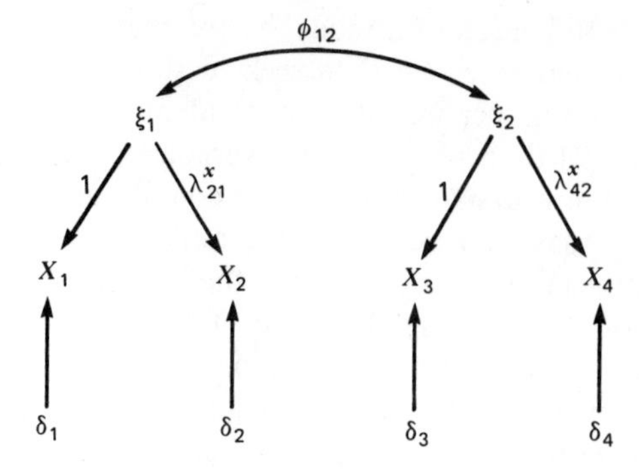

FIGURE 4.17. Two indicators for each of two latent variables.

Solving first for the λ's, we get

$$\lambda_{21}^x = \frac{(i)}{(g)} = \frac{\sigma_{23}}{\sigma_{13}} = \frac{(j)}{(h)} = \frac{\sigma_{24}}{\sigma_{14}}$$

$$\lambda_{42}^x = \frac{(h)}{(g)} = \frac{\sigma_{14}}{\sigma_{13}} = \frac{(j)}{(i)} = \frac{\sigma_{24}}{\sigma_{23}}$$

showing that these parameters are overidentified. With knowledge of the λ's, we derive expressions for the other parameters by successive substitution:

$$(e') \quad \phi_{11} = \frac{\sigma_{12}}{\lambda_{21}^x}$$

$$(f') \quad \phi_{22} = \frac{\sigma_{34}}{\lambda_{42}^x}$$

$$(g') \quad \phi_{12} = \frac{\sigma_{13}}{\phi_{11}\phi_{22}}$$

$$(a') \quad \theta_{11}^{\delta} = \sigma_{11} - \phi_{11}$$

$$(b') \quad \theta_{22}^{\delta} = \sigma_{22} - \phi_{11}\lambda_{21}^x$$

$$(c') \quad \theta_{33}^{\delta} = \sigma_{33} - \phi_{22}$$

$$(d') \quad \theta_{44}^{\delta} = \sigma_{44} - \phi_{22}\lambda_{42}^x$$

Next, let us analyze the model shown in Figure 4.18, where one latent variable is measured without error and the other has two indicators with uncorrelated measurement errors. There are $3(4)/2 = 6$ observed covariances and an equal number of free parameters: λ^x_{32}, ϕ_{11}, ϕ_{12}, ϕ_{22}, θ^δ_{22}, and θ^δ_{33}. Expressing the observed covariances as functions of the parameters, we get:

$$(a)\quad \sigma_{11} = \phi_{11}$$

$$(b)\quad \sigma_{22} = \phi_{22} + \theta^\delta_{22}$$

$$(c)\quad \sigma_{33} = \lambda^x_{32}\phi_{22} + \theta^\delta_{33}$$

$$(d)\quad \sigma_{12} = \phi_{11}\phi_{12}\phi_{22}$$

$$(e)\quad \sigma_{13} = \phi_{11}\phi_{12}\phi_{22}\lambda^x_{32}$$

$$(f)\quad \sigma_{23} = \phi_{22}\lambda^x_{32}$$

Solving for the parameters, which are just identified, completes the demonstration:

$$\lambda_{32} = \frac{(e)}{(d)} = \frac{\sigma_{13}}{\sigma_{12}}$$

$$(a')\quad \phi_{11} = \sigma_{11}$$

$$(f')\quad \phi_{22} = \frac{\sigma_{23}}{\lambda^x_{32}}$$

$$(d')\quad \phi_{12} = \frac{\sigma_{12}}{\phi_{11}\phi_{22}}$$

$$(b')\quad \theta^\delta_{22} = \sigma_{22} - \phi_{22}$$

$$(c')\quad \theta^\delta_{33} = \sigma_{33} - \lambda^x_{32}\phi_{22}$$

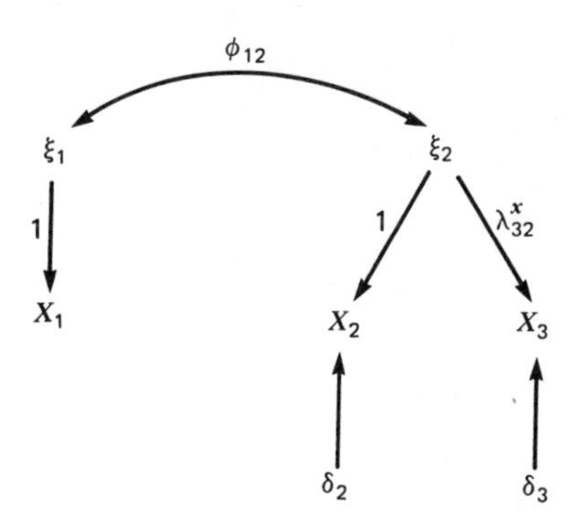

FIGURE 4.18. Two indicators for one latent variable and another variable measured without error.

PROBLEMS

4.17. The model in Figure 4.19 is adapted from work by Bielby, Hauser, and Featherman (1977) on response errors in models of the stratification process. Alternative measures collected on different occasions were available for each of the latent variables in the model:

ξ_1 Father's occupational status

ξ_2 Father's education

ξ_3 Parents' income

η_1 Respondent's education

η_2 Occupational status of the respondent's first job

η_3 Current occupational status

(a) Comment on the specification of the model, paying attention to the measurement-model assumptions as well as to the structural model.

(b) Show that the model is identified.

(c) Using the correlations for a sample of 578 nonblack males in the civilian labor force, given in Table 4.12, estimate the parameters of the model. Is it important to take measurement error into account in estimating this model?

4.18. The data in Table 4.13 were compiled by Inverarity (1976) as part of a study of the relationship between "mechanical solidarity" and "repres-

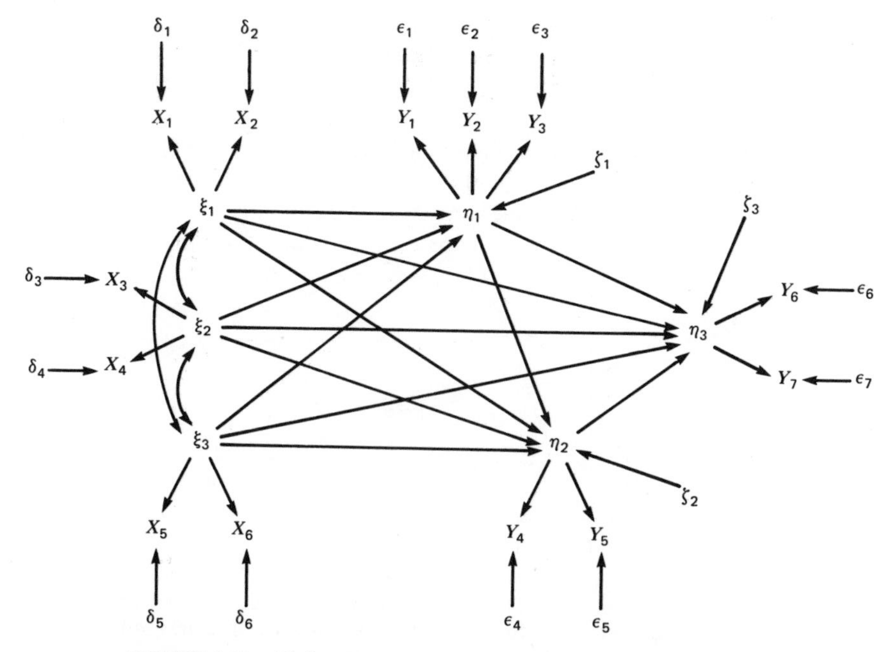

FIGURE 4.19. Bielby, Hauser, and Featherman's stratification model.

sive justice." Both concepts, and the hypothesized link between them, are due to the 19th-century French sociologist, Emile Durkheim, who distinguished between two forms of social solidarity: mechanical solidarity based on social similarity, and organic solidarity based on a detailed social division of labor. Inverarity examines the incidence of lynching in the period 1889–1896 for 59 Louisiana parishes, relating this measure of repressive justice to characteristics of the parishes. He argues that proportion black (X_1) and religious homogeneity (X_3) should exert a positive influence on mechanical solidarity among whites, and that urbanization (X_2, coded one for parishes with some urban population, and zero otherwise) should have a negative impact. Inverarity predicts that the number of lynchings in a parish (Y_3) is directly related to the level of mechanical solidarity (η_1) and to the size of the black population (X_4, in thousands) in the parish. The proportion of the Democratic Party vote in the 1892 presidential election (Y_1) and in the 1896 gubernatorial election (Y_2) are taken as indicators of mechanical solidarity. The model in Figure 4.20 was specified by Bagozzi (1977) for Inverarity's data. (Also see Wasserman, 1977; Pope and Ragin, 1977; Bohrnstedt, 1977; and Inverarity, 1977).

TABLE 4.12. Bielby, Hauser, and Featherman's Stratification Data

	X_1	X_2	X_3	X_4	X_5	X_6	Y_1	Y_2
X_1	1.000							
X_2	0.869	1.000						
X_3	0.585	0.589	1.000					
X_4	0.597	0.599	0.939	1.000				
X_5	0.422	0.437	0.477	0.467	1.000			
X_6	0.426	0.450	0.436	0.478	0.913	1.000		
Y_1	0.428	0.430	0.448	0.445	0.426	0.439	1.000	
Y_2	0.445	0.443	0.483	0.492	0.485	0.502	0.838	1.000
Y_3	0.419	0.419	0.467	0.467	0.486	0.501	0.801	0.921
Y_4	0.398	0.410	0.290	0.300	0.370	0.358	0.581	0.644
Y_5	0.409	0.409	0.325	0.322	0.363	0.348	0.578	0.642
Y_6	0.340	0.369	0.280	0.284	0.291	0.296	0.504	0.563
Y_7	0.364	0.390	0.291	0.308	0.307	0.301	0.519	0.603

	Y_3	Y_4	Y_5	Y_6	Y_7
Y_3	1.000				
Y_4	0.637	1.000			
Y_5	0.631	0.847	1.000		
Y_6	0.534	0.585	0.599	1.000	
Y_7	0.566	0.618	0.620	0.797	1.000

Source: Bielby, Hauser, and Featherman, in Aigner and Goldberger, eds., *Latent Variables in Socio-economic Models*, © Elsevier North-Holland, Inc., 1977. Reprinted with permission from Bielby, Hauser, and Featherman (1977: Table 2).

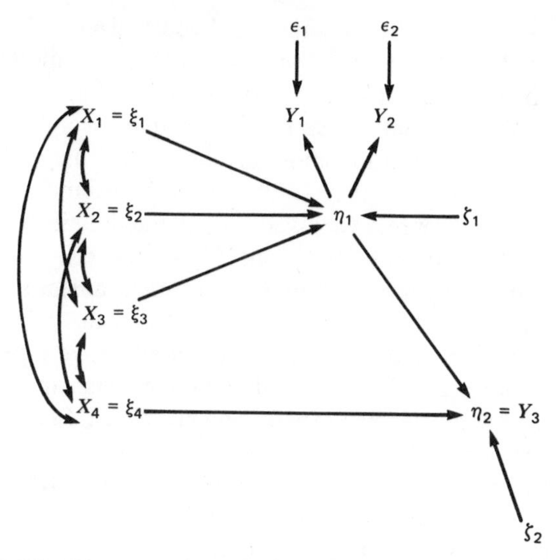

FIGURE 4.20. Bagozzi's model for Inverarity's data on lynchings.

(a) Comment on the specification of Bagozzi's model. Does it adequately capture Inverarity's argument?

(b) Show that the model is identified. What happens to the identification status of the model if Y_3 is specified to be measured with error (that is, if Y_3 is taken as an imperfect indicator of repressive justice)?

(c) Estimate the model using the covariances in Table 4.13.

4.19. (a) Derive equations (4.68) for Σ_{yx}, Σ_{xy}, and Σ_{xx} from the LISREL model.

(b) Derive the log likelihood for the LISREL model [equation (4.69)].

TABLE 4.13. Covariances for Inverarity's Lynching Data

	X_1	X_2	X_3	X_4	Y_1	Y_2	Y_3
X_1	0.04537						
X_2	−0.01029	0.19272					
X_3	0.00403	−0.01566	0.00672				
X_4	0.53791	1.94136	−0.20256	121.61678			
Y_1	0.01449	0.00903	0.00424	0.25706	0.04410		
Y_2	0.03166	−0.00583	0.00334	0.58084	0.02385	0.05244	
Y_3	0.09991	0.11866	−0.00183	7.32943	0.08448	0.07756	2.52810
Mean	0.508	0.254	0.902	11.031	0.766	0.589	1.41

Source: Adapted with permission from Inverarity (1976: Table 2).

4.7. EVALUATION OF STRUCTURAL-EQUATION MODELS

As is general with statistical models, having fit a structural-equation model to data, it is desirable to determine, to the extent possible, whether the model adequately represents the data. Certain checks on the adequacy of the model are implicit in the process of interpretation and testing, as when we ask whether an estimated coefficient assumes a reasonable value, or is statistically distinguishable from zero.

Some procedures for analysis of residuals, presented in Section 3.2 for single-equation linear models, may be extended to structural-equation models (see Belsley, Kuh, and Welsch, 1980: 266–269). For recursive models, the application of these procedures is straightforward, since a recursive model is simply a collection of related regression equations, each estimated by OLS. For nonrecursive models, estimated say by 2SLS, endogenous independent variables are in general correlated with structural residuals. In examining residual plots for nonlinearity, therefore, we must discount visually whatever linear relation is present. An alternative is to use second-stage regression residuals for certain residual analyses, for example, in the detection of outliers.

Two topics related to the question of model quality are dealt with at greater length in this section. First, we develop measures of fit for structural-equation models; and second, we assess the adequacy of overidentifying restrictions.

We should not conclude, however, that all assumptions underlying a structural-equation model may be examined in light of the data. As pointed out in Section 4.5, we cannot in general expect data to mediate issues of causal priority. Nor can we generally assess the crucial assumption of independence between exogenous variables and disturbances: Indeed, in a just-identified model, exogenous variables and structural residuals are perfectly uncorrelated in the sample, much as the independent variables are uncorrelated with the residuals in OLS regression. In an overidentified equation, however, structural residuals may have nonzero correlations with excluded exogenous variables. If these correlations are substantial, we should question the adequacy of the overidentifying restrictions, a topic pursued in Section 4.7.2.

4.7.1. Indices of Fit for Structural-Equation Models

For endogenous variables measured in a meaningful metric, the estimated standard error, S_{E_j}, provides an interpretable measure of fit, as in linear regression analysis. Since recursive models are fit by OLS, there is a multiple correlation coefficient for each structural equation; these R^2 values may be interpreted in the usual manner.

Although several R^2 analogs have been proposed for structural equations in nonrecursive models, these statistics do not have certain properties that we associate with the multiple correlation coefficient. In OLS regression

$$R^2 = \frac{\text{TSS} - \text{RSS}}{\text{TSS}} = 1 - \frac{\text{RSS}}{\text{TSS}} = 1 - \frac{\dfrac{\text{RSS}}{n}}{\dfrac{\text{TSS}}{n}} \simeq 1 - \frac{S_E^2}{S_Y^2}$$

Though the statistic $1 - (S_{E_j}^2/S_{Y_j}^2)$ may be computed for a structural equation in a nonrecursive model, and though this statistic seems sensible, Basmann (1962) has shown that it is unbounded below. Likewise, in OLS regression, $R = r_{Y\hat{Y}}$. In a nonrecursive structural-equation model, the correlation between observed and fitted endogenous-variable values may be negative (Basmann, 1962). Nevertheless, statistics such as $1 - (S_{E_j}^2/S_{Y_j}^2)$ and $r_{Y_j\hat{Y}_j}$ are frequently reported because of their simplicity and intuitive appeal.

Since the reduced form may be estimated consistently by OLS regression, we may justifiably report the multiple correlation for each reduced-form equation (i.e., for each endogenous variable). Hooper (1959) has extended this approach, formulating a correlational index for nonrecursive structural-equation models that assesses the degree of joint dependence of the endogenous on the exogenous variables. This dependence is expressed in the reduced-form equation

$$\underset{(n \times q)}{\mathbf{Y}} = \underset{(n \times m)}{\mathbf{X}} \underset{(m \times q)}{\mathbf{\Pi}'} + \underset{(n \times q)}{\mathbf{\Delta}}$$

Hooper estimates the reduced-form parameters by OLS regression:

$$\mathbf{P}' = (\mathbf{X}'\mathbf{X})^{-1}\mathbf{X}'\mathbf{Y}$$

(An alternative would be to obtain $\mathbf{P} = -\mathbf{B}^{-1}\mathbf{C}$, but in an overidentified model this procedure would change the ultimate interpretation of Hooper's correlational measure.) Fitted endogenous-variable values are then given by

$$\hat{\mathbf{Y}} = \mathbf{X}\mathbf{P}' = \mathbf{X}(\mathbf{X}'\mathbf{X})^{-1}\mathbf{X}'\mathbf{Y}$$

The multivariate analog of total variation (TSS) is the sum-of-squares-and-products matrix $\mathbf{Y}'\mathbf{Y}$; similarly, $\hat{\mathbf{Y}}'\hat{\mathbf{Y}}$ is the analog of "explained" variation (RegrSS). $\hat{\mathbf{Y}}'\hat{\mathbf{Y}}(\mathbf{Y}'\mathbf{Y})^{-1}$, therefore, is a multivariate version of explained "divided by" total variation. Hooper's *trace correlation* statistic is defined as

$$\overline{R}^2 = \frac{1}{q}\text{trace}\left[\hat{\mathbf{Y}}'\hat{\mathbf{Y}}(\mathbf{Y}'\mathbf{Y})^{-1}\right] \tag{4.71}$$

Hooper demonstrates that $\overline{R}^2$ is the mean squared canonical correlation[20]

[20] Canonical correlations assess the strength of linear dependencies between two sets of variables—here, the exogenous and endogenous variables in a structural-equation model. For details of canonical-correlation theory, see Morrison (1976: 259–263).

between the sets of endogenous and exogenous variables. Notice that when $q = 1$, equation (4.71) specializes to the usual R^2 statistic.

For the Duncan, Haller, and Portes peer-influences model in Figure 4.1, $\overline{R}^2 = .230$. Roughly, then, 23 percent of the joint variation of Y_4 and Y_6 is accounted for by their relation to X_1, X_2, X_3, and X_4. The individual R^2's for the two reduced-form regressions are .264 and .319.

4.7.2. Overidentification Tests

We have mentioned at several points in this chapter that an overidentified structural-equation model may be inconsistent with the observed data. One descriptive method for tracing the consequences of overidentification is to calculate model-implied covariances among observed variables, comparing these to sample covariances computed directly from the data. Implied covariances may be calculated by equations (4.51), or for LISREL models, by the sample analogs of equations (4.67) and (4.68). Implied covariances (correlations) for the standardized Blau and Duncan model were shown in Table 4.9. It is clear that this model closely reproduces covariances among observed variables.[21]

A formal test of overidentifying restrictions may be constructed by the likelihood-ratio principle. We develop this test for the LISREL model, because of that model's generality.[22]

The log likelihood for the LISREL model is given by equation (4.69). Suppose that a model with r free parameters is overidentified and has a likelihood of L_0. If the overidentifying restrictions are removed, $\mathbf{S}$ and the model-implied estimate of $\boldsymbol{\Sigma}$ become identical, yielding log likelihood

$$\log L_1 = -\frac{N(p+q)}{2}\log(2\pi) - \frac{N}{2}(\log|\mathbf{S}| + p + q)$$

The number of free parameters in the just-identified model is equal to the number of observed covariances: $(p+q)(p+q+1)/2$. The likelihood-ratio test statistic for the overidentifying restrictions is therefore

$$\begin{aligned} G_0^2 &= -2(\log L_0 - \log L_1) \\ &= N\left[\log|\hat{\boldsymbol{\Sigma}}| + \operatorname{trace}(\mathbf{S}\hat{\boldsymbol{\Sigma}}^{-1}) - \log|\mathbf{S}| - (p+q)\right] \qquad (4.72) \end{aligned}$$

Here $\hat{\boldsymbol{\Sigma}}$ is the estimate of $\boldsymbol{\Sigma}$ implied by the maximum-likelihood estimates of

[21] Even an overidentified recursive model necessarily reproduces certain covariances precisely. See Fox (1980) for further discussion of this point. For the Blau and Duncan model, in fact, only S_{14}^* and S_{15}^* may depart from the corresponding S_{ij}'s.

[22] Overidentification tests are also available for single-equation methods such as 2SLS. See, for example, Fox (1979a).

the model parameters. As a likelihood-ratio statistic, G_0^2 is asymptotically distributed as χ^2 with $[(p+q)(p+q+1)/2] - r$ degrees of freedom (the degree of overidentification of the model). Notice that G_0^2 will be small when the reproduced covariances $\hat{\boldsymbol{\Sigma}}$ are similar to their directly observed counterparts $\mathbf{S}$. Since observed-variable structural-equation models estimated by FIML are special cases of LISREL estimation, the test in equation (4.72) is applicable to these models as well.

We have noted that OLS and FIML are identical for recursive models. For these models, the overidentification test statistic in equation (4.72) specializes to (Land, 1973):

$$G_0^2 = n \sum_{j=1}^{q} \log \frac{S_{E_j}^2}{S_{E_j^*}^2}$$

where n is the sample size, q is the number of equations in the model, $S_{E_j}^2$ is the estimated residual variance from the jth structural equation, and $S_{E_j^*}^2$ is the estimated residual variance for the jth equation respecified so that no prior variables are excluded. If the $S_{E_j}^2$ are appreciably larger than the $S_{E_j^*}^2$, then we should suspect that prior variables have been falsely excluded; when this is the case, the test statistic will be large. The degrees of freedom for G_0^2 are equal to the number of excluded prior variables in all structural equations of the model.

Overidentification test statistics for the models discussed in this chapter are shown in Table 4.14. Notice that the extremely large sample for the Blau and Duncan study results in a statistically significant overidentification test even though the model fits the data very closely: Given a large enough sample, virtually any overidentified model may be rejected. (Indeed, for the Blau and Duncan model, the difference between the calculated G_0^2 and zero may be the

TABLE 4.14. Overidentification Tests

Model	n	G_0^2	df	p
Duncan et al. Nonrecursive (Figure 4.1)	329	2.81	2	.25
Blau and Duncan Recursive (Figure 4.2)	20,700	30.9	2	$\ll .001$
Duncan et al. Block-Recursive (Figure 4.4)	329	3.81	2	.15
Duncan et al. LISREL (Figure 4.16)	329	26.70	15	.03

result of rounding errors.) The significant overidentification test for the LISREL model fit to the peer-influences data is more distressing, yet even here we should be loath to reject the model on this basis alone.

A statistically significant overidentification test may motivate model respecification, perhaps by removing one or more overidentifying restrictions, perhaps by more drastic reformulation. The data may be helpful in suggesting which overidentifying restrictions to remove, but we must be careful always to guide our model-building activity by substantive criteria.

The magnitude of differences between model-implied and directly calculated sample covariances is an uncertain guide to model respecification (see Costner and Schoenberg, 1973; Sörbom, 1975). Sörbom (1975) has suggested examining the partial derivatives of the likelihood function with respect to the *fixed* (i.e., constrained) parameters. These derivatives are available as by-products of the model-fitting procedure. When the likelihood is maximized, its partial derivatives with respect to the *free* parameters are, of course, zero; a steep gradient with respect to a *fixed* parameter, however, indicates that the likelihood might be substantially increased if that parameter were permitted to take on a different value.

PROBLEMS

4.20. Compute trace correlations and reduced-form correlations for the nonrecursive models estimated in Problems 4.11 and 4.12.

4.21. Calculate model-implied covariances and likelihood-ratio overidentification tests for the structural-equation models fit in Problems 4.12, 4.13, 4.14, 4.17, and 4.18. In each case, if the model appears to be inadequate, consider how it might be respecified.

5

Logit and Log-Linear Models for Qualitative Data

All of the statistical models described in the previous chapters are for quantitative dependent variables. It is unnecessary to document the prevalence of qualitative, or nominal, data in the social sciences. In developing the general linear model, we introduced qualitative *independent* variables through the device of coding dummy-variable regressors. There is, of course, no substantive reason why qualitative variables should not also appear as dependent variables, affected by other factors, both qualitative and quantitative.

The first section of this chapter develops linear logit models for qualitative dependent variables. Contingency tables arise when all variables under consideration are qualitative (or at least categorical). There has been much recent interest in log-linear models for contingency tables. We consider these models in some detail in Section 5.2, and demonstrate their relationship to the logit models discussed in Section 5.1.

5.1. LINEAR LOGIT MODELS

The general linear logit model expresses a qualitative dependent variable as a function of several independent variables. We begin our discussion of logit models by attempting to apply linear regression to qualitative dependent variables. The difficulties encountered point the way to more satisfactory models for qualitative data. We initially develop the linear logit model for dichotomous dependent variables, subsequently extending the model to polychotomous data in two different ways. Finally, we describe diagnostic methods for assessing the adequacy of a fitted logit model.

5.1.1. The Linear Probability Model

When we first encountered qualitative independent variables in Section 2.1, we dealt with them by coding zero/one dummy regressors. A dichotomous variable, we learned, gives rise to a single dummy regressor. Because the use of dummy regressors created no difficulties for the general linear model, it seems reasonable to ask whether dummy dependent variables may be similarly employed.

For simplicity, let us examine the relationship between a dichotomous dependent variable and a single quantitative independent variable. Suppose, for instance, that we are interested in the relationship of married women's participation in the work force to their husbands' income. Let Y be a dummy variable coded one for women who are employed and zero for those who are not. Let X represent the husband's income. It then appears natural to write Y as a linear function of X, at least as a first approximation, and to make the usual assumptions of the simple-regression model. That is

$$Y_i = \alpha + \beta x_i + \varepsilon_i \tag{5.1}$$

where $\varepsilon_i \sim N(0, \sigma_\varepsilon^2)$, and ε_i and ε_j are independent for $i \neq j$. If X is random, then we assume that it is independent of ε.

This model is untenable, as we shall see presently, but first we pursue its interpretation. Y, we know, can assume only the values zero and one. Let

$$\pi_i \equiv \Pr(Y = 1 | X = x_i) = \Pr(Y_i = 1)$$

and thus $\Pr(Y_i = 0) = 1 - \pi_i$. Under the assumption that $E(\varepsilon_i) = 0$, it follows that $E(Y_i) = \alpha + \beta x_i$. The expectation of Y_i may be expanded in the following manner:

$$E(Y_i) = \pi_i(1) + (1 - \pi_i)(0) = \pi_i$$

Model (5.1) therefore implies that

$$\pi_i = \alpha + \beta x_i \tag{5.2}$$

which accounts for the *linear probability model* designation.

We previously alluded to difficulties in the specification of the linear probability model (5.1). One such problem concerns the distribution of the error ε. The assumption that $E(\varepsilon_i) = 0$ may reasonably be maintained, at least over a limited range of X-values (see below). If Y_i can take on only two values, however, then ε_i is dichotomous as well: if $Y_i = 1$, then

$$\varepsilon_i = 1 - E(Y_i) = 1 - (\alpha + \beta x_i) = 1 - \pi_i$$

(which occurs with probability π_i); and if $Y_i = 0$, then $\varepsilon_i = 0 - (\alpha + \beta x_i) = -\pi_i$ (with probability $1 - \pi_i$). Because the error is dichotomous, it cannot be even approximately normally distributed. Moreover, the variance of ε is not constant, as we may readily demonstrate: Recalling that $E(\varepsilon_i) = 0$, and using the relations just noted,

$$V(\varepsilon_i) = \pi_i(1 - \pi_i)^2 + (1 - \pi_i)(-\pi_i)^2$$

$$= \pi_i(1 - \pi_i)$$

Dummy *regressor* variables caused no comparable difficulties because the general linear model does not make distributional assumptions about the regressors.

The striking nonnormality and heteroscedasticity of the errors bode ill for ordinary-least-squares estimation of the linear probability model. Goldberger (1964: 248–250) has proposed a correction for heteroscedasticity employing weighted least squares. (Weighted least squares was discussed in Problem 3.13, Chapter 3.) Because the variances $V(\varepsilon_i)$ depend upon the π_i, however, which in turn are functions of the unknown parameters α and β, we require preliminary estimates for the model in order to define weights. Goldberger obtains ad hoc estimates from a preliminary OLS regression; that is, he takes $\widehat{V(\varepsilon_i)} = \hat{Y}_i(1 - \hat{Y}_i)$. Because the fitted values from an OLS regression are not constrained to the interval (0, 1), some of these "variances" may be negative.

This last remark suggests a further difficulty with the linear-probability model as formulated in equation (5.2): If we observe X over a suitably wide range, then π will not remain in the unit interval (as shown in Figure 5.1) and therefore cannot be interpreted as a probability. One solution to this problem is simply to constrain π to the unit interval while retaining the linear relation

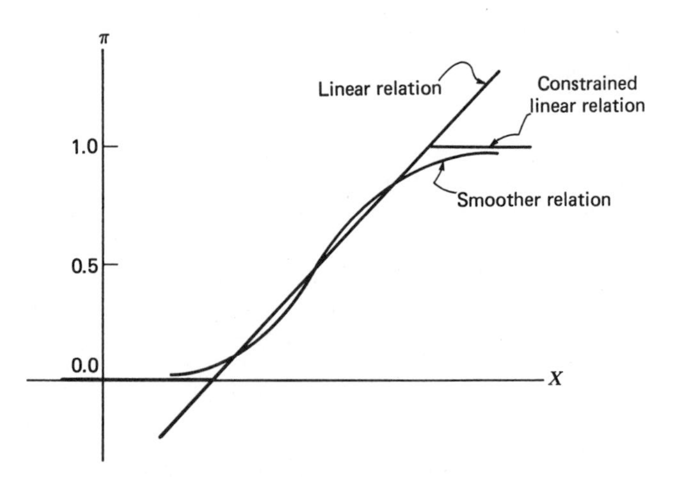

FIGURE 5.1. Linear and constrained-linear probability models. (Adapted with permission from Nerlove and Press, 1973: Fig. 1, The Rand Corporation, Santa Monica, R-1306-EDA-NIH.)

between π and X within this interval:

$$\pi = \begin{cases} 0 & 0 > \alpha + \beta X \\ \alpha + \beta X & 0 \leq \alpha + \beta X \leq 1 \\ 1 & \alpha + \beta X > 1 \end{cases} \tag{5.3}$$

The constrained linear probability model is also shown in Figure 5.1. Although it cannot be dismissed on logical grounds, model (5.3) has certain unattractive features: (1) Under certain circumstances, the model generates unstable estimators (Nerlove and Press, 1973: 5–7); (2) the abrupt changes in slope at $\pi = 0$ and $\pi = 1$ make the model hard to fit to data; and (3) these abrupt changes are unlikely to be substantively reasonable. A smoother relation between π and X (again, see Figure 5.1) is generally more sensible.

We have pointed to several problems with the linear probability model, in both its constrained and its unconstrained versions. Yet, for values of π not too close to zero or one, the linear model may adequately approximate a more reasonable general specification. Furthermore, there is some evidence that least-squares estimates of the linear probability model frequently provide results similar to those produced by the more formally justifiable methods that are the subject of the remainder of this chapter (see Knoke, 1975; Goodman, 1976; and Magidson, 1978; cf. Gillespie, 1977).

5.1.2. Transformations of π

We noted that one difficulty with the unconstrained linear probability model is the implication that π may lie outside the unit interval. What we require to correct this problem is a positive monotone (i.e., nondecreasing) function that will transform $\alpha + \beta x_i$ to the unit interval. Any cumulative probability distribution function P meets this requirement. That is, we may respecify the model as

$$\pi_i = P(\alpha + \beta x_i) \tag{5.4}$$

If we choose P as the cumulative rectangular distribution, for example, we obtain the constrained linear probability model (5.3). An *a priori* reasonable P should be both smooth and symmetric, and should approach $\pi = 0$ and $\pi = 1$ as asymptotes. Moreover, it is advantageous if P is strictly increasing, for then the transformation (5.4) is one-to-one, permitting us to rewrite the model as

$$P^{-1}(\pi_i) = \alpha + \beta x_i$$

where P^{-1} is the inverse of the cumulative distribution function P. Thus, we have a linear model for a monotone transformation of π, or, equivalently, a nonlinear model for π itself.

P is generally chosen as the CDF of the unit-normal distribution

$$\Phi(z) = \frac{1}{\sqrt{2\pi}} \int_{-\infty}^{z} e^{-1/2Z^2}\, dZ$$

or of the *standardized logistic distribution*

$$\Lambda(z) = \frac{1}{1 + e^{-z}}$$

(In these equations, π and e are the mathematical constants.) $\Phi(z)$ yields the *linear probit model* (see Finney, 1971):

$$\pi_i = \Phi(\alpha + \beta x_i)$$
$$= \frac{1}{\sqrt{2\pi}} \int_{\infty}^{\alpha + \beta x_i} e^{-1/2Z^2}\, dZ$$

$\Lambda(z)$ produces the *linear logistic* or *logit model*:

$$\pi_i = \Lambda(\alpha + \beta x_i)$$
$$= \frac{1}{1 + e^{-(\alpha + \beta x_i)}} \tag{5.5}$$

Once variances are equated, the two transformations are so similar that it is not possible in practice to distinguish between them, as is apparent in Figure 5.2. It is also clear from this figure that either function is nearly linear over

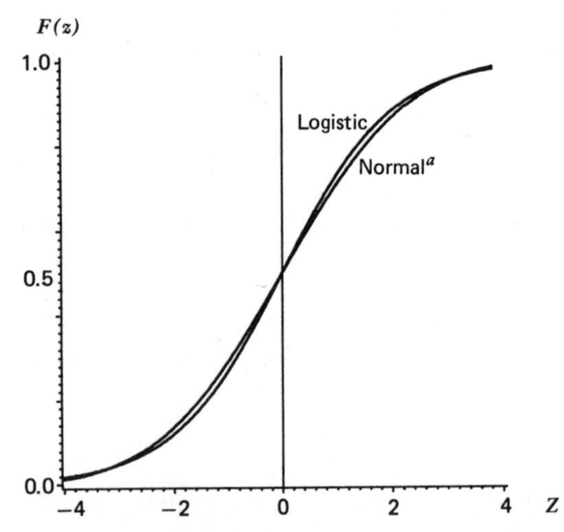

FIGURE 5.2. Logistic and normal cumulative probability distribution functions.
[a] $N(0, \pi^2/3)$ to match variance of standardized logistic.

much of its range. This is why the linear probability model produces results similar to the logit and probit models, except for extreme values of π.

Theoretical arguments may be adduced for both the logit and probit specifications (Nerlove and Press, 1973: 14–15; McFadden, 1974; Finney, 1971: Chapter 2). It is hard, however, to think of formal justifications for the linear probability model in either its constrained or its unconstrained versions. As was pointed out, the linear model may be an adequate approximation to a more reasonable specification; additionally, as Swafford (1980: 667) mentions, because the model specifies a linear relation between π and X, it may be easier to describe than the logit or probit models, although this is largely a result of habit.

Two practical advantages of the logit, compared to the probit, specification are the logit's relative simplicity (it is unnecessary to evaluate an integral) and the interpretability of the inverse transformation $\Lambda^{-1}(\pi)$ as a *log odds*.[1] Rearranging equation (5.5), we get

$$\frac{1}{\pi_i} = 1 + e^{-(\alpha+\beta x_i)}$$

$$\frac{1}{\pi_i} - 1 = e^{-(\alpha+\beta x_i)}$$

$$\frac{\pi_i}{1-\pi_i} = e^{\alpha+\beta x_i}$$

Then, taking the log of both sides,

$$\log\frac{\pi_i}{1-\pi_i} = \alpha + \beta x_i$$

$\Lambda^{-1}(\pi) = \log[\pi/(1-\pi)]$, called the *logit* of π, is the log of the odds that Y is one rather than zero. If the odds are even, that is if $\pi = .5$, then the logit is zero; if $\pi < .5$, then the logit is negative; and if $\pi > .5$, then the logit is positive.

5.1.3. The Linear Logit Model for Dichotomous Data

The simple logit regression model of the previous section may easily be extended in the manner of the general linear model. Suppose, as before, that Y is a dichotomous dependent variable, taking on the values one and zero with probabilities π and $1-\pi$, respectively. For the ith observation, we have

[1]An advantage of the probit model is that it may more readily be extended to simultaneous-equation models that include both qualitative and quantitative endogenous variables and that permit nonrecursive structures. See Heckman (1978); cf. Section 5.2.6.

$\Pr(Y_i = 1) = \pi_i$. Let $\underset{(n \times k)}{\mathbf{X}}$ be a design matrix of full column rank, with one row, $\mathbf{x}_i'$, for each observation. As in the general linear model, $\mathbf{X}$ can contain a constant regressor, interval-level regressors, functions of interval-scale independent variables, dummy regressors, and so on. The linear logit model may then be written

$$\pi_i = \Lambda(\mathbf{x}_i'\boldsymbol{\beta}) = \frac{1}{1 + e^{-\mathbf{x}_i'\boldsymbol{\beta}}} \tag{5.6}$$

or, alternatively

$$\Lambda^{-1}(\pi_i) = \log\left(\frac{\pi_i}{1 - \pi_i}\right) = \mathbf{x}_i'\boldsymbol{\beta}$$

where $\underset{(k \times 1)}{\boldsymbol{\beta}}$ is the vector of parameters relating π to $\mathbf{X}$. For the logit simple-regression model of the previous section, $\mathbf{x}_i' = (1, x_i)$ and $\boldsymbol{\beta} = (\alpha, \beta)'$.

Since each Y_i takes on the values one or zero with probabilities π_i and $1 - \pi_i$, and if the observations are independent, the joint probability for the observations is given by

$$\begin{aligned} p(y_1, \ldots, y_n) &= \prod_{i=1}^{n} \pi_i^{y_i}(1 - \pi_i)^{1-y_i} \\ &= \prod_{i=1}^{n} \left(\frac{\pi_i}{1 - \pi_i}\right)^{y_i}(1 - \pi_i) \end{aligned} \tag{5.7}$$

From equation (5.6) we have

$$\frac{\pi_i}{1 - \pi_i} = e^{\mathbf{x}_i'\boldsymbol{\beta}}$$

and

$$1 - \pi_i = \frac{1}{1 + e^{\mathbf{x}_i'\boldsymbol{\beta}}}$$

Using these expressions in equation (5.7) produces

$$p(y_1, \ldots, y_n | \mathbf{X}) = \prod_{i=1}^{n} \left(e^{\mathbf{x}_i'\boldsymbol{\beta}}\right)^{y_i} \left(\frac{1}{1 + e^{\mathbf{x}_i'\boldsymbol{\beta}}}\right)$$

and the log likelihood function is

$$\log L(\boldsymbol{\beta}) = \sum_{i=1}^{n} Y_i \mathbf{x}_i'\boldsymbol{\beta} - \sum_{i=1}^{n} \log\left(1 + e^{\mathbf{x}_i'\boldsymbol{\beta}}\right)$$

The partial derivatives of the log likelihood with respect to $\boldsymbol{\beta}$ are

$$\frac{\partial \log L(\boldsymbol{\beta})}{\partial \boldsymbol{\beta}} = \sum_{i=1}^{n} Y_i \mathbf{x}_i - \sum_{i=1}^{n} \left(\frac{e^{\mathbf{x}_i'\boldsymbol{\beta}}}{1 + e^{\mathbf{x}_i'\boldsymbol{\beta}}} \right) \mathbf{x}_i$$

$$= \sum_{i=1}^{n} Y_i \mathbf{x}_i - \sum_{i=1}^{n} \left(\frac{1}{1 + e^{-\mathbf{x}_i'\boldsymbol{\beta}}} \right) \mathbf{x}_i \tag{5.8}$$

Setting the vector of partial derivatives to zero to maximize the likelihood yields estimating equations

$$\sum_{i=1}^{n} \left(\frac{1}{1 + e^{-\mathbf{x}_i'\mathbf{b}}} \right) \mathbf{x}_i = \sum_{i=1}^{n} Y_i \mathbf{x}_i \tag{5.9}$$

The estimating equations (5.9) have an intuitive justification: $P_i \equiv 1/(1 + e^{-\mathbf{x}_i'\mathbf{b}})$ is the fitted value of π_i; and thus the estimating equations set the fitted sum $\Sigma P_i \mathbf{x}_i$ equal to the corresponding observed sum $\Sigma Y_i \mathbf{x}_i$. In matrix form, we may write the estimating equations as $\mathbf{X}'\mathbf{p} = \mathbf{X}'\mathbf{y}$, where $\mathbf{p} = (P_1, P_2, \ldots, P_n)'$. Note the similarity to the least-squares estimating equations for the general linear model, which may be written as $\mathbf{X}'\hat{\mathbf{y}} = \mathbf{X}'\mathbf{y}$.

Because $\mathbf{b}$ is a maximum-likelihood estimator, its estimated asymptotic covariance matrix may be obtained from the inverse of the information matrix

$$\mathbf{I}(\boldsymbol{\beta}) = -E\left[\frac{\partial^2 \log L(\boldsymbol{\beta})}{\partial \boldsymbol{\beta}\, \partial \boldsymbol{\beta}'} \right]$$

evaluated at $\boldsymbol{\beta} = \mathbf{b}$. Differentiating equation (5.8) and making the appropriate substitutions,

$$\widehat{\mathscr{V}(\mathbf{b})} = \left[\sum_{i=1}^{n} \frac{e^{-\mathbf{x}_i'\mathbf{b}}}{\left(1 + e^{-\mathbf{x}_i'\mathbf{b}}\right)^2} \mathbf{x}_i \mathbf{x}_i' \right]^{-1}$$

$$= \left[\sum_{i=1}^{n} P_i(1 - P_i) \mathbf{x}_i \mathbf{x}_i' \right]^{-1}$$

$$= (\mathbf{X}'\mathbf{V}\mathbf{X})^{-1}$$

where $\underset{(n \times n)}{\mathbf{V}} \equiv \text{diag}\{P_i(1 - P_i)\}$. Note that $P_i(1 - P_i)$ is the estimated variance of the observation Y_i. The square roots of the diagonal entries of $\widehat{\mathscr{V}(\mathbf{b})}$ are estimated asymptotic standard errors for the logit-model coefficients, which may be used to construct large-sample normal-distribution tests and confidence intervals for the coefficients.

A likelihood-ratio chi-square test may be employed for contrasting two models, when one model is a restricted version of the other. Suppose, for example, that L_1 is the maximized likelihood for the model

$$\Lambda^{-1}(\pi_i) = \beta_1 X_{1i} + \cdots + \beta_p X_{pi} + \beta_{p+1} X_{p+1,i} + \cdots + \beta_k X_{ki}$$

and L_0 is the maximized likelihood for a model that sets the last $k - p$ coefficients to zero; that is

$$\Lambda^{-1}(\pi_i) = \beta_1 X_{1i} + \cdots + \beta_p X_{pi}$$

Then the likelihood-ratio test statistic

$$G_0^2 = -2 \log \frac{L_0}{L_1} = 2(\log L_1 - \log L_0) \tag{5.10}$$

is asymptotically distributed as χ^2 with $k - p$ degrees of freedom, under the null hypothesis $H_0: \beta_{p+1} = \cdots = \beta_k = 0$. Tests of this form are analogous to incremental sum of squares F-tests for linear models (Section 1.2.6).

The logit-model estimating equations (5.9) are nonlinear functions of $\mathbf{b}$ and therefore require numerical solution. The most commonly employed procedure for solving the estimating equations is the iterative *Newton–Raphson method*, which may be described as follows:

1. We start with estimates $\mathbf{b}_0$—one particularly simple choice of initial estimates is $\mathbf{b}_0 = \mathbf{0}$.
2. At each iteration $l + 1$ we compute new estimates

$$\mathbf{b}_{l+1} = \mathbf{b}_l + (\mathbf{X}'\mathbf{V}_l\mathbf{X})^{-1}\mathbf{X}'(\mathbf{y} - \mathbf{p}_l) \tag{5.11}$$

where $\mathbf{p}_l = \{1/(1 + e^{-\mathbf{x}_i'\mathbf{b}_l})\}$ and $\mathbf{V}_l = \text{diag}\{P_{li}/(1 - P_{li})\}$.

3. Iterations continue until $\mathbf{b}_{l+1} \simeq \mathbf{b}_l$.

Notice that when convergence takes place,

$$(\mathbf{X}'\mathbf{V}_l\mathbf{X})^{-1}\mathbf{X}'(\mathbf{y} - \mathbf{p}_l) \simeq \mathbf{0}$$

and thus the estimating equations $\mathbf{X}'\mathbf{y} = \mathbf{X}'\mathbf{p}$ are approximately satisfied. Conversely, if $\mathbf{X}'\mathbf{y}$ is very different from $\mathbf{X}'\mathbf{p}_l$, there will be a large adjustment in $\mathbf{b}$ from one iteration to the next.

5.1.4. Extension of the Linear Logit Model to Polychotomous Data

The logit model developed in the previous section is appropriate for dichotomous data, and therefore has limited application. In this section, we show how the logit model may be generalized to accommodate polychotomous dependent variables.

Suppose that a dependent variable Y may take on any of m qualitative values, which, for convenience, we number $1, 2, \dots, m$. A married woman, for example, may (1) work full time, (2) work part time, or (3) not work outside the home. Although the categories of Y are numbered, we do not in general attribute ordinal properties to these numbers. Let π_{ij} represent the probability that the ith observation falls in the jth dependent-variable category; that is $\pi_{ij} = \Pr(Y_i = j)$.

We have available a set of k regressors $\underset{(1\times k)}{\mathbf{x}'}$, on which π depends. More specifically, we assume that the relationship of π to $\mathbf{x}'$ is given by the symmetric form of the *multivariate logistic distribution function*:

$$\pi_{ij} = \frac{e^{\mathbf{x}_i'\boldsymbol{\gamma}_j}}{\sum_{l=1}^{m} e^{\mathbf{x}_i'\boldsymbol{\gamma}_l}} \tag{5.12}$$

where $\mathbf{x}_i'$ is the regressor vector for the ith observation and $\boldsymbol{\gamma}_j$ is the parameter vector for the jth dependent-variable category. Because $\sum_{j=1}^{m}\pi_{ij} = 1$, we need to impose a linear constraint on the $\boldsymbol{\gamma}_j$ to define them uniquely, much as in the analysis-of-variance model discussed in Chapter 2. It is convenient to impose the constraint $\sum_{j=1}^{m}\boldsymbol{\gamma}_j = \mathbf{0}$.

To gain insight into model (5.12), suppose that this model is specialized to a dichotomous dependent variable, an application familiar from the previous section. Then, $m = 2$, $\boldsymbol{\gamma}_1 = -\boldsymbol{\gamma}_2$, $\pi_{i2} = 1 - \pi_{i1}$, and

$$\log\frac{\pi_{i1}}{\pi_{i2}} = \log\frac{\pi_{i1}}{1 - \pi_{i1}} = \mathbf{x}_i'\boldsymbol{\gamma}_1 - \mathbf{x}_i'\boldsymbol{\gamma}_2 = \mathbf{x}_i'(\boldsymbol{\gamma}_1 - \boldsymbol{\gamma}_2) = \mathbf{x}_i'(2\boldsymbol{\gamma}_1)$$

Thus $2\boldsymbol{\gamma}_1 = \boldsymbol{\beta}$, where $\boldsymbol{\beta}$ is the parameter vector of the logit model (5.6) for dichotomous data defined in the previous section. Notice that if we impose the alternative constraint $\boldsymbol{\gamma}_2 = \mathbf{0}$, employing category two as a "baseline" category, then $\boldsymbol{\gamma}_1 = \boldsymbol{\beta}$, and model (5.12) becomes identical to model (5.6).

More generally, when $m \geq 2$, if we use the standard constraint $\sum\boldsymbol{\gamma}_j = \mathbf{0}$, then the log odds for any pair of categories is a linear function of the difference between their parameter vectors: $\log(\pi_{ij}/\pi_{il}) = \mathbf{x}_i'(\boldsymbol{\gamma}_j - \boldsymbol{\gamma}_l)$. Alternatively, if we employ the constraint $\boldsymbol{\gamma}_m = \mathbf{0}$, then this last relation holds and, additionally, $\log(\pi_{ij}/\pi_{im}) = \mathbf{x}_i'\boldsymbol{\gamma}_j$. Under this constraint, then, $\boldsymbol{\gamma}_j$ implicitly represents the effect of $\mathbf{x}$ on the odds of being in the jth category relative to the last.

To fit model (5.12) to data, we may again invoke the maximum-likelihood method. First, we note that each Y_i takes on its possible values $1, 2, \ldots, m$ with probabilities $\pi_{i1}, \pi_{i2}, \ldots, \pi_{im}$. Following Nerlove and Press (1973), we define dummy variables $W_{i1}, \ldots, W_{im}$, so that $W_{ij} = 1$ if $Y_i = j$ and $W_{ij} = 0$ if $Y_i \neq j$; thus

$$p(y_i) = \prod_{j=1}^{m} \pi_{ij}^{w_{ij}}$$

If the observations are sampled independently, then their joint probability distribution is given by

$$p(y_1, \ldots, y_n) = \prod_{i=1}^{n} \prod_{j=1}^{m} \pi_{ij}^{w_{ij}}$$

Employing model (5.12),

$$p(y_1, \ldots, y_n | \mathbf{X}) = \prod_{i=1}^{n} \prod_{j=1}^{m} \left(\frac{e^{\mathbf{x}_i' \boldsymbol{\gamma}_j}}{\sum_{l=1}^{m} e^{\mathbf{x}_i' \boldsymbol{\gamma}_l}} \right)^{w_{ij}}$$

and the log likelihood is

$$\begin{aligned} \log L(\boldsymbol{\gamma}_1, \ldots, \boldsymbol{\gamma}_m) &= \sum_{i=1}^{n} \sum_{j=1}^{m} W_{ij} \left[\mathbf{x}_i' \boldsymbol{\gamma}_j - \log\left(\sum_{l=1}^{m} e^{\mathbf{x}_i' \boldsymbol{\gamma}_l} \right) \right] \\ &= \sum_{i=1}^{n} \sum_{j=1}^{m} W_{ij} \mathbf{x}_i' \boldsymbol{\gamma}_j - \sum_{i=1}^{n} \log\left(\sum_{l=1}^{m} e^{\mathbf{x}_i' \boldsymbol{\gamma}_l} \right) \end{aligned}$$

(since $\sum_{j=1}^{m} W_{ij} = 1$). Differentiating the log likelihood with respect to the parameters, and setting the partial derivatives to zero, produces the nonlinear estimating equations

$$\sum_{i=1}^{n} W_{ij} \mathbf{x}_i = \sum_{i=1}^{n} \left(\frac{e^{\mathbf{x}_i' \mathbf{c}_j}}{\sum_{l=1}^{m} e^{\mathbf{x}_i' \mathbf{c}_l}} \right) \mathbf{x}_i \tag{5.13}$$

for $j = 1, \ldots, m$. These equations may be solved numerically subject to the constraint $\sum_{j=1}^{m} \mathbf{c}_j = \mathbf{0}$ (see Nerlove and Press, 1973; McFadden, 1974; Haberman, 1979: Chapter 6). The resulting vectors $\mathbf{c}_1, \ldots, \mathbf{c}_m$ share the usual properties of maximum-likelihood estimators.

As in the dichotomous case, the estimating equations set observed sums equal to fitted sums: Using the model (5.12), the fitted probabilities are given by

$$P_{ij} = \frac{e^{\mathbf{x}_i'\mathbf{c}_j}}{\sum_{l=1}^{m} e^{\mathbf{x}_i'\mathbf{c}_l}}$$

Thus, the estimating equations (5.13) may be written $\Sigma W_{ij}\mathbf{x}_i = \Sigma P_{ij}\mathbf{x}_i$, for $j = 1, \ldots, m$.

5.1.5. Modeling Polychotomous Data Employing Nested Dichotomies

As we have seen, the parameters $\gamma_1, \ldots, \gamma_m$ of equation (5.12) may be used to model log odds ratios for pairs of dependent-variable categories. This sort of comparison between pairs of response categories is not, unfortunately, always of substantive interest. An alternative approach that frequently produces more readily interpreted results is to fit separate logit models to each of a set of dichotomies; the dichotomies are constructed so that the likelihood for the polychotomous dependent variable is the product of the likelihoods for the dichotomies. We shall show that the likelihood is separable in this manner if the set of dichotomies is *nested*.

A nested set of $m - 1$ dichotomies is produced from a polychotomous variable by successive binary partitions of the categories of the variable. Two examples for a four-category variable are shown in Figure 5.3. In part (a) of this figure, the dichotomies produced are $\{12, 34\}$, $\{1, 2\}$, and $\{3, 4\}$; in part (b) the nested dichotomies are $\{1, 234\}$, $\{2, 34\}$, and $\{3, 4\}$.

For example, in examining the employment status of married women using the trichotomy (1) working full time, (2) working part time, and (3) not

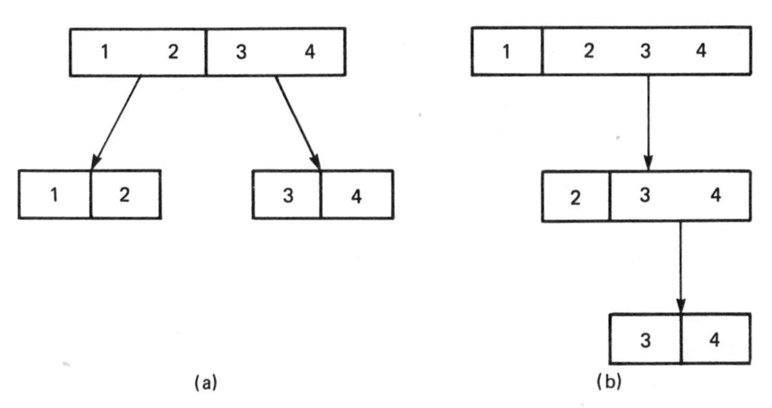

FIGURE 5.3. Illustrative nested dichotomies.

working outside the home, it seems natural to define the nested dichotomies (i) working versus not working $\{12, 3\}$, and (ii) working full time versus working part time $\{1, 2\}$.

Fienberg (1977: 86–90) considers a special case of nested dichotomies that is often of interest when the dependent-variable categories are ordered. Here the jth of $m - 1$ dichotomies contrasts category j of the dependent variable with categories $j + 1, \ldots, m$ [as in Figure 5.3(b)]. The odds ratios formed from these dichotomies are termed "continuation ratios."

We shall demonstrate the separability of the likelihood function for the nested dichotomies $\{12, 3\}$ and $\{1, 2\}$. Let Y_i and W_{ij} be defined as before. Let $Y_i' = 1$ for $Y_i = 1$ or 2, and zero for $Y_i = 3$. Let $Y_i'' = 1$ for $Y_i = 1$, and zero for $Y_i = 2$; Y_i'' is undefined for $Y_i = 3$. Note that Y_i' corresponds to the dichotomy $\{12, 3\}$ and Y_i'' to $\{1, 2\}$. We wish to show that $p(y_i) = p(y_i')p(y_i'')$. (For $Y_i = 3$, we take $p(y_i'') = 1$.)

The probability distribution of Y_i' is given by

$$
\begin{aligned}
p(y_i') &= (\pi_{i1} + \pi_{i2})^{y_i'} \pi_{i3}^{(1-y_i')} \\
&= (\pi_{i1} + \pi_{i2})^{(w_{i1} + w_{i2})} \pi_{i3}^{w_{i3}}
\end{aligned}
\tag{5.14}
$$

To derive the probability distribution of Y_i'', we note that

$$
\Pr(Y_i'' = 1) = \Pr(Y_i = 1 \mid Y_i \neq 3) = \frac{\pi_{i1}}{\pi_{i1} + \pi_{i2}}
$$

$$
\Pr(Y_i'' = 0) = \Pr(Y_i = 2 \mid Y_i \neq 3) = \frac{\pi_{i2}}{\pi_{i1} + \pi_{i2}}
$$

and thus

$$
\begin{aligned}
p(y_i'') &= \left(\frac{\pi_{i1}}{\pi_{i1} + \pi_{i2}}\right)^{y_i''} \left(\frac{\pi_{i2}}{\pi_{i1} + \pi_{i2}}\right)^{(1-y_i'')} \\
&= \left(\frac{\pi_{i1}}{\pi_{i1} + \pi_{i2}}\right)^{w_{i1}} \left(\frac{\pi_{i2}}{\pi_{i1} + \pi_{i2}}\right)^{w_{i2}}
\end{aligned}
\tag{5.15}
$$

Multiplying equation (5.14) by equation (5.15), we obtain

$$
p(y_i')p(y_i'') = \pi_{i1}^{w_{i1}} \pi_{i2}^{w_{i2}} \pi_{i3}^{w_{i3}} = p(y_i)
$$

which is the required result. Although we have demonstrated this result in the context of a specific example, the argument applies to any binary division of a set of categories, and, therefore, may be extended by repeated application to nested dichotomies generally.

The implication of this argument is that we may fit separate logit models to each of a set of nested dichotomies. The resulting maximum-likelihood estimates are identical to those that would be produced by maximizing the likelihood simultaneously with respect to the combined parameters in all of the models. Moreover, since the log of the likelihood for the combined model is the sum of the log likelihoods for the separate models, likelihood-ratio chi-square statistics may be summed to produce tests for the model as a whole. This is not to say, however, that the models for a set of nested dichotomies are equivalent to the model (5.12) developed in the previous section. In general, the two models are not identical, although they usually yield similar results. We have simply shown that, having specified a set of models for nested dichotomies, we may fit the models all at once or separately to each dichotomy.

The data in Table 5.1 are drawn from a sample survey of the Canadian population conducted in 1977. We shall use these data to examine how the labor-force participation of young married women (21 to 30 years of age) is affected by the presence of children in the household, by husband's income, and by the region of the country in which they reside. As described in this section, labor-force participation is treated as a pair of nested dichotomies: (W) working versus not working outside of the home; and, for those who work, (F) working full time versus working part time. Presence of children and husband's income are expected to have negative effects on both dichotomies. Husband's annual income (I), measured in thousands of dollars, is determined by subtracting each woman's income from her reported family income. Presence of children is represented by a dummy variable (K) coded one if minor children are present in the household and zero otherwise. Since husband's income might well have a greater effect among women without children, an interaction regressor ($I \times K$) is included in the model. Finally, four dummy variables (R_1—R_4) are constructed to represent five regions of Canada: the Atlantic provinces, Quebec, Ontario, the prairie provinces, and British Columbia.

Several models are fit to the data in order to provide tests of the independent-variable effects; the log likelihood and degrees of freedom for each model are shown in Table 5.2. The tests, constructed according to the likelihood-ratio approach (equation (5.10)), are given in Table 5.3. The first row of this table reports a simultaneous test for all effects in the model. This test is analogous to the omnibus F-test for linear models. The income $\times$ children interactions and the region effects are nonsignificant. There are statistically significant main effects of income and presence of children. The general pattern of results is similar for the two dichotomies.

Fitted logit models, deleting the region effects and the $I \times K$ interaction, are given in equations (5.16), where the ratio of each coefficient to its standard error is shown in brackets:

$$\begin{aligned} \Lambda^{-1}(P_W) &= 1.336 - \underset{[-2.14]}{0.04231}\, I - \underset{[-5.39]}{1.576}\, K \\ \Lambda^{-1}(P_F) &= 3.478 - \underset{[-2.74]}{0.1073}\, I - \underset{[-4.90]}{2.651}\, K \end{aligned} \tag{5.16}$$

TABLE 5.1. Labor-Force Participation, Husband's Income, Presence of Children Under 18 in the Household, and Region of Residence, for Married Canadian Women 21 – 30 Years Old in 1977

Observation	L^a	I^b	K^c	R^d
1	3	15	1	3
2	3	13	1	3
3	3	45	1	3
4	3	23	1	3
5	3	19	1	3
6	3	7	1	3
7	3	15	1	3
8	1	7	1	3
9	3	15	1	3
10	3	23	1	3
11	3	23	1	3
12	1	13	1	3
13	3	9	1	4
14	3	9	1	4
15	3	45	1	1
16	3	15	1	1
17	3	5	1	3
18	3	9	1	3
19	3	13	1	3
20	3	13	0	3
21	2	19	0	3
22	3	23	1	4
23	1	10	0	4
24	1	11	0	3
25	3	23	1	3
26	3	23	1	3
27	3	19	1	3
28	3	19	1	3
29	3	17	1	4
30	1	14	1	4
31	3	13	1	3
32	3	13	1	3
33	3	15	1	3
34	3	9	0	3
35	3	9	0	3
36	3	19	0	3
37	3	15	1	3
38	1	20	0	3
39	3	9	1	1
40	2	6	0	1
41	3	9	1	5
42	2	4	1	3
43	2	28	0	3
44	3	23	1	3
45	2	5	1	3
46	3	28	1	3
47	3	7	1	3
48	3	7	1	3
49	3	23	1	4
50	1	15	0	4
51	2	10	1	4
52	2	10	1	4
53	3	9	0	3
54	3	9	0	3
55	2	9	1	1
56	3	17	0	1
57	3	23	1	1
58	3	23	1	1
59	3	9	1	3
60	3	9	1	3
61	1	9	0	3
62	1	28	0	3
63	2	10	1	3
64	2	23	0	4
65	3	11	1	4
66	3	15	1	3
67	3	15	1	3
68	3	19	1	3
69	3	19	1	3
70	3	23	1	3
71	3	17	1	3
72	3	17	1	3
73	3	17	1	3
74	3	17	1	3
75	3	17	1	3
76	2	38	1	3
77	2	38	1	3
78	3	7	1	1
79	3	19	1	4
80	2	19	1	5

TABLE 5.1. (*Continued*)

Observation	L^a	I^b	K^c	R^d
81	1	13	0	3
82	2	15	1	3
83	1	17	1	3
84	1	17	1	3
85	2	23	1	3
86	1	27	0	5
87	1	16	1	5
88	1	27	0	3
89	3	35	0	3
90	3	35	0	3
91	3	35	0	3
92	2	9	1	3
93	2	9	1	3
94	2	9	1	3
95	3	13	1	3
96	3	17	1	3
97	3	17	1	3
98	1	15	0	3
99	1	15	0	3
100	3	15	1	3
101	1	11	0	1
102	3	23	1	1
103	3	15	1	1
104	3	15	0	5
105	2	12	0	5
106	2	12	0	5
107	3	13	1	4
108	3	19	1	3
109	3	19	1	1
110	3	3	1	1
111	3	9	1	1
112	1	17	1	1
113	3	1	1	1
114	3	1	1	1
115	2	13	1	4
116	3	13	1	4
117	3	19	0	5
118	3	19	0	5
119	1	15	0	5
120	2	30	1	3
121	3	9	1	1
122	3	23	1	1
123	1	9	0	3
124	1	9	0	3
125	3	13	1	4
126	2	13	1	3
127	3	17	1	1
128	2	13	1	4
129	2	13	1	4
130	2	19	1	3
131	2	19	1	3
132	3	3	1	3
133	1	14	0	3
134	1	14	0	3
135	1	11	1	3
136	1	11	1	3
137	2	14	1	3
138	3	13	1	3
139	3	28	1	3
140	3	28	1	3
141	3	14	1	3
142	3	14	1	3
143	3	11	1	4
144	3	13	1	4
145	3	13	1	4
146	2	11	1	1
147	2	11	1	1
148	3	19	1	5
149	1	6	0	5
150	3	28	0	5
151	1	13	0	5
152	1	13	0	5
153	3	5	0	5
154	2	28	1	5
155	2	11	1	5
156	3	23	1	5
157	2	15	1	5
158	3	13	1	5
159	1	22	0	3
160	1	15	0	3
161	3	15	1	3
162	3	15	1	1
163	1	5	1	1
164	1	1	0	4

TABLE 5.1. (*Continued*)

Observation	L^a	I^b	K^c	R^d
165	1	1	0	4
166	3	9	1	1
167	3	15	1	3
168	1	13	0	3
169	3	19	1	1
170	2	8	1	5
171	1	7	1	4
172	3	19	1	3
173	3	7	1	3
174	1	9	0	3
175	1	9	0	3
176	1	24	0	3
177	3	15	1	3
178	1	13	0	3
179	3	13	0	5
180	1	13	0	5
181	1	17	1	1
182	1	16	0	1
183	1	18	0	3
184	1	18	0	3
185	3	13	0	3
186	2	15	1	5
187	3	13	1	5
188	3	7	1	5
189	1	9	1	1
190	3	23	1	5
191	3	17	1	4
192	3	15	1	5
193	3	11	1	4
194	3	17	1	4
195	3	17	1	4
196	1	5	1	4
197	1	5	1	4
198	3	26	1	3
199	1	10	0	2
200	1	11	0	2
201	1	20	1	2
202	3	13	1	2
203	3	15	1	2
204	3	28	1	2
205	2	9	1	2
206	3	19	1	2
207	3	11	1	2
208	1	11	0	2
209	3	9	1	2
210	1	10	0	2
211	3	19	1	2
212	3	13	1	2
213	1	3	0	2
214	3	15	1	2
215	3	15	1	2
216	2	17	1	2
217	3	7	1	2
218	2	15	0	2
219	3	19	1	2
220	1	16	0	2
221	3	5	0	2
222	3	11	1	2
223	3	11	1	2
224	3	19	1	2
225	3	15	1	2
226	3	15	1	2
227	3	11	1	2
228	1	5	0	2
229	2	23	1	2
230	2	23	1	2
231	3	7	1	2
232	3	13	1	2
233	1	15	0	2
234	1	5	0	2
235	3	7	1	2
236	1	6	0	2
237	1	5	1	2
238	1	5	1	2
239	3	13	1	2
240	3	13	1	2
241	3	13	1	2
242	3	13	0	2
243	3	17	1	2
244	1	6	1	2
245	3	5	1	2
246	2	19	1	2
247	1	3	1	2
248	3	23	0	2
249	3	23	0	2
250	1	15	0	2

TABLE 5.1. (*Continued*)

Observation	L^a	I^b	K^c	R^d
251	3	11	0	2
252	3	23	0	2
253	3	13	1	2
254	2	23	1	2
255	1	11	0	2
256	3	9	0	2
257	1	2	0	2
258	3	15	1	2
259	3	15	0	2
260	3	15	1	2
261	3	11	1	2
262	3	11	0	2
263	3	15	1	2

Source: These data were collected as part of the Social Change in Canada Project, directed by T. Atkinson, B. Blishen, M. Ornstein, and H. Stevenson of York University. The research was supported by SSHRCC grant #S75-0332. The data were made available by the Institute for Behavioural Research of York University. Neither the principal investigators nor the disseminating archive are responsible for the interpretations presented here.

[a]Labor-force participation:
1. working full time
2. working part time
3. not working

[b]Husband's income, in thousands of dollars.

[c]Presence of children in the household:
0. children absent
1. children present

[d]Region:
1. Atlantic provinces
2. Quebec
3. Ontario
4. prairie provinces
5. British Columbia

TABLE 5.2. Log Likelihood for Fitted Logit Models

			Log Likelihood	
Model	Regressors[a]	*df*	W^b	F^c
0	C	1	−178.08	−72.17
1	C, I, K, R	7	−158.65	−50.92
2	C, I, K	3	−159.87	−52.25
3	C, I, R	6	−173.93	−66.88
4	C, K, R	6	−161.22	−54.84
5	$C, I, K, R, I \times K$	8	−158.27	−50.80

[a]C: constant
I: husband's income
K: presence of children
R: region (four dummy variables).

[b]W: working versus not working (n = 263).

[c]F: working full time versus part time (n = 108).

Notice that both husband's income and presence of children have greater effects on the decision to work full time than on the decision to work.

For purposes of comparison, equations (5.17) show the results of fitting linear probability models by OLS regression to the women's labor force participation data:

$$\begin{aligned} P_W &= \underset{}{0.7939} - \underset{[-2.17]}{0.008538\, I} - \underset{[-5.93]}{0.3674\ K} \\ P_F &= 1.085 - \underset{[-3.01]}{0.01644\, I} - \underset{[-6.30]}{0.4931\ K} \end{aligned} \tag{5.17}$$

Although the coefficients of the fitted models in equations (5.16) and (5.17) cannot be directly compared, since the functional forms of the models are different, the signs of the coefficients agree, and the ratios of the coefficients to their standard errors are not grossly different; we should therefore make similar inferences in the two cases. The fitted models are graphed in Figures 5.4(a) and (b). The income ranges in these plots are the ranges observed in the sample. The results for the logit and linear probability models are generally similar, especially for the working/not-working dichotomy. Notice, however, that the linear probability model for the full-time/part-time dichotomy produces some fitted values outside of the unit interval.

Other approaches to modeling polychotomous data are described in McFadden (1974) and Maddala (1983: Chapters 2, 3).

5.1.6. Diagnostic Methods for Logit Models

As in the case of linear models fit by least squares, it is desirable to assess the adequacy of a fitted linear logit model. Many of the diagnostic methods developed in Chapter 3 may be extended to logit models. The discussion in this section is based on the work of Pregibon (1981) and Landwehr, Pregibon, and Shoemaker (1980).

Fitting a logit model by maximum likelihood is intrinsically more complex than fitting a linear model by least squares. One consequence of this relative

TABLE 5.3. Likelihood-Ratio Tests for Terms in the Logit Models

			W		F		$W+F$		
Source	Models Contrasted[a]	df	G^2	p	G^2	p	G^2	df	p
$I, K, R, I \times K$	5–0	7	39.62	< .001	42.74	< .001	82.36	14	< .001
$I \times K$	5–1	1	0.76	.38	0.24	.62	1.00	2	.60
I (Income)	1–4	1	5.14	.02	7.84	.005	12.98	2	.001
K (Children)	1–3	1	30.56	< .001	31.92	< .001	62.48	2	< .001
R (Region)	1–2	4	2.44	.66	2.66	.62	5.10	8	.75

[a] From Table 5.2.

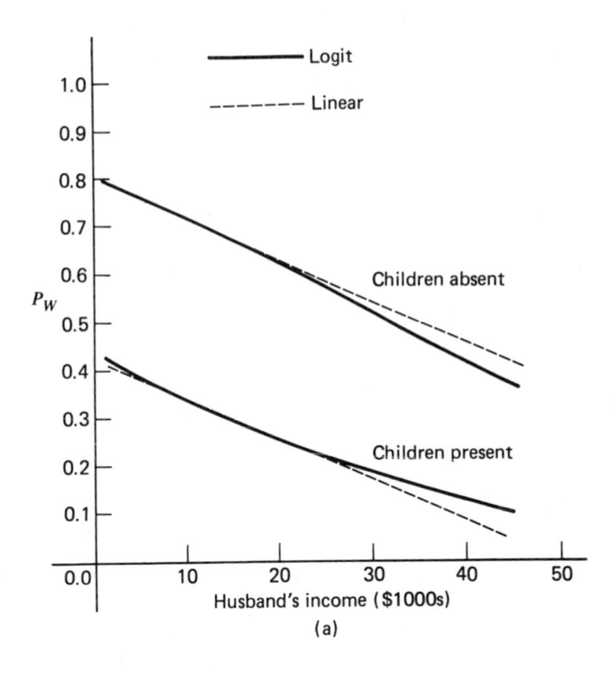

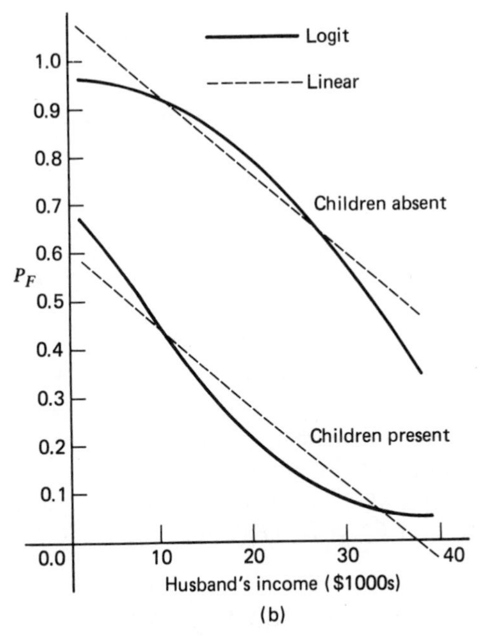

FIGURE 5.4. Models fit to women's employment data. (a) Working versus not working outside the home. (b) Working full time versus part time.

complexity is that it becomes more difficult to assess how the fit is affected—both potentially and actually—by variations in the data. The reader should be cautioned that the parallels between the methods described in this section and those of Chapter 3 are less than perfect. Nevertheless, diagnostic methods for logit models are valuable data-analytic tools, and are likely to be improved and extended in the future.

There are several different ways of defining residuals for logit models. Most straightforwardly, we take $E_i \equiv Y_i - P_i$, where, we recall, $P_i = 1/(1 + e^{-\mathbf{x}_i'\mathbf{b}})$ is the fitted probability for the ith observation.

It is also possible to define residuals by developing parallels to the residual sum of squares in linear models. We shall pursue two such possibilities, both based upon chi-square statistics. The *goodness-of-fit statistic*

$$Z^2 \equiv \sum_{i=1}^{n} \frac{E_i^2}{P_i(1 - P_i)}$$

is asymptotically distributed as $\chi^2(n - k)$ under the model. Large values of this statistic, therefore, indicate a poor fit, although a hypothesis test of the adequacy of the model based on Z^2 has low power because of its lack of specificity. The ith component of Z^2, $Z_i \equiv E_i/\sqrt{P_i(1 - P_i)}$, however, is a useful indicator of lack of fit for observation i. If n is large, then each Z_i follows an approximate unit-normal distribution.

A likelihood-ratio statistic similar to Z^2 is the *deviance*

$$G^2 \equiv -2\log L(\mathbf{b}) = -2\sum[Y_i \log P_i + (1 - Y_i)\log(1 - P_i)]$$

which implicitly contrasts the fitted model with one that predicts each observation perfectly (and hence for which $L = 1$ and $\log L = 0$). G^2, like Z^2, follows an asymptotic chi-square distribution with $n - k$ degrees of freedom. The ith component of the deviance is

$$G_i \equiv \pm\sqrt{-2[Y_i \log P_i + (1 - Y_i)\log(1 - P_i)]}$$

where the sign is chosen to agree with that of E_i.

In Problem 5.5 we show that there is an analogy between logistic regression and weighted-least-squares regression, according to which $\mathbf{y}^* \equiv \mathbf{Xb} + \mathbf{V}^{-1}\mathbf{e}$ plays the role of the "dependent variable." As in Section 5.1.3, $\mathbf{V} = \text{diag}\{P_i(1 - P_i)\}$. Landwehr, Pregibon, and Shoemaker suggest that the Y_i^* may be thought of as *pseudo-observations* on the logit scale; thus $\hat{Y}_i^* \equiv \mathbf{x}_i'\mathbf{b}$ is the logit fitted value, and $E_i^* \equiv E_i/[P_i(1 - P_i)]$ is the logit residual.

Pregibon shows that

$$\mathbf{H} \equiv \mathbf{V}^{1/2}\mathbf{X}(\mathbf{X}'\mathbf{V}\mathbf{X})^{-1}\mathbf{X}'\mathbf{V}^{1/2}$$

is in many respects analogous to the hat matrix in least-squares regression. For instance, $\mathbf{z} = (\mathbf{I}_n - \mathbf{H})\mathbf{z}$ (while in least-squares regression $\mathbf{e} = (\mathbf{I}_n - \mathbf{H})\mathbf{e}$). Furthermore, $\mathbf{H}$ is symmetric and idempotent; the diagonal values h_{ii} are bounded by zero and one; and the average of the h_{ii} is $\bar{h} = k/n$.[2] The h_{ii} are often useful for detecting high-leverage observations. Note that unlike in linear models, however, the hat values depend upon the dependent-variable scores (through the fitted probabilities $\mathbf{p}$, which determine $\mathbf{V}$) as well as upon the design matrix $\mathbf{X}$.

A type of studentized residual may be defined by dedicating a parameter to the ith observation to produce the modified model

$$\Lambda^{-1}(\pi) = \mathbf{x}'\boldsymbol{\beta} + \gamma C_i \tag{5.18}$$

in which the dummy variable C_i is coded one for observation i and zero for all other observations. This procedure is equivalent to deleting the ith observation from the sample. If L represents the maximized log likelihood for the original model and $L_{(-i)}$ the log likelihood for model (5.18), then $G^2_{(-i)} \equiv -2\log(L/L_{(-i)})$ is asymptotically distributed as $\chi^2(1)$ under the hypothesis $H_0: \gamma = 0$. Thus $G_{(-i)}$ (with the sign chosen from the sign of E_i) follows an asymptotic unit-normal distribution. Because the estimating equations for the logit model are nonlinear, to calculate the $G^2_{(-i)}$ exactly requires refitting the model for each observation. Although this is not a practical procedure, Pregibon shows that

$$G^2_{(-i)} \simeq G_i^2 + \frac{Z_i^2 h_{ii}}{1 - h_{ii}} \tag{5.19}$$

Using G_i in place of Z_i in equation (5.19) and taking the square root produces the approximate studentized residual $G_i^* \equiv G_i/\sqrt{1 - h_{ii}}$. Normal-probability plots of G_i^* are useful for revealing outliers, although interpretative problems are sometimes caused by discreteness.[3]

In Section 3.2 we developed the topic of influence in linear models by examining how the deletion of an observation affects the least-squares regression coefficients. Analogously, for the linear logit model, we may define $\mathbf{d}_i \equiv \mathbf{b} - \mathbf{b}_{(-i)}$, where $\mathbf{b}_{(-i)}$ represents the maximum-likelihood estimator of $\boldsymbol{\beta}$ omitting observation i. To find $\mathbf{b}_{(-i)}$ requires refitting the model, but an approximation, $\mathbf{b}^*_{(-i)}$, may be obtained by performing a single Newton–Raphson iteration starting at $\mathbf{b}_0 = \mathbf{b}$. Pregibon proves that

$$\mathbf{d}_i^* \equiv \mathbf{b} - \mathbf{b}^*_{(-i)} = (\mathbf{X}'\mathbf{V}\mathbf{X})^{-1}\mathbf{x}_i \frac{E_i}{1 - h_{ii}} \tag{5.20}$$

(A similar one-step approximation was the basis of equation (5.19).)

[2] Remember that in this chapter, k counts all regressors, including the constant regressor.

[3] Imagine, for instance, a logit simple-regression model $\Lambda^{-1}(\pi) = \alpha + \beta X$ fit to a sample in which there are replicated observations at each X-value. Since Y must either be zero or one, there are only two possible G^*s for each X. More generally, there are two G^*s for each *combination* of independent-variable values. Landwehr, Pregibon, and Shoemaker suggest a simulation procedure that provides an empirical sampling distribution for the residuals.

An analog of Cook's influence statistic follows from the generalized likelihood-ratio test statistic. To test H_0: $\boldsymbol{\beta} = \boldsymbol{\beta}_0$, we would compute

$$G_0^2 = -2\log\frac{L(\boldsymbol{\beta}_0)}{L(\mathbf{b})}$$
$$\simeq (\mathbf{b} - \boldsymbol{\beta}_0)'\widehat{\mathscr{V}(\mathbf{b})}^{-1}(\mathbf{b} - \boldsymbol{\beta}_0)$$
$$= (\mathbf{b} - \boldsymbol{\beta}_0)'\mathbf{X}'\mathbf{V}\mathbf{X}(\mathbf{b} - \boldsymbol{\beta}_0)$$

(The approximation is justified by the asymptotic normality of **b**.) We may use G_0^2 for the hypothesis H_0: $\boldsymbol{\beta} = \mathbf{b}^*_{(-i)}$ as a scale-invariant scalar index of the distance between $\mathbf{b}^*_{(-i)}$ and **b**. Substituting from equation (5.20), we obtain

$$D_i^* \equiv \left(\mathbf{b} - \mathbf{b}^*_{(-i)}\right)'\mathbf{X}'\mathbf{V}\mathbf{X}\left(\mathbf{b} - \mathbf{b}^*_{(-i)}\right)$$
$$= \frac{E_i^2}{(1 - h_{ii})^2}\mathbf{x}_i'(\mathbf{X}'\mathbf{V}\mathbf{X})^{-1}\mathbf{x}_i$$
$$= \frac{Z_i^2 h_{ii}}{(1 - h_{ii})^2}$$

As in linear models, therefore, the influence of an observation is an increasing function of its residual and its leverage.

We explained previously that

$$\mathbf{y}^* = \mathbf{X}\mathbf{b} + \mathbf{V}^{-1}\mathbf{e} = \hat{\mathbf{y}}^* + \mathbf{e}^*$$

represents pseudo-observations on the logit scale. Nonlinearity can be detecting by plotting the logit residuals $\mathbf{e}^*$ against the columns of **X**. Following Landwehr, Pregibon, and Shoemaker, logit partial residuals for the jth regressor are defined by $\mathbf{e}_j^* \equiv \mathbf{e}^* + B_j\mathbf{x}_j$, where $\mathbf{x}_j$ is the jth *column* of the design matrix **X**, and B_j is the estimated coefficient for the jth regressor. Because of discreteness, plots of logit residuals and partial residuals may require smoothing.

In Section 5.1.5, we fit linear logit models to the women's labor-force data from Table 5.1, defining nested dichotomies for the dependent variable. The final model fitted to the full-time/part-time-work dichotomy F was (from equation (5.16))

$$\Lambda^{-1}(P_F) = 3.478 - 0.1073I - 2.651K \tag{5.21}$$

Recall that I is husband's income and K is a dummy variable recording the presence of children in the household.

Figure 5.5 presents index plots for a variety of diagnostic statistics derived from the fitted logit model. To construct an index plot, we employ the

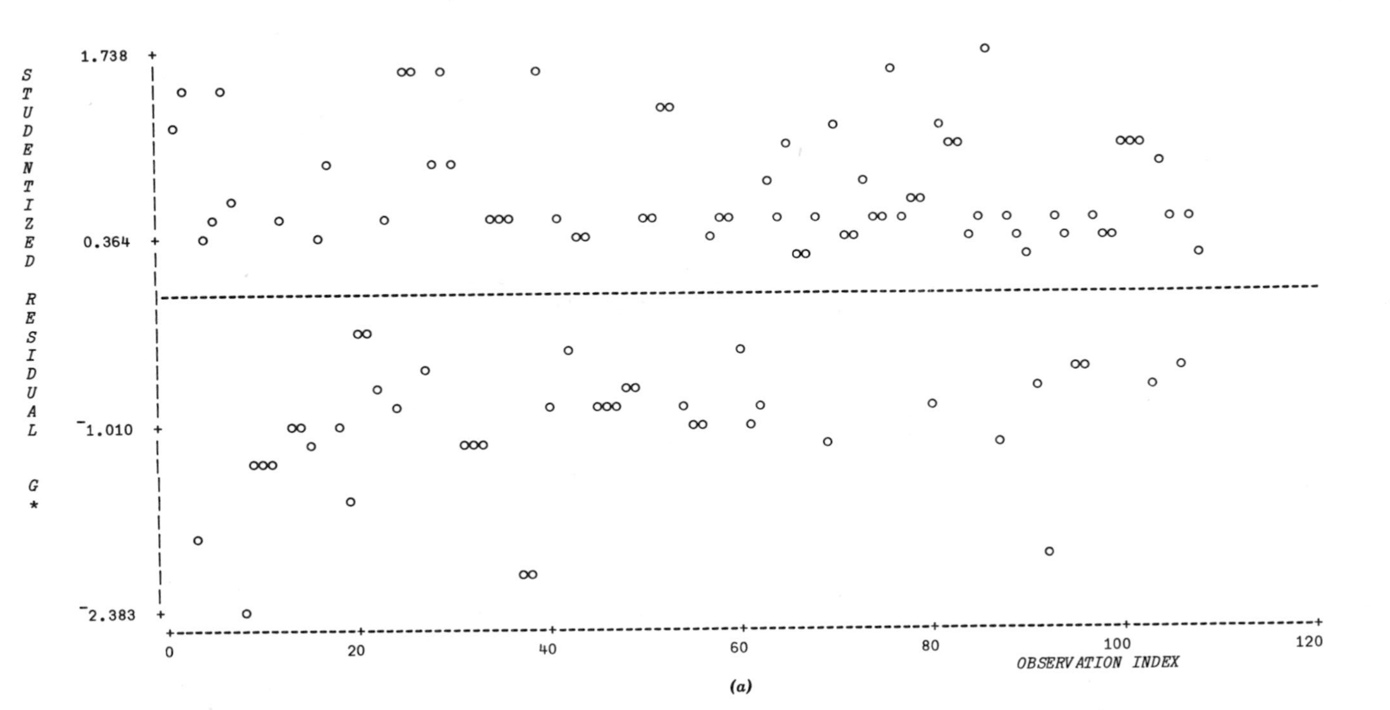

(a)

FIGURE 5.5. Index plots for logit-regression diagnostics, women's employment data. (a) Approximate studentized residuals, G_i^*. (b) Hat values, h_{ii}. (c) Cook's influence statistics, D_i^*. (d) Approximate influence on intercept. (e) Approximate influence on husband's-income coefficient. (f) Approximate influence on presence-of-children coefficient.

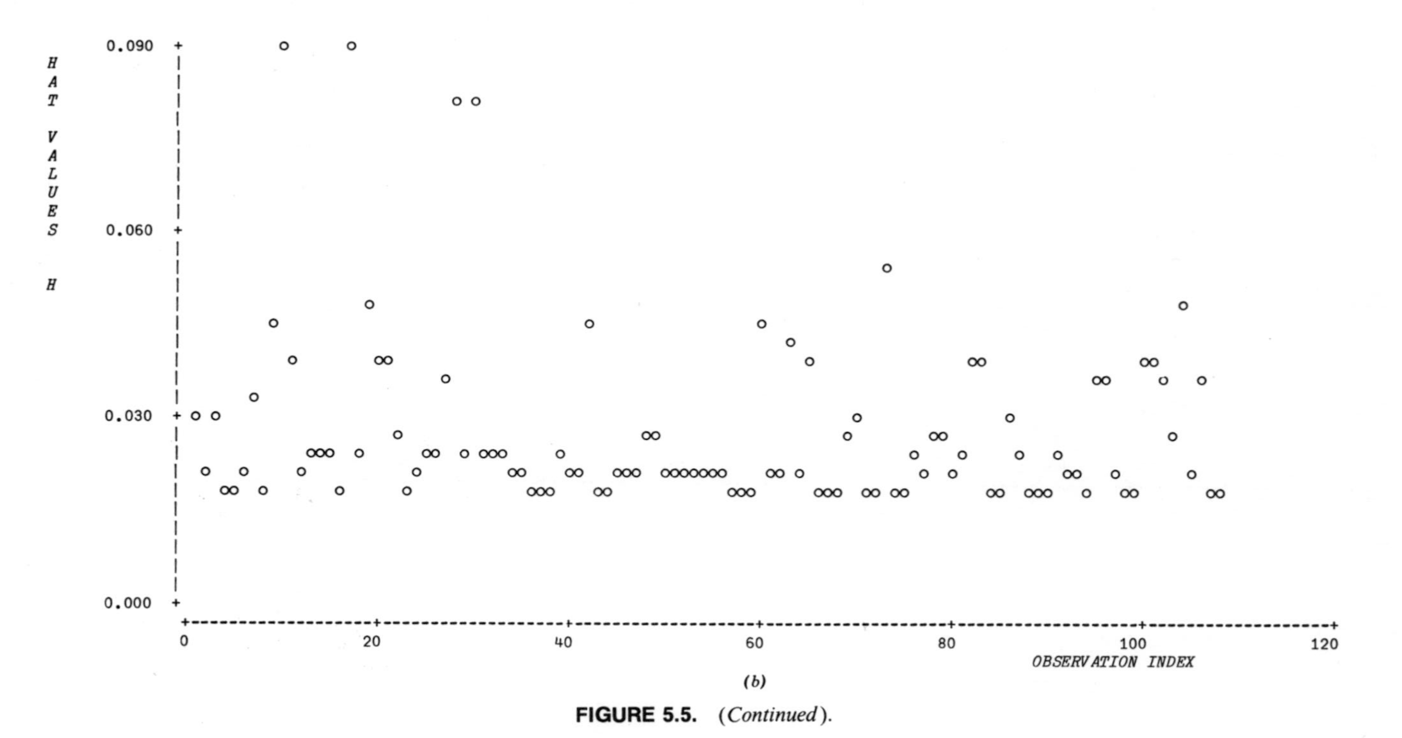

(b)

FIGURE 5.5. *(Continued).*

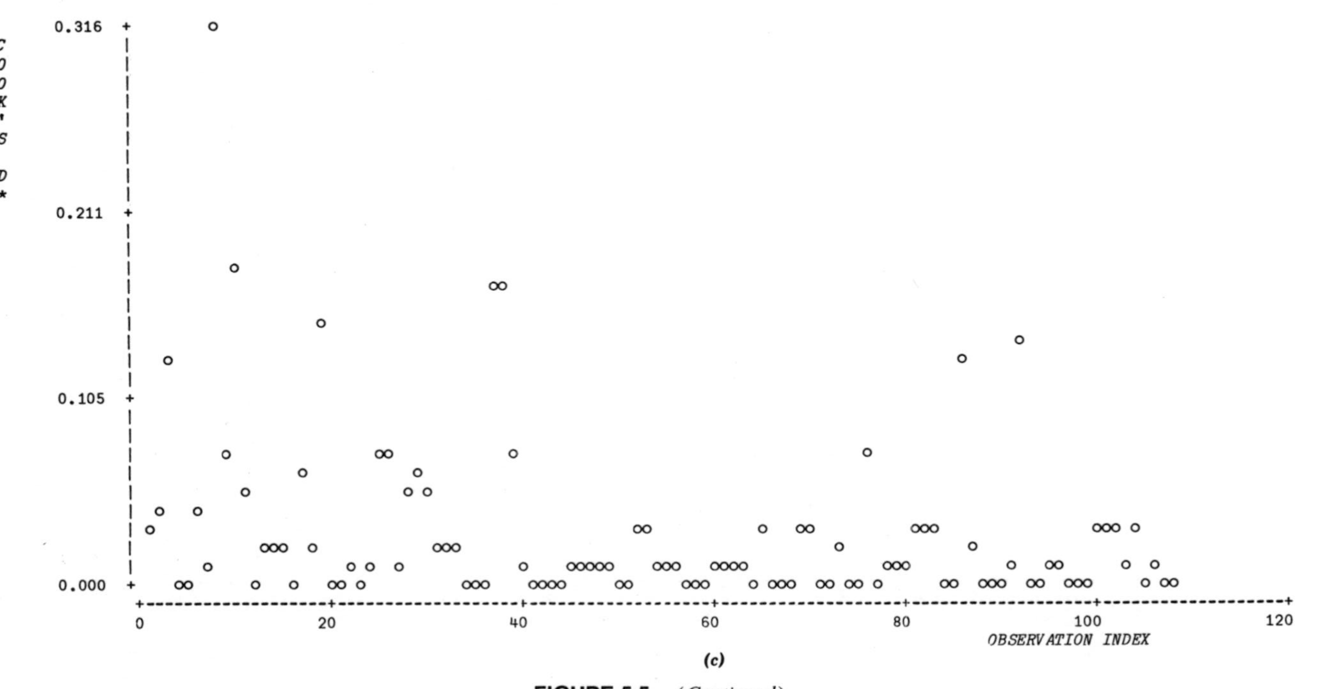

(c)

FIGURE 5.5. (*Continued*).

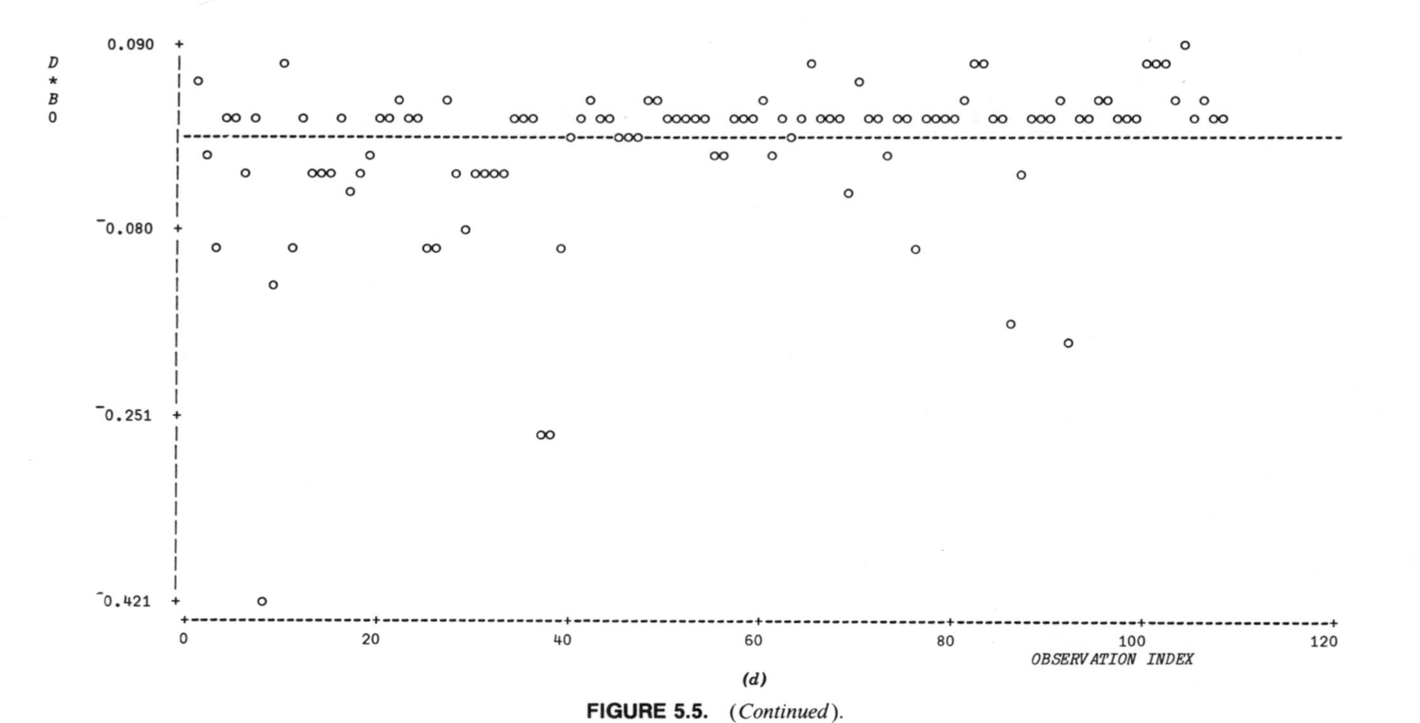

(d)

FIGURE 5.5. (*Continued*).

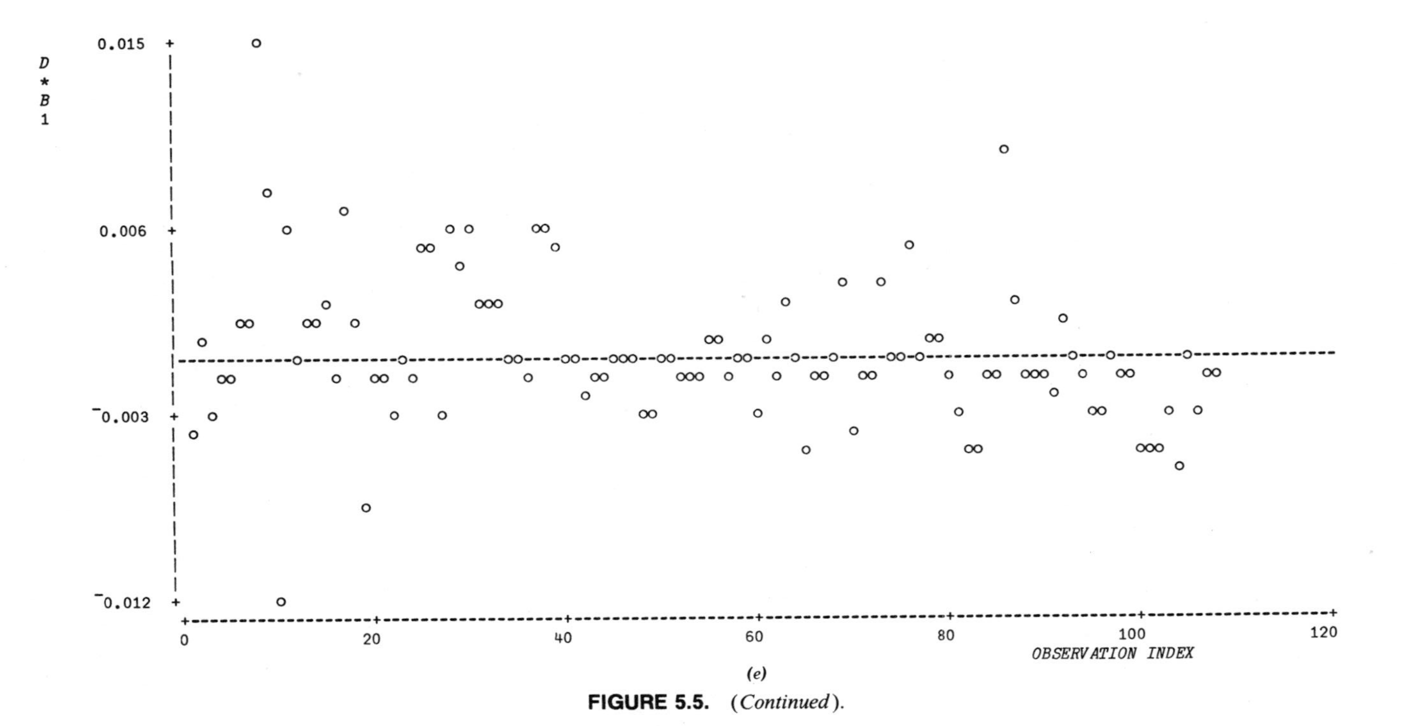

(e)

FIGURE 5.5. (*Continued*).

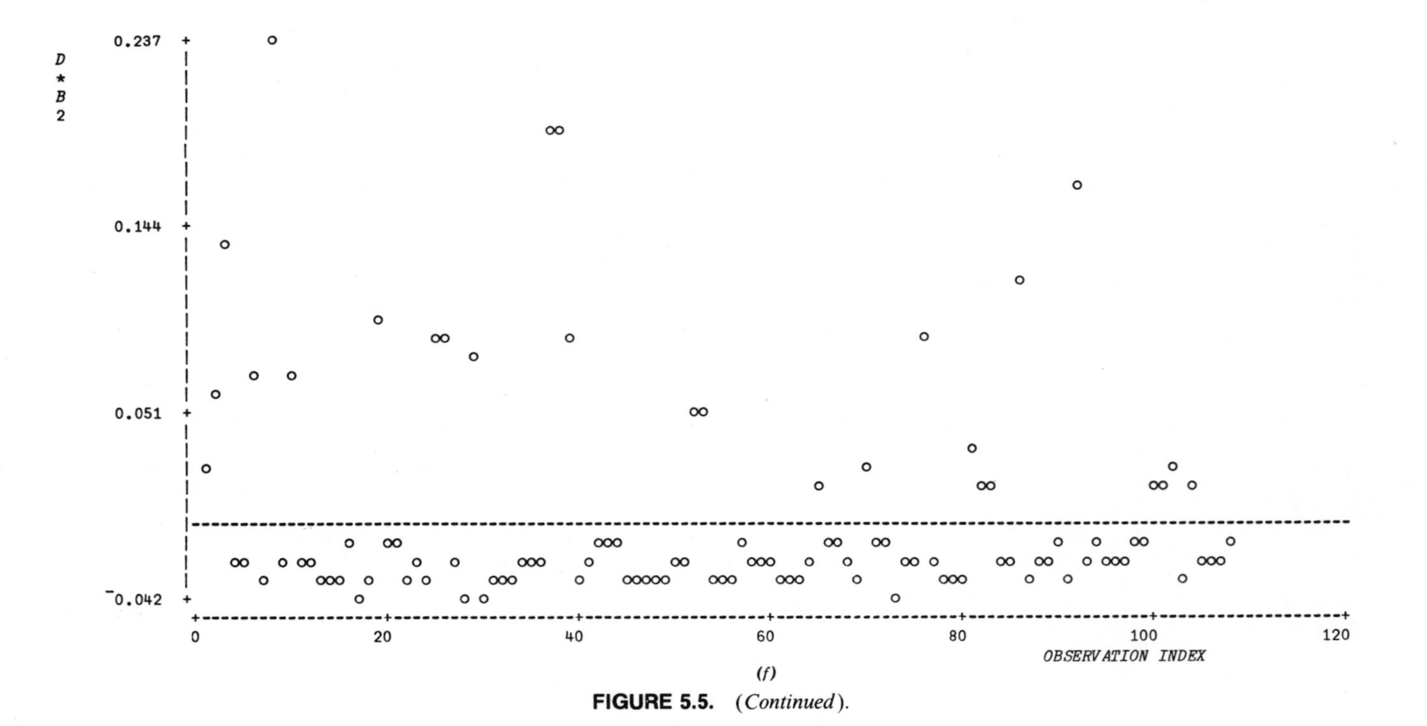

(f)

FIGURE 5.5. (*Continued*).

observation indices $(1, 2, \ldots, n)$ as horizontal coordinates (recall Problem 3.5, Chapter 3). These plots are useful for examining the general distribution of diagnostic statistics and for identifying noteworthy observations.

Figure 5.5(a) shows the approximate studentized residuals G_i^* for the 108 working women in the sample. The largest residual (-2.383) is for the eighth case, but none of the observations appears to be an outlier. Notice that there are few residuals with values near zero. This gap in the residual distribution is a consequence of the discreteness of the observations: Since the dependent variable takes on scores of zero and one, a residual near zero can be obtained only when the fitted probabiiity is itself close to zero or one.

Figure 5.5(b) indicates that several of the observations have hat values h_{ii} in excess of $2\bar{h} = 2(3)/108 = .0556$. The most influential observation according to Cook's statistic D_i^* (graphed in Figure 5.5(c)), however, is observation eight which, curiously, has quite a small hat value. This anomaly points up the complexity of logit-regression diagnostics in comparison with their linear-model counterparts. Figure 5.5(d), (e), and (f) indicate the approximate effect $\mathbf{d}_i^*$ on the fitted coefficients of deleting each observation. Observation eight has the largest influence on all three coefficients, decreasing the intercept, increasing the income slope, and increasing the coefficient for presence of children. (Since the last two coefficients are negative, observation eight *decreases* their magnitude.)

If the eighth observation is omitted, the fitted model becomes

$$\Lambda^{-1}(P_F) = 3.941 - 0.1241I - 2.914K$$

Notice that the approximate influence values given in $\mathbf{d}_8^* = (-0.421, 0.015, 0.237)'$ are quite accurate, although as one-step approximations they slightly understate the difference between $\mathbf{b}$ and $\mathbf{b}_{(-i)}$. On the whole, then, omitting the eighth observation has a small but noticeable impact on the model coefficients. Further investigation reveals that no other observation has as substantial an influence on the fit.

Figure 5.6 shows a partial residual plot for husband's income I, using the full-sample fit of equation (5.21). The filled circles and the line superimposed on the plot represent a moving-median smooth of the sort described in Section 3.2.8. Before we interpret the plot, two of its features require comment. First, note that the partial residuals are highly discrete: For each value of husband's income there are only four possible residuals, corresponding to the four combinations of presence of children and the dependent variable. Second, there are some very large negative residuals in the plot that are detached from the body of the data. The biggest of these, -18.66, is for the eighth observation—a woman without children whose husband earns \$6000, but who works part time rather than full time. Since the fitted value for this woman is near one ($P_8 = .9445$, see Figure 5.4), the logit residual $E^* = (Y - P)/[P(1 - P)]$ is very big ($E_8^* = -18.02$), producing the large negative partial residual shown in the plot. There is some evidence, then, that the logit model produces unrealistically large fitted probabilities of working full time for women with no

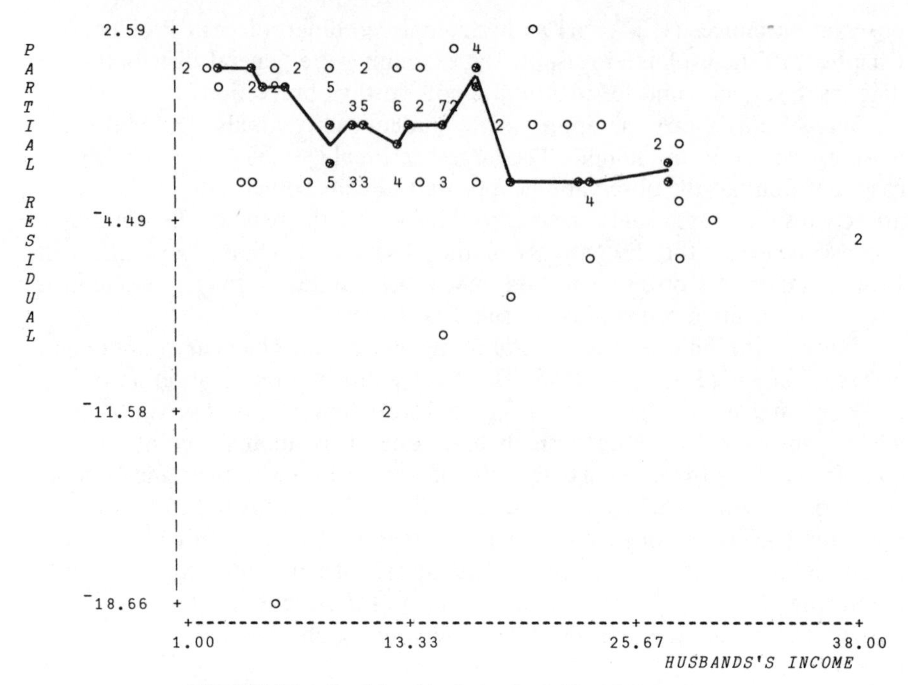

FIGURE 5.6. Partial residual plot for husband's income.

children whose husbands have relatively low incomes. Recall from Figure 5.4 that the linear probability model produced even more extreme results here.

Even the smoothed plot in Figure 5.6 is fairly irregular: The relationship between the logit dependent variable and husband's income appears reasonably straight, although there is some suggestion of nonlinearity. A little experimentation shows that a slight improvement in the fit may be obtained by moving down the ladder of powers, replacing I with $\log I$.

PROBLEMS

5.1. By the June 30, 1982 deadline, 35 states had voted to ratify the Equal Rights Amendment (guaranteeing equal rights under the law to women and men), three short of the 38 states necessary for ratification. Five of these 35 states voted to rescind an earlier ratification vote. Table 5.4 shows the ERA vote by state, along with information on the percent of women in each state legislature, and the percent popular vote for Reagan (as opposed to Carter or Anderson) in the 1980 presidential election. This last variable is intended as a measure of political conservatism.

(a) Define the nested dichotomies: (i) ratified versus rescinded or did not ratify; and (ii) rescinded versus did not ratify. Using a linear

TABLE 5.4. ERA Ratification Data

State	ERA Ratification[a]	Percent Women[b] in Legislature	Percent Vote for Reagan
AL	3	4.3	50.0
AK	1	10.0	61.9
AZ	3	17.0	62.1
AR	3	3.7	48.9
CA	1	10.0	54.2
CO	1	23.0	56.7
CT	1	21.4	48.7
DE	1	14.5	47.7
FL	3	10.0	56.0
GA	3	7.2	41.4
HI	1	18.4	43.7
ID	2	9.5	67.9
IL	3	13.6	50.3
IN	1	8.0	56.8
IA	1	12.0	52.0
KS	1	13.3	59.0
KY	2	7.2	49.5
LA	3	1.4	51.9
ME	1	22.8	46.5
MD	1	14.9	44.6
MA	1	9.5	42.4
MI	3	10.8	49.7
MN	1	11.9	43.6
MS	1	1.1	50.0
MO	3	11.2	51.6
MT	1	11.3	58.4
NE	2	10.2	66.5
NV	3	11.7	64.8
NH	1	29.0	58.3
NJ	1	6.7	52.8
NM	1	6.3	56.0
NY	1	8.6	47.5
NC	3	12.9	49.6
ND	1	12.0	65.3
OH	1	7.6	52.4
OK	3	8.1	61.2
OR	1	22.2	50.1
PA	1	4.3	50.3
RI	1	10.0	37.5
SC	3	6.5	49.8
SD	2	10.5	61.3
TN	2	4.5	49.0
TX	1	6.6	55.8
UT	3	6.7	74.0
VT	1	21.7	45.4
VA	3	6.4	53.9
WA	1	23.8	50.9
WV	1	11.9	45.6
WI	1	15.2	48.8
WY	1	18.5	64.3

[a]Coding:
1 ratified
2 ratified but rescinded
3 did not ratify

[b]*Source:* Adapted with permission from National Women's Political Caucus (1981: Table 1).

logit model, regress each dichotomy on percent of women legislators and percent vote for Reagan.

(b) Repeat the analysis in part (a) using a linear probability model fit by least squares. Compare the results obtained here with those of part (a).

(c) Use the diagnostic procedures discussed in Section 5.1.6 to assess the adequacy of the models fit in part (a), attempting to correct any problems that are revealed.

5.2. Use the diagnostic procedures of Section 5.1.6 to assess the adequacy of the logit model (from (5.16)) fit to the working/not-working dichotomy for the women's labor-force data in Table 5.1.

5.3. Fit a polychotomous linear logit model (equation (5.12)) to the women's labor-force data given in Table 5.1. How do the results of your analysis compare with those reported in Section 5.1.5, where labor-force participation was treated as a pair of nested dichotomies? Which analysis do you prefer?

5.4.† Efficient Computation for Discrete Independent Variables and Testing Nonlinearity: The data shown in Table 5.5 are drawn from a 1977

TABLE 5.5. Perceived Need by Income

	Perceived Need	
Family Income	No	Yes
$0	8	5
$1–$1999	17	16
$2000–$3999	88	76
$4000–$5999	125	108
$6000–$7999	134	75
$8000–$9999	130	94
$10,000–$11,999	168	79
$12,000–$13,999	178	65
$14,000–$15,999	240	60
$16,000–$17,999	141	45
$18,000–$19,999	160	45
$20,000–$24,999	299	71
$25,000–$29,999	199	29
$30,000–$39,999	162	23
$40,000–$49,999	61	7
$50,000–$74,999	36	3
$75,000–$99,999	4	0
$100,000 or more	7	1

Source: Social Change in Canada Project (see Table 5.1).

survey of the Canadian population. Respondents to the survey are classified by family income (represented by 18 categories) and by their responses to the question, "During the past year, have there been any major things you or your family *really* needed to buy but have not been able to afford?" We shall refer to the second variable as perceived need.

(a) Calculate the sample logit, log(Yes/No), for perceived need within each income category, and plot these logits against family income. (Since there is a zero frequency in one of the categories, you may add 0.5 to each frequency prior to computing the logits.) Does the relationship between the perceived-need logit and income appear to be roughly linear? In drawing the graph, use the category midpoints as income scores (employing an arbitrary figure, say $125,000, for the final, open category).

(b) The logit simple-regression model could be fit to the data in Table 5.5 by regenerating the observations on which the table is based. There are, for example, eight observations with $X = 0$ and $Y = 0$; five with $X = 0$ and $Y = 1$; and so on. It is more efficient, however, to leave the data in tabular form and simply to weight calculations by the number of observations in each cell of the table. Using this approach:

(i) Show that the logit-regression estimating equations may be written

$$\sum_{i=1}^{18} P_i(N_{i1} + N_{i2}) = \sum_{i=1}^{18} N_{i2}$$

$$\sum_{i=1}^{18} P_i X_i(N_{i1} + N_{i2}) = \sum_{i=1}^{18} X_i N_{i2}$$

where i indexes the categories of X; N_{ij} is the number of observations in cell i, j of the table; and $P_i = 1/[1 + e^{-(A+Bx_i)}]$.

(ii)* How may this approach be generalized to any number of discrete regressors, or whenever specific combinations of independent-variable values are replicated?

(c) Fit a linear logit regression of perceived need on family income, once again using category midpoints as income scores. Graph the logit regression line on the plot constructed in part (a). Compute the likelihood-ratio chi-square statistic to test the hypothesis that the slope in the logit model is zero. Test the same hypothesis by dividing the estimated slope by its estimated asymptotic standard error. Do the two tests agree?

(d) Construct a likelihood-ratio test for nonlinearity in the relationship between the perceived-need logit and income by fitting a model that represents income as a set of 17 dummy variables.

5.5.*† Logit Regression and Weighted Least Squares: In Problem 3.13 we demonstrated that when the errors in the general linear model $\mathbf{y} = \mathbf{X\beta} + \boldsymbol{\varepsilon}$ are normally and independently distributed but heteroscedastic, the maximum-likelihood estimator of $\boldsymbol{\beta}$ is obtained by minimizing the weighted sum of squares $\Sigma_{i=1}^{n} E_i^2/\sigma_i^2$, in which $\sigma_i^2 \equiv V(\varepsilon_i)$.

(a) Show that the weighted-least-squares estimator may be written in matrix form as $\mathbf{b} = (\mathbf{X}'\boldsymbol{\Sigma}\mathbf{X})^{-1}\mathbf{X}'\boldsymbol{\Sigma}\mathbf{y}$ where $\boldsymbol{\Sigma} \equiv \text{diag}(\sigma_1^2, \ldots, \sigma_n^2)$. (*Hint*: Differentiate the weighted sum of squares $\Sigma E_i^2/\sigma_i^2 = \mathbf{e}'\boldsymbol{\Sigma}\mathbf{e}$ and solve the resulting estimating equations.)

(b) In logit regression, when the Newton–Raphson method converges to a solution $\mathbf{b}$, we have (from equation (5.11)): $\mathbf{b} = \mathbf{b} + (\mathbf{X}'\mathbf{V}\mathbf{X})^{-1}\mathbf{X}'(\mathbf{y} - \mathbf{p})$. Show that this equation may be written in the following form: $\mathbf{b} = (\mathbf{X}'\mathbf{V}\mathbf{X})^{-1}\mathbf{X}'\mathbf{V}\mathbf{y}^*$, where $\mathbf{y}^* \equiv \mathbf{Xb} + \mathbf{V}^{-1}(\mathbf{y} - \mathbf{p})$.

(c) On the basis of parts (a) and (b), develop the analogy between logit regression and weighted least squares. Is there any sense in which this analogy is weak?

5.2. LOG-LINEAR MODELS FOR CONTINGENCY TABLES

When all variables are categorical (i.e., discrete), their joint sample distribution defines a *cross classification* or *contingency table*, where, in general, each combination of variable categories is observed more than once. Log-linear models are models for the association among variables in a contingency table. Although they bear many formal affinities to linear models, unlike linear models (and the logit models treated earlier in this chapter), *log-linear models do not distinguish a dependent variable*. There is, however, a relationship between log-linear and logit models that we shall develop later in this section. Indeed, since most applications treat one variable as the dependent variable, we generally employ log-linear models as a convenient means of fitting an equivalent logit model when all independent variables are qualitative (or categorical).

5.2.1. Two-Way Tables

We shall examine contingency tables for two variables in some detail, for this is the simplest case, and many of the results we establish here extend straightforwardly to tables of higher dimension. Consider the example shown in Table 5.6, a two-way table constructed from data reported in *The American Voter* (Campbell et al., 1960), a classic survey-research study of voting behavior. The table relates intensity of partisan preference to voting turnout in the 1956 U.S. presidential election. To anticipate our analysis, the data indicate that voting turnout increases with increasing intensity of party preference.

TABLE 5.6. Voter Turnout by Intensity of Partisan Preference, 1956 U.S. Presidential Election

(1) Intensity of of Preference	(2) Voter Turnout	
	Voted	Did Not Vote
Weak	305	126
Medium	405	125
Strong	265	49

Source: Adapted with permission from Campbell et al. (1960: Table 5-3).

More generally, two qualitative variables (or classifications) with r and c categories, respectively, define an $r \times c$ contingency table, as shown in Table 5.7. N_{ij} is the *observed frequency* in the i, jth cell of the table. We use a "+" to represent summation over a coordinate; thus $N_{i+} = \Sigma_{j=1}^{c} N_{ij}$ is the *marginal frequency* in the ith row; likewise $N_{+j} = \Sigma_{i=1}^{r} N_{ij}$ is the jth column marginal frequency. $n \equiv N_{++} = \Sigma_{i=1}^{r}\Sigma_{j=1}^{c} N_{ij}$ is the total number of observations in the sample.

We assume that the observations in Table 5.7 are produced by choosing an independent random sample of size n from a population characterized by probability π_{ij} of selecting an observation in cell i, j. We may define marginal probability distributions π_{i+} and π_{+j} as above; note that $\pi_{++} = 1$. If the row and column classifications are probabilistically independent, then the joint probability π_{ij} is the product of the marginal probabilities: $\pi_{ij} = \pi_{i+}\pi_{+j}$.

We have mentioned that the observed frequency N_{ij} results from drawing a random sample, and it is therefore a random variable, taking on different values in different samples. The expected value of N_{ij}—that is, the *expected frequency* in cell i, j of the table—is given by $\mu_{ij} \equiv E(N_{ij}) = n\pi_{ij}$. In the case of independence, we have $\mu_{ij} = n\pi_{i+}\pi_{+j}$. Since $\mu_{i+} = \Sigma_{j=1}^{c} n\pi_{ij} = n\pi_{i+}$ and $\mu_{+j} = n\pi_{+j}$, we may write $\mu_{ij} = \mu_{i+}\mu_{+j}/n$. Taking the log of both sides of this last equation produces

$$\lambda_{ij} \equiv \log \mu_{ij} = \log \mu_{i+} + \log \mu_{+j} - \log n \qquad (5.22)$$

TABLE 5.7. General Two-Way Frequency Table

Variable 1	Variable 2				Total
	1	2	$\cdots$	c	
1	N_{11}	N_{12}	$\cdots$	N_{1c}	N_{1+}
2	N_{21}	N_{22}	$\cdots$	N_{2c}	N_{2+}
$\vdots$	$\vdots$	$\vdots$		$\vdots$	$\vdots$
r	N_{r1}	N_{r2}	$\cdots$	N_{rc}	N_{r+}
Total	N_{+1}	N_{+2}	$\cdots$	N_{+c}	n

That is, under independence, the log expected frequencies depend additively upon the logs of the row marginal expected frequencies, the column marginals, and the sample size. As Fienberg (1977: 13–14) points out, equation (5.22) is reminiscent of a main-effects two-way analysis-of-variance model, where $-\log n$ plays the role of the constant, $\log \mu_{i+}$ and $\log \mu_{+j}$ are analogous to "main-effect" parameters, and λ_{ij} appears in place of the dependent-variable expectation. If we impose ANOVA-like constraints on the model, we may reparameterize equation (5.22) in the following manner:

$$\lambda_{ij} = \mu + \alpha_i + \beta_j \tag{5.23}$$

where $\alpha_+ = \Sigma\alpha_i = 0$ and $\beta_+ = \Sigma\beta_j = 0$. Equation (5.23) is the *log-linear model for independence* in the two-way table. Solving the model for the parameters, we obtain

$$\begin{aligned} \mu &= \frac{\lambda_{++}}{rc} \\ \alpha_i &= \frac{\lambda_{i+}}{c} - \mu \\ \beta_j &= \frac{\lambda_{+j}}{r} - \mu \end{aligned} \tag{5.24}$$

It is vital to stress that although the log-linear model is formally similar to an ANOVA model, the *meaning* of the models differs importantly: In analysis of variance, the α_i and β_j are main-effect parameters, specifying the separate relation of the dependent variable to the categories of each independent variable. The log-linear model (5.23) does not distinguish a dependent variable, and, because it is a model for independence, it specifies that the row and column variables are unrelated; for this model, the α_i and β_j merely express the relation of the log expected cell frequencies to the row and column marginals.

Under the model of independence, we describe rc expected frequencies in terms of

$$1 + (r - 1) + (c - 1) = r + c - 1$$

independent parameters. By analogy with the two-way ANOVA model, we may extend the log-linear model to cases where the row and column classifications are not independent, but rather are related in an arbitrary manner:

$$\lambda_{ij} = \mu + \alpha_i + \beta_j + \gamma_{ij} \tag{5.25}$$

where $\alpha_+ = \beta_+ = \gamma_{i+} = \gamma_{+j} = 0$, for all i and j. As before, we may write the parameters of the model in terms of the λ_{ij}. Indeed, the solution for μ, α_i, and β_j is the same as in equations (5.24), while

$$\gamma_{ij} = \lambda_{ij} - \mu - \alpha_i - \beta_j$$

Although by analogy with the ANOVA model the γ_{ij} are typically referred to as "interactions," this usage is potentially confusing, and therefore we shall refer to the γ_{ij} as *association parameters*, since they represent deviations from independence. Note that under model (5.25), called the *saturated model* for the two-way table, the number of independent parameters is equal to the number of cells in the table:

$$1 + (r-1) + (c-1) + (r-1)(c-1) = rc$$

Thus far, we have considered two versions of the log-linear model for the two-way table, but we have yet to indicate how this model may be fit to data. In estimating log-linear models, we shall employ maximum-likelihood methods, although other approaches are available as well (see Grizzle, Starmer, and Koch, 1969; Ku, Varner, and Kullback, 1971; and Bishop, Fienberg, and Holland, 1975: Chapter 10). The most straightforward route to the maximum-likelihood estimates is through maximization of the likelihood with respect to the model parameters. An indirect but generally computationally advantageous approach first finds maximum-likelihood estimates of the expected frequencies, and, from these, computes estimates of the model parameters.

The observed cell frequencies N_{ij} follow a multinomial distribution:

$$\begin{aligned} p(n_{11}, \ldots, n_{rc}) &= \frac{n!}{\prod_{i=1}^{r} \prod_{j=1}^{c} n_{ij}!} \prod_{i=1}^{r} \prod_{j=1}^{c} \pi_{ij}^{n_{ij}} \\ &= \frac{n!}{\prod_{i=1}^{r} \prod_{j=1}^{c} n_{ij}!} \prod_{i=1}^{r} \prod_{j=1}^{c} \left(\frac{\mu_{ij}}{n} \right)^{n_{ij}} \end{aligned} \qquad (5.26)$$

and the log likelihood for the saturated model is therefore

$$\begin{aligned} \log L(\boldsymbol{\alpha}, \boldsymbol{\beta}, \boldsymbol{\Gamma}) &= \log\left(\frac{n!}{\prod_i \prod_j N_{ij}!} \right) + \sum_i \sum_j N_{ij} \lambda_{ij} - n \log n \\ &= \log\left(\frac{n!}{\prod_i \prod_j N_{ij}!} \right) + \sum_i \sum_j N_{ij} (\mu + \alpha_i + \beta_j + \gamma_{ij}) - n \log n \end{aligned} \qquad (5.27)$$

We wish to maximize the likelihood subject to the constraints on equation (5.25). Notice that the first and third terms in equation (5.27) do not depend upon the parameters and thus may be ignored in maximizing L. The log likelihood for the model of independence (5.23) may be obtained from equation (5.27) by deleting the γ_{ij} (i.e., by setting these parameters to zero).

We have mentioned that it is generally simpler first to estimate the expected frequencies μ_{ij} than to maximize the likelihood directly.[4] Under the saturated model, it is possible to show (Bishop, Fienberg, and Holland, 1975: Chapter 5) that maximum-likelihood estimators of the μ_{ij} are given by the observed cell frequencies; that is $M_{ij} = N_{ij}$. Under the model for independence, the observed marginals provide maximum-likelihood estimators of the expected marginals, $M_{i+} = N_{i+}$, and $M_{+j} = N_{+j}$, which, in turn may be used to estimate the expected cell frequencies: $M_{ij} = M_{i+}M_{+j}/n$. With the M_{ij} in hand, we may calculate maximum-likelihood estimates of the model parameters:

$$L_{ij} \equiv \log M_{ij}$$

$$M = \frac{L_{++}}{rc}$$

$$A_i = \frac{L_{i+}}{c} - M$$

$$B_j = \frac{L_{+j}}{r} - M$$

$$C_{ij} = L_{ij} - M - A_i - B_j$$

Under the model for independence, of course, $C_{ij} = 0$.

The constraints $\gamma_{ij} = 0$ imposed on the model for independence may be tested by the likelihood-ratio approach, contrasting the model for independence (5.23) with the more general saturated model (5.25). Let L_0 represent the likelihood for model (5.23) and L_1 the likelihood for the saturated model. Then under $H_0: \gamma_{ij} = 0$, using equation (5.27) we get

$$\begin{aligned} G_0^2 &= -2(\log L_0 - \log L_1) \\ &= -2\left(\sum_i \sum_j N_{ij} L_{ij} - \sum_i \sum_j N_{ij} \log N_{ij}\right) \\ &= 2\sum_i \sum_j N_{ij} \log \frac{N_{ij}}{M_{ij}} \end{aligned} \tag{5.28}$$

[4] One advantage of directly maximizing the likelihood with respect to the parameters (e.g., by the Newton–Raphson method) is that asymptotic standard errors may be obtained in the usual manner from the information matrix. If, instead, expected frequencies are computed first, there is in general no simple way of obtaining standard errors. For certain models, however (including the saturated model), standard errors may be simply computed. See Goodman (1970), and Bishop, Fienberg, and Holland (1975: 141–146).

TABLE 5.8. Estimated Expected Frequencies for the Model of Independence Fit to Table 5.6

	M_{ij}		
i	$j = 1$	$j = 2$	M_{i+}
1	329.59	101.41	431
2	405.29	124.71	530
3	240.12	73.88	314
M_{+j}	975	300	1275

As is usual, the likelihood-ratio test statistic G_0^2 is asymptotically distributed as χ^2; here, G_0^2 has

$$rc - [1 + (r-1) + (c-1)] = (r-1)(c-1)$$

degrees of freedom, corresponding to the number of independent γ_{ij} parameters set to zero by the model of independence. In equation (5.28), the M_{ij} are estimated expected frequencies under the model of independence.[5]

Table 5.8 shows estimated expected frequencies under the model of independence for the data in Table 5.6. These expected frequencies differ markedly from the observed frequencies, leading us to question the adequacy of the model. Indeed, $G_0^2 = 19.43$, with two degrees of freedom. Since the p-value for this G^2 is less than .001, we reject $H_0: \gamma_{ij} = 0$ in favor of the saturated model, concluding that intensity of preference and turnout are related in the sampled population.

Log expected frequencies for the saturated model ($L_{ij} = \log N_{ij}$) are given in Table 5.9. Table 5.10 shows parameter estimates for the model: The C_{ij} appear in the body of the table, the A_i and B_j are in the margins, and M is shown in the corner. It is apparent from the C_{ij} values that there is a positive relationship between intensity of preference and turning out to vote.

5.2.2. Three-Way Tables

The full or saturated log-linear model for the three-way ($a \times b \times c$) table is defined in analogy with the three-way ANOVA model, although, as in the case

[5]The usual test for independence in two-way tables employs the same expected frequencies, but calculates a different test statistic: $Z_0^2 = \Sigma\Sigma(N_{ij} - M_{ij})^2/M_{ij}$. Z^2 is called the *Pearson* or *goodness-of-fit chi-square statistic*, to distinguish it from G^2, the likelihood-ratio chi-square statistic. Although the two statistics are asymptotically equivalent, we shall use G^2 here, since it has some desirable properties not shared with Z^2 (see Bishop, Fienberg, and Holland, 1975: 124–130, 513–518). When a G^2 with several degrees of freedom is partitioned, for example, its components sum exactly to the overall G^2, a property that does not hold for Z^2.

TABLE 5.9. Log Estimated Expected Frequencies for the Saturated Model Fit to Table 5.6

	$L_{ij} = \log M_{ij} = \log N_{ij}$	
i	$j = 1$	$j = 2$
1	5.720	4.836
2	6.004	4.828
3	5.580	3.892

of two-way tables, the meaning of the parameters is different:

$$\lambda_{ijk} = \mu + \alpha_{1(i)} + \alpha_{2(j)} + \alpha_{3(k)} + \alpha_{12(ij)} + \alpha_{13(ik)} + \alpha_{23(jk)} + \alpha_{123(ijk)} \tag{5.29}$$

with ANOVA-like constraints specifying that each set of parameters sums to zero over every coordinate; for example, $\alpha_{1(+)} = \alpha_{12(i+)} = \alpha_{123(ij+)} = 0$. Given the constraints, we may solve for the parameters in terms of the log expected frequencies. This solution follows the usual ANOVA pattern; for example

$$\begin{aligned}
\mu &= \frac{\lambda_{+++}}{abc} \\
\alpha_{1(i)} &= \frac{\lambda_{i++}}{bc} - \mu \\
\alpha_{12(ij)} &= \frac{\lambda_{ij+}}{c} - \mu - \alpha_{1(i)} - \alpha_{2(j)} \\
\alpha_{123(ijk)} &= \lambda_{ijk} - \mu - \alpha_{1(i)} - \alpha_{2(j)} - \alpha_{3(k)} - \alpha_{12(ij)} - \alpha_{13(ik)} - \alpha_{23(jk)}
\end{aligned} \tag{5.30}$$

TABLE 5.10. Estimated Parameters for the Saturated Model Fit to Table 5.6

	C_{ij}		
i	$j = 1$	$j = 2$	A_i
1	−0.183	0.183	0.135
2	−0.037	0.037	0.273
3	0.219	−0.219	−0.408
B_j	0.625	−0.625	5.143 = M

The presence of the three-way term[6] $\boldsymbol{\alpha}_{123}$ in the model implies that the relationship between any pair of variables (say, one and two) depends upon the category of the third variable (say, three).

Other log-linear models are defined by suppressing certain terms in the saturated model; that is, by setting parameters to zero. In specifying a restricted log-linear model, we shall always be guided by the principle of marginality, discussed in Section 2.1.3: Whenever a high-order term is included in the model, its lower-order relatives are included as well. As in the case of linear models, models of this sort are called *hierarchical*. Non-hierarchical log-linear models may be suitable for special applications, but they are not sensible in general (see Fienberg, 1979). According to the principle of marginality, if $\boldsymbol{\alpha}_{12}$ appears in a model, for example, so do $\boldsymbol{\alpha}_1$ and $\boldsymbol{\alpha}_2$.

If we set $\boldsymbol{\alpha}_{123} = \boldsymbol{\alpha}_{12} = \boldsymbol{\alpha}_{13} = \boldsymbol{\alpha}_{23} = 0$, we produce the *model of mutual independence*, specifying that the variables in the table are completely unrelated:

$$\lambda_{ijk} = \mu + \alpha_{1(i)} + \alpha_{2(j)} + \alpha_{3(k)}$$

Setting $\boldsymbol{\alpha}_{123} = \boldsymbol{\alpha}_{13} = \boldsymbol{\alpha}_{23} = 0$ yields the model

$$\lambda_{ijk} = \mu + \alpha_{1(i)} + \alpha_{2(j)} + \alpha_{3(k)} + \alpha_{12(ij)}$$

which specifies (1) that variables one and two are related, controlling for (i.e., within categories of) variable three; (2) that this partial relationship is constant across the categories of variable three; and (3) that variable three is independent of variables one and two taken jointly. Note that there are two other models of this sort: one in which $\boldsymbol{\alpha}_{13}$ is nonzero; and another in which $\boldsymbol{\alpha}_{23}$ is nonzero.

A third sort of model has two nonzero two-way terms; for example, setting $\boldsymbol{\alpha}_{123} = \boldsymbol{\alpha}_{23} = 0$, we obtain

$$\lambda_{ijk} = \mu + \alpha_{1(i)} + \alpha_{2(j)} + \alpha_{3(k)} + \alpha_{12(ij)} + \alpha_{13(ik)}$$

This model implies that: (1) variables one and two have a constant partial relationship across the categories of variable three; (2) variables one and three have a constant partial relationship across the categories of variable two; and (3) variables two and three are independent, controlling for variable one. Again, there are two other models of this type.

Finally, let us examine the model that sets only the three-way term $\boldsymbol{\alpha}_{123}$ to zero:

$$\lambda_{ijk} = \mu + \alpha_{1(i)} + \alpha_{2(j)} + \alpha_{3(k)} + \alpha_{12(ij)} + \alpha_{13(ik)} + \alpha_{23(jk)}$$

[6]For compactness, we represent a set of α's as a vector (regardless of its dimension), deleting the parenthetical subscript. $\boldsymbol{\alpha}_{123}$, therefore, contains the $\alpha_{123(ijk)}$. We used the same convention in Chapter 2.

This model specifies that each pair of variables (for example, one and two) has a constant partial association across the several categories of the remaining variable (e.g., three).

These descriptions are relatively complex because the log-linear models are models of association among variables. We shall see later (Section 5.2.4) that if one of the variables in a table is taken as the dependent variable, the log-linear model for the table is equivalent to a logit model with a simpler interpretation.

As in the two-way table treated earlier, it is generally simpler to estimate expected frequencies and to obtain estimates of model parameters from equations (5.30) than it is directly to determine maximum-likelihood estimates of the parameters. The observed marginal frequencies corresponding to the terms in the log-linear model that have no higher-order relatives constitute a set of minimal sufficient statistics under the model, from which the maximum-likelihood estimates of expected frequencies may be calculated (Bishop, Fienberg, and Holland, 1975: Chapter 3).

We applied this rule implicitly in our consideration of log-linear models for the two-way table. There, we estimated expected frequencies under the model for independence (with high-order terms α_i and β_j) from the marginals N_{i+} and N_{+j}; for the saturated model (with high-order term γ_{ij}), we employed N_{ij}. In general, the maximum-likelihood estimates of the expected frequencies are constrained to have the same marginals as the observed frequencies, for each marginal corresponding to a high-order term of the model. Table 5.11 shows the set of marginals to be fit under each model described earlier in this section.

It is unfortunately the case that in tables of dimension three and higher, the estimated expected frequencies for certain log-linear models cannot be written in closed form. There is, however, an algorithm known as *iterative proportional fitting*, described in the next section, which determines estimated expected frequencies to a prespecified degree of accuracy for any hierarchical log-linear model. In the three-way contingency table, only model 5 in Table 5.11 does not have closed-form estimates. The model of mutual independence (model 2), for example, has expected frequencies

$$\mu_{ijk} = n\pi_{i++}\pi_{+j+}\pi_{++k}$$

TABLE 5.11. Representative Log-Linear Models for the Three-Way Table, Showing Marginals Fit Under Each Model

Model	Parameters	Marginals Fit
1	$\mu, \alpha_1, \alpha_2, \alpha_3, \alpha_{12}, \alpha_{13}, \alpha_{23}, \alpha_{123}$	N_{ijk}
2	$\mu, \alpha_1, \alpha_2, \alpha_3$	$N_{i++}, N_{+j+}, N_{++k}$
3	$\mu, \alpha_1, \alpha_2, \alpha_3, \alpha_{12}$	N_{ij+}, N_{++k}
4	$\mu, \alpha_1, \alpha_2, \alpha_3, \alpha_{12}, \alpha_{13}$	N_{ij+}, N_{i+k}
5	$\mu, \alpha_1, \alpha_2, \alpha_3, \alpha_{12}, \alpha_{13}, \alpha_{23}$	$N_{ij+}, N_{i+k}, N_{+jk}$

and, hence, estimated expected frequencies

$$M_{ijk} = n\left(\frac{N_{i++}}{n}\right)\left(\frac{N_{+j+}}{n}\right)\left(\frac{N_{++k}}{n}\right)$$

$$= \frac{N_{i++}N_{+j+}N_{++k}}{n^2}$$

Once expected frequencies have been obtained for a model, a likelihood-ratio chi-square test for the model may be calculated according to equation (5.28) (suitably modifying sums and subscripts to conform to the dimension of the table at hand). This test implicitly contrasts a model with the saturated model, for which $M_{ijk} = N_{ijk}$, and thus tests the hypothesis that the parameters absent from the unsaturated model are equal to zero. The degrees of freedom for the G^2 statistic are equal to the number of independent parameters set to zero by the model. Similarly, a likelihood-ratio test for the hypothesis that any set of parameters is zero may be obtained by contrasting two models (calculating the log likelihood ratio, or, equivalently, taking the difference in their G^2 values) that are identical save that one sets to zero the parameters in question. Much as in the general linear model, this procedure is justified as long as the larger model includes all nonzero parameters. For example, if the three-way terms $\boldsymbol{\alpha}_{123}$ are nil, we may test $H_0: \boldsymbol{\alpha}_{23} = \mathbf{0}$ by contrasting models 4 and 5 from Table 5.11. The degrees of freedom for such a test are equal to the number of independent parameters by which the two models differ, here $(b - 1)(c - 1)$.

Table 5.12 shows a three-way contingency table elaborating the example presented earlier in Table 5.6. The sample from *The American Voter* study is cross classified by perceived closeness of the election, as well as by intensity of preference and voting turnout. We have fit all hierarchical log-linear models to this three-way table, displaying the results in Table 5.13. Here we employ a compact notation for marginals fit under a model: for example, {12} repre-

TABLE 5.12. Voter Turnout by Perceived Closeness of the Election and Intensity of Partisan Preference, 1956 U.S. Presidential Election

(1) Perceived Closeness	(2) Intensity of Preference	(3) Turnout	
		Voted	Did Not Vote
One-Sided	Weak	91	39
	Medium	121	49
	Strong	64	24
Close	Weak	214	87
	Medium	284	76
	Strong	201	25

Source: Reprinted with permission from Campbell et al. (1960: Table 5-3).

TABLE 5.13. Log-Linear Models Fit to Table 5.12

Model	Marginals Fit	Degrees of Freedom: General	Degrees of Freedom: Table 5.12	G^2	p
1	{1}{2}{3}	$(a-1)(b-1)+(a-1)(c-1)$ $+(b-1)(c-1)+(a-1)(b-1)(c-1)$	7	36.39	.000
2	{12}{3}	$(a-1)(c-1)+(b-1)(c-1)$ $+(a-1)(b-1)(c-1)$	5	34.83	.000
3	{13}{2}	$(a-1)(b-1)+(b-1)(c-1)$ $+(a-1)(b-1)(c-1)$	6	27.78	.000
4	{1}{23}	$(a-1)(b-1)+(a-1)(c-1)$ $+(a-1)(b-1)(c-1)$	5	16.96	.005
5	{12}{13}	$(b-1)(c-1)+(a-1)(b-1)(c-1)$	4	26.22	.000
6	{12}{23}	$(a-1)(c-1)+(a-1)(b-1)(c-1)$	3	15.40	.001
7	{13}{23}	$(a-1)(b-1)+(a-1)(b-1)(c-1)$	4	8.35	.079
8	{12}{13}{23}	$(a-1)(b-1)(c-1)$	2	7.12	.028
9	{123}	0	0	0.0	—

sents the two-way N_{ij+} marginal table, and {3} represents the one-way N_{++k} marginal table. We shall discuss the results of this analysis in Section 5.2.4, after explaining the relationship between log-linear and logit models.

The methods presented in this section generalize readily, and in an obvious manner, to tables of higher dimension. The interpretation of models including high-order associations can become complex, however, and for this reason we do not pursue the topic here. Interpretation is simpler when one classification is treated as the dependent variable, as we shall see shortly. Discussion of log-linear models for higher-order tables may be found in many places, including Goodman (1970), Bishop, Fienberg, and Holland (1975), Fienberg (1977), Haberman (1978), and Upton (1978).

5.2.3. Estimating Expected Frequencies by Iterative Proportional Fitting

We mentioned in the previous section that the iterative-proportional-fitting algorithm provides maximum-likelihood estimates of the expected frequencies for any hierarchical log-linear model. Beginning with a table of ones, the method successively adjusts the estimated expected frequencies to agree with each marginal table fit under a model. Adjustment for one such marginal generally disturbs agreement with the others. This procedure is repeated, however, until the estimated expected frequencies agree simultaneously with all marginals to be fit. We recognize that convergence has taken place when the estimates stabilize (to some preset level of precision) from one cycle of adjustments to the next.

We shall illustrate the algorithm by fitting model 2 from Table 5.13 to the data in Table 5.12. This model, which fits marginals {12} and {3}, has closed-form estimates available, and therefore the iterative algorithm converges after one cycle.[7] For model 2, then, each cycle consists of two iterations, one to adjust for {12}, and the other for {3}. We denote by $M_{ijk}^{(l,p)}$ the values calculated for the pth iteration of the lth cycle, calling the initial values $M_{ijk}^{(0,2)}$. We proceed in the following manner (adapted from Fienberg, 1977: 33–36):

1. Set initial values $M_{ijk}^{(0,2)} = 1$, for all cells of the table;

2. For each cycle of iterations, l

2.1. Adjust for {12}:

$$M_{ijk}^{(l,1)} = \frac{N_{ij+}}{M_{ij+}^{(l-1,2)}} M_{ijk}^{(l-1,2)}$$

2.2. Adjust for {3}:

$$M_{ijk}^{(l,2)} = \frac{N_{++k}}{M_{++k}^{(l,1)}} M_{ijk}^{(l,1)}$$

3. Stop when the largest absolute difference between the $M_{ijk}^{(l,2)}$ and the $M_{ijk}^{(l-1,2)}$ is less than a pre-established convergence criterion, say 0.01.

The course of this iterative procedure for the example is shown in Table 5.14(a). Table 5.14(b) shows the course of iterations for model 8, which fits {12}{13}{23} and which does not have closed-form estimates. The method of iterative proportional fitting is obviously generalizable to hierarchical models fit to tables of any dimension.

5.2.4. Log-Linear Models and Logit Models

We have mentioned more than once that the interpretation of log-linear models is complicated by the absence of a dependent variable. When one of the variables in a contingency table is regarded as the dependent variable, the log-linear model for the table implies a logit model, the parameters of which bear a simple relation to the parameters of the log-linear model. This fact

[7]Because iterative proportional fitting may always be employed to estimate expected frequencies for hierarchical log-linear models, we have not bothered to state the rules for determining whether closed-form estimates are also available. For discussion of this point see Bishop, Fienberg, and Holland (1975: 76–78). Haberman (1978) presents clear descriptions of different methods for fitting log-linear models to contingency tables.

TABLE 5.14. Obtaining Estimated Expected Frequencies by Iterative Proportional Fitting

(a) Fitting Marginals {12}{3} to Table 5.12

Cell			Estimates				
i	j	k	$M_{ijk}^{(0,2)}$	$M_{ijk}^{(1,1)}$	$M_{ijk}^{(1,2)}$	$M_{ijk}^{(2,1)}$	$M_{ijk} = M_{ijk}^{(2,2)}$
1	1	1	1.0	65.00	99.41	99.41	99.41
1	1	2	1.0	65.00	30.59	30.59	30.59
1	2	1	1.0	85.00	130.00	130.00	130.00
1	2	2	1.0	85.00	40.00	40.00	40.00
1	3	1	1.0	44.00	67.29	67.29	67.29
1	3	2	1.0	44.00	20.71	20.71	20.71
2	1	1	1.0	150.50	230.18	230.18	230.18
2	1	2	1.0	150.50	70.82	70.82	70.82
2	2	1	1.0	180.00	275.29	275.29	275.29
2	2	2	1.0	180.00	84.71	84.71	84.71
2	3	1	1.0	113.00	172.82	172.82	172.82
2	3	2	1.0	113.00	53.18	53.18	53.18

(b) Fitting Marginals {12}{13}{23} to Table 5.12

Cell			Estimates							
i	j	k	$M_{ijk}^{(0,3)}$	$M_{ijk}^{(1,1)}$	$M_{ijk}^{(1,2)}$	$M_{ijk}^{(1,3)}$	$M_{ijk}^{(2,1)}$	$M_{ijk}^{(2,2)}$	$M_{ijk}^{(2,3)}$	$M_{ijk} = M_{ijk}^{(3,3)}$
1	1	1	1.0	65.00	92.47	85.55	84.12	84.18	84.14	84.14
1	1	2	1.0	65.00	37.53	46.67	45.88	45.80	45.85	45.86
1	2	1	1.0	85.00	120.93	121.04	121.06	121.16	121.17	121.17
1	2	2	1.0	85.00	49.07	48.93	48.94	48.84	48.83	48.83
1	3	1	1.0	44.00	62.60	68.92	70.60	70.66	70.70	70.69
1	3	2	1.0	44.00	25.40	16.98	17.40	17.36	17.31	17.31
2	1	1	1.0	150.50	237.20	219.45	221.08	220.99	220.86	220.86
2	1	2	1.0	150.50	63.80	79.33	79.92	80.04	80.15	80.14
2	2	1	1.0	180.00	283.70	283.96	283.93	283.82	283.83	283.83
2	2	2	1.0	180.00	76.30	76.07	76.07	76.18	76.17	76.17
2	3	1	1.0	113.00	178.10	196.08	194.28	194.20	194.30	194.31
2	3	2	1.0	113.00	47.90	32.02	31.72	31.77	31.69	31.69

makes log-linear models a convenient, though indirect, means of fitting logit models when all independent variables are qualitative.[8] We begin with a discussion of dichotomous dependent variables, later generalizing our findings to polychotomous dependent variables.

It seems natural to regard voter turnout in Table 5.12 as a dichotomous dependent variable potentially affected by perceived closeness of the election and by intensity of partisan preference. With this example in mind, let us return to the saturated log-linear model for the three-way table (repeating equation (5.29)):

$$\lambda_{ijk} = \mu + \alpha_{1(i)} + \alpha_{2(j)} + \alpha_{3(k)} + \alpha_{12(ij)} + \alpha_{13(ik)} + \alpha_{23(jk)} + \alpha_{123(ijk)} \quad (5.31)$$

For convenience, we suppose that the dependent variable is number three, as in the illustration. Let Ω_{ij} symbolize the dependent-variable logit within categories i, j of the independent variables; that is

$$\Omega_{ij} \equiv \log\frac{\pi_{ij1}}{\pi_{ij2}} = \log\frac{n\pi_{ij1}}{n\pi_{ij2}} = \log\frac{\mu_{ij1}}{\mu_{ij2}} = \lambda_{ij1} - \lambda_{ij2}$$

Then, from the saturated log-linear model (5.31),

$$\Omega_{ij} = \left(\alpha_{3(1)} - \alpha_{3(2)}\right) + \left(\alpha_{13(i1)} - \alpha_{13(i2)}\right) + \left(\alpha_{23(j1)} - \alpha_{23(j2)}\right) + \left(\alpha_{123(ij1)} - \alpha_{123(ij2)}\right) \quad (5.32)$$

Noting that the first parenthetical term in equation (5.32) does not depend upon the independent variables, that the second depends only upon variable one, and so forth, we rewrite this equation in the following manner:

$$\Omega_{ij} = \omega + \omega_{1(i)} + \omega_{2(j)} + \omega_{12(ij)} \quad (5.33)$$

where, because of the ANOVA-like constraints on the α's:

$$\omega \equiv \alpha_{3(1)} - \alpha_{3(2)} = 2\alpha_{3(1)}$$

$$\omega_{1(i)} \equiv \alpha_{13(i1)} - \alpha_{13(i2)} = 2\alpha_{13(i1)}$$

$$\omega_{2(j)} \equiv \alpha_{23(j1)} - \alpha_{23(j2)} = 2\alpha_{23(j1)}$$

$$\omega_{12(ij)} \equiv \alpha_{123(ij1)} - \alpha_{123(ij2)} = 2\alpha_{123(ij1)}$$

[8] A tabular approach is also computationally more efficient when quantitative independent variables are discrete. See McFadden (1974), Bock (1975: Chapter 8), and Problem 5.4.

TABLE 5.15. Logit Models, Corresponding Log-Linear Models, and Fitted Marginals for the Three-Way Table

Logit Model	Log-linear Model	Marginals Fit
$\Omega_{ij} = \omega$	$\lambda_{ijk} = \mu + \alpha_{1(i)} + \alpha_{2(j)} + \alpha_{3(k)} + \alpha_{12(ij)}$	{12}{3}
$\Omega_{ij} = \omega + \omega_{1(i)}$	$\lambda_{ijk} = \mu + \alpha_{1(i)} + \alpha_{2(j)} + \alpha_{3(k)} + \alpha_{12(ij)} + \alpha_{13(ik)}$	{12}{13}
$\Omega_{ij} = \omega + \omega_{2(j)}$	$\lambda_{ijk} = \mu + \alpha_{1(i)} + \alpha_{2(j)} + \alpha_{3(k)} + \alpha_{12(ij)} + \alpha_{23(jk)}$	{12}{23}
$\Omega_{ij} = \omega + \omega_{1(i)} + \omega_{2(j)}$	$\lambda_{ijk} = \mu + \alpha_{1(i)} + \alpha_{2(j)} + \alpha_{3(k)} + \alpha_{12(ij)} + \alpha_{13(ik)} + \alpha_{23(jk)}$	{12}{13}{23}
$\Omega_{ij} = \omega + \omega_{1(i)} + \omega_{2(j)} + \omega_{12(ij)}$	$\lambda_{ijk} = \mu + \alpha_{1(i)} + \alpha_{2(j)} + \alpha_{3(k)} + \alpha_{12(ij)} + \alpha_{13(ik)} + \alpha_{23(jk)} + \alpha_{123(ijk)}$	{123}

Furthermore, because they are defined as twice the α's, the ω's are also constrained to sum to zero over any coordinate: $\omega_{1(+)} = \omega_{2(+)} = \omega_{12(i+)} = \omega_{12(+j)} = 0$, for all i and j.

Notice that the log-linear-model parameters for the association of the independent variables do not appear in equation (5.32). This equation (or, equivalently, equation (5.33)), the saturated logit model for the table, therefore shows how the dependent-variable log odds depend upon the independent variables and their interactions: In light of the constraints that they satisfy, the ω's are interpretable as ANOVA-like effect parameters.

A similar argument may be pursued with respect to any unsaturated log-linear model: Each such model implies a model for the dependent-variable logits. Because, however, our purpose is to examine the effects of the independent variables on the dependent variable, and not to explore the relations between the independent variables, we generally include $\boldsymbol{\alpha}_{12}$ and its lower-order relatives in any model that we fit, thereby treating the associations between the independent variables as given. Indeed, in experimental research, the independent-variable marginal table frequently is fixed by design.[9] Fitted marginals corresponding to logit models for the three-way table are shown in Table 5.15.

Sample logits are analogous to cell means in ANOVA, and, therefore, deserve examination: It is just as unenlightening to report significance tests for

[9] Fienberg (1977: 80) points out that when data are sparse it may be advantageous to smooth the relationships among independent variables by not treating their marginal table as fixed. As we have mentioned, *any* log-linear model implies a logit model. In what follows, however, we shall always suppose that the independent-variable marginal table is fit.

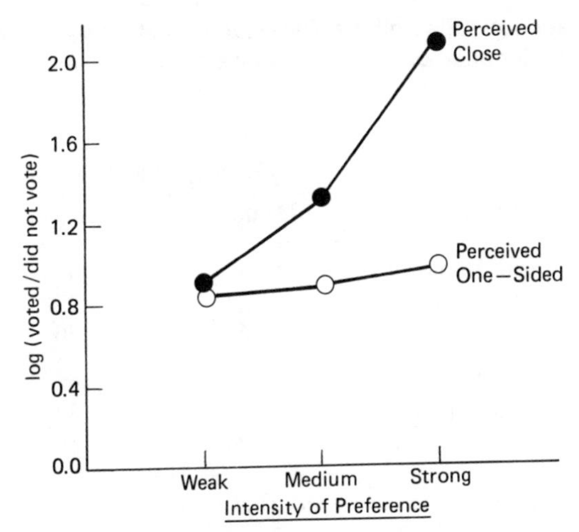

FIGURE 5.7. Plotted logits for the data in Table 5.12.

the parameters of a logit model in the absence of cell logits (or parameter estimates) as it is to report an analysis-of-variance table without cell means.

The sample logits for the *American Voter* data are plotted in Figure 5.7. Perceived closeness of the election and intensity of preference appear to interact in affecting turnout: Both among those who perceive the election to be one-sided, and among those who perceive it to be close, turnout increases with strength of preference; the increase is much more dramatic, however, when the election is perceived to be close.

In Table 5.13 we fitted all hierarchical log-linear models to the *American Voter* data; some of these log-linear models correspond to logit models, as we have explained in this section, and may be used to construct likelihood-ratio tests for the parameters of the logit model. These tests are shown in Table 5.16, which is similar in format to an ANOVA table. We conclude that the

TABLE 5.16. Tests for Logit-Model Parameters, Data From Table 5.12

Source	Logit Parameters	Models Contrasted[a]	G^2	df	p
Perceived Closeness	ω_1	6–8	8.28	1	.004
Intensity of Preference	ω_2	5–8	19.10	2	.000
Closeness × Intensity	ω_{12}	8–9	7.12	2	.028

[a] From Table 5.13.

TABLE 5.17. Respondent's Political Party Identification by Parents' Party Identification and Parents' Level of Political Activity, 1958

(1) Parents' Level of Political Activity	(2) Parents' Party Identification	(3) Respondent's Party Identification		
		Democrat	Independent	Republican
One or both active	Both Democrats	263	43	27
	Both Republicans	27	29	138
	Other	63	36	35
Neither active	Both Democrats	234	62	12
	Both Republicans	32	32	123
	Other	70	66	64

Source: Adapted with permission from Campbell et al. (1960: Table 7-1).

perceived-closeness × intensity-of-preference interaction noted in Figure 5.7 is statistically significant, and we therefore do not interpret the tests for main effects.

The relationship between log-linear and logit models may be extended to polychotomous dependent variables. Consider, for example, Table 5.17, also adapted from *The American Voter* (Campbell et al., 1960). In this table, the political-party identification of respondents to a 1958 sample survey is cross classified by parental level of political activity and parents' party identification. Treating respondent's party identification as the dependent variable, we may restrict attention to models that fit the $\{12\}$ (parents' level-of-activity by parents' party-identification) marginal table. These log-linear models are equivalent to logit models for the three-category dependent variable, as developed in Section 5.1.4.

Alternatively, we may fit log-linear/logit models for nested dichotomies. In the present example, this approach seems preferable, for the following nested dichotomies appear natural: $\{2, 13\}$ independents versus party identifiers; and $\{1, 3\}$ Democratic versus Republican identifiers. The logits for these dichotomies are graphed in Figures 5.8(a) and (b). In Figure 5.8(a), the odds of being independent appear to increase when there is no consistent pattern of parental affiliation and when parents are politically inactive; interaction effects seem slight. Figure 5.8(b) shows that children's party identification appears to follow that of parents; the effect of parental level of activity is virtually absent, and once more the interactions seem relatively small.

Table 5.18 reports the results of fitting log-linear/logit models to the table as a whole and to the nested dichotomies. Tests of interactions and main effects in the logit models are given in Table 5.19. The last column of this table pools the G^2 statistics for the nested dichotomies; note that the resulting test statistics are nearly identical to those for the table as a whole. The tests in Table 5.18 largely confirm the descriptive findings apparent from Figure 5.8.

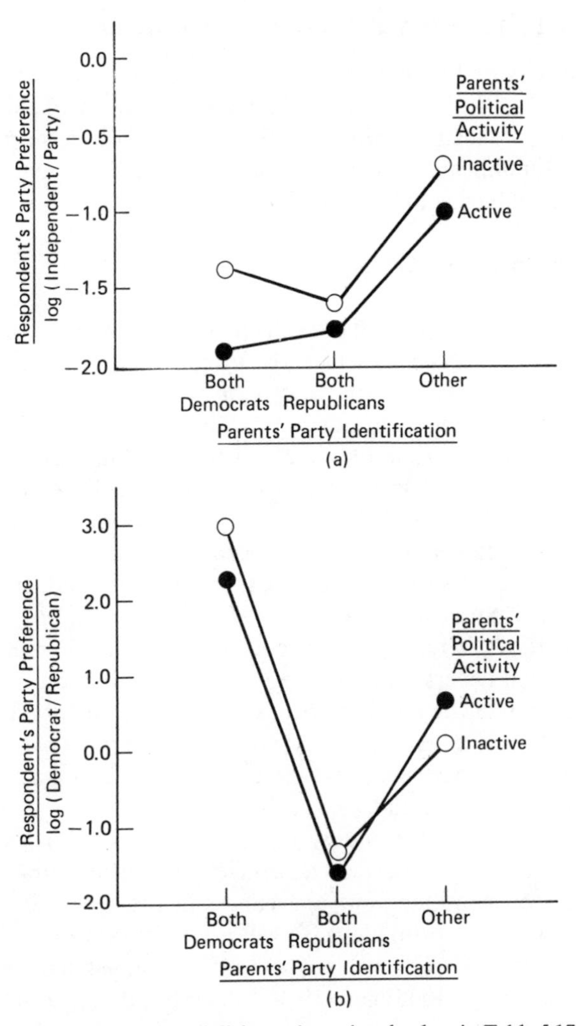

FIGURE 5.8. Plotted logits for nested dichotomies, using the data in Table 5.17. (a) Independent versus party. (b) Democrat versus Republican.

TABLE 5.18. Log-Linear / Logit Models Fit to Table 5.17

		Full Table		Nested Dichotomies			
				$\{I, DR\}$		$\{D, R\}$	
Model	Marginals Fit	G^2	*df*	G^2	*df*	G^2	*df*
1	{12}{3}	574.52	10	38.13	5	536.39	5
2	{12}{13}	564.80	8	28.54	4	536.27	4
3	{12}{23}	16.22	6	7.87	3	8.35	3
4	{12}{13}{23}	9.55	4	1.20	2	8.17	2
5	{123}	0.0	0	0.0	0	0.0	0

Both parents' activity and parents' party exert main effects on whether or not the respondent is an independent. Although the activity × parents' party interaction has a statistically significant effect on whether the respondent is a Democrat or a Republican, this interaction is so small compared with the parents' party main effect as to be negligible.

5.2.5. Tables With Empty Cells

It sometimes happens that one or more cells in a contingency table are empty; that is, have observed frequencies of zero. Empty cells occur for two quite different reasons: (1) Certain combinations of categories may be empty because of logical or definitional considerations. In a society in which intra-clan marriage is forbidden, for example, a contingency table classifying clan of husband by clan of wife necessarily has an empty main diagonal (cf. Bishop, Fienberg, and Holland, 1975: 191–192). Empty cells of this sort are called *structural zeroes*. Log-linear models have been adapted to deal with tables containing structural zeroes (see Bishop, Fienberg, and Holland, 1975: Chapter 5; Fienberg, 1977: Chapter 8; Haberman, 1979: Chapter 7), but we shall not describe these methods here. (2) Zero frequencies may also arise by chance, when a particular cell occurs rarely in the population, when the sample size is not large compared with the number of cells in the table, or for a combination of these reasons. Empty cells of this second type are termed *sampling zeroes*.

The log-linear models we have considered in this chapter are appropriate for tables with sampling zeroes, and any such model may be fit to these tables without difficulty, as long as no *marginal* fit under the model has a zero entry. Indeed, we may regard the expected frequencies for a well-fitting unsaturated model as a means of smoothing out unevenness in the observed table due to sampling. If we wish to estimate a model that fits a zero marginal (the saturated model, for example), we may proceed to adjust the cell counts to make them all positive. Goodman (1970) suggests adding a small positive

TABLE 5.19. Tests for Logit-Model Parameters, Data From Table 5.17

				Nested Dichotomies								
	Full Table			$\{I, DR\}$			$\{D, R\}$			Sum		
Source	G^2	*df*	*p*	G^2	*df*	*p*	G^2	*df*	*p*	G^2	*df*	*p*
Parents' activity	6.67	2	.035	6.67	1	.010	0.18	1	.671	6.85	2	.032
Parents' party	555.25	4	.000	27.34	2	.000	528.10	2	.000	555.44	4	.000
Activity × Party	9.55	4	.049	1.20	2	.546	8.17	2	.016	9.37	4	.052

quantity, say one half, to each cell of the observed table. Bishop, Fienberg, and Holland (1975: Chapter 12) consider more complex methods for replacing sampling zeroes.

5.2.6. Log-Linear / Logit Models for Recursive Causal Structures

Goodman (1972, 1973a, 1973b) has shown how logit models for contingency tables may be extended to situations in which a recursive causal structure is specified among the variables in a table.[10] As Fienberg (1977: 91–92) has pointed out, however, the analogy between these recursive logit systems and the structural-equation models of Chapter 4 is not perfect: (1) For polychotomous causes or effects, there are several logit parameters associated with each causal relation. (2) Logit equations frequently incorporate interactions. (3) Perhaps most important, the overall statistical association between pairs of variables cannot be straightforwardly decomposed into components that are functions of the logit parameters. As we shall show, however, a system of logit equations specifying recursive causal relations among a set of variables implies a structure for the contingency table relating the variables.

Consider a four-way contingency table for variables A, B, C, and D, in which variables A and B are taken to be causally prior to C, which in turn is prior to D. A and B, therefore, are analogous to exogenous variables in a structural-equation model, while C and D are analogous to endogenous variables. The recursive logit system consists of two models: one for variable C in the $\{ABC\}$ three-way marginal table, and the other for variable D in the full four-way table $\{ABCD\}$.

As we know, each logit model implies a set of expected frequencies for the table to which it is fit. We shall show that the logit system implies a set of expected frequencies for the full four-way table. The particular structure of the logit models comprising the system is irrelevant to this demonstration. For simplicity and concreteness, however, imagine that all four variables are dichotomous, that only main effects are present in the models, and that B exerts no direct effect on D; these conditions are reflected in the logit equations:

$$\begin{aligned} \Omega_{ij}^{AB\overline{C}} &= \omega^{AB\overline{C}} + \omega_{A(i)}^{AB\overline{C}} + \omega_{B(j)}^{AB\overline{C}} \\ \Omega_{ijk}^{ABC\overline{D}} &= \omega^{ABC\overline{D}} + \omega_{A(i)}^{ABC\overline{D}} + \omega_{C(k)}^{ABC\overline{D}} \end{aligned} \tag{5.34}$$

[10] Goodman, and Nerlove and Press (1973), have also suggested analogies between log-linear/logit models and nonrecursive structural-equation models. These analogies break down, however, because the parameters purporting to represent reciprocal effects either (1) measure association rather than causal impact, or (2) are estimated ignoring the nonrecursive structure of the model. For further criticism, see Fienberg (1977: 104–105), Rosenthal (1980), and Brier (1979). Heckman (1978) shows how the probit model can be extended to nonrecursive structures. Also see Maddala (1983: Chapter 5).

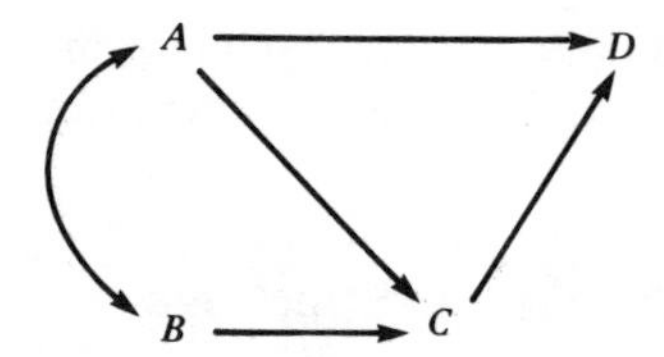

FIGURE 5.9. Illustrative recursive causal structure for a four-way table.

Here, the superscripts indicate the table to which the logit equation applies, with the barred letter specifying the dependent variable; thus, for example, $\Omega_{ij}^{AB\overline{C}}$ is the logit for variable C within categories i and j of variables A and B. The model in (5.34) is diagrammed heuristically in Figure 5.9.

As mentioned, the first equation in (5.34) (or any logit model fit to the $\{ABC\}$ table) implies expected frequencies μ_{ijk}^{ABC}. From these expected frequencies, we may calculate conditional probabilities under the model for C given A and B:

$$\pi_{ijk}^{AB\overline{C}} = \frac{\mu_{ijk}^{ABC}}{\mu_{ij+}^{ABC}} \tag{5.35}$$

where $\pi_{ijk}^{AB\overline{C}}$ stands for $\Pr(C = k | A = i,\ B = j)$. Likewise, from the second equation in (5.34) (or from any logit model fit to the full table), we have

$$\pi_{ijkl}^{ABC\overline{D}} = \frac{\mu_{ijkl}^{ABCD}}{\mu_{ijk+}^{ABCD}} \tag{5.36}$$

Using equations (5.35) and (5.36), we may derive the unconditional cell probabilities under the combined logit models:

$$\pi_{ijkl} = \pi_{ijkl}^{ABC\overline{D}} \pi_{ijk}^{AB\overline{C}} \pi_{ij}^{AB}$$

where π_{ij}^{AB}, from the marginal table for the "exogenous" variables, is taken as given. The expected frequencies under the combined model, then, are given by

$$\mu_{ijkl}^{*} = n\pi_{ijkl} = n\pi_{ijkl}^{ABC\overline{D}} \pi_{ijk}^{AB\overline{C}} \pi_{ij}^{AB}$$

Maximum-likelihood estimates of the expected frequencies are therefore

$$\begin{aligned} M_{ijkl}^{*} &= n \left(\frac{M_{ijkl}^{ABCD}}{N_{ijk+}^{ABCD}} \right) \left(\frac{M_{ijk}^{ABC}}{N_{ij+}^{ABC}} \right) \left(\frac{N_{ij}^{AB}}{n} \right) \\ &= \frac{M_{ijkl}^{ABCD} M_{ijk}^{ABC}}{N_{ijk+}^{ABCD}} \end{aligned}$$

since $M_{ijk+}^{ABCD} = N_{ijk+}^{ABCD}$ under the logit model for D, $M_{ij+}^{ABC} = N_{ij+}^{ABC}$ $(= N_{ij}^{AB})$ under the logit model for C, and $M_{ij}^{AB} = N_{ij}^{AB}$ implicitly for the $\{AB\}$ marginal table. In general, the expected frequencies for the combined model cannot be obtained from a single, hierarchical log-linear model for the full (here, four-way) table (see Goodman, 1973b, for exceptions).

The likelihood-ratio test statistic for the combined model may be partitioned into additive components due to the logit models comprising the combined model. For the example,

$$G_*^2 = 2\sum_i \sum_j \sum_k \sum_l N_{ijkl}^{ABCD} \log \frac{N_{ijkl}^{ABCD}}{M_{ijkl}^*}$$

$$= 2\sum_i \sum_j \sum_k \sum_l N_{ijkl}^{ABCD} \log \frac{N_{ijkl}^{ABCD}}{M_{ijkl}^{ABCD} M_{ijk}^{ABC} / N_{ijk+}^{ABCD}}$$

$$= 2\sum_i \sum_j \sum_k \sum_l \left[N_{ijkl}^{ABCD} \log\left(\frac{N_{ijkl}^{ABCD}}{M_{ijkl}^{ABCD}} \right) + N_{ijkl}^{ABCD} \log\left(\frac{N_{ijk+}^{ABCD}}{M_{ijk}^{ABC}} \right) \right]$$

$$= G_{ABC\overline{D}}^2 + G_{AB\overline{C}}^2$$

The degrees of freedom for the separate logit models add similarly. An illustrative application of these methods appears in Problem 5.10.

PROBLEMS

5.6.# The data in Table 5.20, drawn from *The American Voter*, are for a sample of black respondents residing in the southern United States. All of the southern states had more or less restrictive election laws, aimed

TABLE 5.20. Electoral Participation of Southern Blacks

(1) Frequency of Voting	(2) State Suffrage Law	
	Restrictive	Moderate
All or most elections	3	18
Some elections	11	18
Never	81	53

Source: Adapted with permission from Campbell et al. (1960: Table 11-5).

TABLE 5.21. Opinion of TV Series by Location and Sex

(1) Location	(2) Sex	(3) Opinion Approve	Disapprove
Urban	Female	6	12
	Male	17	1
Rural	Female	3	7
	Male	5	15

Source: Reprinted with permission from Lee (1978: 69).

primarily at limiting the political participation of blacks. The table relates the restrictiveness of the law for the respondent's state to frequency of voting.

(a) Test the null hypothesis that frequency of voting and state suffrage law are independent in the population sampled.

(b) Calculate the log observed frequencies; then, using the log frequencies, compute parameter estimates for the saturated log-linear model for the two-way table.

5.7.# Lee (1978) employs log-linear models to analyze a three-way contingency table (Table 5.21) relating respondent's location (urban or rural), sex, and opinion (approval or disapproval) of a TV series.

(a) Find estimated expected frequencies under the model of independence, and calculate the likelihood-ratio chi-square test statistic for the model. Is the hypothesis of independence a substantively meaningful one for these data?

(b) Compute the log observed frequencies and from them calculate parameter estimates for the saturated log-linear model.

(c) Treating opinion as the dependent variable in the table, find the odds for this variable within combinations of sex and location categories. Compute and graph the sample log odds (logits), commenting on the results.

(d) Compute parameter estimates for the saturated logit model (i) directly from the log odds found in part (c), and (ii) indirectly from the log-linear-model parameter estimates calculated in part (b). Do the two methods of calculation produce the same results?

(e) Use the method of iterative proportional fitting to find estimated expected frequencies for models that fit the following sets of marginals: (i) $\{12\}\{3\}$; (ii) $\{12\}\{13\}$; (iii) $\{12\}\{23\}$; and (iv) $\{12\}\{13\}\{23\}$. (Recall that in all but case (iv) the estimated expected frequencies converge in one cycle.) Performing appropriate

TABLE 5.22. Reactions to Invasions of Personal Space

(1) Density	(2) Sex of Subject	(3) Sex of Intruder	(4) Response	
			Yes	No
Low	Male	Male	18	1
		Female	15	8
	Female	Male	17	5
		Female	12	7
High	Male	Male	13	6
		Female	16	4
	Female	Male	10	9
		Female	14	6

Source: Reprinted with permission from Harris, Luginbuhl, and Fishbein (1978: Table 1).

G^2 tests, and keeping in mind the descriptive results in parts (c) and (d), what conclusions would you draw?

5.8. Harris, Luginbuhl, and Fishbein (1978) conducted a social-psychological experiment that examined reactions to invasions of personal space. The research took place in a field setting provided by a public escalator. The primary results of the study were presented in the form of a contingency table (Table 5.22). Three of the variables in the table were design or independent variables: (1) density of people on the escalator, rated either high or low; (2) the sex of the subject; and (3) the sex of the intruder. The fourth variable was a dichotomous response or dependent variable: whether or not the subject reacted in some manner to the intrusion.

The authors analyzed the data by separately examining the two-way (partial) tables relating density to response within combinations of categories of the other variables. Since there is a statistically significant relationship between density and response in only one of the four partial tables, the authors concluded that "males in the present study were more likely to react to a personal space invasion under low-density conditions than high-density conditions, but only when the intruder was another male. Density had no effect on responses by female subjects" (Harris, Luginbuhl, and Fishbein, 1978: 352–353). The implication is that there is a three-way interaction among the independent variables in determining response.

(a) Calculate the dependent-variable odds within combinations of independent-variable categories. Compute and graph the log odds (as in Figure 5.7), commenting on the results.

(b) Test the statistical significance of the various interactions and main effects of the independent variables on response. Summarize the likelihood-ratio tests for terms in the log-linear/logit model in a table similar to Table 5.16. Do these tests square with the descriptive findings in part (a)?

(c) On the basis of the tests in part (b), fit a final log-linear/logit model that incorporates only those effects shown to be important (and, of course, effects marginal to them). From the estimated expected frequencies for the model, calculate and graph fitted logits (log odds). Finally, report parameter estimates for the logit model.

(d) Test for independence between density and response separately in each of four partial tables. Do you obtain the results reported by Harris, Luginbuhl, and Fishbein?

(e) Do the results of your log-linear/logit analysis support the authors' conclusions (replicated in part (d))? Which analysis do you prefer? (Cf. Fox, 1979b.)

5.9. Table 5.23 elaborates the *American Voter* data from Problem 5.6, (i) by including information on white as well as black southerners, and (ii) by classifying respondents into two categories based upon the percent of nonwhites in their counties of residence. Analyze these data employing log-linear/logit models, treating frequency of voting as the dependent variable.

(a) Define nested dichotomies for the dependent variable, fitting separate models to each dichotomy.

(b) Model the dependent variable as a polychotomy.

(c) Compare the results of the two analyses. Which approach do you prefer for these data?

TABLE 5.23. Electoral Participation in the South

			(4) Frequency of Voting		
(1) Race	(2) County Percent Nonwhite	(3) State Suffrage Law	All or Most Elections	Some Elections	Never
White	< 30	Restrictive	108	58	64
		Moderate	190	74	108
	> 30	Restrictive	46	18	21
		Moderate	41	10	16
Black	< 30	Restrictive	3	5	5
		Moderate	15	8	26
	> 30	Restrictive	0	6	76
		Moderate	3	10	27

Source: Reprinted with permission from Campbell et al. (1960: Table 11-5).

TABLE 5.24. Intention to Vote by Education and Political Interest

(1) Education	(2) Political Interest	(3) Intention to Vote: Will Not Vote	Will Vote
Some High School	Great	5	490
	Medium	69	917
	Low	58	74
No High School	Great	6	270
	Medium	67	602
	Low	100	145

Source: Rosenberg, after Lazarsfeld, Berelson, and Gaudet, *The People's Choice*, © Columbia University Press, 1948. Reprinted with permission from Rosenberg (1968: Table 3-4) after Lazarsfeld, Berelson, and Gaudet (1948: 47).

5.10. The data in Table 5.24 on intention to vote, which originate from Lazarsfeld, Berelson, and Gaudet (1948), are discussed by Rosenberg (1968). Rosenberg hypothesizes that education affects political interest which, in turn, affects intention to vote, but he expects no direct effect of education on intention to vote when interest is controlled:

$$\text{Education} \rightarrow \text{Political Interest} \rightarrow \text{Intention to Vote}$$

Use the method discussed in Section 5.2.6 to determine whether Rosenberg's hypothesis is consistent with the data.

5.11.† Cell Residuals: Bishop, Fienberg, and Holland (1975: 136–141) define *standardized cell residuals* based upon components of the goodness-of-fit and likelihood-ratio chi-square statistics. In the former case, the residual for the **i**th cell is given by $Z_{\mathbf{i}} \equiv (N_{\mathbf{i}} - M_{\mathbf{i}})/\sqrt{M_{\mathbf{i}}}$, where **i** must be replaced by a cell subscript appropriate to the dimension of the table at hand. An examination of residuals may point to cells that are poorly fit, even when the overall G^2 for the model is acceptable. A normal-probability plot of the $Z_{\mathbf{i}}$ is useful for detecting outlying cells. In addition, if an important set of terms has been omitted from the model, this omission may often be discerned in the pattern of the cell residuals. To detect such a pattern, it is helpful to construct a table containing the signs (+ or −) of the cell residuals. Examining residuals is most useful for tables with a large number of cells. Calculate cell residuals for the model fit to Harris, Luginbuhl, and Fishbein's personal-space data in part (c) of Problem 5.8. Can you detect a systematic pattern in these residuals?

A

Notation

Specific notation is introduced at various points in the appendices and chapters. The following general conventions, with few exceptions, are adhered to throughout the text [examples are shown in brackets]:

Known scalar constants (including subscripts) are represented by lower-case italic letters [a, b, i, x_i, y_1^*].

Observable scalar random variables are represented by upper-case italic letters [X, Y_i, B_0'] or, if their names contain more than one character, by Roman letters, the first of which is upper case [RegrSS, RSS_0]. Values of random variables are represented as constants [x, y_i, b_0'].

Scalar parameters are represented by lower-case Greek letters [α, β, β_j^*, γ_2]. (Their estimators are generally denoted by "corresponding" italic characters [A, B, B_j^*, C_2] or by Greek letters with diacritics [$\hat{\alpha}$, $\tilde{\beta}$].)

Unobservable scalar random variables are also represented by lower-case Greek letters [ε_i, δ].

Vectors and matrices are represented by bold-face characters—lower case for vectors [$\mathbf{x}_1$, $\boldsymbol{\gamma}$], upper case for matrices [$\mathbf{X}$, $\boldsymbol{\Gamma}$]. Roman letters are used for constants and observable random variables [$\mathbf{a}$, $\mathbf{A}$, $\mathbf{y}_1$, $\mathbf{X}$]. Greek letters are used for parameters and unobservable random variables [$\boldsymbol{\beta}$, $\boldsymbol{\varepsilon}$, $\boldsymbol{\Gamma}$, $\boldsymbol{\xi}_i$]. Some upper-case Greek and Roman letters—most notably **B**—are typographically indistinguishable. (See the Greek alphabet given in Table A.1 at the end of this appendix.)

Diacritics and symbols such as * and ′ are used freely as modifiers to denote alternative forms [$\mathbf{X}^*$, β', $\tilde{\boldsymbol{\varepsilon}}$, $\hat{B}_1$].

The symbol $\equiv$ may be read as "is defined by," or "is equal to by definition" [$\bar{X} \equiv (\Sigma X_i)/n$].

The symbol $\sim$ means "is distributed as" [$\varepsilon \sim N(0, \sigma_\varepsilon^2)$].

The operator $E(\ )$ denotes the expectation of a scalar, vector, or matrix random variable [$E(Y_i)$, $E(\boldsymbol{\varepsilon})$, $E(\mathbf{X})$]. The operator $V(\ )$ denotes the variance of a scalar random variable, or the variance-covariance matrix of a vector random variable [$V(\varepsilon_i)$, $V(\mathbf{b})$]. Estimated variances or variance-covariance matrices are indicated by a circumflex ("hat") placed over these expressions:

$[\widehat{V(\varepsilon_i)}, \widehat{V(\mathbf{b})}]$. The operator $C(\ )$ gives the covariance of two scalar random variables or the covariance matrix of two vector random variables $[C(X, Y), C(\mathbf{x}_i, \mathbf{y}_i)]$. The operators $\mathscr{E}(\ )$ and $\mathscr{V}(\ )$ denote asymptotic expectation and variance, respectively. Their usage is similar to that of $E(\ )$ and $V(\ )$ $[\mathscr{E}(B), \mathscr{V}(\hat{\boldsymbol{\beta}}), \widehat{\mathscr{V}(A)}]$.

Standard mathematical functions are shown in lower case $[\cos W, \mathrm{trace}(\mathbf{A})]$. The log function denotes the natural logarithm unless otherwise indicated.

The summation sign Σ is used to denote continued addition $[\Sigma_{i=1}^{n} X_i \equiv X_1 + X_2 + \cdots + X_n]$. Often the range of the index is deleted if it is clear from the context $[\Sigma_i X_i]$, and the index may be deleted as well $[\Sigma X_i]$. The symbol Π similarly indicates continued multiplication $[\Pi_{i=1}^{n} X_i \equiv X_1 X_2 \cdots X_n]$.

To avoid awkward and repetitive phrasing in the statement of definitions and results, the words "if" and "when" are understood to mean "if and only if," unless explicitly indicated to the contrary. ["Two matrices are equal if they are of the same order and all corresponding entries are equal."] The general style of presentation in the text is nonrigorous.

TABLE A.1. Greek Alphabet

Greek Letter			Roman Equivalent	
Lower Case	Upper Case		Phonetic	Other
α	A	alpha	a	
β	B	beta	b	
γ	Γ	gamma	g, n	c
δ, ∂	Δ	delta	d	
ε	E	epsilon	e	
ζ	Z	zeta	z	
η	H	eta	e	y
θ	Θ	theta	th	
ι	I	iota	i	
κ	K	kappa	k	
λ	Λ	lambda	l	
μ	M	mu	m	
ν	N	nu	n	
ξ	Ξ	xi	x	
o	O	omicron	o	
π	Π	pi	p	
ρ	P	rho	r	
σ	Σ	sigma	s	
τ	T	tau	t	
υ	Υ	upsilon	y, u	
ϕ	Φ	phi	ph	
χ	X	chi	ch	x
ψ	Ψ	psi	ps	
ω	Ω	omega	o	w

B

Matrices, Linear Algebra, and Vector Geometry

Matrices furnish a natural notation for linear statistical models; the algebra of these models is linear algebra; and vector geometry provides a powerful conceptual tool for visualizing many aspects of linear models. The purpose of this appendix is to present basic concepts and results concerning matrices, linear algebra, and vector geometry. The focus is on topics that are employed in the main body of the book, and the style of presentation is informal rather than rigorous: At points, results are stated without proof; at other points, proofs are outlined or results are justified intuitively. The reader wholly unfamiliar with the subject matter of this appendix might profitably make reference to one of the many available texts on linear algebra, each of which develops in greater depth and detail most of the topics presented here.

The first section of the appendix develops elementary matrix algebra. Sections 2 and 3 introduce vector geometry and vector spaces. Section 4 discusses the related topics of matrix rank and the solution of linear simultaneous equations. Sections 5 and 6 deal with eigenvalues, eigenvectors, quadratic forms, and positive-definite matrices. Section 7 describes differential calculus in matrix form. Although none of the exposition in this appendix is essentially original, Sections B.2 and B.3 owe a special debt to Wonnacott and Wonnacott (1979: 376–391), while sections B.5 and B.6 owe a similar debt to Johnston (1972: 102–106).

B.1. MATRICES

B.1.1. Basic Definitions

A *matrix* is a rectangular table of numbers, or of numerical variables; for example

$$\underset{(m\times n)}{\mathbf{A}} = \begin{pmatrix} a_{11} & a_{12} & \cdots & a_{1n} \\ a_{21} & a_{22} & \cdots & a_{2n} \\ \vdots & \vdots & & \vdots \\ a_{m1} & a_{m2} & \cdots & a_{mn} \end{pmatrix} \tag{B.1}$$

A matrix with m rows and n columns is said to be of *order* m by n, written $(m \times n)$. For purposes of clarity, we shall at times indicate the order of a matrix below the matrix, as in equation (B.1). Each *entry* or *element* of a matrix may be subscripted by its row and column indices: a_{ij} is the entry in the ith row and jth column of the matrix $\mathbf{A}$. Individual numbers, such as the entries of a matrix, are termed *scalars*. Sometimes, for compactness, we specify a matrix by enclosing its typical element in braces; for example, $\underset{(m\times n)}{\mathbf{A}} = \{a_{ij}\}$ is equivalent to equation (B.1).

A matrix consisting of one column is called a *column vector*; for example

$$\underset{(m\times 1)}{\mathbf{a}} = \begin{pmatrix} a_1 \\ a_2 \\ \vdots \\ a_m \end{pmatrix}$$

A matrix consisting of one row is called a *row vector*,

$$\underset{(1\times n)}{\mathbf{b}'} = (b_1 \quad b_2 \quad \cdots \quad b_n)$$

In specifying a row vector, we often place commas between its elements for clarity.

The *transpose* of a matrix $\mathbf{A}$, denoted $\mathbf{A}'$, is formed from $\mathbf{A}$ so that the ith row of $\mathbf{A}'$ consists of the elements of the ith column of $\mathbf{A}$; thus (using the matrix in equation (B.1))

$$\underset{(n\times m)}{\mathbf{A}'} = \begin{pmatrix} a_{11} & a_{21} & \cdots & a_{m1} \\ a_{12} & a_{22} & \cdots & a_{m2} \\ \vdots & \vdots & & \vdots \\ a_{1n} & a_{2n} & \cdots & a_{mn} \end{pmatrix}$$

Note that $(\mathbf{A}')' = \mathbf{A}$. In this text, a vector is a column vector (such as $\mathbf{a}$) unless it is explicitly transposed (such as $\mathbf{b}'$).

A *square matrix* of *order* n, as the name implies, has n rows and n columns. The entries a_{ii} (that is, $a_{11}, a_{22}, \ldots, a_{nn}$) of a square matrix $\mathbf{A}$ comprise the *main diagonal* of the matrix. The sum of the diagonal elements is the *trace* of the matrix:

$$\text{trace}(\mathbf{A}) = \sum_{i=1}^{n} a_{ii}$$

A square matrix $\mathbf{A}$ is *symmetric* if $\mathbf{A} = \mathbf{A}'$; that is, when $a_{ij} = a_{ji}$ for all i and j. An *upper triangular* matrix is a square matrix with zeroes below its main diagonal:

$$\underset{(n \times n)}{\mathbf{U}} = \begin{pmatrix} u_{11} & u_{12} & \cdots & u_{1n} \\ 0 & u_{22} & \cdots & u_{2n} \\ \vdots & \vdots & \ddots & \vdots \\ 0 & 0 & \cdots & u_{nn} \end{pmatrix}$$

Similarly, a *lower triangular* matrix is a square matrix of the form

$$\underset{(n \times n)}{\mathbf{L}} = \begin{pmatrix} l_{11} & 0 & \cdots & 0 \\ l_{21} & l_{22} & \cdots & 0 \\ \vdots & \vdots & \ddots & \vdots \\ l_{n1} & l_{n2} & \cdots & l_{nn} \end{pmatrix}$$

A square matrix is *diagonal* if all entries off its main diagonal are zero; thus

$$\underset{(n \times n)}{\mathbf{D}} = \begin{pmatrix} d_1 & 0 & \cdots & 0 \\ 0 & d_2 & \cdots & 0 \\ \vdots & \vdots & \ddots & \vdots \\ 0 & 0 & \cdots & d_n \end{pmatrix}$$

For compactness, we write $\mathbf{D} = \text{diag}(d_1, d_2, \ldots, d_n)$. A *scalar matrix* is a diagonal matrix all of whose diagonal entries are equal: $\mathbf{S} = \text{diag}(s, s, \ldots, s)$. A particularly important scalar matrix is the *identity matrix* $\mathbf{I}$, which has ones on the main diagonal:

$$\underset{(n \times n)}{\mathbf{I}} = \begin{pmatrix} 1 & 0 & \cdots & 0 \\ 0 & 1 & \cdots & 0 \\ \vdots & \vdots & \ddots & \vdots \\ 0 & 0 & \cdots & 1 \end{pmatrix}$$

We write $\mathbf{I}_n$ for $\underset{(n \times n)}{\mathbf{I}}$.

Two other special matrices are the *zero matrix* $\mathbf{0}$, all of whose entries are zero, and the *unit vector* $\mathbf{1}$, all of whose entries are one.

A *partitioned matrix* is a matrix whose elements are organized into *submatrices*; for example

$$\underset{(4\times 3)}{\mathbf{A}} = \left(\begin{array}{cc|c} a_{11} & a_{12} & a_{13} \\ a_{21} & a_{22} & a_{23} \\ a_{31} & a_{32} & a_{33} \\ \hline a_{41} & a_{42} & a_{43} \end{array}\right) = \left(\begin{array}{c|c} \underset{(3\times 2)}{\mathbf{A}_{11}} & \underset{(3\times 1)}{\mathbf{A}_{12}} \\ \hline \underset{(1\times 2)}{\mathbf{A}_{21}} & \underset{(1\times 1)}{\mathbf{A}_{22}} \end{array}\right)$$

where the submatrix

$$\mathbf{A}_{11} = \begin{pmatrix} a_{11} & a_{12} \\ a_{21} & a_{22} \\ a_{31} & a_{32} \end{pmatrix}$$

and $\mathbf{A}_{12}$, $\mathbf{A}_{21}$, and $\mathbf{A}_{22}$ are similarly defined. When there is no possibility of confusion, we omit the broken lines separating the submatrices. If a matrix is partitioned vertically but not horizontally, then we separate its submatrices by commas; for example $\underset{(m\times n+p)}{\mathbf{C}} = [\underset{(m\times n)}{\mathbf{C}_1}, \underset{(m\times p)}{\mathbf{C}_2}]$

B.1.2. Simple Matrix Arithmetic

Two matrices are equal if they are of the same order and all corresponding entries are equal (a definition used implicitly in Section B.1.1).

Two matrices may be added only if they are of the same order; then their sum is formed by adding corresponding elements. Thus if $\mathbf{A}$ and $\mathbf{B}$ are of order $(m \times n)$, then $\mathbf{C} = \mathbf{A} + \mathbf{B}$ is also of order $(m \times n)$ with $c_{ij} = a_{ij} + b_{ij}$. Likewise, if $\mathbf{D} = \mathbf{A} - \mathbf{B}$, then $\mathbf{D}$ is of order $(m \times n)$ with $d_{ij} = a_{ij} - b_{ij}$. The negative of a matrix $\mathbf{A}$, $\mathbf{E} = -\mathbf{A}$, is of the same order as $\mathbf{A}$ with elements $e_{ij} = -a_{ij}$. Matrix addition, subtraction, and negation follow essentially the same rules as the corresponding scalar operations; in particular

$$\mathbf{A} + \mathbf{B} = \mathbf{B} + \mathbf{A}$$

$$\mathbf{A} + (\mathbf{B} + \mathbf{C}) = (\mathbf{A} + \mathbf{B}) + \mathbf{C}$$

$$\mathbf{A} - \mathbf{B} = \mathbf{A} + (-\mathbf{B}) = -(\mathbf{B} - \mathbf{A})$$

$$\mathbf{A} - \mathbf{A} = \mathbf{0}$$

$$\mathbf{A} + \mathbf{0} = \mathbf{A}$$

$$-(-\mathbf{A}) = \mathbf{A}$$

$$(\mathbf{A} + \mathbf{B})' = \mathbf{A}' + \mathbf{B}'$$

The product of a scalar c and an $(m \times n)$ matrix $\mathbf{A}$ is an $(m \times n)$ matrix $\mathbf{B} = c\mathbf{A}$ where $b_{ij} = ca_{ij}$. The product of a scalar and a matrix obeys the following rules:

$$c\mathbf{A} = \mathbf{A}c$$

$$\mathbf{A}(b + c) = \mathbf{A}b + \mathbf{A}c$$

$$c(\mathbf{A} + \mathbf{B}) = c\mathbf{A} + c\mathbf{B}$$

$$0\mathbf{A} = \mathbf{0}$$

$$1\mathbf{A} = \mathbf{A}$$

$$(-1)\mathbf{A} = -\mathbf{A}$$

The *inner product* of two vectors (each with n entries), say $\underset{(1\times n)}{\mathbf{a}'}$ and $\underset{(n\times 1)}{\mathbf{b}}$, written $\mathbf{a}' \cdot \mathbf{b}$, is a scalar formed by multiplying corresponding entries of the vectors and summing the resulting products:

$$\mathbf{a}' \cdot \mathbf{b} = \sum_{i=1}^{n} a_i b_i$$

Two matrices $\mathbf{A}$ and $\mathbf{B}$ are *conformable for multiplication* in the order given (i.e., $\mathbf{AB}$) if the number of columns of the first factor ($\mathbf{A}$) is equal to the number of rows of the second ($\mathbf{B}$). Thus $\mathbf{A}$ and $\mathbf{B}$ are conformable for multiplication if $\mathbf{A}$ is of order $(m \times n)$ and $\mathbf{B}$ is of order $(n \times p)$. Let $\mathbf{C} = \mathbf{AB}$ be the matrix product; and let $\mathbf{a}'_i$ represent the ith row of $\mathbf{A}$ and $\mathbf{b}_j$ represent the jth column of $\mathbf{B}$. Then $\mathbf{C}$ is a matrix of order $(m \times p)$ in which

$$c_{ij} = \mathbf{a}'_i \cdot \mathbf{b}_j = \sum_{k=1}^{n} a_{ik} b_{kj}$$

Matrix multiplication is associative, $\mathbf{A}(\mathbf{BC}) = (\mathbf{AB})\mathbf{C}$, and distributive with respect to addition:

$$(\mathbf{A} + \mathbf{B})\mathbf{C} = \mathbf{AC} + \mathbf{BC}$$

$$\mathbf{A}(\mathbf{B} + \mathbf{C}) = \mathbf{AB} + \mathbf{AC}$$

but it is not in general commutative: If $\mathbf{A}$ is $(m \times n)$ and $\mathbf{B}$ is $(n \times p)$, then the product $\mathbf{AB}$ is defined but $\mathbf{BA}$ is defined only if $m = p$. Even so, $\mathbf{AB}$ and $\mathbf{BA}$ are of different orders unless $m = n$. And even if $\mathbf{A}$ and $\mathbf{B}$ are square, $\mathbf{AB}$ and $\mathbf{BA}$, though of the same order, may be unequal. If $\mathbf{AB} = \mathbf{BA}$, then $\mathbf{A}$ and $\mathbf{B}$ are said to *commute* with one another. A scalar factor, however, may be moved anywhere within a matrix product: $c\mathbf{AB} = \mathbf{A}c\mathbf{B} = \mathbf{AB}c$.

The identity and zero matrices play roles with respect to matrix multiplication analogous to those of the numbers one and zero in scalar algebra:

$$\underset{(m\times n)}{\mathbf{A}}\mathbf{I}_n = \mathbf{I}_m \underset{(m\times n)}{\mathbf{A}} = \mathbf{A}$$

$$\underset{(m\times n)}{\mathbf{A}}\ \underset{(n\times p)}{\mathbf{0}} = \underset{(m\times p)}{\mathbf{0}}$$

$$\underset{(q\times m)}{\mathbf{0}}\ \underset{(m\times n)}{\mathbf{A}} = \underset{(q\times n)}{\mathbf{0}}$$

A further property of matrix multiplication is $(\mathbf{AB})' = \mathbf{B}'\mathbf{A}'$.

The *powers* of a square matrix are the products of the matrix with itself. Thus $\mathbf{A}^2 = \mathbf{AA}$, $\mathbf{A}^3 = \mathbf{AAA} = \mathbf{AA}^2 = \mathbf{A}^2\mathbf{A}$, and so on. If $\mathbf{B}^2 = \mathbf{A}$, then we call $\mathbf{B}$ the *square root* of $\mathbf{A}$, written, $\mathbf{A}^{1/2}$. If $\mathbf{A}^2 = \mathbf{A}$, then $\mathbf{A}$ is said to be *idempotent.*

For purposes of matrix addition, subtraction, and multiplication, the submatrices of partitioned matrices may be treated as elements, as long as the factors are partitioned conformably. For example, if

$$\mathbf{A} = \left(\begin{array}{ccc|cc} a_{11} & a_{12} & a_{13} & a_{14} & a_{15} \\ a_{21} & a_{22} & a_{23} & a_{24} & a_{25} \\ \hline a_{31} & a_{32} & a_{33} & a_{34} & a_{35} \end{array}\right) = \begin{pmatrix} \mathbf{A}_{11} & \mathbf{A}_{12} \\ \mathbf{A}_{21} & \mathbf{A}_{22} \end{pmatrix}$$

and

$$\mathbf{B} = \left(\begin{array}{ccc|cc} b_{11} & b_{12} & b_{13} & b_{14} & b_{15} \\ b_{21} & b_{22} & b_{23} & b_{24} & b_{25} \\ \hline b_{31} & b_{32} & b_{33} & b_{34} & b_{35} \end{array}\right) = \begin{pmatrix} \mathbf{B}_{11} & \mathbf{B}_{12} \\ \mathbf{B}_{21} & \mathbf{B}_{22} \end{pmatrix}$$

then

$$\mathbf{A} + \mathbf{B} = \left(\begin{array}{c|c} \mathbf{A}_{11} + \mathbf{B}_{11} & \mathbf{A}_{12} + \mathbf{B}_{12} \\ \hline \mathbf{A}_{21} + \mathbf{B}_{21} & \mathbf{A}_{22} + \mathbf{B}_{22} \end{array}\right)$$

Similarly, if

$$\underset{(m+n\times p+q)}{\mathbf{A}} = \begin{pmatrix} \underset{(m\times p)}{\mathbf{A}_{11}} & \underset{(m\times q)}{\mathbf{A}_{12}} \\ \underset{(n\times p)}{\mathbf{A}_{21}} & \underset{(n\times q)}{\mathbf{A}_{22}} \end{pmatrix}$$

and

$$\underset{(p+q \times r+s)}{\mathbf{B}} = \begin{pmatrix} \underset{(p \times r)}{\mathbf{B}_{11}} & \underset{(p \times s)}{\mathbf{B}_{12}} \\ \underset{(q \times r)}{\mathbf{B}_{21}} & \underset{(q \times s)}{\mathbf{B}_{22}} \end{pmatrix}$$

then

$$\underset{(m+n \times r+s)}{\mathbf{AB}} = \left(\begin{array}{c|c} \mathbf{A}_{11}\mathbf{B}_{11} + \mathbf{A}_{12}\mathbf{B}_{21} & \mathbf{A}_{11}\mathbf{B}_{12} + \mathbf{A}_{12}\mathbf{B}_{22} \\ \hline \mathbf{A}_{21}\mathbf{B}_{11} + \mathbf{A}_{22}\mathbf{B}_{21} & \mathbf{A}_{21}\mathbf{B}_{12} + \mathbf{A}_{22}\mathbf{B}_{22} \end{array}\right)$$

B.1.3. Matrix Inverses

The *inverse* of a square matrix[1] $\mathbf{A}$ is a square matrix of the same order, written $\mathbf{A}^{-1}$, with the property that $\mathbf{AA}^{-1} = \mathbf{A}^{-1}\mathbf{A} = \mathbf{I}$. If a square matrix has an inverse, then the matrix is termed *nonsingular*; a square matrix without an inverse is *singular*. If the inverse of a matrix exists, then it is unique; moreover, if for a square matrix $\mathbf{A}$, $\mathbf{AB} = \mathbf{I}$, then necessarily $\mathbf{BA} = \mathbf{I}$, and thus $\mathbf{B} = \mathbf{A}^{-1}$.

In scalar algebra, only the number zero has no inverse. It is simple to show by example that there exist singular nonzero matrices: Let us hypothesize that $\mathbf{B}$ is the inverse of the matrix

$$\mathbf{A} = \begin{pmatrix} 1 & 0 \\ 0 & 0 \end{pmatrix}$$

But

$$\mathbf{AB} = \begin{pmatrix} 1 & 0 \\ 0 & 0 \end{pmatrix}\begin{pmatrix} b_{11} & b_{12} \\ b_{21} & b_{22} \end{pmatrix} = \begin{pmatrix} b_{11} & b_{12} \\ 0 & 0 \end{pmatrix} \neq \mathbf{I}_2$$

which contradicts the hypothesis.

There are many methods for finding the inverse of a nonsingular square matrix. We shall briefly and informally describe a procedure called *Gaussian elimination*. Although there are methods that tend to produce more accurate numerical results when implemented on a digital computer, elimination has the virtue of relative simplicity. To illustrate the method of elimination, we shall employ the matrix

$$\begin{pmatrix} 2 & -2 & 0 \\ 1 & -1 & 1 \\ 4 & 4 & -4 \end{pmatrix} \tag{B.2}$$

[1]It is possible to define various sorts of *generalized inverses* for rectangular matrices and for square matrices that do not have a conventional inverse. Though generalized inverses have statistical applications, we shall not require them in this text. See Rao and Mitra (1971).

We begin by adjoining to this matrix an identity matrix; that is, we form the partitioned or *augmented* matrix

$$\left(\begin{array}{ccc|ccc} 2 & -2 & 0 & 1 & 0 & 0 \\ 1 & -1 & 1 & 0 & 1 & 0 \\ 4 & 4 & -4 & 0 & 0 & 1 \end{array}\right)$$

Then we attempt to reduce the original matrix to an identity matrix by applying operations of three sorts:

E_I: Multiply each entry in a row of the matrix by a nonzero scalar constant.

E_{II}: Add a scalar multiple of one row to another, replacing the other row.

E_{III}: Exchange two rows of the matrix.

E_I, E_{II}, and E_{III} are called *elementary row operations.*

Dealing with each row in turn, we insure that there is a nonzero entry in the diagonal position, employing a row interchange (for a lower row) if necessary. Then we divide the row through by its diagonal element (called the *pivot*) to obtain an entry of one in the diagonal position. Finally, we add multiples of the row in question to the other rows so as to *sweep out* the nonzero elements in the pivot column. For the illustration:

divide row 1 by 2

$$\left(\begin{array}{ccc|ccc} 1 & -1 & 0 & \frac{1}{2} & 0 & 0 \\ 1 & -1 & 1 & 0 & 1 & 0 \\ 4 & 4 & -4 & 0 & 0 & 1 \end{array}\right)$$

subtract row 1 from row 2

$$\left(\begin{array}{ccc|ccc} 1 & -1 & 0 & \frac{1}{2} & 0 & 0 \\ 0 & 0 & 1 & -\frac{1}{2} & 1 & 0 \\ 4 & 4 & -4 & 0 & 0 & 1 \end{array}\right)$$

subtract 4 × row 1 from row 3

$$\left(\begin{array}{ccc|ccc} 1 & -1 & 0 & \frac{1}{2} & 0 & 0 \\ 0 & 0 & 1 & -\frac{1}{2} & 1 & 0 \\ 0 & 8 & -4 & -2 & 0 & 1 \end{array}\right)$$

interchange rows 2 and 3

$$\left(\begin{array}{ccc|ccc} 1 & -1 & 0 & \frac{1}{2} & 0 & 0 \\ 0 & 8 & -4 & -2 & 0 & 1 \\ 0 & 0 & 1 & -\frac{1}{2} & 1 & 0 \end{array}\right)$$

divide row 2 by 8

$$\left(\begin{array}{ccc|ccc} 1 & -1 & 0 & \frac{1}{2} & 0 & 0 \\ 0 & 1 & -\frac{1}{2} & -\frac{1}{4} & 0 & \frac{1}{8} \\ 0 & 0 & 1 & -\frac{1}{2} & 1 & 0 \end{array}\right)$$

add row 2 to row 1

$$\left(\begin{array}{ccc|ccc} 1 & 0 & -\frac{1}{2} & \frac{1}{4} & 0 & \frac{1}{8} \\ 0 & 1 & -\frac{1}{2} & -\frac{1}{4} & 0 & \frac{1}{8} \\ 0 & 0 & 1 & -\frac{1}{2} & 1 & 0 \end{array}\right)$$

add $\frac{1}{2} \times$ row 3 to row 1

$$\left(\begin{array}{ccc|ccc} 1 & 0 & 0 & 0 & \frac{1}{2} & \frac{1}{8} \\ 0 & 1 & -\frac{1}{2} & -\frac{1}{4} & 0 & \frac{1}{8} \\ 0 & 0 & 1 & -\frac{1}{2} & 1 & 0 \end{array}\right)$$

add $\frac{1}{2} \times$ row 3 to row 2

$$\left(\begin{array}{ccc|ccc} 1 & 0 & 0 & 0 & \frac{1}{2} & \frac{1}{8} \\ 0 & 1 & 0 & -\frac{1}{2} & \frac{1}{2} & \frac{1}{8} \\ 0 & 0 & 1 & -\frac{1}{2} & 1 & 0 \end{array}\right)$$

Once the original matrix is reduced to the identity matrix, the final columns of the augmented matrix contain the inverse, as we may verify for the example:

$$\begin{pmatrix} 2 & -2 & 0 \\ 1 & -1 & 1 \\ 4 & 4 & -4 \end{pmatrix} \begin{pmatrix} 0 & \frac{1}{2} & \frac{1}{8} \\ -\frac{1}{2} & \frac{1}{2} & \frac{1}{8} \\ -\frac{1}{2} & 1 & 0 \end{pmatrix} = \begin{pmatrix} 1 & 0 & 0 \\ 0 & 1 & 0 \\ 0 & 0 & 1 \end{pmatrix}$$

It is simple to demonstrate why the elimination method works. Each elementary row operation may be represented as multiplication on the left by an appropriately formulated square matrix. Thus, for example, to interchange the second and third rows we may multiply on the left by

$$\mathbf{E}_{III} = \begin{pmatrix} 1 & 0 & 0 \\ 0 & 0 & 1 \\ 0 & 1 & 0 \end{pmatrix}$$

The elimination procedure applies a sequence of (say p) elementary row operations to the augmented matrix $[\underset{(n\times n)}{\mathbf{A}}, \mathbf{I}_n]$, which we may write as

$$\mathbf{E}_p \cdots \mathbf{E}_2\mathbf{E}_1[\mathbf{A}, \mathbf{I}_n] = [\mathbf{I}_n, \mathbf{B}]$$

using $\mathbf{E}_i$ to represent the ith operation in the sequence. Defining $\mathbf{E} \equiv \mathbf{E}_p \cdots \mathbf{E}_2\mathbf{E}_1$, we have $\mathbf{E}[\mathbf{A}, \mathbf{I}_n] = [\mathbf{I}_n, \mathbf{B}]$; that is, $\mathbf{EA} = \mathbf{I}_n$ (implying that $\mathbf{E} = \mathbf{A}^{-1}$), and $\mathbf{EI}_n = \mathbf{B}$. Consequently, $\mathbf{B} = \mathbf{E} = \mathbf{A}^{-1}$. If $\mathbf{A}$ is singular, we cannot reduce $\mathbf{A}$ to $\mathbf{I}$ by elementary row operations: At some point in the process we shall find that no nonzero pivot is available.

The matrix inverse obeys the following rules:

$$\mathbf{I}^{-1} = \mathbf{I}$$

$$(\mathbf{A}^{-1})^{-1} = \mathbf{A}$$

$$(\mathbf{A}')^{-1} = (\mathbf{A}^{-1})'$$

$$(\mathbf{AB})^{-1} = \mathbf{B}^{-1}\mathbf{A}^{-1}$$

$$(c\mathbf{A})^{-1} = c^{-1}\mathbf{A}^{-1}$$

(where $\mathbf{A}$ and $\mathbf{B}$ are order-n nonsingular matrices, and c is a nonzero scalar). If $\mathbf{D} = \text{diag}(d_1, d_2, \ldots, d_n)$, and if all $d_i \neq 0$, then $\mathbf{D}$ is nonsingular and $\mathbf{D}^{-1} = \text{diag}(1/d_1, 1/d_2, \ldots, 1/d_n)$. Finally, the inverse of a nonsingular symmetric matrix is itself symmetric.

B.1.4. Determinants

Each square matrix $\mathbf{A}$ is associated with a scalar called its *determinant* and written $|\mathbf{A}|$. For a (2×2) matrix $\mathbf{A}$, $|\mathbf{A}| = a_{11}a_{22} - a_{12}a_{21}$. For a (3×3) matrix the determinant is given by

$$\begin{aligned} |\mathbf{A}| = {} & a_{11}a_{22}a_{33} - a_{11}a_{23}a_{32} + a_{12}a_{23}a_{31} \\ & - a_{12}a_{21}a_{33} + a_{13}a_{21}a_{32} - a_{13}a_{22}a_{31} \end{aligned}$$

Although there is a general definition of the determinant for a square matrix of order n, we find it simpler here to define the determinant implicitly by specifying the following properties (or axioms):

1. Multiplying a row of a square matrix by a scalar constant multiplies the determinant of the matrix by the same constant.
2. Adding a multiple of one row to another leaves the determinant unaltered.
3. Interchanging two rows changes the sign of the determinant.
4. $|\mathbf{I}| = 1$.

Properties (1), (2), and (3) specify the effects of elementary row operations on the determinant. Since the Gaussian elimination method described in Section B.1.3 reduces a square matrix to the identity matrix, these properties, along with (4), are sufficient for establishing the value of the determinant. Indeed, the determinant is simply the product of the pivot elements, with the sign of the product reversed if, in the course of elimination, an odd number of row interchanges is employed. For the illustrative matrix (B.2), then, the determinant is $-(2)(8)(1) = -16$. If a matrix is singular, then one or more of the pivots are zero, and the determinant is zero. Conversely, a nonsingular matrix has a nonzero determinant.

B.2. BASIC VECTOR GEOMETRY

Thus far, we have treated vectors algebraically as one-column (or one-row) matrices. Vectors also have a geometric interpretation: The vector $\mathbf{x} = (x_1 \quad x_2 \quad \cdots \quad x_n)'$ is represented as a directed line segment extending from the origin of an n-dimensional Cartesian coordinate space to the point defined by the entries (called the *coordinates*) of the vector. Some examples of geometric vectors in two and three-dimensional space are shown in Figures B.1(a) and (b).

The basic arithmetic operations defined for vectors have simple geometric interpretations. To add two vectors $\mathbf{x}_1$ and $\mathbf{x}_2$ is, in effect, to place the tail of one at the tip of the other. When a vector is shifted from the origin in this manner, it retains its length and its orientation (the angles it makes with respect to the coordinate axes); indeed, length and orientation serve to define a

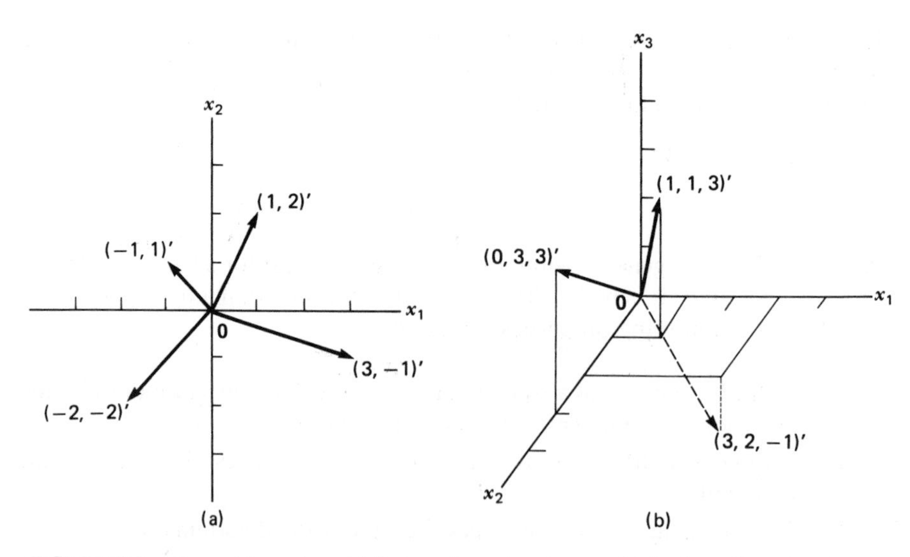

FIGURE B.1. Geometric vectors. (a) In two-dimensional space. (b) In three-dimensional space.

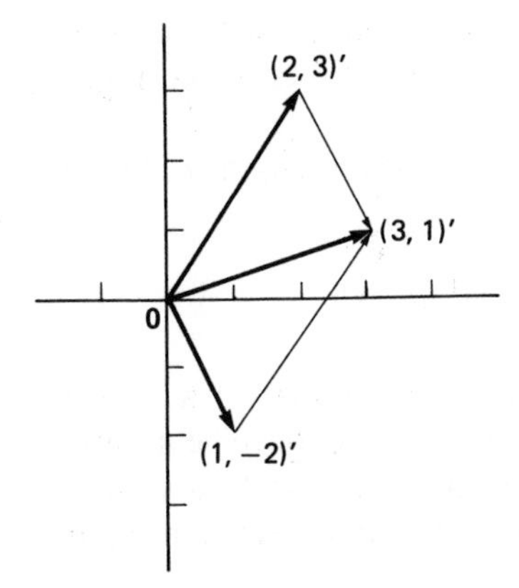

FIGURE B.2. Vector addition.

vector uniquely. The operation of vector addition, illustrated in two dimensions in Figure B.2, is equivalent to completing a parallelogram in which $\mathbf{x}_1$ and $\mathbf{x}_2$ are two adjacent sides; the vector sum is the diagonal of the parallelogram starting at the origin.

As shown in Figure B.3, the difference $\mathbf{x}_1 - \mathbf{x}_2$ is a vector whose length and orientation are obtained by proceeding from the tip of $\mathbf{x}_2$ to the tip of $\mathbf{x}_1$. Likewise, $\mathbf{x}_2 - \mathbf{x}_1$ proceeds from $\mathbf{x}_1$ to $\mathbf{x}_2$.

The length of a vector $\mathbf{x}$, denoted by $\|\mathbf{x}\|$, is the square root of its sum of squared coordinates:

$$\|\mathbf{x}\| = \sqrt{\sum_{i=1}^{n} x_i^2} = \sqrt{\mathbf{x} \cdot \mathbf{x}}$$

This result follows from the Pythagorean theorem in two dimensions, as shown in Figure B.4(a). The result may be extended one dimension at a time to higher-dimensional coordinate spaces, as shown for a three-dimensional space

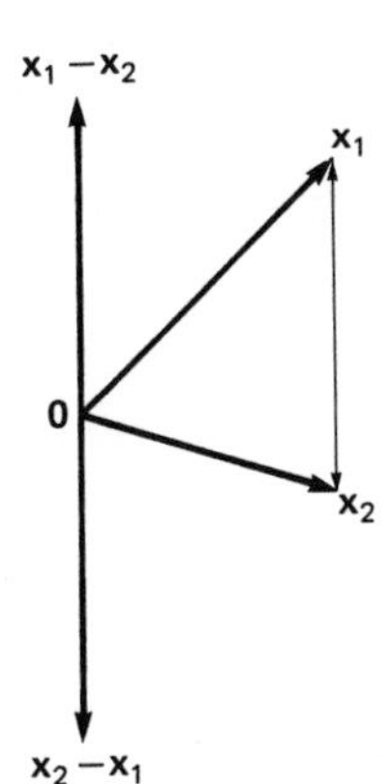

FIGURE B.3. Vector subtraction.

in Figure B.4(b). The distance between two vectors $\mathbf{x}_1$ and $\mathbf{x}_2$, defined as the distance separating their tips, is given by $\|\mathbf{x}_1 - \mathbf{x}_2\| = \|\mathbf{x}_2 - \mathbf{x}_1\|$ (see Figure B.3).

The product $a\mathbf{x}$ of a scalar a and a vector $\mathbf{x}$ is a vector of length $|a|\ \|\mathbf{x}\|$, as is readily verified:

$$\|a\mathbf{x}\| = \sqrt{\sum (ax_i)^2} = \sqrt{a^2 \sum x_i^2} = |a|\ \|\mathbf{x}\|$$

If a is positive, then the orientation of $a\mathbf{x}$ is the same as that of $\mathbf{x}$; if a is negative, then $a\mathbf{x}$ is collinear with $\mathbf{x}$ but in the opposite direction. The negative $-\mathbf{x} = (-1)\mathbf{x}$ of a vector $\mathbf{x}$ is therefore a vector of the same length as $\mathbf{x}$ but of opposite orientation. These results are illustrated for two dimensions in Figure B.5.

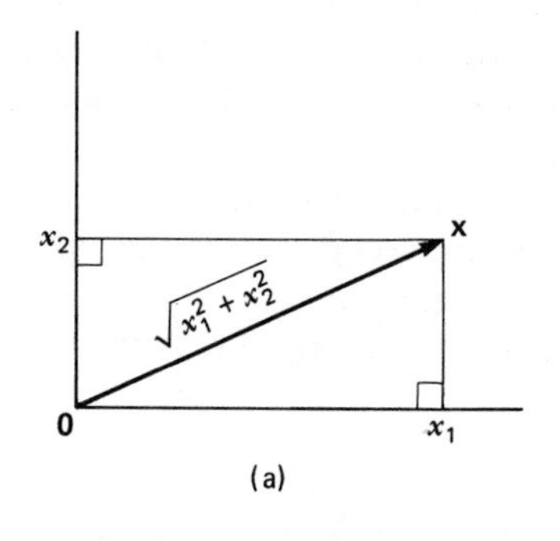

(a)

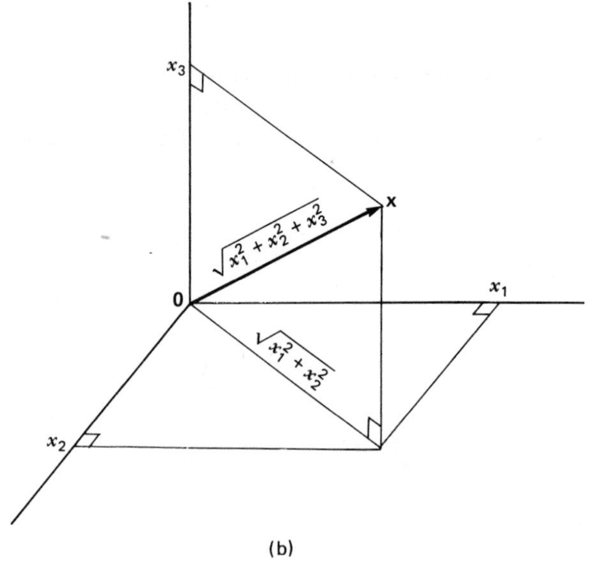

(b)

FIGURE B.4. Vector length. (a) In two dimensions. (b) In three dimensions.

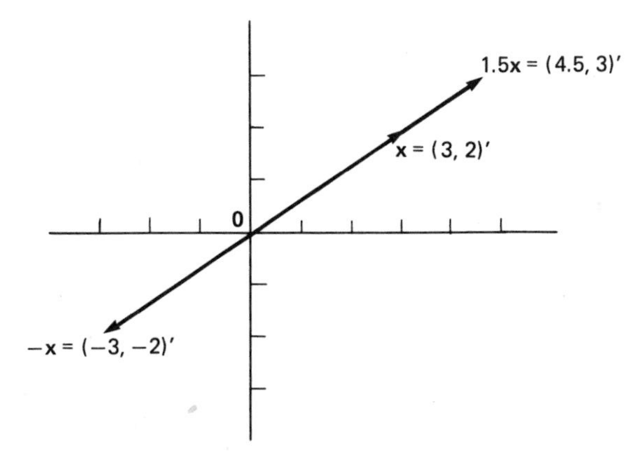

FIGURE B.5. Product of a scalar and a vector.

B.3. VECTOR SPACES AND SUBSPACES

The *vector space of dimension n* is the set of all vectors $\mathbf{x} = (x_1, x_2, \ldots, x_n)'$; the coordinates x_i may be any real numbers. The vector space of dimension one is therefore the real line; the vector space of dimension two is the plane; and so on.

The *subspace* of the n-dimensional vector space *generated* by a set of vectors $\{\mathbf{x}_1, \mathbf{x}_2, \ldots, \mathbf{x}_k\}$ is the subset of vectors $\mathbf{y}$ in the space that can be expressed as linear combinations of the generating set:

$$\mathbf{y} = a_1\mathbf{x}_1 + a_2\mathbf{x}_2 + \cdots + a_k\mathbf{x}_k$$

The set of vectors $\{\mathbf{x}_1, \mathbf{x}_2, \ldots, \mathbf{x}_k\}$ is said to *span* the subspace that it generates.

A set of vectors $\{\mathbf{x}_1, \mathbf{x}_2, \ldots, \mathbf{x}_k\}$ is *linearly independent* if no vector in the set may be expressed as a linear combination of the other vectors

$$\mathbf{x}_i = a_1\mathbf{x}_1 + \cdots + a_{i-1}\mathbf{x}_{i-1} + a_{i+1}\mathbf{x}_{i+1} + \cdots + a_k\mathbf{x}_k \qquad \text{(B.3)}$$

(where some of the constants a_j may be zero). Equivalently, the set of vectors is linearly independent if there are no constants $b_1, b_2, \ldots, b_k$, not all zero, for which $b_1\mathbf{x}_1 + b_2\mathbf{x}_2 + \cdots + b_k\mathbf{x}_k = \mathbf{0}$. Equation (B.3) is called a *linear dependency* or *collinearity*, and if this equation holds the vectors comprise a *linearly dependent* set. Note that the zero vector is linearly dependent upon every other vector, inasmuch as $\mathbf{0} = 0\mathbf{x}$.

The *dimension* of the subspace spanned by a set of vectors is the number of vectors in the largest linearly independent subset. The dimension of the subspace spanned by $\{\mathbf{x}_1, \mathbf{x}_2, \ldots, \mathbf{x}_k\}$ therefore cannot exceed the smaller of k and n. These relations are illustrated for three dimensions in Figure B.6. Figure B.6(a) shows the one-dimensional subspace (i.e., the line) generated by a single nonzero vector; Figure B.6(b) shows the one-dimensional subspace generated

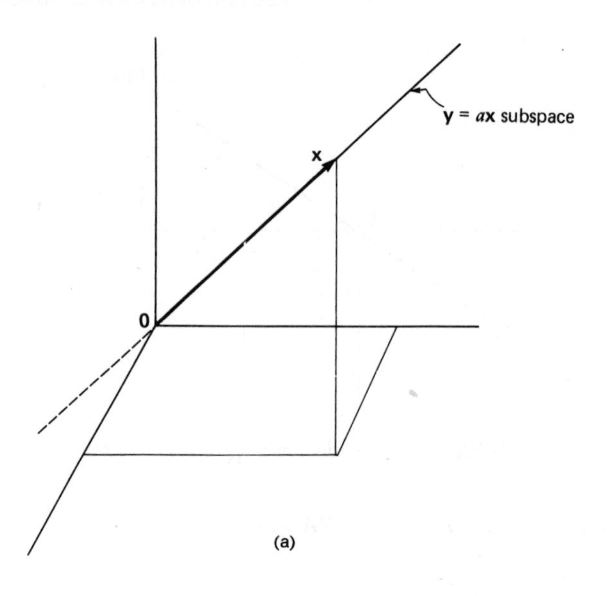

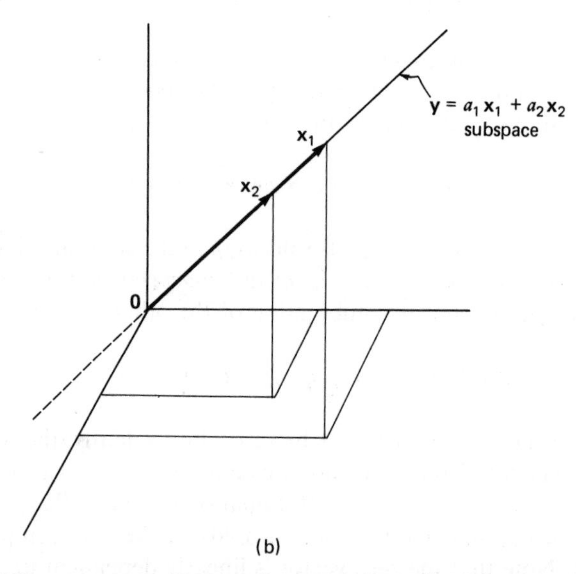

FIGURE B.6. Subspaces generated by sets of vectors in three-dimensional space. (a) One vector. (b) Two collinear vectors. (c) Two linearly independent vectors. (d) Three linearly dependent vectors.

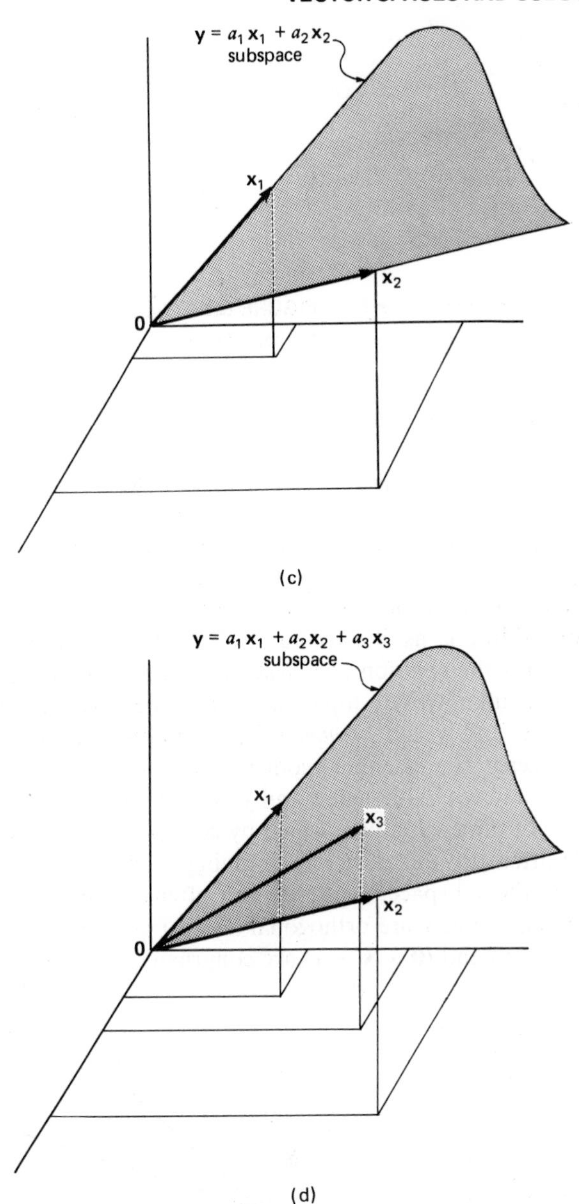

FIGURE B.6. (*Continued*).

by two collinear vectors; Figure B.6(c) shows the two-dimensional subspace (the plane) generated by two linearly independent vectors; and Figure B.6(d) shows the plane generated by three linearly dependent vectors no two of which are collinear.

A linearly independent set of vectors $\{\mathbf{x}_1, \mathbf{x}_2, \ldots, \mathbf{x}_k\}$ is said to provide a *basis* for the subspace that it spans. Any vector $\mathbf{y}$ in the subspace may be

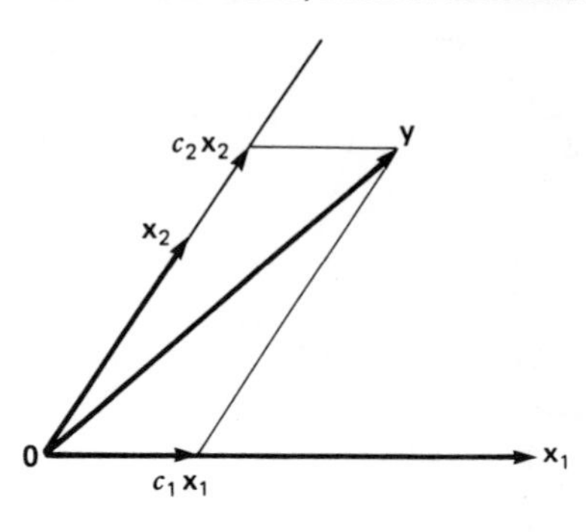

FIGURE B.7. The coordinates of $\mathbf{y}$ with respect to the basis $\{\mathbf{x}_1, \mathbf{x}_2\}$.

written *uniquely* as a linear combination of the basis vectors:

$$\mathbf{y} = c_1\mathbf{x}_1 + c_2\mathbf{x}_2 + \cdots + c_k\mathbf{x}_k$$

The constants $c_1, c_2, \ldots, c_k$ are called the *coordinates of* $\mathbf{y}$ *with respect to the basis* $\{\mathbf{x}_1, \mathbf{x}_2, \ldots, \mathbf{x}_k\}$. The coordinates of a vector with respect to a basis of a two-dimensional subspace may be found geometrically by the parallelogram rule of vector addition, as illustrated in Figure B.7. Finding coordinates algebraically entails the solution of a system of linear simultaneous equations (where the c_i's are unknowns), a topic taken up in the next section.

Two vectors $\mathbf{x}$ and $\mathbf{y}$ are *orthogonal* (i.e., perpendicular) if their inner product $\mathbf{x} \cdot \mathbf{y}$ is zero. The essential geometry of vector orthogonality is shown in Figures B.8(a) and (b). Although $\mathbf{x}$ and $\mathbf{y}$ lie in an n-dimensional space, they span a subspace of dimension two, which by convention we make the plane of the paper. We often use this device in applying vector geometry to statistical problems, where the subspace of interest can often be limited to two or three dimensions. When $\mathbf{x}$ and $\mathbf{y}$ are orthogonal (as in Figure B.8(a)), the two right triangles $(\mathbf{0}, \mathbf{x}, \mathbf{x} + \mathbf{y})$ and $(\mathbf{0}, \mathbf{x}, \mathbf{x} - \mathbf{y})$ are congruent; consequently, $||\mathbf{x} + \mathbf{y}|| =$

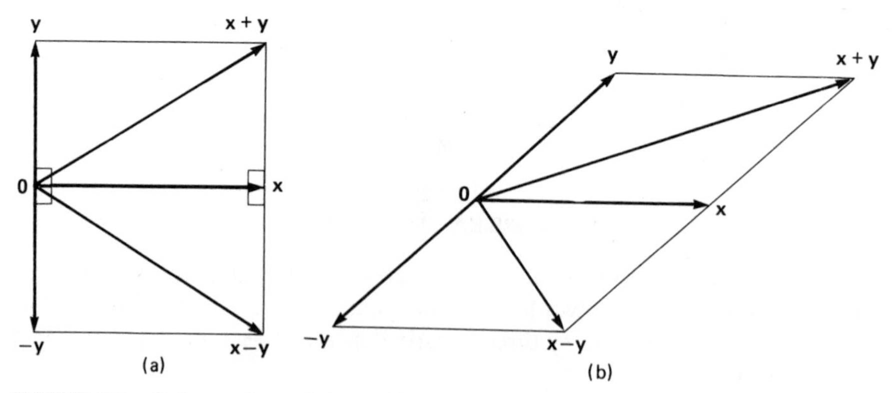

FIGURE B.8. Orthogonality and the inner product. (a) Orthogonal vectors. (b) Nonorthogonal vectors.

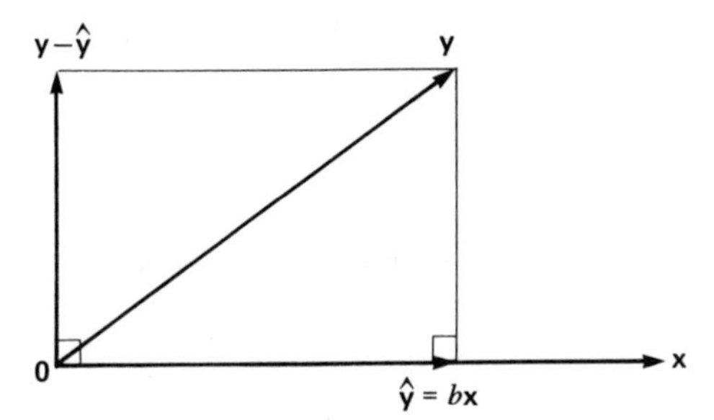

FIGURE B.9. The orthogonal projection of $\mathbf{y}$ on $\mathbf{x}$.

$\|\mathbf{x} - \mathbf{y}\|$. Since the squared length of a vector is the inner product of the vector with itself, we have

$$\begin{aligned}(\mathbf{x} + \mathbf{y}) \cdot (\mathbf{x} + \mathbf{y}) &= (\mathbf{x} - \mathbf{y}) \cdot (\mathbf{x} - \mathbf{y}) \\ \mathbf{x} \cdot \mathbf{x} + 2\mathbf{x} \cdot \mathbf{y} + \mathbf{y} \cdot \mathbf{y} &= \mathbf{x} \cdot \mathbf{x} - 2\mathbf{x} \cdot \mathbf{y} + \mathbf{y} \cdot \mathbf{y} \\ 4\mathbf{x} \cdot \mathbf{y} &= 0 \\ \mathbf{x} \cdot \mathbf{y} &= 0\end{aligned}$$

When $\mathbf{x}$ and $\mathbf{y}$ are not orthogonal (as in Figure B.8(b)), then $\|\mathbf{x} + \mathbf{y}\| \neq \|\mathbf{x} - \mathbf{y}\|$, and $\mathbf{x} \cdot \mathbf{y} \neq 0$.

The definition of orthogonality may be extended to matrices in the following manner: The matrix $\underset{(n \times k)}{\mathbf{X}}$ is orthogonal if each pair of its columns is orthogonal, that is, if $\mathbf{X}'\mathbf{X}$ is diagonal. $\mathbf{X}$ is said to be *orthonormal* if $\mathbf{X}'\mathbf{X} = \mathbf{I}_k$.

The *orthogonal projection* of one vector $\mathbf{y}$ on another vector $\mathbf{x}$ is a scalar multiple $\hat{\mathbf{y}} = b\mathbf{x}$ of $\mathbf{x}$ such that $(\mathbf{y} - \hat{\mathbf{y}})$ is orthogonal to $\mathbf{x}$. The geometry of orthogonal projection is illustrated in Figure B.9. Note that by the Pythagorean theorem (see Figure B.10), $\hat{\mathbf{y}}$ is the point along the line through $\mathbf{x}$ that is closest to $\mathbf{y}$. To find b we use the fact that

$$\mathbf{x} \cdot (\mathbf{y} - \hat{\mathbf{y}}) = \mathbf{x} \cdot (\mathbf{y} - b\mathbf{x}) = 0$$

Thus, $\mathbf{x} \cdot \mathbf{y} - b\mathbf{x} \cdot \mathbf{x} = 0$ and $b = (\mathbf{x} \cdot \mathbf{y})/(\mathbf{x} \cdot \mathbf{x})$.

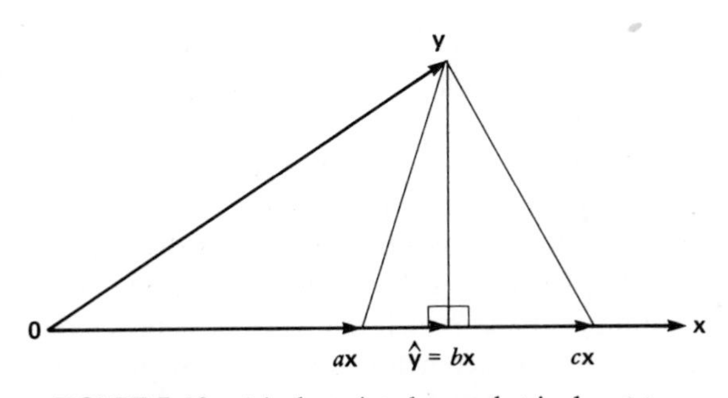

FIGURE B.10. $\hat{\mathbf{y}}$ is the point along $\mathbf{x}$ that is closest to $\mathbf{y}$.

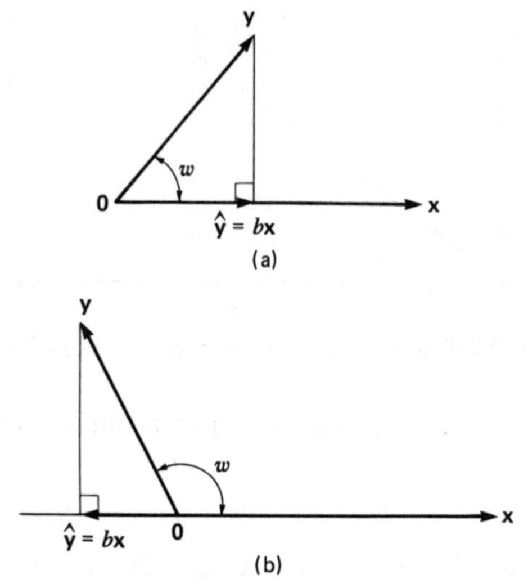

FIGURE B.11. The angle w separating two vectors. (a) $0° < w < 90°$. (b) $90° < w < 180°$.

The orthogonal projection of **y** on **x** may be used to define the cosine of the angle separating the two vectors. We distinguish two cases: In Figure B.11(a) the angle separating the vectors is between 0° and 90°; while in Figure B.11(b) this angle is between 90° and 180°. (By convention, we examine the smaller of the two angles separating a pair of vectors, and therefore do not encounter angles greater than 180°; this is of no consequence since $\cos(360 - w) = \cos w$.) In the first instance (Figure B.11(a)) we have

$$\cos w = \frac{||\hat{\mathbf{y}}||}{||\mathbf{y}||} = \frac{b||\mathbf{x}||}{||\mathbf{y}||} = \frac{\mathbf{x} \cdot \mathbf{y}}{||\mathbf{x}||^2} \times \frac{||\mathbf{x}||}{||\mathbf{y}||}$$

$$= \frac{\mathbf{x} \cdot \mathbf{y}}{||\mathbf{x}|| \, ||\mathbf{y}||}$$

and, likewise, in the second instance (Figure B.11(b))

$$\cos w = \frac{-||\hat{\mathbf{y}}||}{||\mathbf{y}||} = \frac{b||\mathbf{x}||}{||\mathbf{y}||} = \frac{\mathbf{x} \cdot \mathbf{y}}{||\mathbf{x}|| \, ||\mathbf{y}||}$$

Note that in both cases the sign of $\cos w$ is correctly reflected in the sign of the coefficient b for the orthogonal projection of **y** on **x**.

The orthogonal projection of a vector **y** onto the subspace spanned by a set of vectors $\{\mathbf{x}_1, \mathbf{x}_2, \ldots, \mathbf{x}_k\}$ is the vector $\hat{\mathbf{y}} = b_1\mathbf{x}_1 + b_2\mathbf{x}_2 + \cdots + b_k\mathbf{x}_k$ formed as a linear combination of the $\mathbf{x}_i$'s such that $(\mathbf{y} - \hat{\mathbf{y}})$ is orthogonal to each and every vector $\mathbf{x}_i$ in the set. The geometry of orthogonal projection for $k = 2$ is illustrated in Figure B.12. Note that $\hat{\mathbf{y}}$ is the point closest to **y** in the subspace

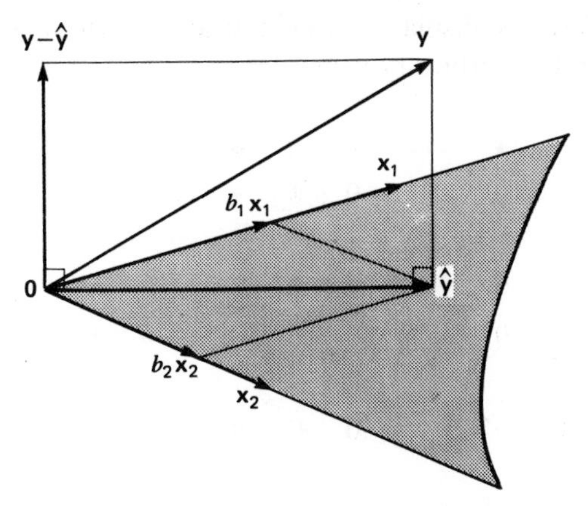

FIGURE B.12. The orthogonal projection of $\mathbf{y}$ on the subspace spanned by $\mathbf{x}_1$ and $\mathbf{x}_2$.

spanned by the $\mathbf{x}_i$'s. Placing the constants b_i into a vector $\mathbf{b}$, and gathering the vectors $\mathbf{x}_i$ into an $(n \times k)$ matrix $\mathbf{X} = [\mathbf{x}_1, \mathbf{x}_2, \ldots, \mathbf{x}_k]$, we have $\hat{\mathbf{y}} = \mathbf{Xb}$. By definition of an orthogonal projection,

$$\mathbf{x}_i \cdot (\mathbf{y} - \hat{\mathbf{y}}) = \mathbf{x}_i \cdot (\mathbf{y} - \mathbf{Xb}) = 0 \qquad \text{for } i = 1, \ldots, k \tag{B.4}$$

Equivalently, $\mathbf{X}'(\mathbf{y} - \mathbf{Xb}) = \mathbf{0}$, or $\mathbf{X}'\mathbf{y} = \mathbf{X}'\mathbf{Xb}$. We may solve this matrix equation uniquely for $\mathbf{b}$ under the supposition that $\underset{(k \times k)}{\mathbf{X}'\mathbf{X}}$ is nonsingular, in which case $\mathbf{b} = (\mathbf{X}'\mathbf{X})^{-1}\mathbf{X}'\mathbf{y}$. $\mathbf{X}'\mathbf{X}$ is nonsingular if $\{\mathbf{x}_1, \mathbf{x}_2, \ldots, \mathbf{x}_k\}$ is a linearly independent set, providing a basis for the subspace that it generates; otherwise, $\mathbf{b}$ is not unique.

B.4. MATRIX RANK AND THE SOLUTION OF LINEAR SIMULTANEOUS EQUATIONS

The *row space* of an $(m \times n)$ matrix $\mathbf{A}$ is the subspace of the n-dimensional vector space spanned by the m rows of $\mathbf{A}$. The *rank* of $\mathbf{A}$ is the dimension of its row space, that is, the maximum number of linearly independent rows in $\mathbf{A}$. It follows immediately that $\text{rank}(\mathbf{A}) \leq \min(m, n)$.

A matrix is said to be in *row-echelon form* if it satisfies the following criteria:

1. All of its nonzero rows (if any) precede all of its zero rows (if any).
2. The first nonzero entry (proceeding from left to right), called the *leading entry*, in each nonzero row is one.
3. The leading entry in each nonzero row after the first is to the right of the leading entry in the previous row.
4. All other entries are zero in a column containing a leading entry.

Row-echelon form is displayed schematically in (B.5), where the asterisks represent elements of arbitrary value:

$$\left(\begin{array}{cccccccccccccccc} 0 & \cdots & 0 & 1 & * & \cdots & * & 0 & * & \cdots & * & 0 & * & \cdots & * \\ 0 & \cdots & 0 & 0 & 0 & \cdots & 0 & 1 & * & \cdots & * & 0 & * & \cdots & * \\ \vdots & & \vdots & \vdots & \vdots & & \vdots & \vdots & \vdots & & \vdots & \vdots & \vdots & & \vdots \\ 0 & \cdots & 0 & 0 & 0 & \cdots & 0 & 0 & 0 & \cdots & 0 & 1 & * & \cdots & * \\ \hdashline 0 & \cdots & 0 & 0 & 0 & \cdots & 0 & 0 & 0 & \cdots & 0 & 0 & 0 & \cdots & 0 \\ \vdots & & \vdots & \vdots & \vdots & & \vdots & \vdots & \vdots & & \vdots & \vdots & \vdots & & \vdots \\ 0 & \cdots & 0 & 0 & 0 & \cdots & 0 & 0 & 0 & \cdots & 0 & 0 & 0 & \cdots & 0 \end{array}\right) \begin{array}{l} \text{nonzero} \\ \text{rows} \\ \\ \\ \text{zero} \\ \text{rows} \\ \end{array} \tag{B.5}$$

The rank of a matrix in row-echelon form is equal to the number of nonzero rows in the matrix: The pattern of leading entries, each located in a column all of whose other elements are zero, insures that no nonzero row can be formed as a linear combination of other rows

A matrix may be placed in row-echelon form by a sequence of elementary row operations, employing the elimination procedure first described in Section B.1.3. For example, starting with the matrix

$$\begin{pmatrix} -2 & 0 & -1 & 2 \\ 4 & 0 & 1 & 0 \\ 6 & 0 & 1 & 2 \end{pmatrix}$$

divide row 1 by -2

$$\begin{pmatrix} 1 & 0 & \frac{1}{2} & -1 \\ 4 & 0 & 1 & 0 \\ 6 & 0 & 1 & 2 \end{pmatrix}$$

subtract $4 \times$ row 1 from row 2

$$\begin{pmatrix} 1 & 0 & \frac{1}{2} & -1 \\ 0 & 0 & -1 & 4 \\ 6 & 0 & 1 & 2 \end{pmatrix}$$

subtract $6 \times$ row 1 from row 3

$$\begin{pmatrix} 1 & 0 & \frac{1}{2} & -1 \\ 0 & 0 & -1 & 4 \\ 0 & 0 & -2 & 8 \end{pmatrix}$$

multiply row 2 by -1

$$\begin{pmatrix} 1 & 0 & \frac{1}{2} & -1 \\ 0 & 0 & 1 & -4 \\ 0 & 0 & -2 & 8 \end{pmatrix}$$

subtract $\frac{1}{2} \times$ row 2 from row 1

$$\begin{pmatrix} 1 & 0 & 0 & 1 \\ 0 & 0 & 1 & -4 \\ 0 & 0 & -2 & 8 \end{pmatrix}$$

add $2 \times$ row 2 to row 3

$$\begin{pmatrix} 1 & 0 & 0 & 1 \\ 0 & 0 & 1 & -4 \\ 0 & 0 & 0 & 0 \end{pmatrix}$$

The rank of a matrix $\mathbf{A}$ is equal to the rank of its row-echelon form $\mathbf{A}_R$, since a zero row in $\mathbf{A}_R$ can only arise if one row of $\mathbf{A}$ is expressible as a linear combination of other rows (or if $\mathbf{A}$ contains a zero row). That is, none of the elementary row operations alters the rank of a matrix. The rank of the matrix reduced to row-echelon form in the example is thus two.

The row-echelon form of a nonsingular square matrix is the identity matrix, and the rank of a nonsingular matrix is therefore equal to its order. Conversely, the rank of a singular matrix is less than its order.

We have defined the rank of a matrix $\mathbf{A}$ as the dimension of its row space. It may be shown that the rank of $\mathbf{A}$ is also equal to the dimension of its *column space*—that is, to the maximum number of linearly independent columns in $\mathbf{A}$.

A system of m linear simultaneous equations in n unknowns may be written in matrix form as

$$\underset{(m \times n)}{\mathbf{A}} \; \underset{(n \times 1)}{\mathbf{x}} = \underset{(m \times 1)}{\mathbf{b}} \tag{B.6}$$

where the elements of $\mathbf{A}$ and $\mathbf{b}$ are specified constants and $\mathbf{x}$ is the vector of unknowns. Suppose that there is an equal number of equations and unknowns—that is, $m = n$. Then if the coefficient matrix $\mathbf{A}$ is nonsingular, the system of equations (B.6) has the *unique solution* $\mathbf{x} = \mathbf{A}^{-1}\mathbf{b}$.

Alternatively, $\mathbf{A}$ may be singular. Then $\mathbf{A}$ can be transformed to row-echelon form by a sequence of (say, p) elementary row operations, representable as successive multiplication on the left by elementary-row-operation matrices:

$$\mathbf{A}_R = \mathbf{E}_p \cdots \mathbf{E}_2\mathbf{E}_1\mathbf{A} = \mathbf{E}\mathbf{A}$$

Applying these operations to both sides of equation (B.6) produces

$$\mathbf{E}\mathbf{A}\mathbf{x} = \mathbf{E}\mathbf{b}$$

$$\mathbf{A}_R\mathbf{x} = \mathbf{b}_R \tag{B.7}$$

where $\mathbf{b}_R \equiv \mathbf{E}\mathbf{b}$. Equation systems (B.6) and (B.7) are *equivalent* in the sense that any solution vector $\mathbf{x} = \mathbf{x}^*$ that satisfies one system also satisfies the other.

Let r represent the rank of $\mathbf{A}$. Because $r < n$, $\mathbf{A}_R$ contains r nonzero rows and $n - r$ zero rows. If any zero row of $\mathbf{A}_R$ is associated with a nonzero entry (say b) in $\mathbf{b}_R$, then the system of equations is inconsistent or *overdetermined*, for it contains the self-contradictory "equation"

$$0x_1 + 0x_2 + \cdots + 0x_n = b \neq 0$$

If, on the other hand, every zero row of $\mathbf{A}_R$ corresponds to a zero entry in $\mathbf{b}_R$, then the equation system is consistent and there is an infinity of solutions satisfying the system: $n - r$ of the unknowns may be given arbitrary values, which determine the values of the remaining r unknowns. Under this circumstance, we say that the equation system is *underdetermined*.

An example of an underdetermined equation system in row-echelon form is

$$\begin{pmatrix} 1\checkmark & 3 & 0 & -1 \\ 0 & 0 & 1\checkmark & -2 \\ 0 & 0 & 0 & 0 \\ 0 & 0 & 0 & 0 \end{pmatrix} \begin{pmatrix} x_1\checkmark \\ x_2 \\ x_3\checkmark \\ x_4 \end{pmatrix} = \begin{pmatrix} 1 \\ -2 \\ 0 \\ 0 \end{pmatrix} \tag{B.8}$$

To express the solutions of this system in general form, we identify the leading entries of $\mathbf{A}_R$ and the unknowns corresponding to them, indicated by arrows in equation (B.8). Separating the unknowns into two sets based upon this identification, we rewrite the system of equations as

$$\begin{pmatrix} 1 & 0 \\ 0 & 1 \end{pmatrix} \begin{pmatrix} x_1 \\ x_3 \end{pmatrix} + \begin{pmatrix} 3 & -1 \\ 0 & -2 \end{pmatrix} \begin{pmatrix} x_2 \\ x_4 \end{pmatrix} = \begin{pmatrix} 1 \\ -2 \end{pmatrix}$$

Then we solve for x_1 and x_3 in terms of x_2 and x_4:

$$\begin{pmatrix} x_1 \\ x_3 \end{pmatrix} = \begin{pmatrix} 1 \\ -2 \end{pmatrix} - \begin{pmatrix} 3 & -1 \\ 0 & -2 \end{pmatrix} \begin{pmatrix} x_2 \\ x_4 \end{pmatrix}$$

Thus (letting x_2^* and x_4^* represent arbitrary values for x_2 and x_4), any vector

$$\begin{pmatrix} x_1 \\ x_2 \\ x_3 \\ x_4 \end{pmatrix} = \begin{pmatrix} 1 - 3x_2^* + x_4^* \\ x_2^* \\ -2 + 2x_4^* \\ x_4^* \end{pmatrix}$$

is a solution to the equation system, and any solution may be written in this form.

Suppose, now, that there are fewer equations than unknowns—that is, $m < n$. Then r is necessarily less than n, and the equations are either overdetermined (if a zero row of $\mathbf{A}_R$ corresponds to a nonzero entry of $\mathbf{b}_R$) or underdetermined (if they are consistent).

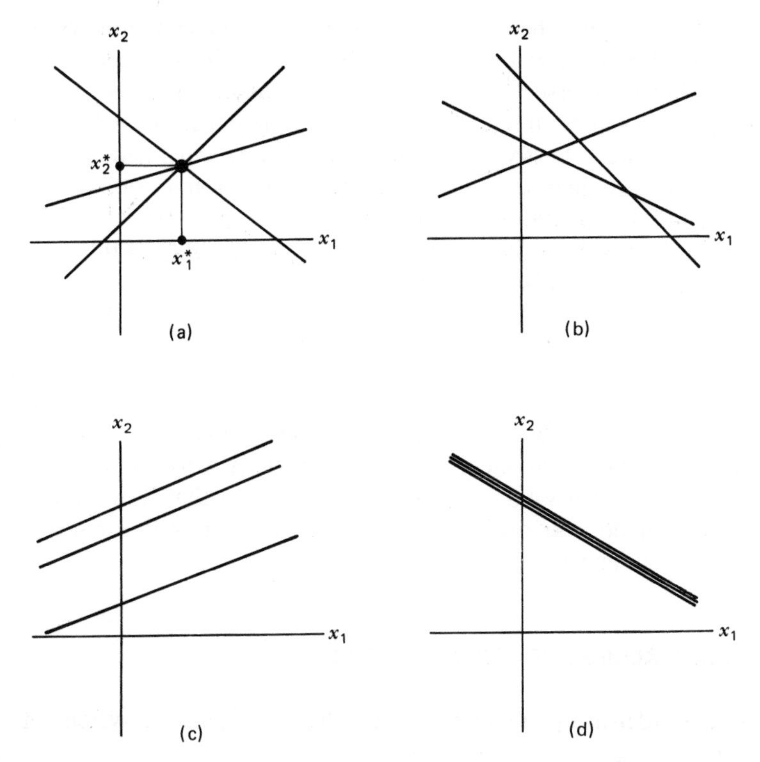

FIGURE B.13. Three linear equations in two unknowns. (a) Unique solution. (b) Overdetermined. (c) Overdetermined. (d) Underdetermined (three coincident lines).

Suppose, finally, that there are more equations than unknowns: $m > n$. If $\mathbf{A}$ is of *full-column rank* (i.e., if $r = n$), then $\mathbf{A}_R$ consists of the order-n identity matrix followed by $m - n$ zero rows: If the equations are consistent, they therefore have a unique solution; otherwise, of course, they are overdetermined. If $r < n$, the equations are either overdetermined (if inconsistent) or underdetermined (if consistent).

To illustrate these results geometrically,[2] consider a system of three linear equations in two unknowns:

$$a_{11}x_1 + a_{12}x_2 = b_1$$
$$a_{21}x_1 + a_{22}x_2 = b_2$$
$$a_{31}x_1 + a_{32}x_2 = b_3$$

Each equation describes a line in two-dimensional coordinate space, as illustrated schematically in Figure B.13. If the three lines intersect at a point, as

[2] This geometric representation of linear equations should not be confused with the geometric vector representation discussed in Section B.2.

in Figure B.13(a), then there is a unique solution to the equation system: Only the pair of values (x_1^*, x_2^*) simultaneously satisfies all three equations. If the three lines fail to intersect at a common point, as in Figure B.13(b) and (c), then no pair of values simultaneously satisfies the equations, which therefore are overdetermined. Lastly, if the three lines are coincident, as in Figure B.13(d), then any pair of values on the common line satisfies all three equations, and the equations are underdetermined.

When the vector **b** in the general equation system (B.6) is the zero vector, the system of equations is said to be *homogeneous*:

$$\underset{(m \times n)}{\mathbf{A}} \; \underset{(n \times 1)}{\mathbf{x}} = \underset{(m \times 1)}{\mathbf{0}}$$

The *trivial solution* $\mathbf{x} = \mathbf{0}$ always satisfies a homogeneous system which, consequently, cannot be inconsistent. From our previous work in this section, we can see that nontrivial solutions exist if $\text{rank}(\mathbf{A}) < n$—that is, when the system is underdetermined. Our results concerning the solution of linear simultaneous equations are summarized in Table B.1.

B.5. EIGENVALUES AND EIGENVECTORS

If **A** is an order-n square matrix, then the homogeneous system of linear equations

$$(\mathbf{A} - l\mathbf{I}_n)\mathbf{x} = \mathbf{0} \tag{B.9}$$

will have nontrivial solutions only for certain values of the scalar l. The results

TABLE B.1. Solutions of Linear Simultaneous Equations

Number of Equations	$m < n$	$m = n$		$m > n$	
Rank	$r < n$	$r < n$	$r = n$	$r < n$	$r = n$
General Equation System					
Consistent	Underdetermined	Underdetermined	Unique solution	Underdetermined	Unique solution
Inconsistent	Overdetermined	Overdetermined	—	Overdetermined	Overdetermined
Homogeneous Equation System	Nontrivial solutions	Nontrivial solutions	Trivial solution	Nontrivial solutions	Trivial solution

m: number of equations
n: number of unknowns
r: rank of coefficient matrix

of the previous section suggest that nontrivial solutions exist when the matrix $(\mathbf{A} - l\mathbf{I}_n)$ is singular; that is when

$$|\mathbf{A} - l\mathbf{I}_n| = 0 \tag{B.10}$$

The determinantal equation (B.10) is called the *characteristic equation* of the matrix $\mathbf{A}$, and values of l for which this equation holds are called *eigenvalues*, *characteristic roots*, or *latent roots* of $\mathbf{A}$. A vector $\mathbf{x}_1$ satisfying the equation (B.9) for a particular eigenvalue l_1 is called an *eigenvector*, *characteristic vector*, or *latent vector* of $\mathbf{A}$ associated with l_1.

We shall examine the (2×2) case in some detail. For this case, the characteristic equation is

$$\begin{vmatrix} a_{11} - l & a_{12} \\ a_{21} & a_{22} - l \end{vmatrix} = 0$$

$$(a_{11} - l)(a_{22} - l) - a_{12}a_{21} = 0 \tag{B.11}$$

$$l^2 - (a_{11} + a_{22})l + a_{11}a_{22} - a_{12}a_{21} = 0$$

Using the quadratic formula to solve equation (B.11) produces the two roots

$$\begin{aligned} l_1 &= \frac{1}{2}\left[a_{11} + a_{22} + \sqrt{(a_{11} + a_{22})^2 - 4(a_{11}a_{22} - a_{12}a_{21})}\right] \\ l_2 &= \frac{1}{2}\left[a_{11} + a_{22} - \sqrt{(a_{11} + a_{22})^2 - 4(a_{11}a_{22} - a_{12}a_{21})}\right] \end{aligned} \tag{B.12}$$

These roots are real if the quantity under the radical is nonnegative. Notice, incidentally, that $l_1 + l_2 = a_{11} + a_{22}$ (the sum of the eigenvalues of $\mathbf{A}$ is the trace of $\mathbf{A}$), and that $l_1 l_2 = a_{11}a_{22} - a_{12}a_{21}$ (the product of the eigenvalues is the determinant of $\mathbf{A}$). Furthermore, when $\mathbf{A}$ is singular, l_2 is zero.

If $\mathbf{A}$ is symmetric, then $a_{12} = a_{21}$, and equations (B.12) become

$$\begin{aligned} l_1 &= \frac{1}{2}\left[a_{11} + a_{22} + \sqrt{(a_{11} - a_{22})^2 + 4a_{12}^2}\right] \\ l_2 &= \frac{1}{2}\left[a_{11} + a_{22} - \sqrt{(a_{11} - a_{22})^2 + 4a_{12}^2}\right] \end{aligned} \tag{B.13}$$

The eigenvalues of a (2×2) symmetric matrix are necessarily real since the quantity under the radical in equations (B.13) is the sum of two squares, which cannot be negative.

We shall use the following (2×2) symmetric matrix for an illustration:

$$\begin{pmatrix} 1 & 0.5 \\ 0.5 & 1 \end{pmatrix}$$

Here

$$l_1 = \frac{1}{2}\left[1 + 1 + \sqrt{(1-1)^2 + 4(0.5)^2}\right] = 1.5$$

$$l_2 = \frac{1}{2}\left[1 + 1 - \sqrt{(1-1)^2 + 4(0.5)^2}\right] = 0.5$$

To find the eigenvectors associated with l_1, we solve the homogeneous system of equations

$$\begin{pmatrix} -0.5 & 0.5 \\ 0.5 & -0.5 \end{pmatrix}\begin{pmatrix} x_{11} \\ x_{21} \end{pmatrix} = \begin{pmatrix} 0 \\ 0 \end{pmatrix}$$

yielding

$$\mathbf{x}_1 = \begin{pmatrix} x_{11} \\ x_{21} \end{pmatrix} = \begin{pmatrix} x_{21}^* \\ x_{21}^* \end{pmatrix}$$

Similarly, for l_2, we solve

$$\begin{pmatrix} 0.5 & 0.5 \\ 0.5 & 0.5 \end{pmatrix}\begin{pmatrix} x_{12} \\ x_{22} \end{pmatrix} = \begin{pmatrix} 0 \\ 0 \end{pmatrix}$$

which produces

$$\mathbf{x}_2 = \begin{pmatrix} x_{12} \\ x_{22} \end{pmatrix} = \begin{pmatrix} -x_{22}^* \\ x_{22}^* \end{pmatrix}$$

The set of eigenvectors associated with each eigenvalue spans a one-dimensional subspace: When one of the entries of the eigenvector is specified, the other entry follows. Notice further that the eigenvectors $\mathbf{x}_1$ and $\mathbf{x}_2$ are orthogonal:

$$\mathbf{x}_1 \cdot \mathbf{x}_2 = -x_{21}^* x_{22}^* + x_{21}^* x_{22}^* = 0$$

Many of the properties of eigenvalues and eigenvectors of (2×2) matrices generalize to $(n \times n)$ matrices. In particular:

1. The characteristic equation $|\mathbf{A} - l\mathbf{I}_n| = 0$ of an $(n \times n)$ matrix $\mathbf{A}$ is an n-th order polynomial in l; there are, consequently, n eigenvalues, not all necessarily distinct.
2. The sum of the eigenvalues of $\mathbf{A}$ is the trace of $\mathbf{A}$.
3. The product of the eigenvalues of $\mathbf{A}$ is the determinant of $\mathbf{A}$.
4. The number of nonzero eigenvalues of $\mathbf{A}$ is the rank of $\mathbf{A}$.

5. If **A** is a symmetric matrix, then the eigenvalues of **A** are all real numbers.
6. If the eigenvalues of **A** are distinct (i.e., all different), then the set of eigenvectors associated with a particular eigenvalue spans a one-dimensional subspace. If, alternatively, k eigenvalues are equal, then their common set of eigenvectors spans a subspace of dimension k.
7. Eigenvectors associated with different eigenvalues are orthogonal.

B.6. QUADRATIC FORMS AND POSITIVE-DEFINITE MATRICES

The expression

$$\underset{(1\times n)}{\mathbf{x}'}\ \underset{(n\times n)}{\mathbf{A}}\ \underset{(n\times 1)}{\mathbf{x}} \tag{B.14}$$

is called a *quadratic form* in **x**. In this section, **A** will always represent a symmetric matrix. **A** is said to be *positive definite* if the quadratic form (B.14) is positive for all nonzero **x**. **A** is *positive semidefinite* if this quadratic form is nonnegative (i.e., positive or zero) for all nonzero vectors **x**. The eigenvalues of a positive-definite matrix are all positive (and, consequently, the matrix is nonsingular); those of a positive-semidefinite matrix are all positive or zero.

Let $\underset{(m\times m)}{\mathbf{C}} = \underset{(m\times n)}{\mathbf{B}'}\ \underset{(n\times n)}{\mathbf{A}}\ \underset{(n\times m)}{\mathbf{B}}$, where **A** is positive definite and **B** is of full-column rank ($m \leq n$). We shall show that **C** is positive definite. Note that **C** is symmetric:

$$\mathbf{C}' = (\mathbf{B}'\mathbf{A}\mathbf{B})' = \mathbf{B}'\mathbf{A}'\mathbf{B} = \mathbf{B}'\mathbf{A}\mathbf{B} = \mathbf{C}$$

If **y** is any $(m \times 1)$ nonzero vector, then $\underset{(n\times 1)}{\mathbf{x}} = \mathbf{B}\mathbf{y}$ is also nonzero: Since **B** is of rank m, we can select m linearly independent rows of **B**, forming the nonsingular matrix $\mathbf{B}^*$; $\mathbf{x}^* = \mathbf{B}^*\mathbf{y}$, which contains a subset of the entries in **x**, is nonzero because $\mathbf{y} = \mathbf{B}^{*-1}\mathbf{x}^* \neq 0$. Consequently

$$\mathbf{y}'\mathbf{C}\mathbf{y} = \mathbf{y}'\mathbf{B}'\mathbf{A}\mathbf{B}\mathbf{y} = \mathbf{x}'\mathbf{A}\mathbf{x}$$

is necessarily positive, and **C** is positive definite. By similar reasoning, if rank(**B**) $< m$, then **C** is positive semidefinite. The matrix $\underset{(m\times n)}{\mathbf{B}'}\ \underset{(n\times m)}{\mathbf{B}} = \mathbf{B}'\mathbf{I}_n\mathbf{B}$ is therefore positive definite if **B** is of full-column rank (since $\mathbf{I}_n$ is clearly positive definite), and positive semidefinite otherwise. (Cf. the geometric discussion following equation (B.4) in Section B.3.)

B.7. DIFFERENTIAL CALCULUS IN MATRIX FORM

Let y be a differentiable function of the entries of an n-element vector **x**; that is

$$y = f(x_1, x_2, \ldots, x_n) = f(\mathbf{x})$$

The *vector partial derivative* of y with respect to $\mathbf{x}$ is defined to be the column vector of partial derivatives of y with respect to the entries of $\mathbf{x}$:

$$\frac{\partial y}{\partial \mathbf{x}} \equiv \left(\frac{\partial y}{\partial x_1}, \frac{\partial y}{\partial x_2}, \ldots, \frac{\partial y}{\partial x_n} \right)'$$

If, therefore, y is a linear function of $\mathbf{x}$,

$$y = \underset{(1\times n)}{\mathbf{a}'} \underset{(n\times 1)}{\mathbf{x}} = a_1x_1 + a_2x_2 + \cdots + a_nx_n$$

then $\partial y/\partial \mathbf{x} = \mathbf{a}$. Suppose, alternatively, that y is a quadratic form in $\mathbf{x}$: $y = \underset{(1\times n)}{\mathbf{x}'} \underset{(n\times n)}{\mathbf{A}} \underset{(n\times 1)}{\mathbf{x}}$, where $\mathbf{A}$ is symmetric. Expanding the matrix product gives us

$$\begin{aligned} y = {} & a_{11}x_1^2 + a_{22}x_2^2 + \cdots + a_{nn}x_n^2 \\ & + 2a_{12}x_1x_2 + \cdots + 2a_{1n}x_1x_n + \cdots + 2a_{n-1,n}x_{n-1}x_n \end{aligned}$$

and thus

$$\frac{\partial y}{\partial x_i} = 2(a_{i1}x_1 + a_{i2}x_2 + \cdots + a_{in}x_n) = 2\mathbf{a}_i'\mathbf{x}$$

where $\mathbf{a}_i'$ represents the ith row of $\mathbf{A}$. Placing these partial derivatives in a vector produces $\partial y/\partial \mathbf{x} = 2\mathbf{A}\mathbf{x}$. The vector partial derivatives of linear and quadratic forms are strikingly similar to the analogous scalar derivatives $d(ax)/dx = a$ and $d(ax^2)/dx = 2ax$.

The so-called *Hessian* matrix of second-order partial derivatives of a function $y = f(\underset{(n\times 1)}{\mathbf{x}})$ is defined in the following manner:

$$\frac{\partial^2 y}{\partial \mathbf{x}\, \partial \mathbf{x}'} \equiv \begin{pmatrix} \dfrac{\partial^2 y}{\partial x_1^2} & \dfrac{\partial^2 y}{\partial x_1\, \partial x_2} & \cdots & \dfrac{\partial^2 y}{\partial x_1\, \partial x_n} \\ \dfrac{\partial^2 y}{\partial x_2\, \partial x_1} & \dfrac{\partial^2 y}{\partial x_2^2} & \cdots & \dfrac{\partial^2 y}{\partial x_2\, \partial x_n} \\ \vdots & \vdots & & \vdots \\ \dfrac{\partial^2 y}{\partial x_n\, \partial x_1} & \dfrac{\partial^2 y}{\partial x_n\, \partial x_2} & \cdots & \dfrac{\partial^2 y}{\partial x_n^2} \end{pmatrix}$$

For instance, $\partial^2(\mathbf{x}'\mathbf{A}\mathbf{x})/\partial \mathbf{x}\, \partial \mathbf{x}' = 2\mathbf{A}$, for a symmetric matrix $\mathbf{A}$.

To minimize a function $y = f(\mathbf{x})$ of several variables, we often equate the vector partial derivative to zero, $\partial y/\partial \mathbf{x} = \mathbf{0}$, and solve the resulting set of simultaneous equations for $\mathbf{x}$, obtaining a solution $\mathbf{x} = \mathbf{x}^*$. This solution represents a (local) minimum of the function in question if the Hessian matrix evaluated at $\mathbf{x} = \mathbf{x}^*$ is positive definite.[3]

[3]A positive-definite Hessian is a sufficient but not necessary condition for a minimum.

C

Probability and Estimation

The purpose of this appendix is to outline basic results in probability and statistical inference that are employed at various points in the body of the text. In Section C.1, we review concepts in elementary probability theory. This material is assumed to be generally familiar; it is covered in more detail in introductory texts on statistical theory (though it is often treated only cursorily in introductory social-statistics texts).

Sections C.2 and C.3 briefly describe several probability distributions that are of particular importance to the study of linear and related models. Section C.4 outlines asymptotic distribution theory, which we shall occasionally require to determine properties of statistical estimators, a subject that is taken up in Section C.5. The concluding section of the appendix (C.6) develops the broadly applicable method of maximum-likelihood estimation.

C.1. ELEMENTARY PROBABILITY THEORY

C.1.1. Basic Definitions

In probability theory, an *experiment* is a repeatable procedure for making an observation; an *outcome* is a possible observation resulting from an experiment; and the *sample space* of the experiment is the set of all possible outcomes. Any *realization* of the experiment produces a particular outcome. Sample spaces may be discrete and finite, discrete and infinite (i.e., countably infinite), or continuous.

If we flip a coin twice and record on each flip whether the coin shows heads (H) or tails (T), then the sample space of the experiment is discrete and finite:

$S = \{\text{HH}, \text{HT}, \text{TH}, \text{TT}\}$. If we flip a coin repeatedly until a head shows and record the number of flips required, then the sample space is discrete and infinite, consisting of the positive integers: $S = \{1, 2, 3, \ldots\}$. If we burn a light bulb until it fails, recording the burning time in hours and fractions of an hour, then the sample space of the experiment is continuous and consists of all positive real numbers (abstracting from the discontinuous quality of real devices for measuring time and not bothering to specify an upper limit for the life of a bulb): $S = \{x : x > 0\}$. In this section we shall limit consideration to discrete, finite sample spaces.

An *event* is a subset of the sample space of an experiment—that is, a set of outcomes. An event is said to occur for a realization of the experiment if one of its constituent outcomes occurs. For example, for $S = \{\text{HH}, \text{HT}, \text{TH}, \text{TT}\}$, the event $E = \{\text{HH}, \text{HT}\}$, representing a head on the first flip of the coin, occurs if we obtain the outcome HH or the outcome HT.

Let $S = \{o_1, o_2, \ldots, o_n\}$ be the sample space of an experiment; let $O_1 = \{o_1\}$, $O_2 = \{o_2\}, \ldots, O_n = \{o_n\}$ be the *simple events*, each consisting of one of the outcomes; and let $E = \{o_a, o_b, \ldots, o_m\}$ be any subset of S. Probabilities are numbers assigned to events in a manner consistent with the following axioms (rules):

1. $\Pr(E) \geq 0$: The probability of an event is zero or positive.
2. $\Pr(E) = \Pr(O_a) + \Pr(O_b) + \cdots + \Pr(O_m)$: The probability of an event is the sum of probabilities of its contituent outcomes.
3. $\Pr(S) = 1$; $\Pr(\varnothing) = 0$, where $\varnothing$ is the *empty event*, which contains no outcomes: Some outcome must occur.

In classical statistics, the perspective adopted in this book, probabilities are interpreted as long-run relative frequencies—that is, as proportions. Thus, if the probability of an event is one half, then the event will occur approximately half the time when the experiment is repeated many times, and the approximation is expected to improve as the number of repetitions increases.

A number of important relations may be defined among events. The *intersection* of two events E_1 and E_2, denoted $E_1 \cap E_2$, contains all outcomes common to the two. $\Pr(E_1 \cap E_2)$ is thus the probability that both E_1 *and* E_2 occur simultaneously. If $E_1 \cap E_2 = \varnothing$, then E_1 and E_2 are said to be *disjoint* or *mutually exclusive*. By extension, the intersection of many events $E_1 \cap E_2 \cap \cdots \cap E_k$ contains all outcomes that are members of each and every event.

The *union* of two events $E_1 \cup E_2$ contains all outcomes in either or both events. Thus $\Pr(E_1 \cup E_2)$ is the probability that E_1 occurs *or* E_2 occurs (or both occur). The union of several events $E_1 \cup E_2 \cup \cdots \cup E_k$ contains all outcomes that are in one or more of the events. If these events are disjoint, then

$$\Pr(E_1 \cup E_2 \cup \cdots \cup E_k) = \sum_{i=1}^{k} \Pr(E_i)$$

otherwise

$$\Pr(E_1 \cup E_2 \cup \cdots \cup E_k) < \sum_{i=1}^{k} \Pr(E_i)$$

(since some outcomes contribute more than once when the probabilities are summed). For two events, for example,

$$\Pr(E_1 \cup E_2) = \Pr(E_1) + \Pr(E_2) - \Pr(E_1 \cap E_2)$$

The *conditional probability* of E_1 given E_2, written $\Pr(E_1|E_2)$, is defined as $\Pr(E_1 \cap E_2)/\Pr(E_2)$. The conditional probability is interpreted as the probability that E_1 occurs if E_2 is known to occur. Two events E_1 and E_2 are said to be *independent* if $\Pr(E_1 \cap E_2) = \Pr(E_1)\Pr(E_2)$. Independence of E_1 and E_2 implies that $\Pr(E_1) = \Pr(E_1|E_2)$ and that $\Pr(E_2) = \Pr(E_2|E_1)$. (Note that independence is different from disjointness.) More generally, a set of events $\{E_1, E_2, \ldots, E_k\}$ is independent if for every subset $\{E_a, E_b, \ldots, E_m\}$ containing two or more of the events

$$\Pr(E_a \cap E_b \cap \cdots \cap E_m) = \Pr(E_a)\Pr(E_b) \cdots \Pr(E_m)$$

The *difference* between two events $E_1 - E_2$ contains all outcomes in the first event that are not in the second. The difference $\overline{E} \equiv S - E$ is called the *complement* of E. Note that $\Pr(\overline{E}) = 1 - \Pr(E)$.

Let $E \equiv E_1 \cap E_2 \cap \cdots \cap E_k$. Then $\overline{E} = \overline{E}_1 \cup \overline{E}_2 \cup \cdots \cup \overline{E}_k$. Applying previous results,

$$\begin{aligned}\Pr(E_1 \cap E_2 \cap \cdots \cap E_k) &= \Pr(E) = 1 - \Pr(\overline{E}) \\ &\geq 1 - \sum_{i=1}^{k} \Pr(\overline{E}_i) \qquad \text{(C.1)}\end{aligned}$$

Suppose that all of the events $E_1, E_2, \ldots, E_k$ have equal probabilities, say $\Pr(E_i) = 1 - b$ (i.e., $\Pr(\overline{E}_i) = b$). Then

$$\begin{aligned}1 - a &\equiv \Pr(E_1 \cap E_2 \cap \cdots \cap E_k) \\ &\geq 1 - kb \qquad \text{(C.2)}\end{aligned}$$

(C.2) and the more general (C.1) are called *Bonferroni inequalities*. Inequality

(C.2) is useful for evaluating the combined Type-I error rate a for k nonindependent statistical tests with equal individual error rates b; the result is that $a \leq kb$. For instance, if we test 20 true statistical hypotheses, each at a significance level of .01, then the probability of rejecting at least one hypothesis is at most 20(.01) = .20.

C.1.2. Random Variables

A *random variable* is a function that assigns a number to each outcome of the sample space of an experiment. For the sample space $S = \{\text{HH}, \text{HT}, \text{TH}, \text{TT}\}$ introduced earlier, a random variable X that counts the number of heads in an outcome assigns a value of two to HH, of one to each of HT and TH, and of zero to TT.

If X is a discrete random variable, then we write $p(x)$ for $\Pr(X = x)$, where the capital letter X represents the variable itself, while the lower-case letter x denotes a *particular value* of the variable. The probabilities $p(x)$ for all values of X comprise the *probability distribution* of the random variable. If, for example, each outcome of the coin-flipping experiment has probability $\frac{1}{4}$ (i.e., if the coin is "fair"), then $p(0) = \frac{1}{4}$, $p(1) = \frac{1}{2}$, and $p(2) = \frac{1}{4}$. The *cumulative distribution function* (*CDF*) of a random variable X, written $P(x)$, indicates the probability of observing a value of the variable less than or equal to a particular value: $P(x) \equiv \Pr(X \leq x) = \sum_{x' \leq x} p(x')$. Thus, for the example $P(0) = \frac{1}{4}$, $P(1) = \frac{3}{4}$, and $P(2) = 1$.

Random variables defined on continuous sample spaces may themselves be continuous. We still take $P(x)$ as $\Pr(X \leq x)$, but it becomes meaningless to refer to the probability of observing individual values of X (as long as $P(x)$ is absolutely continuous). We may, however, introduce the *probability density function* $p(x)$ as the continuous analog of the discrete probability distribution, defining $p(x) = dP(x)/dx$. Reversing this relation, $P(x) = \int_{-\infty}^{x} p(x)\,dx$; and

$$\Pr(x_0 \leq x \leq x_1) = P(x_1) - P(x_0) = \int_{x_0}^{x_1} p(x)\,dx$$

Thus, areas under the probability density function are interpreted as probabilities.

A particularly simple continuous probability distribution is the *rectangular distribution*

$$p(x) = \begin{cases} 0 & a > x \\ \dfrac{1}{b - a} & a \leq x \leq b \\ 0 & x > b \end{cases}$$

This density function is pictured in Figure C.1(a), and the corresponding

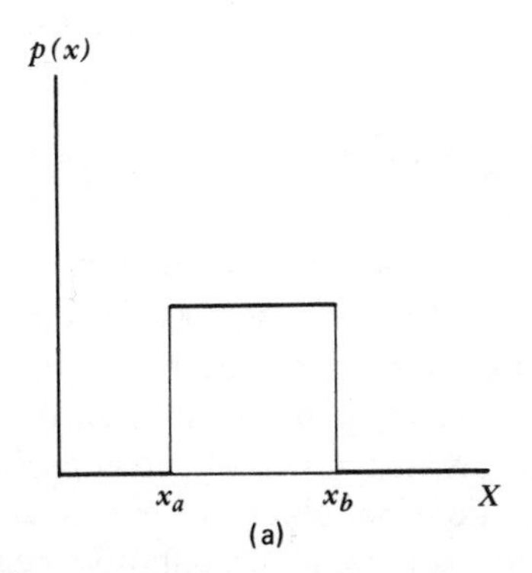

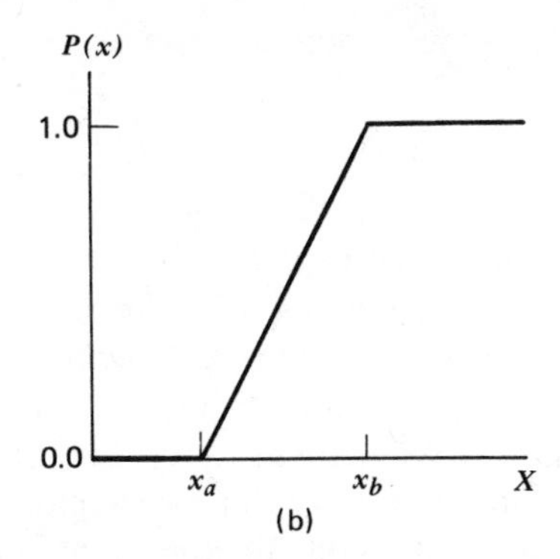

FIGURE C.1. The rectangular distribution. (a) Probability density function. (b) Cumulative distribution function.

cumulative distribution function is shown in Figure C.1(b). Note that

$$\int_{-\infty}^{\infty} p(x)\,dx = \int_{a}^{b} p(x)\,dx = \frac{1}{b-a}(b-a) = 1$$

Two fundamental properties of a random variable are its *expected value* (or *mean*) and its *variance*.[1] The expected value specifies the center of the probability distribution of the random variable (in the same sense as the arithmetic mean of a set of scores specifies the center of the frequency distribution of the scores), while the variance indicates how spread out the distribution is about its expectation. The expectation is interpretable as the mean score of the random variable that would be observed over many repetitions of the experiment, while the variance is the mean-squared distance between the scores and their expectation.

In the discrete case, the expectation of a random variable X, indicated by $E(X)$ or μ_X, is given by

$$E(X) = \mu_X \equiv \sum_{\text{all } x} xp(x)$$

The analogous formula for the continuous case is

$$E(X) \equiv \int_{-\infty}^{\infty} xp(x)\,dx$$

The variance of a random variable X, written $V(X)$ or σ_X^2, is defined as $E[(X - \mu_X)^2]$. Thus, in the discrete case

$$V(X) \equiv \sum_{\text{all } x} (x - \mu_X)^2 p(x)$$

[1]The expectation and variance are undefined for some random variables. We shall ignore this possibility here.

while in the continuous case

$$V(X) \equiv \int_{-\infty}^{\infty} (x - \mu_X)^2 p(x)\, dx$$

The variance is expressed in the squared units of the random variable, but the standard deviation $\sigma \equiv +\sqrt{\sigma^2}$ is measured in the same units as the variable.

The *joint probability distribution* of two discrete random variables X_1 and X_2 gives the probability of simultaneously observing any pair of values for the two variables. We write $p(x_1, x_2)$ for $\Pr(X_1 = x_1 \text{ and } X_2 = x_2)$. The *joint probability density* $p(x_1, x_2)$ of two continuous random variables is defined analogously. Extension to the joint probability or joint probability density $p(x_1, x_2, \ldots, x_n)$ of several random variables is straightforward. To distinguish it from the joint probability distribution, we call $p(x_i)$ the *marginal probability distribution* of X_i. Note that $p(x_1) = \Sigma_{x_2} p(x_1, x_2)$.

The *conditional probability* (or probability density) $p(x_1|x_2)$ (read, "the conditional probability that $X_1 = x_1$ given that $X_2 = x_2$") is equal to $p(x_1, x_2)/p(x_2)$. The random variables X_1 and X_2 are said to be *independent* if $p(x_1) = p(x_1|x_2)$ for all values of X_1 and X_2; that is, when X_1 and X_2 are independent, the conditional and marginal distributions of X_1 are identical. Equivalent conditions for independence are $p(x_2) = p(x_2|x_1)$ and $p(x_1, x_2) = p(x_1)p(x_2)$ (i.e., when X_1 and X_2 are independent, their joint probability is the product of their marginal probabilities): If X_1 and X_2 are independent, then by the original definition

$$p(x_1|x_2) \equiv \frac{p(x_1, x_2)}{p(x_2)} = p(x_1)$$

thus, $p(x_1, x_2) = p(x_1)p(x_2)$; and

$$p(x_2|x_1) \equiv \frac{p(x_1, x_2)}{p(x_1)} = \frac{p(x_1)p(x_2)}{p(x_1)} = p(x_2)$$

More generally, the set of n random variables $X_1, X_2, \ldots, X_n$ is independent, if for every subset $X_a, X_b, \ldots, X_m$ of size n or smaller, $p(x_a, x_b, \ldots, x_m) = p(x_a)p(x_b) \cdots p(x_m)$.

The *covariance* of two random variables is a measure of the *linear* dependence between them; the covariance is given by

$$C(X_1, X_2) = \sigma_{12} \equiv E[(X_1 - \mu_1)(X_2 - \mu_2)]$$

When large values of X_1 are associated with large values of X_2 (and, conversely, small values with small values), the covariance is positive; when large values of X_1 are associated with small values of X_2 (and vice-versa), the covariance is negative; the covariance is zero otherwise, for instance (but not exclusively) when the random variables are independent. Note that $C(X, X) = V(X)$.

The correlation ρ_{12} between two random variables X_1 and X_2 is a normalized version of their covariance, defined by the formula $\rho_{12} \equiv \sigma_{12}/\sigma_1\sigma_2$. The smallest possible value of the correlation, $\rho = -1$, is indicative of a perfect inverse linear relationship between the random variables, while the largest value, $\rho = 1$, is indicative of a perfect direct linear relationship; $\rho = 0$ corresponds to a covariance of zero and indicates the absence of a linear relationship.

It is often convenient to write a collection of random variables as a *vector random variable*: for example, $\underset{(n\times 1)}{\mathbf{x}} = (X_1, X_2, \ldots, X_n)'$. The expectation of a vector random variable is the vector of expectations of its elements; thus

$$E(\mathbf{x}) = \boldsymbol{\mu}_x \equiv [E(X_1), E(X_2), \ldots, E(X_n)]'$$

The *variance* (or *variance-covariance*) *matrix* of a vector random variable $\mathbf{x}$ is defined by analogy with the scalar variance as

$$V(\mathbf{x}) = \underset{(n\times n)}{\boldsymbol{\Sigma}_{xx}} \equiv E[(\mathbf{x} - \boldsymbol{\mu}_x)(\mathbf{x} - \boldsymbol{\mu}_x)']$$

$$= \begin{pmatrix} \sigma_1^2 & \sigma_{12} & \cdots & \sigma_{1n} \\ \sigma_{21} & \sigma_2^2 & \cdots & \sigma_{2n} \\ \vdots & \vdots & \ddots & \vdots \\ \sigma_{n1} & \sigma_{n2} & \cdots & \sigma_n^2 \end{pmatrix}$$

$V(\mathbf{x})$ is symmetric and positive semidefinite. The *covariance matrix* for two vector random variables $\underset{(n\times 1)}{\mathbf{x}}$ and $\underset{(m\times 1)}{\mathbf{y}}$ is

$$C(\mathbf{x}, \mathbf{y}) = \underset{(n\times m)}{\boldsymbol{\Sigma}_{xy}} \equiv E[(\mathbf{x} - \boldsymbol{\mu}_x)(\mathbf{y} - \boldsymbol{\mu}_y)']$$

C.1.3. Transformations of Random Variables

Suppose that a random variable Y is a linear function $a + bX$ of a discrete random variable X, with expectation μ_X and variance σ_X^2. Then

$$E(Y) = \mu_Y = \sum_x (a + bx)p(x)$$

$$= a\sum p(x) + b\sum xp(x) = a + b\mu_X$$

and (employing this property of the expectation operator)

$$V(Y) \equiv E[(Y - \mu_Y)^2] = E\{[(a + bX) - (a + b\mu_X)]^2\}$$

$$= b^2 E[(X - \mu_X)^2] = b^2\sigma_X^2$$

Now, let Y be a linear function $a_1X_1 + a_2X_2$ of two discrete random variables X_1 and X_2, with expectations μ_1 and μ_2, variances σ_1^2 and σ_2^2, and covariance σ_{12}. Then

$$\begin{aligned}
E(Y) = \mu_Y &= \sum_{x_1}\sum_{x_2}(a_1x_1 + a_2x_2)p(x_1, x_2) \\
&= \sum_{x_1}\sum_{x_2} a_1x_1p(x_1, x_2) + \sum_{x_1}\sum_{x_2} a_2x_2p(x_1, x_2) \\
&= a_1\sum_{x_1} x_1p(x_1) + a_2\sum_{x_2} x_2p(x_2) \\
&= a_1\mu_1 + a_2\mu_2
\end{aligned}$$

and

$$\begin{aligned}
V(Y) &\equiv E\left[(Y - \mu_Y)^2\right] \\
&= E\left\{\left[(a_1X_1 + a_2X_2) - (a_1\mu_1 + a_2\mu_2)\right]^2\right\} \\
&= E\left\{\left[(a_1X_1 - a_1\mu_1) + (a_2X_2 - a_2\mu_2)\right]^2\right\} \\
&= a_1^2E\left[(X_1 - \mu_1)^2\right] + a_2^2E\left[(X_2 - \mu_2)^2\right] \\
&\quad + 2a_1a_2E\left[(X_1 - \mu_1)(X_2 - \mu_2)\right] \\
&= a_1^2\sigma_1^2 + a_2^2\sigma_2^2 + 2a_1a_2\sigma_{12}
\end{aligned}$$

Note that when X_1 and X_2 are independent, and consequently $\sigma_{12} = 0$, $V(Y) = a_1^2\sigma_1^2 + a_2^2\sigma_2^2$.

These results generalize to vector random variables in the following manner: Let $\underset{(m\times 1)}{\mathbf{y}}$ be a linear transformation $\underset{(m\times n)}{\mathbf{A}}\ \underset{(n\times 1)}{\mathbf{x}}$ of the vector random variable $\mathbf{x}$, which has expectation $E(\mathbf{x}) = \boldsymbol{\mu}_x$ and variance-covariance matrix $V(\mathbf{x}) = \boldsymbol{\Sigma}_{xx}$. It may be shown (in a manner analogous to the scalar proofs given previously) that

$$E(\mathbf{y}) = \underset{(m\times 1)}{\boldsymbol{\mu}_y} = \mathbf{A}\boldsymbol{\mu}_x$$

$$V(\mathbf{y}) = \underset{(m\times m)}{\boldsymbol{\Sigma}_{yy}} = \mathbf{A}\boldsymbol{\Sigma}_{xx}\mathbf{A}'$$

If the entries of $\mathbf{x}$ are pairwise independent, then the off-diagonal elements of $\boldsymbol{\Sigma}_{xx}$ are zero, and consequently

$$\sigma_{Y_i}^2 = \sum_{j=1}^{n} a_{ij}^2\sigma_{X_j}^2$$

Although we have developed these rules for discrete random variables, they apply equally to the continuous case. For instance, if $Y = a + bX$ is a linear function of the continuous random variable X, then

$$E(Y) = \int_{-\infty}^{\infty} (a + bx) p(x)\, dx$$

$$= a\int_{-\infty}^{\infty} p(x)\, dx + b\int_{-\infty}^{\infty} xp(x)\, dx$$

$$= a + bE(X)$$

At times, when $\mathbf{y} = f(\mathbf{x})$, we need to know not merely $E(\mathbf{y})$ and $V(\mathbf{y})$ but the probability distribution of $\mathbf{y}$. Indeed f may be a nonlinear function. Suppose that there is the same number of elements (n) in $\mathbf{y}$ and $\mathbf{x}$; that the function f is differentiable; and that f is one-to-one over the domain of $\mathbf{x}$-values under consideration. The last property implies that we may write the reverse transformation $\mathbf{x} = f^{-1}(\mathbf{y})$. The probability density for $\mathbf{y}$ is given by

$$p(\mathbf{y}) = p(\mathbf{x})\left|\frac{\partial \mathbf{x}}{\partial \mathbf{y}}\right|_{+} = p(\mathbf{x})\left|\frac{\partial \mathbf{y}}{\partial \mathbf{x}}\right|_{+}^{-1}$$

where $|\partial \mathbf{x}/\partial \mathbf{y}|_{+}$, called the *Jacobian* of the transformation, is the absolute value of the determinant

$$\begin{vmatrix} \dfrac{\partial X_1}{\partial Y_1} & \cdots & \dfrac{\partial X_n}{\partial Y_1} \\ \vdots & \ddots & \vdots \\ \dfrac{\partial X_1}{\partial Y_n} & \cdots & \dfrac{\partial X_n}{\partial Y_n} \end{vmatrix}$$

and $|\partial \mathbf{y}/\partial \mathbf{x}|_{+}$ is similarly defined.

C.2. TWO DISCRETE PROBABILITY DISTRIBUTIONS

In this section we define two important discrete probability distributions: the binomial distribution and its generalization, the multinomial distribution.

C.2.1. The Binomial Distribution

The coin-flipping experiment described at the beginning of Section C.1.2 gives rise to a binomial random variable. To extend this example, let the random variable X count the number of heads that appear in n independent flips of a coin. Let π denote the probability of obtaining a head on any given flip; $1 - \pi$ is then the probability of obtaining a tail. The probability of observing exactly x heads and $n - x$ tails (i.e., $\Pr(X = x)$), is given by the *binomial distribution*:

$$p(x) = \binom{n}{x}\pi^x(1 - \pi)^{n-x} \tag{C.3}$$

where x is any integer between zero and n, inclusive; $\pi^x(1 - \pi)^{n-x}$ is the probability of observing x heads and $n - x$ tails in a *particular* arrangement (e.g., x heads followed by $n - x$ tails); and $\binom{n}{x} \equiv n!/[x!(n - x)!]$, called the *binomial coefficient*, is the number of *different* arrangements of x heads and $n - x$ tails. The expectation of X is $E(X) = n\pi$, and its variance is $V(X) = n\pi(1 - \pi)$.

C.2.2. The Multinomial Distribution

Imagine n repeated, independent trials of a process that on each trial can give rise to one of k different categories of outcomes. Let the random variable X_i count the number of outcomes in category i. Let π_i denote the probability of obtaining an outcome in category i on any given trial. Note that $\sum_{i=1}^{k}\pi_i = 1$, and similarly that $\sum_{i=1}^{k} X_i = n$.

Suppose, for instance, that we toss a die n times, letting X_1 count the number of ones, X_2 the number of twos,..., X_6 the number of sixes. Then $k = 6$, and π_1 is the probability of obtaining a one on any toss, π_2 is the probability of obtaining a two, and so on.

Returning to the general case, the vector random variable $\mathbf{x} \equiv (X_1, X_2, \ldots, X_k)'$ follows the *multinomial distribution*

$$p(\mathbf{x}) = p(x_1, x_2, \ldots, x_k) = \frac{n!}{x_1!x_2! \cdots x_k!}\pi_1^{x_1}\pi_2^{x_2} \cdots \pi_k^{x_k}$$

Note that $\pi_1^{x_1}\pi_2^{x_2}\ldots\pi_k^{x_k}$ gives the probability of obtaining x_1 outcomes in category one, x_2 in category two, and so on, in a particular arrangement; and $n!/(x_1!x_2! \cdots x_k!)$ counts the number of different arrangements. Finally, if $k = 2$, then $x_2 = n - x_1$, and the multinomial distribution reduces to the binomial distribution of equation (C.3).

C.3. THE NORMAL DISTRIBUTION AND ITS RELATIVES

In this section we describe five families of continuous random variables that play central roles in the development of linear statistical models: the univariate normal, chi-square, t, and F distributions; and the multivariate-normal distribution.

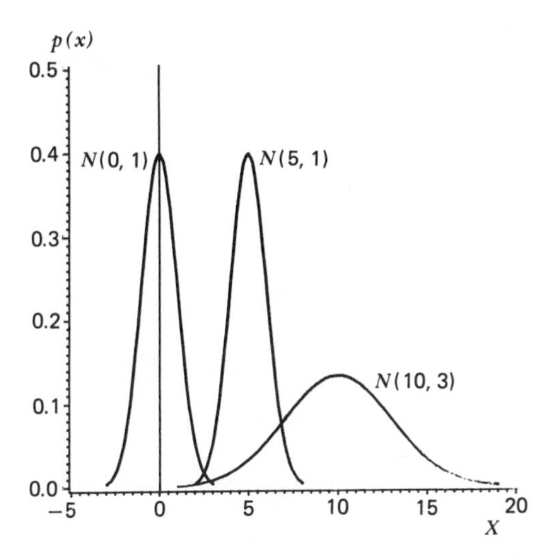

FIGURE C.2. Normal probability density functions.

C.3.1. The Normal Distribution

A *normally distributed* (or *Gaussian*) random variable X has probability density function

$$p(x) = \frac{1}{\sigma\sqrt{2\pi}} \exp\left[-\frac{(x-\mu)^2}{2\sigma^2}\right]$$

where the *parameters* of the distribution μ and σ^2 are, respectively, the mean and variance of X. There is, therefore, a different normal distribution for each choice of μ and σ^2; several examples are shown in Figure C.2. On occasion, we employ the abbreviation $X \sim N(\mu, \sigma^2)$, meaning that X is distributed nomally with expectation μ and variance σ^2. Of particular importance is the *unit-normal* random variable $Z \sim N(0, 1)$, with density function

$$\phi(z) = \frac{1}{\sqrt{2\pi}} e^{-z^2/2}$$

The CDF $\Phi(z)$ of the unit-normal distribution is shown in Figure C.3.

C.3.2. The Chi-Square (χ^2) Distribution

If $Z_1, Z_2, \ldots, Z_n$ are independently distributed unit-normal random variables then

$$X^2 = Z_1^2 + Z_2^2 + \cdots + Z_n^2$$

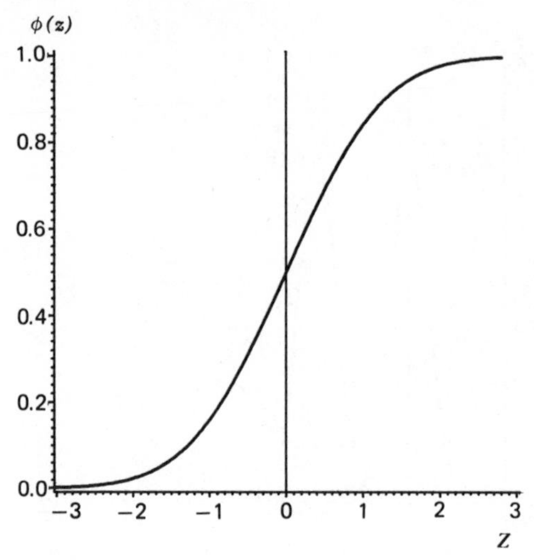

FIGURE C.3. The cumulative unit-normal distribution.

follows a *chi-square distribution* with *n degrees of freedom*, abbreviated $\chi^2(n)$. The probability density function of the chi-square variable is specified by

$$p(x^2) = \frac{1}{2^{n/2}\Gamma\left(\frac{n}{2}\right)}(x^2)^{(n-2)/2}e^{-x^2/2}$$

where Γ is the *gamma function*

$$\Gamma\left(\frac{n}{2}\right) = \begin{cases} \left(\frac{n}{2} - 1\right)! & \text{for } n \text{ even} \\ \left(\frac{n}{2} - 1\right)\left(\frac{n}{2} - 2\right)\ldots\left(\frac{3}{2}\right)\left(\frac{1}{2}\right)\sqrt{\pi} & \text{for } n \text{ odd} \end{cases}$$

The expectation and variance of a chi-square random variable are $E(X^2) = n$, and $V(X^2) = 2n$. Several chi-square distributions are graphed in Figure C.4.

C.3.3. The *t*-Distribution

If Z follows a unit-normal distribution, and X^2 independently follows a chi-square distribution with n degrees of freedom, then $t = Z/\sqrt{X^2/n}$ is a t random variable with n degrees of freedom, abbreviated $t(n)$.[2] The probability

[2] We use a lower-case t for the random variable in deference to nearly universal usage.

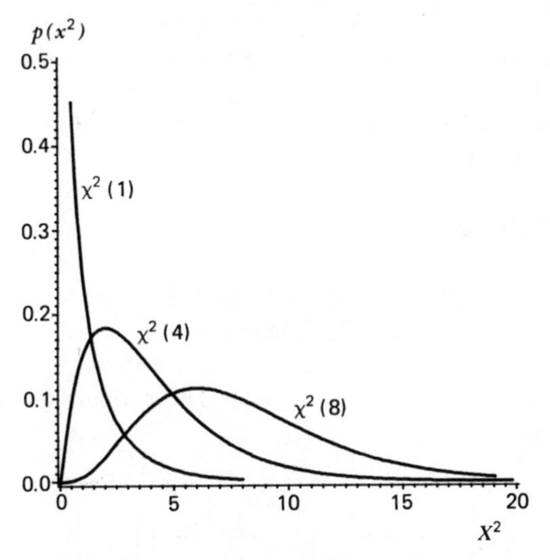

FIGURE C.4. Chi-square probability density functions.

density function of t is

$$p(t) = \frac{\Gamma\left(\dfrac{n+1}{2}\right)}{\sqrt{\pi n}\,\Gamma\left(\dfrac{n}{2}\right)} \times \frac{1}{\left(1 + \dfrac{t^2}{n}\right)^{(n+1)/2}} \tag{C.4}$$

From the symmetry of the formula for $p(t)$ around $t = 0$, it is clear that

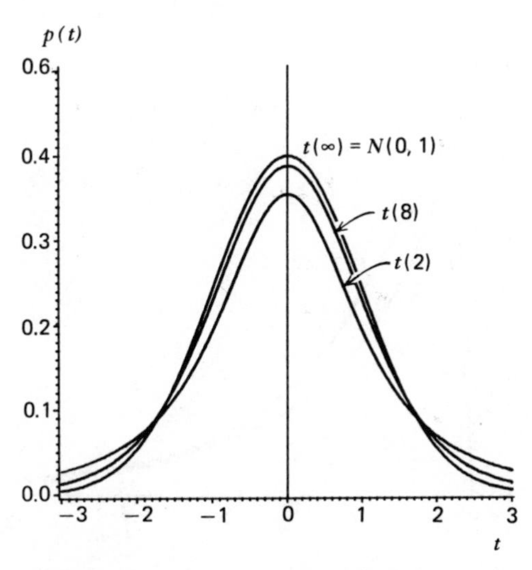

FIGURE C.5. t probability density functions.

$E(t) = 0$. It may be shown that $V(t) = n/(n-2)$, for $n > 2$; thus the variance of t is large for small degrees of freedom, and approaches one as n increases. Several t-distributions are shown in Figure C.5. As degrees of freedom grow, the t-distribution more and more closely approximates the unit-normal distribution; in the limit, $t(\infty) = N(0, 1)$.

C.3.4. The *F*-Distribution

Let X_1^2 and X_2^2 be independently distributed chi-square variables with n_1 and n_2 degrees of freedom respectively. Then $F = (X_1^2/n_1)/(X_2^2/n_2)$ follows an *F-distribution* with n_1 numerator degrees of freedom and n_2 denominator degrees of freedom, abbreviated $F(n_1, n_2)$. The probability density for F is given by the equation

$$p(f) = \frac{\Gamma\left(\frac{n_1+n_2}{2}\right)}{\Gamma\left(\frac{n_1}{2}\right)\Gamma\left(\frac{n_2}{2}\right)} \left(\frac{n_1}{n_2}\right)^{n_1/2} f^{(n_1-2)/2} \left(1 + \frac{n_1}{n_2} f\right)^{-(n_1+n_2)/2} \tag{C.5}$$

Note that (comparing equations (C.4) and (C.5)) $t^2(n) = F(1, n)$. As n_2 grows larger, $F(n_1, n_2)$ approaches $\chi^2(n_1)/n_1$; in the limit $F(n, \infty) = \chi^2(n)/n$. For $n_2 > 2$, $E(F) = n_2/(n_2 - 2)$; and for $n_2 > 4$,

$$V(F) = \frac{2n_2^2(n_1 + n_2 - 2)}{n_1(n_2 - 2)^2(n_2 - 4)}$$

Figure C.6 shows several F probability density functions.

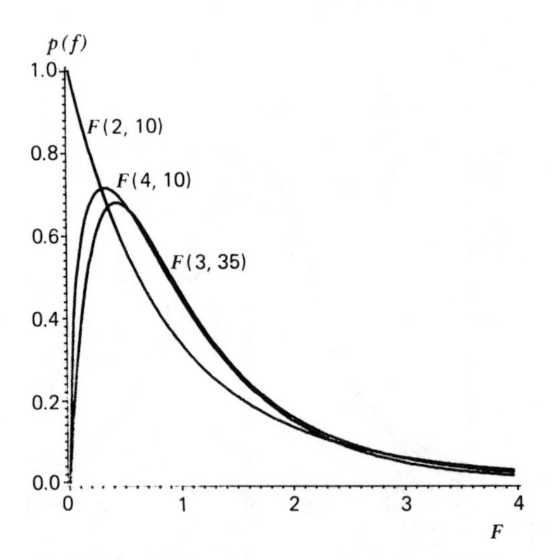

FIGURE C.6. F probability density functions.

C.3.5. The Multivariate-Normal Distribution

The joint probability density for a *multivariate-normal* vector random variable $\mathbf{x} = (X_1, X_2, \ldots, X_n)'$ with mean vector $\boldsymbol{\mu}$ and positive-definite variance-covariance matrix $\boldsymbol{\Sigma}$ is given by

$$p(\mathbf{x}) = \frac{1}{(2\pi)^{n/2}|\boldsymbol{\Sigma}|^{1/2}} \exp\left[-\frac{1}{2}(\mathbf{x} - \boldsymbol{\mu})'\boldsymbol{\Sigma}^{-1}(\mathbf{x} - \boldsymbol{\mu})\right]$$

which we abbreviate as $\mathbf{x} \sim N_n(\boldsymbol{\mu}, \boldsymbol{\Sigma})$. If $\mathbf{x}$ is multivariately normally distributed, then the marginal distribution of each of its components is univariate normal, $X_i \sim N(\mu_i, \sigma_i^2)$. Further, if $\mathbf{x} \sim N_n(\boldsymbol{\mu}, \boldsymbol{\Sigma})$ and $\underset{(m\times 1)}{\mathbf{y}} = \underset{(m\times n)}{\mathbf{A}} \underset{(n\times 1)}{\mathbf{x}}$ is a linear transformation of $\mathbf{x}$ where $\text{rank}(\mathbf{A}) = m \leq n$, then $\mathbf{y} \sim N_m(\mathbf{A}\boldsymbol{\mu}, \mathbf{A}\boldsymbol{\Sigma}\mathbf{A}')$. We say that a vector random variable $\mathbf{x}$ follows a *singular normal* distribution if the covariance matrix $\boldsymbol{\Sigma}$ of $\mathbf{x}$ is singular, but if a maximal linearly independent subset of $\mathbf{x}$ is multivariately normally distributed. A bivariate-normal density function for $\mu_1 = 5$, $\mu_2 = 6$, $\sigma_1 = 1.5$, $\sigma_2 = 3$, and $\rho_{12} = .5$ (i.e., $\sigma_{12} = (.5)(1.5)(3) = 2.25$) is depicted in Figure C.7.

C.4. ASYMPTOTIC DISTRIBUTION THEORY: AN INTRODUCTION

Partly because it is at times difficult to determine the small-sample properties of statistical estimators, it is of interest to investigate how an estimator behaves as the sample size grows. *Asymptotic distribution theory* provides tools for this investigation. We shall merely outline the theory here; more complete accounts may be found in many sources, including Theil (1971: Chapter 8).

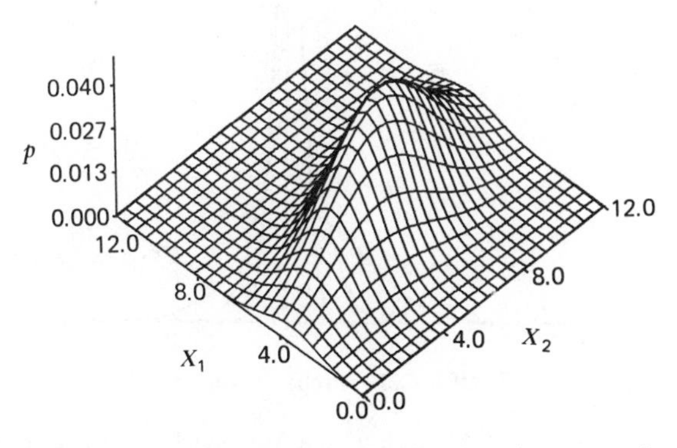

FIGURE C.7. A bivariate-normal probability density function ($\mu_1 = 5$, $\mu_2 = 6$, $\sigma_1 = 1.5$, $\sigma_2 = 3$, $\rho_{12} = .5$).

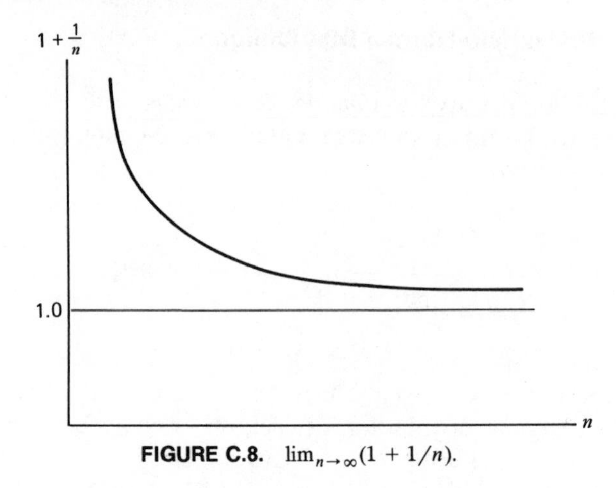

FIGURE C.8. $\lim_{n \to \infty}(1 + 1/n)$.

C.4.1. Probability Limits

Although asymptotic distribution theory applies to sequences of random variables, we first consider the nonstochastic[3] *infinite sequence* $\{a_1, a_2, \ldots, a_n, \ldots\}$. As the reader may be aware, this sequence has a *limit* a when, given any positive number ε no matter how small, there is a positive integer $n(\varepsilon)$ such that $|a_n - a| < \varepsilon$ for all $n > n(\varepsilon)$. In words, a_n can be made arbitrarily close to a by picking n sufficiently large. To describe this state of affairs compactly, we write $\lim_{n \to \infty} a_n = a$. If, for example, $a_n = 1 + 1/n$, then $\lim_{n \to \infty} a_n = 1$; this sequence and its limit are graphed in Figure C.8.

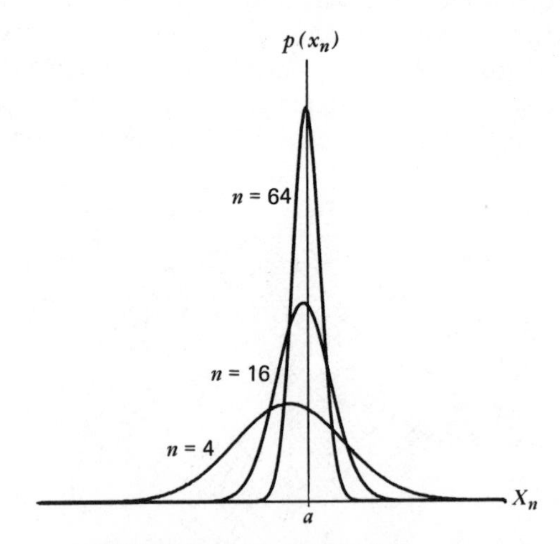

FIGURE C.9. Probability limit.

[3] By "nonstochastic" we mean that each a_n is a fixed number rather than a random variable.

Consider now a sequence of random variables $\{X_1, X_2, \ldots, X_n, \ldots\}$. In a typical application, X is some estimator and n is the size of the sample for which the estimator is calculated. Let $p_n \equiv \Pr(|X_n - a| < \delta)$ where a is a constant and δ is a small positive number. Suppose that the nonstochastic sequence of probabilities $\{p_1, p_2, \ldots, p_n, \ldots\}$ approaches a limit of one; that is $\lim_{n\to\infty}\Pr(|X_n - a| < \delta) = 1$. Then the random variable X_n concentrates more and more of its probability in a small region around a as n grows, a situation illustrated in Figure C.9. If this result holds regardless of how small δ is, then we say that a is the *probability limit* of X_n, denoted plim $X_n = a$. We generally drop the subscript n to write the even more compact expression, plim $X = a$.

Probability limits have the following very useful property: If plim $X = a$, and if $Y = f(X)$ is some continuous function of X, then plim $Y = f(a)$. Moreover, if plim $X = a$ and plim $Y = b$, and if $Z = f(X, Y)$ is a continuous function of X and Y, then plim $Z = f(a, b)$.

C.4.2. Asymptotic Expectation and Variance

We return to the sequence of random variables $\{X_1, X_2, \ldots, X_n, \ldots\}$. Let μ_n denote the expectation of X_n; $\{\mu_1, \mu_2, \ldots, \mu_n, \ldots\}$ is a nonstochastic sequence. If this sequence approaches a limit μ, then we call μ the *asymptotic expectation* of X, also written $\mathscr{E}(X)$.

Although it seems natural to define an asymptotic variance analogously as the limit of the sequence of variances, this definition is not satisfactory since (as the following example illustrates) $\lim_{n\to\infty}V(X_n)$ is zero in most interesting cases. Suppose that we calculate the mean $\overline{X}_n$ for a sample of size n drawn from a population with mean μ and variance σ^2. From elementary statistics, we know that $E(\overline{X}_n) = \mu$ and that $V(\overline{X}_n) \equiv E[(\overline{X}_n - \mu)^2] = \sigma^2/n$. Consequently, $\lim_{n\to\infty}V(\overline{X}_n) = 0$. Inserting the factor $\sqrt{n}$ within the square, however, produces the expectation $E\{[\sqrt{n}(\overline{X}_n - \mu)]^2\} = \sigma^2$. Then, dividing by n and taking the limit yields the *asymptotic variance* of the sample mean:

$$\mathscr{V}(\overline{X}) \equiv \lim_{n\to\infty} \frac{1}{n} E\left\{\left[\sqrt{n}\left(\overline{X}_n - \mu\right)\right]^2\right\}$$

$$= \frac{1}{n}\mathscr{E}\left\{\left[\sqrt{n}\left(\overline{X}_n - \mu\right)\right]^2\right\} = \sigma^2/n$$

This result is uninteresting for the present illustration because $\mathscr{V}(\overline{X}) = V(\overline{X})$, but in certain applications it is possible to find the asymptotic variance of a statistic when the finite-sample variance is intractable. Then we may apply the asymptotic result as an approximation in large samples.

In the general case, where X_n has expectation μ_n, the asymptotic variance of X is defined to be[4]

$$\mathscr{V}(X) \equiv \frac{1}{n}\mathscr{E}\{[\sqrt{n}(X_n - \mu_n)]^2\} \tag{C.6}$$

C.4.3. Asymptotic Distribution

Let $\{P_1, P_2, \ldots, P_n, \ldots\}$ represent the CDFs of a sequence of random variables $\{X_1, X_2, \ldots, X_n, \ldots\}$. The CDF of X converges to the *asymptotic distribution P* if, given any positive number ε, however small, we can find a sufficiently large $n(\varepsilon)$ such that $|P_n(x) - P(x)| < \varepsilon$ for $n > n(\varepsilon)$ and for all values x of the random variable. A familiar illustration is provided by the *central limit theorem*, which (in one of its versions) states that the mean of a set of independent and identically distributed random variables with finite expectations and variances follows an approximate normal distribution, the approximation improving as the number of random variables increases.

The results in this section extend straightforwardly to vectors and matrices. We say that $\text{plim} \underset{(m\times 1)}{\mathbf{x}} = \underset{(m\times 1)}{\mathbf{a}}$ when $\text{plim}\, X_i = a_i$ for $i = 1, \ldots, m$. Likewise, $\text{plim} \underset{(m\times p)}{\mathbf{X}} = \underset{(m\times p)}{\mathbf{A}}$ means that $\text{plim}\, X_{ij} = a_{ij}$ for all i and j. The asymptotic expectation of a vector random variable $\underset{(m\times 1)}{\mathbf{x}}$ is defined as the vector of asymptotic expectations of its entries: $\boldsymbol{\mu} = \mathscr{E}(\mathbf{x}) \equiv [\mathscr{E}(X_1), \mathscr{E}(X_2), \ldots, \mathscr{E}(X_m)]'$. The asymptotic variance-covariance matrix of $\mathbf{x}$ is given by (but recall footnote 4)

$$\mathscr{V}(\mathbf{x}) \equiv \frac{1}{n}\mathscr{E}\{[\sqrt{n}(\mathbf{x} - \boldsymbol{\mu})][\sqrt{n}(\mathbf{x} - \boldsymbol{\mu})]'\}$$

C.5. PROPERTIES OF ESTIMATORS[5]

An *estimator* is a sample statistic (i.e., a function of the observations of a sample) used to estimate an unknown population parameter. Since its value varies from one sample to the next, an estimator is a random variable. An *estimate* is the value of an estimator in a particular sample. The probability distribution of an estimator is called its *sampling distribution*; the variance of this distribution is called the *sampling variance* of the estimator; and the

[4]It is generally preferable to define asymptotic expectation and variance in terms of the asymptotic distribution (Section C.4.3), since the sequences used for this purpose here do not exist in all cases (see Theil, 1971: 375–376; also see McCallum, 1973). Our use of the symbols $\mathscr{E}$ and $\mathscr{V}$ for asymptotic expectation and variance extends a notation suggested by Johnston (1972); the reader should be aware that $\mathscr{E}$ and $\mathscr{V}$ are often used in place of E and V to denote *ordinary* expectation and variance.

[5]Most of the material in this and the following section may be traced directly to a seminal paper on estimation by Fisher (1922).

standard deviation of the sampling distribution is often called the *standard error* of the estimator.

C.5.1. Bias

An estimator A of the parameter α is said to be *unbiased* if $E(A) = \alpha$. The difference $E(A) - \alpha$ (which, of course, is zero for an unbiased estimator) is the *bias* of A. Suppose, for example, that we draw n independent observations from a population with mean μ and variance σ^2. Then the sample mean $\overline{X} \equiv \Sigma X_i/n$ is an unbiased estimator of μ, while

$$S_*^2 \equiv \frac{\Sigma(X_i - \overline{X})^2}{n} \tag{C.7}$$

is a biased estimator of σ^2, since $E(S_*^2) = [(n-1)/n]\sigma^2$; the bias of S_*^2 is $-\sigma^2/n$. Sampling distributions of unbiased and biased estimators are illustrated in Figure C.10. The *asymptotic bias* of an estimator A of α is $\mathscr{E}(A) - \alpha$, and the estimator is *asymptotically unbiased* if $\mathscr{E}(A) = \alpha$.

C.5.2. Mean-Squared Error and Efficiency

To say that an estimator is unbiased means that its average value over repeated samples is equal to the parameter being estimated. This is clearly a desirable property for an estimator to possess, but it is cold comfort if the estimator does not provide estimates that are close to the parameter: In forming the expectation, large negative estimation errors for some samples could offset large positive errors for others.

The *mean-squared error* of an estimator A of the parameter α is the average squared difference between the estimator and the parameter: $\text{MSE}(A) \equiv$

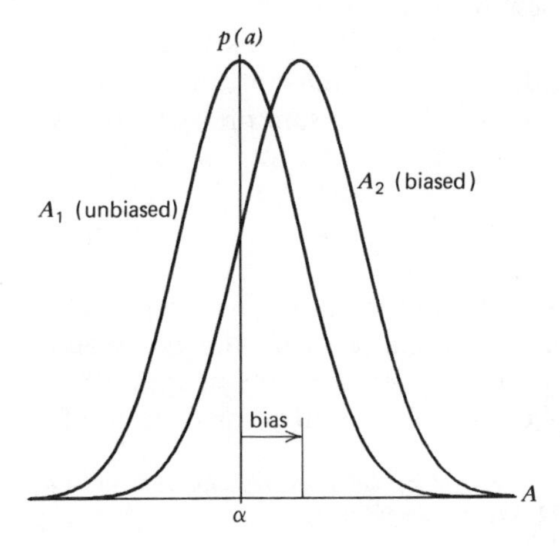

FIGURE C.10. Biased and unbiased estimators.

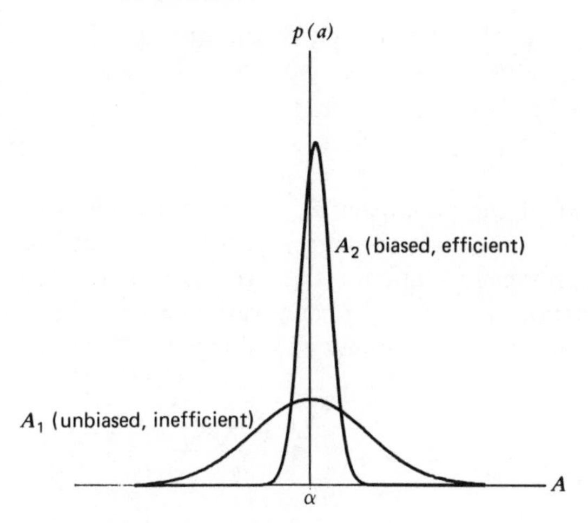

FIGURE C.11. Relative efficiency of estimators.

$E[(A - \alpha)^2]$. The *efficiency* of an estimator is inversely proportional to its mean-squared error. We should generally prefer a more efficient estimator to a less efficient one. Note that the mean-squared error of an unbiased estimator is equal to its sampling variance. For a biased estimator, we may write

$$\begin{aligned}
\text{MSE}(A) &= E\{[A - E(A) + E(A) - \alpha]^2\} \\
&= E\{[A - E(A)]^2\} + [E(A) - \alpha]^2 \\
&\quad + 2[E(A) - E(A)][E(A) - \alpha] \\
&= V(A) + [\text{bias}(A)]^2
\end{aligned}$$

The efficiency of an estimator increases, therefore, as its sampling variance and bias decline. In comparing two estimators, an advantage in sampling variance can more than offset a disadvantage due to bias, as illustrated in Figure C.11. *Asymptotic efficiency* is inversely proportional to the sum of asymptotic variance and asymptotic squared bias.

C.5.3. Consistency

An estimator A of the parameter α is *consistent* if plim $A = \alpha$. A sufficient (but not necessary[6]) condition for consistency is that an estimator be asymptotically unbiased and that the sampling variance of the estimator approach zero as n increases; this condition implies that the mean-squared error of the

[6]There are cases in which plim $A = \alpha$ but the variance and asymptotic expectation of A (as defined here) do not exist. See Johnston (1972: 272) for an example.

estimator approaches a limit of zero. Figure C.9 illustrates consistency, if we construe X as an estimator of a. The estimator S_*^2 given in equation (C.7) is a consistent estimator of the population variance σ^2 even though it is biased in finite samples.

C.5.4. Sufficiency

Sufficiency is a more difficult property to grasp than unbias, efficiency, or consistency. A statistic S based on a sample of observations is *sufficient* for a parameter α if the statistic exhausts all information about α that is present in the sample. More formally, let the observations $X_1, X_2, \ldots, X_n$ be drawn from a probability distribution with parameter α, and let $S = f(X_1, X_2, \ldots, X_n)$. Then S is a sufficient statistic for α if the probability distribution of the observations conditional on S, $p(x_1, x_2, \ldots, x_n | S)$ does not depend upon α. Notice that S need not be an estimator of α.

To illustrate the idea of sufficiency, suppose that n observations are independently sampled, and that each observation X_i takes on a value of one with probability π and a value of zero with probability $1 - \pi$. We shall show that the sample sum $S = \sum_{i=1}^{n} X_i$ is a sufficient statistic for π. If we know the value s of S, then there are $\binom{n}{s}$ different possible arrangements of the s ones and $n - s$ zeroes, each with probability $1/\binom{n}{s}$. Because this probability does not depend upon the parameter π, S is a sufficient statistic. By a similar argument, the sample proportion $P = S/n$ is also a sufficient statistic. The proportion P, but not the sum S, is an estimator of π.

The concept of sufficiency can be extended to sets of parameters and statistics: Given a sample of (possibly multivariate) observations $\mathbf{x}_1, \mathbf{x}_2, \ldots, \mathbf{x}_n$, a vector of statistics $\mathbf{s} = (S_1, S_2, \ldots, S_p)' = f(\mathbf{x}_1, \mathbf{x}_2, \ldots, \mathbf{x}_n)$ is *jointly sufficient* for the parameters $\boldsymbol{\alpha} = (\alpha_1, \alpha_2, \ldots, \alpha_k)'$ if the conditional distribution of the observations given $\mathbf{s}$ does not depend upon $\boldsymbol{\alpha}$. It may be shown, for example, that the mean $\overline{X}$ and variance S^2 calculated from an independent random sample are sufficient statistics for the parameters μ and σ^2 of a normal distribution (as are the sample sum $\sum X_i$ and sum of squares $\sum X_i^2$, which jointly contain the same information as $\overline{X}$ and S^2). A set of sufficient statistics is called *minimally sufficient* if there is no smaller sufficient set.

C.6. MAXIMUM-LIKELIHOOD ESTIMATION

The *maximum-likelihood* method provides estimators that have both a reasonable intuitive basis and many desirable statistical properties. A disadvantage of the method, however, is that it frequently requires strong assumptions about the structure of the data.

Suppose, for example, that we wish to estimate the mean of a normally distributed variable that (unrealistically) has a known variance of one. We select an independent random sample of n observations; each observation X_i then has the probability density function

$$p(x_i) = \frac{1}{\sqrt{2\pi}} e^{-(1/2)(x_i - \mu)^2}$$

where μ is the unknown expectation we wish to estimate. Because the observations are independent, their joint probability density is the product of their marginal densities:

$$p(x_1, x_2, \ldots, x_n) = \prod_{i=1}^{n} p(x_i)$$

$$= (2\pi)^{-n/2} e^{-(1/2)\Sigma_{i=1}^{n}(x_i - \mu)^2}$$

(where the operator $\prod$ indicates continued multiplication).

The parameter μ is a fixed though unknown value, and the observations $X_1, X_2, \ldots, X_n$ are random variables. In a particular sample, however, the observed or realized values $x_1, x_2, \ldots, x_n$ are constants, and the probability density $p(x_1, x_2, \ldots, x_n)$ for the observed values depends upon μ. We may therefore change our point of view and write the probability density as a function of μ:

$$L(\mu) = (2\pi)^{-n/2} e^{-(1/2)\Sigma_{i=1}^{n}(x_i - \mu)^2} \qquad \text{(C.8)}$$

$L(\mu)$ is called the *likelihood function*: It gives the probability density (or likelihood) of observing the obtained data for different possible values of the unknown parameter μ.

We proceed to select an estimate of μ that maximizes the likelihood—that is, we choose an estimated parameter value that, were it the true value, would make our observed data as probable as they can be. It would, after all, be intuitively unreasonable to choose an estimate of μ that is unlikely to give rise to the data that we in fact obtained.

In the present case, and in most applications, it is simpler to work with the natural logarithm of the likelihood function, $\log L(\mu)$, than with $L(\mu)$ itself. Because $\log L$ is a monotone, increasing function of L, maximizing $\log L$ is equivalent to maximizing L. Taking the log of equation (C.8) for our illustration produces

$$\log L(\mu) = -\frac{n}{2}\log 2\pi - \frac{1}{2}\sum_{i=1}^{n}(x_i - \mu)^2$$

To maximize $\log L$, we find its derivative with respect to μ,

$$\frac{d \log L(\mu)}{d\mu} = -\frac{1}{2}\sum 2(x_i - \mu)(-1)$$

set the derivative to zero, $\Sigma(x_i - \mu) = 0$; and solve for the estimate $\hat{\mu}$ of μ that satisfies the equation, producing $\hat{\mu} = \Sigma x_i/n = \bar{x}$. Expressed in general terms, the maximum-likelihood *estimator* is $\hat{\mu} = \overline{X}$. We generally avoid this awkward substitution of the estimator for the estimate by replacing x_i with X_i in the likelihood function [equation (C.8)].

Under very broad conditions, maximum-likelihood estimators have the following general properties (see, e.g., Theil, 1971: Chapter 8; Kendall and Stuart, 1979: Chapter 18):

1. Maximum-likelihood estimators are consistent.
2. They are asymptotically unbiased, through they may be biased in finite samples.
3. They are asymptotically efficient—no asymptotically unbiased estimator has a smaller asymptotic sampling variance.
4. They are asymptotically normally distributed.
5. If there is a sufficient statistic for a parameter, then the maximum-likelihood estimator of the parameter is a function of a sufficient statistic.
6. The asymptotic sampling variance of the maximum-likelihood estimator $\hat{\alpha}$ of a parameter α may be obtained from the second derivative of the log likelihood:

$$\mathscr{V}(\hat{\alpha}) = \frac{1}{-E\left[\dfrac{d^2 \log L(\alpha)}{d\alpha^2}\right]} \tag{C.9}$$

   In practice, we substitute the maximum-likelihood estimate $\hat{\alpha}$ into the right-hand size of equation (C.9) to obtain an *estimated* asymptotic sampling variance $\widehat{\mathscr{V}(\hat{\alpha})}$.

7. Suppose that $L(\hat{\alpha})$ is the value of the likelihood function for the maximum-likelihood estimator $\hat{\alpha}$, while $L(\alpha)$ is the likelihood for the true parameter α. Then the *log likelihood-ratio statistic*

$$G^2 \equiv -2 \log \frac{L(\alpha)}{L(\hat{\alpha})} = 2[\log L(\hat{\alpha}) - \log L(\alpha)] \tag{C.10}$$

   follows an asymptotic chi-square distribution with one degree of freedom. The likelihood-ratio statistic may be used to test the hypothesis $H_0: \alpha = \alpha_0$ by substituting the hypothesized value α_0 for α in equation (C.10).

The maximum-likelihood approach may be generalized to simultaneous estimation of several parameters. Let $p(\underset{(n \times m)}{\mathbf{X}})$ represent the probability density for n possibly multivariate observations ($m \geq 1$), and suppose that this density depends upon a vector $\boldsymbol{\alpha}$ of k independent parameters. (To say that the parameters are independent means that the value of none can be obtained from the values of the others.) The likelihood $L(\boldsymbol{\alpha})$ is a function of the parameter vector $\boldsymbol{\alpha}$, and we seek the values $\hat{\boldsymbol{\alpha}}$ that maximize the function. As before, it is generally more convenient to work with $\log L(\boldsymbol{\alpha})$ in place of $L(\boldsymbol{\alpha})$. We find the vector partial derivative $\partial \log L(\boldsymbol{\alpha})/\partial \boldsymbol{\alpha}$, set this derivative to $\mathbf{0}$, and solve the resulting matrix equation for $\hat{\boldsymbol{\alpha}}$; if there is more than one root, we choose the solution that produces the largest likelihood.

As in the case of a single parameter, the maximum-likelihood estimator is consistent, asymptotically unbiased, asymptotically efficient, asymptotically multivariate-normal, and based upon sufficient statistics. Its asymptotic variance-covariance matrix is given by

$$\mathscr{V}(\boldsymbol{\alpha}) = \left\{ -E\left[\frac{\partial^2 \log L(\boldsymbol{\alpha})}{\partial \boldsymbol{\alpha}\, \partial \boldsymbol{\alpha}'} \right] \right\}^{-1} \tag{C.11}$$

The matrix in braces in equation (C.11) is called the *information matrix* $\mathbf{I}(\boldsymbol{\alpha})$ (not to be confused with an identity matrix); $\mathscr{V}(\boldsymbol{\alpha})$ is thus the inverse of the information matrix.

The likelihood-ratio statistic of equation (C.10) generalizes in the following manner: Suppose that we wish to test the hypothesis H_0 that p of the k elements of $\boldsymbol{\alpha}$ are equal to particular values. Let $L(\hat{\boldsymbol{\alpha}}_0)$ represent the maximized likelihood under the constraint represented by the hypothesis (i.e., setting the p parameters equal to their hypothesized values, but leaving the other parameters free to be estimated); $L(\hat{\boldsymbol{\alpha}}_1)$ represents the globally maximized likelihood when the constraint is relaxed. Then, under the hypothesis H_0,

$$G_0^2 = -2 \log \left[\frac{L(\hat{\boldsymbol{\alpha}}_0)}{L(\hat{\boldsymbol{\alpha}}_1)} \right]$$

has an asymptotic chi-square distribution with p degrees of freedom.

The following example (adapted from Theil, 1971: 389–390) illustrates these results. Suppose that a sample of n independent observations is drawn from a normally distributed population with unknown mean μ and variance σ^2; we wish to estimate μ and σ^2. The likelihood function is [cf. equation (C.8)]

$$L(\mu, \sigma^2) = (2\pi\sigma^2)^{-n/2} \exp\left[-\frac{1}{2\sigma^2} \sum_{i=1}^{n} (X_i - \mu)^2 \right]$$

and the log likelihood is

$$\log L(\mu, \sigma^2) = -\frac{n}{2} \log 2\pi - \frac{n}{2} \log \sigma^2 - \frac{1}{2\sigma^2} \sum (X_i - \mu)^2$$

with partial derivatives

$$\frac{\partial \log L}{\partial \mu} = \frac{1}{\sigma^2} \sum (X_i - \mu)$$

$$\frac{\partial \log L}{\partial \sigma^2} = -\frac{n}{2\sigma^2} + \frac{1}{2\sigma^4} \sum (X_i - \mu)^2$$

Setting the partial derivatives to zero and solving simultaneously for the maximum-likelihood estimators of μ and σ^2 produces

$$\hat{\mu} = \frac{\Sigma X_i}{n} = \overline{X}$$

$$\hat{\sigma}^2 = \frac{\Sigma (X_i - \overline{X})^2}{n}$$

The matrix of second derivatives is

$$\begin{pmatrix} \dfrac{\partial^2 \log L}{\partial \mu^2} & \dfrac{\partial^2 \log L}{\partial \mu \, \partial \sigma^2} \\ \dfrac{\partial^2 \log L}{\partial \sigma^2 \, \partial \mu} & \dfrac{\partial^2 \log L}{\partial (\sigma^2)^2} \end{pmatrix} = \begin{pmatrix} -\dfrac{n}{\sigma^2} & -\dfrac{1}{\sigma^4} \sum (X_i - \mu) \\ -\dfrac{1}{\sigma^4} \sum (X_i - \mu) & \dfrac{n}{2\sigma^4} - \dfrac{1}{4\sigma^6} \sum (X_i - \mu)^2 \end{pmatrix} \tag{C.12}$$

Taking expectations in equation (C.12), noting that $E(X_i - \mu) = 0$ and that $E[(X_i - \mu)^2] = \sigma^2$, produces the negative of the information matrix

$$-\mathbf{I}(\mu, \sigma^2) = \begin{pmatrix} -\dfrac{n}{\sigma^2} & 0 \\ 0 & -\dfrac{n}{2\sigma^4} \end{pmatrix}$$

Finally, the asymptotic covariance marix of the maximum-likelihood estimators is given by the inverse of the information matrix:

$$\mathscr{V}(\mu, \sigma^2) = \mathbf{I}^{-1}(\mu, \sigma^2) = \begin{pmatrix} \dfrac{\sigma^2}{n} & 0 \\ 0 & \dfrac{2\sigma^4}{n} \end{pmatrix}$$

Note that the maximum-likelihood estimator of σ^2 is biased but consistent (and indeed is the estimator S_*^2 given in equation (C.7)).

D

Approximate Critical Values for the Largest Studentized Residual from a Linear Model

n: sample size

k: number of regressors, excluding constant

a: significance level (two-tail)

n	a	k: 0	1	2	3	4	5	6	7	8	9	10
4	0.25	3.8100										
	0.10	6.2053										
	0.05	8.8602										
	0.01	19.9625										
	0.001	63.2337										
5	0.25	3.1824	4.3026									
	0.10	4.5407	6.9646									
	0.05	5.8409	9.9248									
	0.01	10.2145	22.3271									
	0.001	22.2037	70.7001									
6	0.25	2.9573	3.4258	4.7445								
	0.10	3.9608	4.8567	7.6488								
	0.05	4.8510	6.2315	10.9859								
	0.01	7.5287	10.9688	24.4643								
	0.001	13.6531	23.6016	77.4500								
7	0.25	2.8526	3.1145	3.6411	5.1487							
	0.10	3.6805	4.1478	5.1377	8.2767							
	0.05	4.3817	5.0675	6.5797	11.7687							
	0.01	6.3518	7.8414	11.4532	26.4292							
	0.001	10.3808	14.1989	24.8512	83.6570							
8	0.25	2.7979	2.9674	3.2542	3.8352	5.5234						
	0.10	3.5212	3.8100	4.3147	5.0920	8.8602						
	0.05	4.1152	4.5257	5.2611	6.9052	12.5897						
	0.01	5.7090	6.5414	8.1216	11.9838	28.2578						
	0.001	8.7294	10.6728	14.6888	25.9866	89.4343						
9	0.25	2.7679	2.8877	3.0702	3.3803	4.0127	5.8743					
	0.10	3.4216	3.6190	3.9263	4.4657	5.6251	9.4076					
	0.05	3.9467	4.2209	4.6553	5.4366	7.1849	13.3604					
	0.01	5.3101	5.8399	6.7126	8.3763	12.4715	29.9750					
	0.001	7.7550	8.9142	10.9366	15.1344	27.0308	94.8604					
10	0.25	2.7515	2.8412	2.9687	3.1634	3.4954	4.1765	6.2053				
	0.10	3.3554	3.4995	3.7074	4.0321	4.6041	5.9409	9.9248				
	0.05	3.8325	4.0293	4.3168	4.7733	5.5976	7.4533	14.0891				
	0.01	5.0413	5.4079	5.9588	6.8688	8.6103	12.9240	31.5991				
	0.001	7.1200	7.8846	9.0824	11.1777	15.5441	28.0001	99.9925				
11	0.25	2.7432	2.8133	2.9079	3.0426	3.2488	3.6015	4.3292	6.5196			
	0.10	3.3095	3.4198	3.5705	3.7883	4.1293	4.7320	6.0423	10.4164			
	0.05	3.7513	3.8999	4.1048	4.4047	4.8819	5.7465	7.7041	14.7818			
	0.01	4.8494	5.1183	5.4973	6.0680	7.0128	8.8271	13.3470	33.1436			
	0.001	6.6768	7.2172	8.0033	9.2369	11.4001	15.9240	28.9067	104.8737			
12	0.25	2.7400	2.7964	2.8699	2.9691	3.1106	3.3276	3.7001	4.4723	6.8195		
	0.10	3.2768	3.3642	3.4789	3.6358	3.8630	4.2193	4.8510	6.2315	10.8859		
	0.05	3.6915	3.8079	3.9618	4.1743	4.4858	4.9825	5.8853	7.9398	15.4435		
	0.01	4.7065	4.9124	5.1892	5.5799	6.1690	7.1464	9.0294	13.7450	34.6194		
	0.001	6.3517	6.7533	7.3069	8.1130	9.3800	11.6067	16.2788	29.7598	109.5377		

n	a	k: 0	1	2	3	4	5	6	7	8	9	10
13	0.25	2.7400	2.7866	2.8455	2.9220	3.0257	3.1736	3.4010	3.7923	4.6072	7.1067	
	0.10	3.2531	3.3242	3.4147	3.5335	3.6963	3.9323	4.3032	4.9625	6.4102	11.3359	
	0.05	3.6462	3.7401	3.8602	4.0191	4.2388	4.5612	5.0764	6.0154	8.1624	16.0780	
	0.01	4.5966	4.7594	4.9706	5.2549	5.6565	6.2630	7.2712	9.2192	14.1214	36.0347	
	0.001	6.1041	6.4141	6.8243	7.3901	8.2150	9.5134	11.7999	16.6119	30.5666	114.0110	
14	0.25	2.7422	2.7815	2.8298	2.8909	2.9705	3.0783	3.2324	3.4696	3.9790	4.7351	7.3827
	0.10	3.2357	3.2949	3.3682	3.4616	3.5844	3.7527	3.9971	4.3817	5.0675	6.5797	11.7687
	0.05	3.6113	3.6887	3.7852	3.9088	4.0724	4.2980	4.6317	5.1644	6.1380	8.3738	16.6883
	0.01	4.5099	4.6420	4.8087	5.0249	5.3162	5.7282	6.3510	7.3884	9.3983	14.4787	37.3965
	0.001	5.9099	6.1564	6.4722	6.9004	7.4678	8.3103	9.6384	11.9815	16.9262	31.3331	118.3153
15	0.25	2.7459	2.7795	2.8200	2.8701	2.9333	3.0158	3.1276	3.2875	3.5341	3.9608	4.8567
	0.10	3.2229	3.2729	3.3338	3.4092	3.5054	3.6319	3.8055	4.0579	4.4557	5.1668	6.7411
	0.05	3.5838	3.6489	3.7283	3.8273	3.9542	4.1224	4.3553	4.6979	5.2474	6.2541	8.5752
	0.01	4.4401	4.5496	4.6845	4.8547	5.0757	5.3737	5.7954	6.4338	7.4990	9.5679	14.8194
	0.001	5.7539	5.9546	6.2052	6.5266	6.9524	7.5407	8.4000	9.7561	12.1529	17.2241	32.0637
16	0.25	2.7506	2.7797	2.8143	2.8561	2.9077	2.9730	3.0582	3.1738	3.3393	3.5950	4.0384
	0.10	3.2135	3.2565	3.3078	3.3702	3.4477	3.5465	3.6766	3.8552	4.1152	4.5257	5.2611
	0.05	3.5621	3.6176	3.6842	3.7654	3.8669	3.9969	4.1693	4.4084	4.7604	5.3259	6.3643
	0.01	4.3830	4.4752	4.5868	4.7244	4.8980	5.1235	5.4278	5.8588	6.5121	7.6037	9.7291
	0.001	5.6261	5.7927	5.9965	6.2512	6.5778	7.0108	7.6094	8.4845	9.8673	12.3153	17.5074
17	0.25	2.7561	2.7815	2.8114	2.8462	2.8900	2.9432	3.0104	3.0982	3.2174	3.3883	3.6527
	0.10	3.2067	3.2440	3.2880	3.3405	3.4045	3.4839	3.5852	3.7187	3.9022	4.1694	4.5921
	0.05	3.5447	3.5926	3.6493	3.7173	3.8004	3.9041	4.0372	4.2136	4.4585	4.8196	5.4005
	0.01	4.3356	4.4144	4.5084	4.6219	4.7620	4.9388	5.1686	5.4789	5.9188	6.5862	7.7032
	0.001	5.5199	5.6604	5.8294	6.0361	6.2945	6.6261	7.0659	7.6743	8.5646	9.9729	12.4697
18	0.25	2.7619	2.7844	2.8106	2.8412	2.8777	2.9220	2.9766	3.0457	3.1360	3.2587	3.4348
	0.10	3.2019	3.2346	3.2727	3.3177	3.3714	3.4368	3.5182	3.6219	3.7586	3.9467	4.2209
	0.05	3.5306	3.5725	3.6214	3.6793	3.7487	3.8334	3.9394	4.0752	4.2556	4.5062	4.8759
	0.01	4.2958	4.3640	4.4442	4.5396	4.6551	4.7975	4.9774	5.2114	5.5274	5.9757	6.6568
	0.001	5.4302	5.5505	5.6928	5.8640	6.0735	6.3355	6.6719	7.1182	7.7360	8.6407	10.0733
19	0.25	2.7681	2.7882	2.8112	2.8380	2.8694	2.9069	2.9522	3.0082	3.0792	3.1718	3.2979
	0.10	3.1986	3.2276	3.2610	3.2999	3.3458	3.4007	3.4675	3.5506	3.6566	3.7965	3.9890
	0.05	3.5193	3.5562	3.5989	3.6487	3.7076	3.7783	3.8648	3.9728	4.1114	4.2955	4.5514
	0.01	4.2620	4.3217	4.3910	4.4724	4.5693	4.6866	4.8312	5.0141	5.2520	5.5735	6.0300
	0.001	5.3538	5.4578	5.5794	5.7236	5.8969	6.1091	6.3745	6.7154	7.1679	7.7947	8.7132
20	0.25	2.7745	2.7925	2.8131	2.8366	2.8640	2.8961	2.9345	2.9809	3.0382	3.1109	3.2060
	0.10	3.1966	3.2224	3.2520	3.2860	3.3257	3.3725	3.4284	3.4966	3.5814	3.6897	3.8325
	0.05	3.5101	3.5430	3.5805	3.6239	3.6746	3.7345	3.8065	3.8945	4.0045	4.1458	4.3335
	0.01	4.2332	4.2858	4.3464	4.4166	4.4992	4.5975	4.7165	4.8633	5.0490	5.2907	5.6174
	0.001	5.2879	5.3788	5.4840	5.6070	5.7528	5.9281	6.1429	6.4116	6.7568	7.2153	7.8507
22	0.25	2.7876	2.8024	2.8190	2.8378	2.8591	2.8837	2.9123	2.9458	2.9857	3.0342	3.0941
	0.10	3.1952	3.2163	3.2399	3.2667	3.2973	3.3325	3.3736	3.4221	3.4801	3.5508	3.6389
	0.05	3.4966	3.5231	3.5530	3.5869	3.6256	3.6704	3.7227	3.7846	3.8590	3.9500	4.0638
	0.01	4.1868	4.2287	4.2761	4.3301	4.3922	4.4643	4.5490	4.6500	4.7722	4.9232	5.1141
	0.001	5.1805	5.2517	5.3326	5.4254	5.5327	5.6583	5.8072	5.9864	6.2060	6.4808	6.8342

n	a	k: 0	1	2	3	4	5	6	7	8	9	10
24	0.25	2.8008	2.8132	2.8269	2.8423	2.8595	2.8780	2.9011	2.9266	2.9562	2.9911	3.0326
	0.10	3.1965	3.2139	3.2333	3.2550	3.2794	3.3070	3.3386	3.3749	3.4173	3.4674	3.5274
	0.05	3.4880	3.5099	3.5342	3.5615	3.5921	3.6270	3.6668	3.7129	3.7667	3.8305	3.9071
	0.01	4.1516	4.1858	4.2240	4.2669	4.3154	4.3706	4.4341	4.5079	4.5947	4.6981	4.8233
	0.001	5.0970	5.1544	5.2186	5.2911	5.3735	5.4680	5.5774	5.7054	5.8572	6.0399	6.2639
26	0.25	2.8138	2.8244	2.8360	2.8488	2.8630	2.8788	2.8966	2.9167	2.9396	2.9660	2.9966
	0.10	3.1994	3.2142	3.2304	3.2483	3.2683	3.2906	3.3156	3.3440	3.3765	3.4139	3.4576
	0.05	3.4827	3.5011	3.5213	3.5438	3.5687	3.5966	3.6281	3.6638	3.7047	3.7520	3.8072
	0.01	4.1244	4.1529	4.1844	4.2193	4.2582	4.3020	4.3515	4.4079	4.4728	4.5481	4.6368
	0.001	5.0307	5.0779	5.1302	5.1884	5.2538	5.3274	5.4112	5.5073	5.6185	5.7488	5.9033
28	0.25	2.8266	2.8357	2.8456	2.8565	2.8684	2.8816	2.8962	2.9126	2.9309	2.9516	2.9753
	0.10	3.2035	3.2162	3.2300	3.2451	3.2618	3.2802	3.3006	3.3235	3.3492	3.3783	3.4116
	0.05	3.4797	3.4955	3.5126	3.5314	3.5522	3.5751	3.6006	3.6292	3.6614	3.6980	3.7398
	0.01	4.1031	4.1272	4.1536	4.1826	4.2147	4.2503	4.2900	4.3346	4.3850	4.4425	4.5086
	0.001	4.9769	5.0165	5.0599	5.1078	5.1609	5.2200	5.2863	5.3612	5.4462	5.5438	5.6568
30	0.25	2.8389	2.8469	2.8555	2.8649	2.8751	2.8863	2.8985	2.9121	2.9271	2.9439	2.9627
	0.10	3.2084	3.2194	3.2313	3.2443	3.2584	3.2739	3.2909	3.3097	3.3306	3.3540	3.3804
	0.05	3.4786	3.4922	3.5069	3.5229	3.5405	3.5597	3.5808	3.6043	3.6303	3.6595	3.6924
	0.01	4.0862	4.1069	4.1294	4.1540	4.1808	4.2103	4.2429	4.2791	4.3195	4.3648	4.4162
	0.001	4.9327	4.9664	5.0031	5.0432	5.0872	5.1357	5.1895	5.2495	5.3167	5.3926	5.4789
32	0.25	2.8509	2.8580	2.8655	2.8737	2.8825	2.8921	2.9026	2.9140	2.9266	2.9405	2.9559
	0.10	3.2138	3.2235	3.2339	3.2451	3.2572	3.2705	3.2849	3.3007	3.3181	3.3373	3.3587
	0.05	3.4787	3.4906	3.5034	3.5173	3.5323	3.5486	3.5665	3.5861	3.6076	3.6315	3.6581
	0.01	4.0728	4.0908	4.1102	4.1312	4.1541	4.1790	4.2063	4.2362	4.2693	4.3061	4.3471
	0.001	4.8959	4.9250	4.9563	4.9904	5.0275	5.0681	5.1127	5.1619	5.2163	5.2771	5.3451
34	0.25	2.8625	2.8688	2.8755	2.8827	2.8904	2.8988	2.9078	2.9176	2.9283	2.9401	2.9529
	0.10	3.2196	3.2282	3.2373	3.2471	3.2577	3.2691	3.2815	3.2950	3.3097	3.3259	3.3436
	0.05	3.4799	3.4904	3.5016	3.5137	3.5267	3.5408	3.5561	3.5727	3.5909	3.6108	3.6328
	0.01	4.0620	4.0778	4.0947	4.1130	4.1327	4.1540	4.1772	4.2025	4.2301	4.2606	4.2942
	0.001	4.8650	4.8902	4.9174	4.9468	4.9785	5.0130	5.0505	5.0916	5.1367	5.1864	5.2416
36	0.25	2.8738	2.8794	2.8854	2.8917	2.8986	2.9059	2.9138	2.9223	2.9316	2.9416	2.9525
	0.10	3.2256	3.2332	3.2414	3.2501	3.2594	3.2694	3.2801	3.2918	3.3044	3.3181	3.3331
	0.05	3.4817	3.4911	3.5010	3.5117	3.5231	3.5354	3.5486	3.5629	3.5785	3.5954	3.6138
	0.01	4.0534	4.0674	4.0823	4.0983	4.1154	4.1339	4.1539	4.1755	4.1990	4.2246	4.2527
	0.001	4.8387	4.8609	4.8847	4.9102	4.9377	4.9673	4.9994	5.0343	5.0722	5.1137	5.1593
38	0.25	2.8846	2.8897	2.8950	2.9008	2.9069	2.9134	2.9203	2.9278	2.9359	2.9446	2.9540
	0.10	3.2318	3.2386	3.2459	3.2536	3.2619	3.2707	3.2802	3.2904	3.3013	3.3132	3.3260
	0.05	3.4842	3.4925	3.5014	3.5109	3.5210	3.5318	3.5434	3.5559	3.5693	3.5839	3.5996
	0.01	4.0465	4.0590	4.0722	4.0863	4.1014	4.1176	4.1350	4.1537	4.1739	4.1958	4.2196
	0.001	4.8161	4.8358	4.8568	4.8793	4.9033	4.9291	4.9568	4.9868	5.0192	5.0544	5.0927
40	0.25	2.8951	2.8997	2.9045	2.9097	2.9152	2.9210	2.9272	2.9338	2.9409	2.9485	2.9568
	0.10	3.2380	3.2442	3.2507	3.2577	3.2650	3.2729	3.2813	3.2903	3.2999	3.3102	3.3214
	0.05	3.4870	3.4946	3.5026	3.5110	3.5200	3.5297	3.5399	3.5509	3.5626	3.5753	3.5889
	0.01	4.0410	4.0522	4.0640	4.0766	4.0900	4.1043	4.1196	4.1359	4.1535	4.1725	4.1929
	0.001	4.7967	4.8143	4.8330	4.8529	4.8741	4.8967	4.9210	4.9470	4.9750	5.0052	5.0379

n	a	k: 0	1	2	3	4	5	6	7	8	9	10
45	0.25	2.9198	2.9235	2.9274	2.9314	2.9357	2.9402	2.9450	2.9500	2.9554	2.9611	2.9672
	0.10	3.2536	3.2585	3.2637	3.2691	3.2748	3.2809	3.2873	3.2941	3.3013	3.3089	3.3171
	0.05	3.4955	3.5015	3.5077	3.5143	3.5213	3.5286	3.5364	3.5446	3.5534	3.5627	3.5726
	0.01	4.0318	4.0405	4.0497	4.0594	4.0697	4.0805	4.0920	4.1041	4.1170	4.1308	4.1455
	0.001	4.7587	4.7723	4.7866	4.8017	4.8177	4.8347	4.8526	4.8717	4.8920	4.9136	4.9368
50	0.25	2.9426	2.9456	2.9488	2.9521	2.9555	2.9592	2.9630	2.9670	2.9712	2.9756	2.9803
	0.10	3.2689	3.2729	3.2771	3.2815	3.2861	3.2909	3.2959	3.3013	3.3069	3.3128	3.3190
	0.05	3.5051	3.5099	3.5150	3.5202	3.5258	3.5316	3.5378	3.5442	3.5510	3.5581	3.5657
	0.01	4.0272	4.0343	4.0416	4.0493	4.0574	4.0659	4.0749	4.0843	4.0942	4.1047	4.1158
	0.001	4.7313	4.7422	4.7535	4.7655	4.7780	4.7911	4.8050	4.8196	4.8350	4.8513	4.8686
55	0.25	2.9637	2.9662	2.9689	2.9716	2.9744	2.9774	2.9805	2.9838	2.9872	2.9908	2.9945
	0.10	3.2837	3.2871	3.2905	3.2942	3.2979	3.3019	3.3060	3.3103	3.3148	3.3195	3.3245
	0.05	3.5151	3.5191	3.5233	3.5277	3.5322	3.5369	3.5419	3.5471	3.5525	3.5582	3.5642
	0.01	4.0257	4.0314	4.0375	4.0438	4.0504	4.0572	4.0644	4.0719	4.0798	4.0881	4.0967
	0.001	4.7113	4.7201	4.7294	4.7391	4.7491	4.7597	4.7707	4.7823	4.7944	4.8071	4.8205
60	0.25	2.9833	2.9854	2.9877	2.9900	2.9924	2.9949	2.9975	3.0002	3.0030	3.0060	3.0090
	0.10	3.2980	3.3008	3.3038	3.3068	3.3100	3.3133	3.3167	3.3202	3.3239	3.3278	3.3318
	0.05	3.5253	3.5287	3.5322	3.5359	3.5397	3.5436	3.5477	3.5520	3.5564	3.5611	3.5659
	0.01	4.0261	4.0310	4.0360	4.0413	4.0467	4.0524	4.0583	4.0644	4.0709	4.0775	4.0845
	0.001	4.6964	4.7038	4.7115	4.7195	4.7278	4.7364	4.7455	4.7548	4.7646	4.7749	4.7855
65	0.25	3.0016	3.0035	3.0054	3.0074	3.0094	3.0115	3.0137	3.0160	3.0184	3.0209	3.0234
	0.10	3.3117	3.3141	3.3166	3.3192	3.3219	3.3247	3.3276	3.3306	3.3337	3.3369	3.3403
	0.05	3.5355	3.5384	3.5414	3.5445	3.5477	3.5511	3.5545	3.5581	3.5618	3.5657	3.5697
	0.01	4.0280	4.0321	4.0364	4.0408	4.0454	4.0502	4.0551	4.0603	4.0656	4.0712	4.0769
	0.001	4.6854	4.6916	4.6981	4.7048	4.7118	4.7190	4.7265	4.7343	4.7424	4.7508	4.7595
70	0.25	3.0187	3.0204	3.0220	3.0238	3.0255	3.0274	3.0293	3.0312	3.0333	3.0354	3.0376
	0.10	3.3248	3.3269	3.3291	3.3313	3.3336	3.3360	3.3385	3.3411	3.3437	3.3465	3.3493
	0.05	3.5456	3.5481	3.5507	3.5534	3.5561	3.5590	3.5619	3.5650	3.5682	3.5714	3.5748
	0.01	4.0308	4.0344	4.0381	4.0419	4.0458	4.0499	4.0541	4.0584	4.0629	4.0676	4.0724
	0.001	4.6771	4.6825	4.6880	4.6938	4.6997	4.7058	4.7122	4.7187	4.7255	4.7326	4.7399
80	0.25	3.0501	3.0513	3.0526	3.0540	3.0554	3.0568	3.0582	3.0597	3.0613	3.0628	3.0645
	0.10	3.3493	3.3510	3.3526	3.3544	3.3562	3.3580	3.3599	3.3618	3.3638	3.3659	3.3680
	0.05	3.5650	3.5670	3.5690	3.5710	3.5732	3.5753	3.5776	3.5799	3.5822	3.5847	3.5872
	0.01	4.0383	4.0411	4.0439	4.0468	4.0498	4.0528	4.0560	4.0593	4.0626	4.0661	4.0696
	0.001	4.6665	4.6706	4.6748	4.6792	4.6836	4.6882	4.6929	4.6978	4.7028	4.7080	4.7133
90	0.25	3.0781	3.0791	3.0802	3.0812	3.0823	3.0835	3.0846	3.0858	3.0870	3.0883	3.0896
	0.10	3.3718	3.3732	3.3745	3.3759	3.3773	3.3787	3.3802	3.3818	3.3833	3.3849	3.3866
	0.05	3.5834	3.5850	3.5866	3.5882	3.5899	3.5916	3.5934	3.5952	3.5970	3.5989	3.6009
	0.01	4.0472	4.0494	4.0516	4.0539	4.0563	4.0587	4.0612	4.0637	4.0663	4.0689	4.0717
	0.001	4.6612	4.6645	4.6678	4.6712	4.6747	4.6782	4.6819	4.6856	4.6895	4.6934	4.6975
100	0.25	3.1035	3.1043	3.1052	3.1060	3.1069	3.1079	3.1088	3.1098	3.1107	3.1118	3.1128
	0.10	3.3926	3.3937	3.3948	3.3959	3.3971	3.3982	3.3994	3.4007	3.4019	3.4032	3.4045
	0.05	3.6008	3.6021	3.6034	3.6047	3.6061	3.6075	3.6089	3.6104	3.6118	3.6134	3.6149
	0.01	4.0568	4.0586	4.0604	4.0622	4.0641	4.0661	4.0681	4.0701	4.0722	4.0743	4.0764
	0.001	4.6593	4.6619	4.6646	4.6673	4.6701	4.6730	4.6759	4.6789	4.6819	4.6851	4.6883

n	a	k: 0	1	2	3	4	5	6	7	8	9	10
120	0.25	3.1478	3.1484	3.1490	3.1497	3.1503	3.1510	3.1516	3.1523	3.1530	3.1537	3.1544
	0.10	3.4297	3.4304	3.4312	3.4320	3.4328	3.4336	3.4345	3.4354	3.4362	3.4371	3.4380
	0.05	3.6325	3.6335	3.6344	3.6353	3.6363	3.6372	3.6382	3.6392	3.6403	3.6413	3.6424
	0.01	4.0764	4.0776	4.0789	4.0802	4.0815	4.0828	4.0842	4.0856	4.0870	4.0885	4.0899
	0.001	4.6613	4.6631	4.6649	4.6668	4.6687	4.6707	4.6727	4.6747	4.6768	4.6789	4.6810
140	0.25	3.1857	3.1862	3.1866	3.1871	3.1876	3.1881	3.1886	3.1891	3.1896	3.1901	3.1906
	0.10	3.4620	3.4625	3.4631	3.4637	3.4643	3.4649	3.4656	3.4662	3.4669	3.4675	3.4682
	0.05	3.6608	3.6615	3.6622	3.6629	3.6636	3.6643	3.6650	3.6658	3.6665	3.6673	3.6681
	0.01	4.0955	4.0964	4.0974	4.0983	4.0993	4.1003	4.1013	4.1023	4.1033	4.1044	4.1054
	0.001	4.6674	4.6687	4.6701	4.6715	4.6729	4.6743	4.6758	4.6772	4.6787	4.6802	4.6818
160	0.25	3.2188	3.2191	3.2195	3.2199	3.2203	3.2206	3.2210	3.2214	3.2218	3.2222	3.2226
	0.10	3.4905	3.4910	3.4914	3.4919	3.4924	3.4929	3.4933	3.4938	3.4943	3.4948	3.4954
	0.05	3.6861	3.6867	3.6872	3.6878	3.6883	3.6889	3.6894	3.6900	3.6906	3.6912	3.6918
	0.01	4.1136	4.1144	4.1151	4.1158	4.1166	4.1173	4.1181	4.1189	4.1197	4.1205	4.1213
	0.001	4.6755	4.6765	4.6775	4.6786	4.6797	4.6808	4.6819	4.6830	4.6841	4.6853	4.6865
180	0.25	3.2481	3.2484	3.2487	3.2490	3.2493	3.2496	3.2499	3.2502	3.2505	3.2508	3.2512
	0.10	3.5161	3.5165	3.5168	3.5172	3.5176	3.5180	3.5184	3.5188	3.5192	3.5196	3.5200
	0.05	3.7091	3.7095	3.7099	3.7104	3.7108	3.7113	3.7117	3.7122	3.7126	3.7131	3.7136
	0.01	4.1307	4.1313	4.1319	4.1325	4.1331	4.1337	4.1343	4.1349	4.1355	4.1362	4.1368
	0.001	4.6844	4.6852	4.6861	4.6869	4.6878	4.6886	4.6895	4.6904	4.6913	4.6922	4.6931
200	0.25	3.2743	3.2746	3.2748	3.2751	3.2753	3.2756	3.2758	3.2761	3.2764	3.2766	3.2769
	0.10	3.5393	3.5396	3.5399	3.5402	3.5405	3.5408	3.5411	3.5414	3.5418	3.5421	3.5424
	0.05	3.7300	3.7303	3.7307	3.7311	3.7314	3.7318	3.7322	3.7325	3.7329	3.7333	3.7337
	0.01	4.1467	4.1472	4.1477	4.1482	4.1487	4.1492	4.1496	4.1502	4.1507	4.1512	4.1517
	0.001	4.6938	4.6944	4.6951	4.6958	4.6965	4.6972	4.6979	4.6986	4.6993	4.7000	4.7008
250	0.25	3.3302	3.3304	3.3305	3.3307	3.3309	3.3310	3.3312	3.3313	3.3315	3.3317	3.3319
	0.10	3.5890	3.5892	3.5894	3.5896	3.5898	3.5900	3.5902	3.5904	3.5906	3.5908	3.5910
	0.05	3.7754	3.7756	3.7758	3.7761	3.7763	3.7765	3.7768	3.7770	3.7773	3.7775	3.7778
	0.01	4.1827	4.1830	4.1833	4.1836	4.1839	4.1842	4.1846	4.1849	4.1852	4.1855	4.1859
	0.001	4.7169	4.7174	4.7178	4.7183	4.7187	4.7192	4.7196	4.7201	4.7205	4.7210	4.7215
300	0.25	3.3759	3.3760	3.3761	3.3762	3.3764	3.3765	3.3766	3.3767	3.3769	3.3770	3.3771
	0.10	3.6301	3.6303	3.6304	3.6305	3.6307	3.6308	3.6310	3.6311	3.6313	3.6314	3.6316
	0.05	3.8133	3.8135	3.8136	3.8138	3.8140	3.8141	3.8143	3.8145	3.8146	3.8148	3.8150
	0.01	4.2137	4.2139	4.2141	4.2143	4.2146	4.2148	4.2150	4.2153	4.2155	4.2157	4.2160
	0.001	4.7388	4.7391	4.7394	4.7397	4.7401	4.7404	4.7407	4.7410	4.7413	4.7417	4.7420
400	0.25	3.4480	3.4481	3.4482	3.4482	3.4483	3.4484	3.4484	3.4485	3.4486	3.4487	3.4487
	0.10	3.6957	3.6958	3.6959	3.6959	3.6960	3.6961	3.6962	3.6963	3.6964	3.6965	3.6966
	0.05	3.8743	3.8744	3.8745	3.8746	3.8747	3.8748	3.8749	3.8750	3.8751	3.8752	3.8753
	0.01	4.2650	4.2651	4.2653	4.2654	4.2655	4.2656	4.2658	4.2659	4.2660	4.2662	4.2663
	0.001	4.7775	4.7777	4.7779	4.7781	4.7782	4.7784	4.7786	4.7788	4.7790	4.7791	4.7793
500	0.25	3.5038	3.5039	3.5039	3.5039	3.5040	3.5040	3.5041	3.5041	3.5042	3.5042	3.5043
	0.10	3.7469	3.7470	3.7470	3.7471	3.7471	3.7472	3.7472	3.7473	3.7474	3.7474	3.7475
	0.05	3.9223	3.9224	3.9225	3.9225	3.9226	3.9227	3.9227	3.9228	3.9229	3.9229	3.9230
	0.01	4.3063	4.3064	4.3065	4.3066	4.3067	4.3067	4.3068	4.3069	4.3070	4.3071	4.3072
	0.001	4.8103	4.8104	4.8105	4.8107	4.8108	4.8109	4.8110	4.8111	4.8112	4.8114	4.8115

n	a	k: 12	14	16	18	20	25	30	40	50	60
16	0.25	7.9059									
	0.10	12.5897									
	0.05	17.8466									
	0.01	39.9812									
	0.001	126.4852									
17	0.25	5.0838									
	0.10	7.0430									
	0.05	8.9521									
	0.01	15.4575									
	0.001	33.4329									
18	0.25	4.1826	8.3966								
	0.10	5.4366	13.3604								
	0.05	6.5697	18.9341								
	0.01	10.0299	42.4087								
	0.001	18.0361	134.1585								
19	0.25	3.7600	5.2930								
	0.10	4.7156	7.3215								
	0.05	5.5393	9.3001								
	0.01	7.8888	16.0471								
	0.001	12.7578	34.6984								
20	0.25	3.5212	4.3147	8.8602							
	0.10	4.3168	5.5976	14.0891							
	0.05	4.9807	6.7582	19.9625							
	0.01	6.7883	10.3062	44.7046							
	0.001	10.2609	18.5224	141.4160							
22	0.25	3.2696	3.6002	4.4366	9.3007						
	0.10	3.8999	4.4047	5.7465	14.7818						
	0.05	4.4046	5.0768	6.9328	20.9404						
	0.01	5.6997	6.9092	10.5625	46.8882						
	0.001	7.9555	10.4332	18.9733	148.3189						

n	a	k: 12	14	16	18	20	25	30	40	50	60
24	0.25	3.1452	3.3279	3.6730	4.5501	9.7213					
	0.10	3.6915	3.9618	4.4858	5.8853	15.4435					
	0.05	4.1181	4.4700	5.1656	7.0957	21.8747					
	0.01	5.1740	5.7755	7.0210	10.8016	48.9745					
	0.001	6.9054	8.0523	10.5929	19.3944	154.9145					
26	0.25	3.0757	3.1924	3.3818	3.7406	4.6564					
	0.10	3.5709	3.7401	4.0191	4.5612	6.0154					
	0.05	3.9514	4.1683	4.5305	5.2482	7.2484					
	0.01	4.8706	5.2294	5.8458	7.1251	11.0260					
	0.001	6.3175	6.9714	8.1421	10.7418	19.7898					
28	0.25	3.0340	3.1155	3.2361	3.4320	3.8038					
	0.10	3.4949	3.6113	3.7852	4.0724	4.6317					
	0.05	3.8448	3.9925	4.2150	4.5869	5.3255					
	0.01	4.6759	4.9144	5.2810	5.9114	7.2226					
	0.001	5.9461	6.3673	7.0329	8.2260	10.8813					
30	0.25	3.0083	3.0688	3.1527	3.2768	3.4789	6.2315				
	0.10	3.4443	3.5296	3.6489	3.8273	4.1224	8.5752				
	0.05	3.7725	3.8798	4.0308	4.2586	4.6398	10.8688				
	0.01	4.5420	4.7123	4.9554	5.3292	5.9729	18.7089				
	0.001	5.6925	5.9861	6.4139	7.0904	8.3047	40.4152				
32	0.25	2.9925	3.0392	3.1013	3.1874	3.3151	4.2868				
	0.10	3.4095	3.4749	3.5621	3.6842	3.8669	5.3259				
	0.05	3.7214	3.8031	3.9126	4.0668	4.2996	6.2270				
	0.01	4.4454	4.5733	4.7465	4.9939	5.3745	8.8118				
	0.001	5.5094	5.7259	6.0237	6.4577	7.1446	14.1947				
34	0.25	2.9829	3.0203	3.0682	3.1318	3.2201	3.7306	11.5975			
	0.10	3.3850	3.4369	3.5036	3.5926	3.7173	4.4585	18.3984			
	0.05	3.6841	3.7486	3.8318	3.9434	4.1007	5.0556	26.0480			
	0.01	4.3731	4.4728	4.6027	4.7787	5.0301	6.6367	58.2967			
	0.001	5.3719	5.5382	5.7574	6.0591	6.4989	9.5273	184.3868			
36	0.25	2.9777	3.0083	3.0466	3.0955	3.1605	3.4795	5.1079			
	0.10	3.3676	3.4098	3.4626	3.5306	3.6214	4.0752	6.5697			
	0.05	3.6564	3.7087	3.7742	3.8589	3.9725	4.5485	7.8998			
	0.01	4.3176	4.3976	4.4987	4.6305	4.8091	5.7458	11.9851			
	0.001	5.2653	5.3971	5.5654	5.7872	6.0925	7.7751	21.4808			
38	0.25	2.9754	3.0010	3.0323	3.0713	3.1213	3.3412	4.0696			
	0.10	3.3552	3.3903	3.4332	3.4870	3.5562	3.8648	4.9295			
	0.05	3.6356	3.6788	3.7319	3.7985	3.8846	4.2722	5.6526			
	0.01	4.2739	4.3397	4.4208	4.5232	4.6568	5.2729	7.6363			
	0.001	5.1807	5.2878	5.4211	5.5911	5.8154	6.8872	11.4738			
40	0.25	2.9752	2.9970	3.0231	3.0550	3.0948	3.2565	3.6766			
	0.10	3.3464	3.3761	3.4118	3.4554	3.5101	3.7345	4.3335			
	0.05	3.6197	3.6561	3.7000	3.7539	3.8215	4.1013	4.8637			
	0.01	4.2391	4.2941	4.3607	4.4427	4.5465	4.9838	6.2340			
	0.001	5.1122	5.2010	5.3092	5.4438	5.6155	6.3590	8.6396			

n	a	k: 12	14	16	18	20	25	30	40	50	60
45	0.25	2.9806	2.9959	3.0137	3.0346	3.0595	3.1486	3.3177	7.1849		
	0.10	3.3350	3.3557	3.3797	3.4080	3.4417	3.5631	3.7965	9.8538		
	0.05	3.5945	3.6197	3.6491	3.6836	3.7249	3.8744	4.1643	12.4715		
	0.01	4.1780	4.2155	4.2592	4.3109	4.3730	4.6000	5.0506	21.4337		
	0.001	4.9881	5.0476	5.1173	5.2002	5.3002	5.6718	6.4338	46.2719		
50	0.25	2.9905	3.0019	3.0150	3.0298	3.0469	3.1040	3.1966	3.8325		
	0.10	3.3326	3.3479	3.3653	3.3852	3.4082	3.4850	3.6105	4.5008		
	0.05	3.5822	3.6007	3.6218	3.6460	3.6739	3.7676	3.9216	5.0413		
	0.01	4.1399	4.1672	4.1983	4.2340	4.2754	4.4152	4.6480	6.4420		
	0.001	4.9063	4.9490	4.9978	5.0540	5.1196	5.3430	5.7223	8.9070		
55	0.25	3.0026	3.0115	3.0214	3.0326	3.0452	3.0852	3.1441	3.4221	7.7041	
	0.10	3.3351	3.3470	3.3602	3.3750	3.3917	3.4450	3.5241	3.9023	10.5517	
	0.05	3.5770	3.5913	3.6073	3.6252	3.6454	3.7101	3.8062	4.2721	13.3470	
	0.01	4.1155	4.1363	4.1595	4.1857	4.2154	4.3106	4.4534	5.1651	22.9239	
	0.001	4.8494	4.8816	4.9178	4.9585	5.0049	5.1544	5.3818	6.5624	49.4764	
60	0.25	3.0156	3.0227	3.0306	3.0394	3.0490	3.0788	3.1199	3.2794	3.9618	
	0.10	3.3404	3.3499	3.3603	3.3718	3.3846	3.4239	3.4786	3.6924	4.6398	
	0.05	3.5763	3.5876	3.6001	3.6140	3.6294	3.6769	3.7430	4.0035	5.1892	
	0.01	4.0995	4.1159	4.1341	4.1541	4.1765	4.2457	4.3426	4.7312	6.6155	
	0.001	4.8085	4.8337	4.8615	4.8925	4.9271	5.0344	5.1863	5.8101	9.1306	
65	0.25	3.0289	3.0348	3.0412	3.0482	3.0559	3.0790	3.1095	3.2142	3.5091	8.1624
	0.10	3.3474	3.3551	3.3636	3.3728	3.3829	3.4132	3.4535	3.5924	3.9908	11.1686
	0.05	3.5782	3.5875	3.5976	3.6087	3.6209	3.6573	3.7057	3.8738	4.3624	14.1214
	0.01	4.0892	4.1025	4.1170	4.1330	4.1505	4.2031	4.2735	4.5204	5.2612	24.2425
	0.001	4.7782	4.7985	4.8206	4.8450	4.8719	4.9528	5.0616	5.4500	6.6706	52.3123
70	0.25	3.0421	3.0471	3.0525	3.0583	3.0646	3.0830	3.1066	3.1811	3.3492	4.0724
	0.10	3.3553	3.3618	3.3688	3.3763	3.3846	3.4087	3.4396	3.5378	3.7616	4.7590
	0.05	3.5820	3.5897	3.5981	3.6071	3.6170	3.6459	3.6830	3.8012	4.0727	5.3162
	0.01	4.0827	4.0937	4.1057	4.1186	4.1327	4.1743	4.2278	4.3994	4.8017	6.7649
	0.001	4.7553	4.7721	4.7901	4.8098	4.8313	4.8946	4.9765	5.2427	5.8846	9.3233
80	0.25	3.0679	3.0716	3.0755	3.0796	3.0841	3.0968	3.1122	3.1561	3.2339	3.4095
	0.10	3.3724	3.3771	3.3822	3.3876	3.3934	3.4098	3.4299	3.4870	3.5889	3.8215
	0.05	3.5925	3.5981	3.6041	3.6105	3.6174	3.6370	3.6609	3.7292	3.8514	4.1327
	0.01	4.0771	4.0850	4.0935	4.1026	4.1124	4.1402	4.1743	4.2720	4.4485	4.8630
	0.001	4.7244	4.7364	4.7491	4.7628	4.7774	4.8193	4.8708	5.0193	5.2916	5.9495
90	0.25	3.0922	3.0950	3.0980	3.1012	3.1045	3.1138	3.1248	3.1540	3.1994	3.2801
	0.10	3.3900	3.3936	3.3975	3.4015	3.4058	3.4178	3.4320	3.4697	3.5286	3.6338
	0.05	3.6050	3.6092	3.6138	3.6186	3.6237	3.6379	3.6547	3.6995	3.7697	3.8956
	0.01	4.0774	4.0834	4.0898	4.0965	4.1037	4.1238	4.1475	4.2109	4.3109	4.4918
	0.001	4.7060	4.7149	4.7244	4.7344	4.7451	4.7750	4.8104	4.9056	5.0570	5.3348
100	0.25	3.1149	3.1172	3.1195	3.1220	3.1246	3.1318	3.1400	3.1610	3.1911	3.2380
	0.10	3.4073	3.4101	3.4132	3.4163	3.4197	3.4288	3.4394	3.4663	3.5051	3.5657
	0.05	3.6181	3.6215	3.6251	3.6288	3.6328	3.6436	3.6561	3.6880	3.7339	3.8059
	0.01	4.0810	4.0857	4.0907	4.0959	4.1014	4.1166	4.1341	4.1788	4.2435	4.3456
	0.001	4.6949	4.7019	4.7092	4.7169	4.7251	4.7475	4.7735	4.8399	4.9366	5.0906

n	a	k: 12	14	16	18	20	25	30	40	50	60
120	0.25	3.1559	3.1574	3.1590	3.1607	3.1624	3.1670	3.1722	3.1847	3.2009	3.2230
	0.10	3.4399	3.4418	3.4439	3.4460	3.4482	3.4541	3.4607	3.4765	3.4972	3.5253
	0.05	3.6446	3.6468	3.6492	3.6517	3.6543	3.6612	3.6690	3.6876	3.7119	3.7451
	0.01	4.0930	4.0961	4.0994	4.1028	4.1064	4.1160	4.1267	4.1526	4.1864	4.2326
	0.001	4.6855	4.6901	4.6949	4.6999	4.7051	4.7191	4.7348	4.7727	4.8224	4.8907
140	0.25	3.1917	3.1928	3.1940	3.1952	3.1964	3.1997	3.2033	3.2117	3.2219	3.2349
	0.10	3.4695	3.4710	3.4724	3.4739	3.4755	3.4796	3.4842	3.4947	3.5077	3.5241
	0.05	3.6697	3.6713	3.6730	3.6748	3.6766	3.6815	3.6868	3.6991	3.7143	3.7336
	0.01	4.1076	4.1099	4.1122	4.1147	4.1172	4.1238	4.1311	4.1481	4.1690	4.1956
	0.001	4.6850	4.6882	4.6916	4.6951	4.6988	4.7084	4.7190	4.7436	4.7740	4.8127
160	0.25	3.2235	3.2243	3.2252	3.2261	3.2270	3.2295	3.2321	3.2382	3.2453	3.2539
	0.10	3.4964	3.4975	3.4986	3.4997	3.5009	3.5040	3.5073	3.5148	3.5238	3.5347
	0.05	3.6930	3.6942	3.6955	3.6969	3.6982	3.7018	3.7057	3.7145	3.7249	3.7376
	0.01	4.1230	4.1247	4.1264	4.1282	4.1301	4.1350	4.1403	4.1523	4.1666	4.1839
	0.001	4.6888	4.6913	4.6938	4.6964	4.6991	4.7061	4.7138	4.7311	4.7517	4.7768
180	0.25	3.2518	3.2525	3.2532	3.2539	3.2546	3.2566	3.2586	3.2632	3.2684	3.2746
	0.10	3.5208	3.5216	3.5225	3.5234	3.5243	3.5267	3.5293	3.5350	3.5416	3.5493
	0.05	3.7145	3.7155	3.7165	3.7176	3.7186	3.7214	3.7244	3.7310	3.7387	3.7477
	0.01	4.1381	4.1394	4.1408	4.1422	4.1437	4.1474	4.1515	4.1604	4.1709	4.1831
	0.001	4.6950	4.6969	4.6988	4.7009	4.7029	4.7083	4.7141	4.7270	4.7419	4.7595
200	0.25	3.2774	3.2780	3.2785	3.2791	3.2797	3.2813	3.2829	3.2865	3.2905	3.2952
	0.10	3.5431	3.5438	3.5445	3.5452	3.5459	3.5479	3.5499	3.5544	3.5594	3.5653
	0.05	3.7345	3.7352	3.7361	3.7369	3.7377	3.7400	3.7423	3.7475	3.7534	3.7602
	0.01	4.1528	4.1538	4.1549	4.1561	4.1572	4.1602	4.1634	4.1704	4.1783	4.1875
	0.001	4.7023	4.7038	4.7054	4.7070	4.7086	4.7129	4.7174	4.7274	4.7387	4.7518
250	0.25	3.3322	3.3326	3.3330	3.3333	3.3337	3.3347	3.3357	3.3379	3.3403	3.3430
	0.10	3.5915	3.5919	3.5924	3.5928	3.5933	3.5945	3.5958	3.5985	3.6015	3.6048
	0.05	3.7783	3.7788	3.7793	3.7798	3.7804	3.7818	3.7832	3.7864	3.7898	3.7937
	0.01	4.1866	4.1873	4.1880	4.1887	4.1894	4.1913	4.1932	4.1974	4.2020	4.2072
	0.001	4.7224	4.7234	4.7244	4.7254	4.7264	4.7290	4.7318	4.7377	4.7443	4.7515
300	0.25	3.3774	3.3776	3.3779	3.3781	3.3784	3.3791	3.3798	3.3813	3.3829	3.3847
	0.10	3.6319	3.6322	3.6325	3.6329	3.6332	3.6340	3.6349	3.6367	3.6387	3.6409
	0.05	3.8154	3.8157	3.8161	3.8164	3.8168	3.8178	3.8188	3.8209	3.8232	3.8257
	0.01	4.2164	4.2169	4.2174	4.2179	4.2184	4.2197	4.2210	4.2238	4.2269	4.2302
	0.001	4.7426	4.7433	4.7440	4.7447	4.7454	4.7472	4.7490	4.7530	4.7573	4.7619
400	0.25	3.4489	3.4490	3.4492	3.4493	3.4495	3.4499	3.4503	3.4511	3.4520	3.4529
	0.10	3.6967	3.6969	3.6971	3.6973	3.6975	3.6979	3.6984	3.6994	3.7005	3.7017
	0.05	3.8755	3.8757	3.8759	3.8761	3.8763	3.8769	3.8775	3.8786	3.8799	3.8812
	0.01	4.2666	4.2668	4.2671	4.2674	4.2677	4.2684	4.2691	4.2707	4.2723	4.2740
	0.001	4.7797	4.7801	4.7805	4.7808	4.7812	4.7822	4.7833	4.7854	4.7876	4.7900
500	0.25	3.5044	3.5045	3.5046	3.5047	3.5048	3.5050	3.5053	3.5058	3.5064	3.5070
	0.10	3.7476	3.7477	3.7478	3.7479	3.7481	3.7484	3.7487	3.7493	3.7500	3.7507
	0.05	3.9231	3.9233	3.9234	3.9235	3.9237	3.9240	3.9244	3.9251	3.9259	3.9267
	0.01	4.3074	4.3075	4.3077	4.3079	4.3081	4.3085	4.3090	4.3100	4.3110	4.3121
	0.001	4.8117	4.8120	4.8122	4.8125	4.8127	4.8134	4.8140	4.8153	4.8167	4.8182

References

Abler, R., J. S. Adams, and P. Gould (1971). *Spatial Organization: The Geographers's View of the World*. Englewood Cliffs: Prentice-Hall.

Afifi, A. A., and R. M. Elashoff (1966). Missing observations in multivariate statistics I: Review of the literature. *Journal of the American Statistical Association*, **61**: 595–604.

Aigner, D. J., and A. S. Goldberger, eds. (1977). *Latent Variables in Socio-Economic Models*. New York: North-Holland.

Alwin, D. F., and R. M. Hauser (1975). The decomposition of effects in path analysis. *American Sociological Review*, **40**: 37–47.

Amemiya, T. (1974). The nonlinear two-stage least-squares estimator. *Journal of Econometrics*, **2**: 105–110.

Amemiya, T. (1977). The maximum likelihood estimator and the nonlinear three-stage least squares estimator in the general nonlinear simultaneous equation model. *Econometrica*, **45**: 955–968.

Andrews, D. F. (1979). The robustness of residual displays. Pages 19–32 in R. L. Launer and G. N. Wilkinson, eds. *Robustness in Statistics*. New York: Academic Press.

Angell, R. C. (1951). The moral integration of American cities. *American Journal of Sociology*, **57** (part 2): 1–140.

Anscombe, F. J. (1960). Rejection of outliers [with commentary]. *Technometrics*, **2**: 123–166.

Anscombe, F. J. (1961). Examination of residuals. *Proceedings of the Fourth Berkeley Symposium on Mathematical Statistics and Probability*, **1**: 1–36.

Anscombe, F. J. (1981). *Computing in Statistical Science Through APL*. New York: Springer-Verlag.

Anscombe, F. J., and J. W. Tukey (1963). The examination and analysis of residuals. *Technometrics*, **5**: 141–160.

Arnold, S. F. (1981). *The Theory of Linear Models and Multivariate Analysis*. New York: Wiley.

Atkinson, A. C. (1982). Regression diagnostics, transformations and constructed variables [with commentary]. *Journal of the Royal Statistical Society, Series B*, **44**: 1–36.

Bailey, J. R. (1977). Tables of the Bonferroni *t* statistic. *Journal of the American Statistical Association*, **72**: 469–478.

Bagozzi, R. P. (1977). Populism and lynching in Louisiana (comment on Inverarity). *American Sociological Review*, **42**: 355–358.

Bard, Y. (1974). *Nonlinear Parameter Estimation*. New York: Academic Press.

Barnett, V., and T. Lewis (1978). *Outliers in Statistical Data*. New York: Wiley.

Bartlett, M. S. (1937). Properties of sufficiency and statistical tests. *Proceedings of the Royal Society, A*, **160**: 268–282.

Basmann, R. L. (1957). A generalized method of linear estimation of coefficients in a structural equation. *Econometrica*, **25**: 77–83.

Basmann, R. L. (1962). Letter to the editor. *Econometrica*, **30**: 824–826.

Beale, E. M. L., and R. J. A. Little (1975). Missing values in multivariate analysis. *Journal of the Royal Statistical Society, Series B*, **37**: 129–145.

Beckman, R. J., and R. D. Cook (1983). Outliers. *Technometrics*, **25**: 119–163.

Beckman, R. J., and H. J. Trussell (1974). The distribution of an arbitrary studentized residual and the effects of updating in multiple regression. *Journal of the American Statistical Association*, **69**: 199–201.

Behnken, D. W., and N. R. Draper (1972). Residuals and their variance patterns. *Technometrics*, **14**: 101–111.

Belsley, D. A., E. Kuh, and R. E. Welsch (1980). *Regression Diagnostics: Identifying Influential Data and Sources of Collinearity*. New York: Wiley.

Berk, R. A., and S. F. Berk (1978). A simultaneous equation model for the division of household labor. *Sociological Methods and Research*, **6**: 431–468.

Bielby, W. T., R. M. Hauser, and D. L. Featherman (1977). Response errors of non-black males in models of the stratification process. Pages 227–251 in D. J. Aigner and A. S. Goldberger, eds. *Latent Variables in Socio-Economic Models*. New York: North-Holland.

Bishop, Y. M. M., S. E. Fienberg, and P. W. Holland (1975). *Discrete Multivariate Analysis: Theory and Practice*. Cambridge: MIT.

Blalock, H. M. Jr. (1979). *Social Statistics, Revised Second Edition*. New York: McGraw-Hill.

Blau, P. M., and O. D. Duncan (1967). *The American Occupational Structure*. New York: Wiley.

Bock, R. D. (1975). *Multivariate Statistical Methods in Behavioral Research*. New York: McGraw-Hill.

Bohrnstedt, G. W. (1977). Use of the multiple indicators-multiple causes (MIMIC) model (comment on Inverarity). *American Sociological Review*, **42**:656–663.

Bollen, K. A., and S. Ward (1979). Ratio variables in aggregate data analysis: Their uses, problems, and alternatives. *Sociological Methods and Research*, **7**: 431–450.

Box, G. E. P. (1953). Non-normality and tests on variances. *Biometrika*, **40**: 318–335.

Box, G. E. P., and D. R. Cox (1964). An analysis of transformations. *Journal of the Royal Statistical Society, Series B*, **26**: 211–252.

Box, G. E. P., and P. W. Tidwell (1962). Transformation of the independent variables. *Technometrics*, **4**: 531–550.

Brier, S. S. (1979). The utility of simultaneous logistic response equations. Pages 119–129 in K. F. Schuessler, ed. *Sociological Methodology 1979*. San Francisco: Jossey-Bass.

Campbell, A., P. E. Converse, W. E. Miller, and D. E. Stokes (1960). *The American Voter*. New York: Wiley.

Canada (1962). *Major Roads (Map)*. Ottawa: Department of Mines and Technical Surveys.

Canada (1971). *Census of Canada*. Ottawa: Statistics Canada.

Canada (1972). *Canada Year Book*. Ottawa: Statistics Canada.

Chatterjee, S., and B. Price (1977). *Regression Analysis by Example*. New York: Wiley.

Chirot, D., and C. Ragin (1975). The market, tradition and peasant rebellion: The case of Romania in 1907. *American Sociological Review*, **40**: 428–444.

Christ, C. F. (1966). *Econometric Models and Methods*. New York: Wiley.

Cleveland, W. S. (1979) Robust locally weighted regression and smoothing scatterplots. *Journal of the American Statistical Association*, **74**: 829–836.

Cleveland, W. S., and B. Kleiner (1975). A graphical technique for enhancing scatterplots with moving statistics. *Technometrics*, **17**: 447–454.

Cook, R. D. (1977a). Detection of influential observations in linear regression. *Technometrics*, **19**: 15–18.

Cook, R. D. (1977b). Letter to the editor. *Technometrics*, **19**: 349.

Cook, R. D. (1979). Influential observations in linear regression. *Journal of the American Statistical Association*, **74**: 169–174.

Cook, R. D., and P. Prescott (1981). On the accuracy of Bonferroni significance levels for detecting outliers in linear models. *Technometrics*, **22**: 59–63.

Cook, R. D., and S. Weisberg (1980). Characterizations of an empirical influence function for detecting influential cases in regression. *Technometrics*, **22**: 495–508.

Cook, R. D., and S. Weisberg (1982). *Residuals and Influence in Regression*. New York: Chapman-Hall.

Costner, H. L., and R. Schoenberg (1973). Diagnosing indicator ills in multiple indicator models. Pages 168–199 in A. S. Goldberger and O. D. Duncan, eds. *Structural Equation Models in the Social Sciences*. New York: Seminar Press.

Daniel, C., and F. S. Wood (1980). *Fitting Equations to Data, Second Edition*. New York: Wiley.

Draper, N. R., and W. G. Hunter (1969). Transformations: Some examples revisited. *Technometrics*, **11**: 23–40.

Draper, N. R., and J. A. John (1981). Influential observations and outliers in regression. *Technometrics*, **23**: 21–26.

Draper, N. R., and H. Smith (1966). *Applied Regression Analysis*. New York: Wiley.

Draper, N. R., and H. Smith (1981). *Applied Regression Analysis, Second Edition*. New York: Wiley.

Draper, N. R., and R. C. Van Nostrand (1979). Ridge regression and James–Stein estimators: Review and comments. *Technometrics*, **21**: 451–466.

Duncan, O. D. (1966). Path analysis: Sociological examples. *American Journal of Sociology*, **72**: 1–16.

Duncan, O. D. (1975). *Introduction to Structural Equation Models*. New York: Academic Press.

Duncan, O. D., D. L. Featherman, and B. Duncan (1972). *Socioeconomic Background and Achievement*. New York: Seminar Press.

Duncan, O. D., A. O. Haller, and A. Portes (1968). Peer influences on aspirations: A reinterpretation. *American Journal of Sociology*, **74**: 119–137.

Efron, B. (1982). *The Jackknife, the Bootstrap, and Other Resampling Plans*. Philadelphia: Society for Industrial and Applied Mathematics.

Ellenberg, J. H. (1973). The joint distribution of the standardized least squares residuals from a general linear regression. *Journal of the American Statistical Association*, **68**: 941–943.

Ellenberg, J. H. (1976). Testing for a single outlier from a general linear regression. *Biometrics*, **32**: 637–645.

Fienberg, S. E. (1977). *The Analysis of Cross-Classified Categorical Data*. Cambridge: MIT.

Fienberg, S. E. (1979). A note on fitting and interpreting parameters in models for categorical data. Pages 112–118 in K. F. Schuessler, ed. *Sociological Methodology 1979*. San Francisco: Jossey-Bass.

Finifter, B. M. (1972). The generation of confidence: Evaluating research findings by random subsample replication. Pages 112–175 in H. L. Costner, ed. *Sociological Methodology 1972*. San Francisco: Jossey-Bass.

Finney, D. J. (1971). *Probit Analysis, Third Edition*. London: Cambridge University Press.

Finney, J. M. (1972). Indirect effects in path analysis. *Sociological Methods and Research*, **1**: 175–186.

Fisher, F. M. (1966). *The Identification Problem in Econometrics.* New York: McGraw-Hill.

Fisher, R. A. (1922). On the mathematical foundations of theoretical statistics. *Philosophical Transactions of the Royal Society of London, Series A*, **222**: 309–368.

Fisher, R. A. (1925). *Statistical Methods for Research Workers.* Edinburgh: Oliver and Boyd.

Fox, B. (1980). *Women's Domestic Labour and their Involvement in Wage Work: Twentieth-Century Changes in the Reproduction of Daily Life.* Unpublished doctoral dissertation.

Fox, J. (1979a). Simultaneous equation models and two-stage least squares. Pages 130–150 in K. F. Schuessler, ed. *Sociological Methodology 1979.* San Francisco: Jossey-Bass.

Fox, J. (1979b). Comment on Harris, Luginbuhl, and Fishbein: Loglinear and logit models for tabular experimental data. *Social Psychology Quarterly*, **42**: 431–433.

Fox, J. (1980). Effect analysis in structural equation models: Extensions and simplified methods of computation. *Sociological Methods and Research*, **9**: 3–28.

Fox, J., and M. Guyer (1978). "Public" choice and cooperation in *n*-person prisoner's dilemma. *Journal of Conflict Resolution*, **22**: 469–481.

Francis, I. (1973). Comparison of several analysis of variance programs. *Journal of the American Statistical Association*, **68**: 860–865.

Friendly, M., and P. Franklin (1980). Interactive presentation in multitrial free recall. *Memory and Cognition*, **8**: 265–270.

Gallant, A. R. (1975). Nonlinear regression. *The American Statistician*, **29**: 73–81.

Gallant, A. R. (1977). Three-stage least-squares estimation for a system of simultaneous, nonlinear, implicit equations. *Journal of Econometrics*, **5**: 71–88.

Gentleman, J. F. (1978). Algorithm AS130: Moving statistics for enhanced scatter plots. *Applied Statistics*, **27**: 354–358.

Geraci, V. J. (1977). Identification of simultaneous equation models with measurement error. Pages 163–185 in D. J. Aigner and A. S. Goldberger, eds. *Latent Variables in Socio-Economic Models.* New York: North-Holland.

Gillespie, M. W. (1977). Log-linear techniques and the regression analysis of dummy dependent variables: Further bases for comparison. *Sociological Methods and Research*, **6**: 103–122.

Gillespie, M. W., and J. Fox (1980). Specification error and negatively correlated disturbances in "parallel" simultaneous-equation models. *Sociological Methods and Research*, **8**: 273–308.

Gnanadesikan, R. (1977). *Methods for Statistical Analysis of Multivariate Data.* New York: Wiley.

Goldberger, A. S. (1964). *Econometric Theory.* New York: Wiley.

Goldberger, A. S. (1971). Econometrics and psychometrics: A survey of communalities. *Psychometrika*, **36**: 83–107.

Goldberger, A. S. (1972). Structural equation methods in the social sciences. *Econometrica*, **40**: 979–1001.

Goldberger, A. S. (1973). Structural equation models: An overview. Pages 1–18 in A. S. Goldberger and O. D. Duncan, eds. *Structural Equation Models in the Social Sciences.* New York: Seminar Press.

Goldberger, A. S., and O. D. Duncan, eds. (1973). *Structural Equation Models in the Social Sciences.* New York: Seminar Press.

Goodman, L. A. (1970). The multivariate analysis of qualitative data: Interactions among multiple classifications. *Journal of the American Statistical Association*, **65**: 226–256.

Goodman, L. A. (1972). A general model for the analysis of surveys. *American Journal of Sociology*, **77**: 1035–1086.

Goodman, L. A. (1973a). Causal analysis of data from panel studies and other kinds of surveys. *American Journal of Sociology*, **78**: 1135–1191.

Goodman, L. A. (1973b). The analysis of multidimensional contingency tables when some variables are posterior to others: A modified path analysis approach. *Biometrika*, **60**: 179–192.

Goodman, L. A. (1976). The relationship between modified and usual multiple-regression approaches to the analysis of dichotomous variables. Pages 83–110 in D. R. Heise, ed. *Sociological Methodology 1976*. San Francisco: Jossey-Bass.

Greene, V. L. (1977). An algorithm for total and indirect causal effects. *Political Methodology*, **4**: 369–381.

Griliches, Z. (1977). Errors in variables and other nonobservables. Pages 1–33 in D. J. Aigner and A. S. Goldberger, eds. *Latent Variables in Socio-Economic Models*. New York: North-Holland.

Grizzle, J. E., C. F. Starmer, and G. G. Koch (1969). Analysis of categorical data by linear models. *Biometrics*, **25**: 489–504.

Haberman, S. J. (1978). *Analysis of Qualitative Data, Volume 1: Introductory Topics*. New York: Academic Press.

Haberman, S. J. (1979). *Analysis of Qualitative Data, Volume 2: New Developments*. New York: Academic Press.

Haitovsky, Y. (1968). Missing data in regression analysis. *Journal of the Royal Statistical Society, Series B*, **30**: 67–82.

Harris, B., J. E. R. Luginbuhl, and J. E. Fishbein (1978). Density and personal space in a field setting. *Social Psychology*, **41**: 350–353.

Hawkins, D. M. (1980). *Identification of Outliers*. New York: Chapman and Hall.

Heckman, J. J. (1978). Dummy endogenous variables in a simultaneous equation system. *Econometrica*, **46**: 931–959.

Hedayat, A., and D. S. Robson (1970). Independent stepwise residuals for testing homoscedasticity. *Journal of the American Statistical Association*, **65**: 1573–1581.

Heise, D. R. (1975). *Causal Analysis*. New York: Wiley.

Hoaglin, D. C., and R. E. Welsch (1978). The hat matrix in regression and ANOVA. *The American Statistician*, **32**: 17–22.

Hocking, R. R. (1976). The analysis and selection of variables in linear regression. *Biometrics*, **32**: 1–49.

Hocking, R. R., and F. M. Speed (1975). A full rank analysis of some linear model problems. *Journal of the American Statistical Association*, **70**: 706–712.

Hoerl, A. E. (1962). Application of ridge analysis to regression problems. *Chemical Engineering Progress*, **58**: 54–59.

Hoerl, A. E., and R. W. Kennard (1970a). Ridge regression: Biased estimation for nonorthogonal problems. *Technometrics*, **12**: 55–67.

Hoerl, A. E., and R. W. Kennard (1970b). Ridge regression: Applications to nonorthogonal problems. *Technometrics*, **12**: 69–82.

Hooper, H. W. (1959). Simultaneous equations and canonical correlation theory. *Econometrica*, **27**: 245–256.

Inverarity, J. (1976). Populism and lynching in Louisiana: 1889–1896: A test of Erickson's theory of the relationship between boundary crises and repressive justice. *American Sociological Review*, **41**: 262–280.

Inverarity, J. M. (1977). Reply to Bagozzi, Berk, Bohrnstedt, Pope and Ragin, and Wasserman. *American Sociological Review*, **42**: 663–667.

Johnston, J. (1972). *Econometric Methods, Second Edition*. New York: McGraw-Hill.

Jöreskog, K. G. (1970). A general method for analysis of covariance structures. *Biometrika*, **57**: 239–251.

Jöreskog, K. G. (1973). A general method for estimating a linear structural equation system. Pages 85–112 in A. S. Goldberger and O. D. Duncan, eds. *Structural Equation Models in the Social Sciences*. New York: Seminar Press.

Jöreskog, K. G., and D. Sörbom (1977). Statistical models and methods for analysis of longitudinal data. Pages 285–325 in D. J. Aigner and A. S. Goldberger, eds. *Latent Variables in Socio-Economic Models*. New York: North-Holland.

Jöreskog, K. G., and D. Sörbom (1978). *LISREL: Analysis of Linear Structural Relations by the Method of Maximum Likelihood: User's Guide (Version IV, Release 2)*. Chicago: National Educational Resources.

Jöreskog, K. G., and D. Sörbom (1981). *LISREL: Analysis of Linear Structural Relations by the Method of Maximum Likelihood: User's Guide (Version V)*. Chicago: National Educational Resources.

Kendall, M., and A. Stuart (1979). *The Advanced Theory of Statistics, Volume 2: Inference and Relationship, Fourth Edition*. New York: Macmillan.

Kim, J.-O., and J. Curry (1977). The treatment of missing data in multivariate analysis. *Sociological Methods and Research*, **6**: 215–240.

Kish, L. (1965). *Survey Sampling*. New York: Wiley.

Kleppner, D., and N. Ramsey (1965). *Quick Calculus*. New York: Wiley.

Kmenta, J. (1971). *Elements of Econometrics*. New York: Macmillan.

Knoke. D. (1975). A comparison of log-linear and regression models for systems of dichotomous variables. *Sociological Methods and Research*, **3**: 416–434.

Koopmans, T. C., H. Rubin, and R. B. Leipnik (1950). Measuring the equations systems of dynamic economics. Pages 53–237 in T. C. Koopmans, ed. *Statistical Inference in Dynamic Economic Models*. New York: Wiley.

Ku, H. H., R. N. Varner, and S. Kullback (1971). Analysis of multidimensional contingency tables. *Journal of the American Statistical Association*, **66**: 55–64.

Land, K. C. (1973). Identification, parameter estimation, and hypothesis testing in recursive sociological models. Pages 19–49 in A. S. Goldberger and O. D. Duncan, eds. *Structural Equation Models in the Social Sciences*. New York: Seminar Press.

Landwehr, J. M., D. Pregibon, and A. C. Shoemaker (1980). Some graphical procedures for studying a logistic regression fit. *1980 Proceedings of the Business and Economic Statistics Section, American Statistical Association*, 15–20.

Larsen, W. A., and S. J. McCleary (1972). The use of partial residual plots in regression analysis. *Technometrics*, **14**: 781–790.

Lazarsfeld, P. F., B. Berelson, and H. Gaudet (1948). *The People's Choice*. New York: Columbia University Press.

Lee, S. K. (1978). An example for teaching some basic concepts in multidimensional contingency table analysis. *The American Statistician*, **32**: 69–71.

Lewis-Beck, M. S. (1974). Determining the importance of an independent variable: A path analytic solution. *Social Science Research*, **3**: 95–107.

Lewis-Beck, M. S., and L. B. Mohr (1976). Evaluating effects of independent variables. *Political Methodology*, **3**: 27–47.

Lincoln, J. R. (1978). Community structure and industrial conflict: An analysis of strike activity in SMSAs. *American Sociological Review*,**43**: 199–220.

Long, S. B. (1979). The continuing debate over the use of ratio variables: Facts and fiction. Pages 37–67 in K. F. Schuessler, ed. *Sociological Methodology 1980*. San Francisco: Jossey-Bass.

Lord, F. M., and M. R. Novick (1968). *Statistical Theories of Mental Test Scores*. Reading: Addison-Wesley.

Lund, R. E. (1975). Tables for an approximate test for outliers in linear models. *Technometrics*, **17**: 473–476.

McCallum, B. T. (1973). A note concerning asymptotic covariance expressions. *Econometrica*, **41**: 581–583.

McFadden, D. (1974). Conditional logit analysis of qualitative choice behavior. Pages 105–142 in P. Zarembka, ed. *Frontiers in Econometrics*. New York: Academic Press.

McNeil, D. R. (1977). *Interactive Data Analysis: A Practical Primer*. New York: Wiley.

Maddala, G. S. (1983). *Limited Dependent and Qualitative Variables in Econometrics*. New York: Cambridge University Press.

Mage, D. T. (1982). An objective graphical method for testing normal distributional assumptions using probability plots. *The American Statistician*, **36**: 116–120.

Magidson, J. (1978). An illustrative comparison of Goodman's approach to logit analysis with dummy variable regression analysis. Pages 27–54 in L. A. Goodman (J. Magidson, ed.) *Analyzing Qualitative/Categorical Data: Log-Linear Models and Latent-Structure Analysis*. Cambridge: Abt.

Malinvaud, E. (1970). *Statistical Methods of Econometrics, Second Revised Edition*. New York: American Elsevier.

Mandel, J. (1982). Use of the singular value decomposition in regression analysis. *The American Statistician*, **36**: 15–24.

Marquardt, D. W., and R. D. Snee (1975). Ridge regression in practice. *The American Statistician*, **29**: 3–20.

Moore, J. C., Jr., and E. Krupat (1971). Relationship between source status, authoritarianism, and conformity in a social setting. *Sociometry*, **34**: 122–134.

Morrison, D. F. (1976). *Multivariate Statistical Methods, Second Edition*. New York: McGraw-Hill.

Mosteller, F., and J. W. Tukey (1977). *Data Analysis and Regression: A Second Course in Statistics*. Reading: Addison-Wesley.

National Women's Political Caucus (1981). *National Directory of Women Elected Officials 1981*. Washington, D.C.: National Women's Political Caucus.

Nelder, J. A. (1974). Letter to the editor. *Applied Statistics*, **23**: 232.

Nelder, J. A. (1976). Letter to the editor. *The American Statistician*, **30**: 103.

Nelder, J. A. (1977). A reformulation of linear models [with commentary]. *Journal of the Royal Statistical Society, Series A*, **140**: 48–76.

Nerlove, M., and S. J. Press (1973). *Univariate and Multivariate Log-Linear and Logistic Models*. Santa Monica: Rand Corporation.

Obenchain, R. L. (1977a). Letter to the editor. *Technometrics*, **19**: 348–349.

Obenchain, R. L. (1977b). Classical *F*-tests and confidence intervals for ridge regression. *Technometrics*, **19**: 429–439.

Ornstein, M. D. (1976). The boards and executives of the largest Canadian corporations: Size, composition, and interlocks. *The Canadian Journal of Sociology*, **1**: 411–437.

Pecknold, J. C., D. J. McClure, L. Appeltauer, L. Wrzesinski, and T. Allan (1982). Treatment of anxiety using fenobam (a nonbenzodiazepine) in a double-blind standard (diazepam) placebo-controlled study. *Journal of Clinical Psychopharmacology*, **2**: 129–133.

Pope, W., and C. Ragin (1977). Mechanical solidarity, repressive justice, and lynchings in Louisiana (comment on Inverarity). *American Sociological Review*, **42**: 363–368.

Pregibon, D. (1981). Logistic regression diagnostics. *The Annals of Statistics*, **9**: 705–724.

Prescott, P. (1975). An approximate test for outliers in linear models. *Technometrics*, **17**: 129–132.

Putter, J. (1967). Orthonormal bases of error spaces and their use for investigating the normality and variance of residuals. *Journal of the American Statistical Association*, **62**: 1022–1036.

Ramsey, J. B. (1969). Tests for specification errors in classical linear least-squares regression analysis. *Journal of the Royal Statistical Society, Series B*, **31**: 350–371.

Rao, C. R. (1973). *Linear Statistical Inference and its Applications, Second Edition*. New York: Wiley.

Rao, C. R., and S. K. Mitra (1971). *Generalized Inverse of Matrices and Its Applications.* New York: Wiley.

Rindfuss, R. R., L. Bumpass, and C. St. John (1980). Education and fertility: Implications for the roles women occupy. *American Sociological Review*, **45**: 431–447.

Riordan, C. A., B. Quigley-Fernandez, and J. T. Tedeschi (1982). Some variables affecting changes in interpersonal attraction. *Journal of Experimental Social Psychology*, **18**: 358–374.

Rosenberg, M. (1968). *The Logic of Survey Analysis.* New York: Basic Books.

Rosenthal, H. (1980). Review of L. Goodman, *Analyzing Qualitative/Categorical Data: Log-Linear Models and Latent Structure Analysis. Contemporary Sociology*, **9**: 207–212.

Sahlins, M. (1972). *Stone Age Economics.* New York: Aldine.

Scheffé, H. (1959). *The Analysis of Variance.* New York: Wiley.

Scudder, T. (1962). *The Ecology of the Gwembe Tonga.* Manchester: Manchester University Press.

Searle, S. R. (1971). *Linear Models.* New York: Wiley.

Searle, S. R., F. M. Speed, and H. V. Henderson (1981). Some computational and model equivalences in analysis of variance of unequal-subclass-numbers data. *The American Statistician*, **35**: 16–33.

Searle, S. R., F. M. Speed, and G. A. Milliken (1980). Population marginal means in the linear model: An alternative to least squares means. *The American Statistician*, **34**: 216–221.

Seber, G. A. F. (1977). *Linear Regression Analysis.* New York: Wiley.

Shryock, H. S., and J. S. Siegel, and associates (1973). *The Methods and Materials of Demography (Two Volumes).* Washington D.C.: U.S. Bureau of the Census.

Smith, G., and F. Campbell (1980). A critique of some ridge regression methods [with commentary]. *Journal of the American Statistical Association*, **75**: 74–103.

Sobel, M. E. (1982). Asymptotic confidence intervals for indirect effects in structural equation models. Pages 290–312 in S. Leinhardt, ed. *Sociological Methodology 1982.* San Francisco: Jossey-Bass.

Sörbom, D. (1975). Detection of correlated errors in longitudinal data. *British Journal of Mathematical and Statistical Psychology*, **28**: 138–151.

Speed, F. M., and R. R. Hocking (1976). The use of the $R(\ )$-notation with unbalanced data. *The American Statistician*, **30**: 30–33.

Speed, F. M., R. R. Hocking, and O. P. Hackney (1978). Methods of analysis of linear models with unbalanced data. *Journal of the American Statistical Association*, **73**: 105–112.

Speed, F. M., and C. J. Monlezun (1979). Exact F tests for the method of unweighted means in a 2^k experiment. *The American Statistician*, **33**: 15–18.

Steinhorst, R. K. (1982). Resolving current controversies in analysis of variance. *The American Statistician*, **36**: 138–139.

Stolzenberg, R. M. (1979). The measurement and decomposition of causal effects in nonlinear and nonadditive models. Pages 459–488 in K. F. Schuessler, ed. *Sociological Methodology 1980.* San Francisco: Jossey-Bass.

Swafford, M. (1980). Three parametric techniques for contingency table analysis: A nontechnical commentary. *American Sociological Review*, **45**: 664–690.

Theil, H. (1971). *Principles of Econometrics.* New York: Wiley.

Tietjen, G. L., R. H. Moore, and R. J. Beckman (1973). Testing for a single outlier in simple linear regression. *Technometrics*, **15**: 717–721.

Torgerson, W. S. (1958). *Theory and Methods of Scaling.* New York: Wiley.

Tukey, J. W. (1977). *Exploratory Data Analysis.* Reading: Addison-Wesley.

Upton, C. J. G. (1978). *The Analysis of Cross-Tabulated Data.* New York: Wiley.

Urquhart, M. C., and K. A. H. Buckley, eds. (1965). *Historical Statistics of Canada*. Toronto: Macmillan.

Velleman, P. F., and R. E. Welsch (1981). Efficient computing of regression diagnostics. *The American Statistician*, **35**: 234–241.

Vinod, H. D. (1978). A survey of ridge regression and related techniques for improvements over ordinary least squares. *The Review of Economics and Statistics*, **60**: 121–131.

Vinod, H. D., and A. Ullah (1981). *Recent Advances in Regression Methods*. New York: Dekker.

Wasserman, I. M. (1977). Southern violence and the political process (comment on Inverarity). *American Sociological Review*, **42**: 359–362.

Weisberg, S. (1980). *Applied Linear Regression*. New York: Wiley.

Wiley, D. E. (1973). The identification problem for structural equation models with unmeasured variables. Pages 69–83 in A. S. Goldberger and O. D. Duncan, eds. *Structural Equation Models in the Social Sciences*. New York: Seminar Press.

Wilk, M. B., and R. Gnanadesikan (1968). Probability plotting methods for the analysis of data. *Biometrika*, **55**: 1–17.

Wonnacott, R. J., and T. H. Wonnacott (1979). *Econometrics, Second Edition*. New York: Wiley.

Wood, F. S. (1973). The use of individual effects and residuals in fitting equations to data. *Technometrics*, **15**: 677–695.

Yates, F. (1934). The analysis of multiple classifications with unequal numbers in the different classes. *Journal of the American Statistical Association*, **29**: 51–66.

Author Index

Subject Index

Data Set Index